Mathematical Engineering of Deep Learning

Mathematical Engineering of Deep Learning provides a complete and concise overview of deep learning using the language of mathematics. The book provides a self-contained background on machine learning and optimization algorithms and progresses through the key ideas of deep learning. These ideas and architectures include deep neural networks, convolutional models, recurrent models, long/short-term memory, the attention mechanism, transformers, variational auto-encoders, diffusion models, generative adversarial networks, reinforcement learning, and graph neural networks. Concepts are presented using simple mathematical equations together with a concise description of relevant tricks of the trade. The content is the foundation for state-of-the-art artificial intelligence applications, involving images, sound, large language models, and other domains. The focus is on the basic mathematical description of algorithms and methods and does not require computer programming. The presentation is also agnostic to neuroscientific relationships, historical perspectives, and theoretical research. The benefit of such a concise approach is that a mathematically equipped reader can quickly grasp the essence of deep learning.

Key Features:

- A perfect summary of deep learning not tied to any computer language, or computational framework.
- An ideal handbook of deep learning for readers that feel comfortable with mathematical notation.
- An up-to-date description of the most influential deep learning ideas that have made an impact on vision, sound, natural language understanding, and scientific domains.
- The exposition is not tied to the historical development of the field or to neuroscience, allowing the reader to quickly grasp the essentials.

Deep learning is easily described through the language of mathematics at a level accessible to many professionals. Readers from fields such as engineering, statistics, physics, pure mathematics, econometrics, operations research, quantitative management, quantitative biology, applied machine learning, or applied deep learning will quickly gain insights into the key mathematical engineering components of the field.

CHAPMAN & HALL/CRC DATA SCIENCE SERIES

Reflecting the interdisciplinary nature of the field, this book series brings together researchers, practitioners, and instructors from statistics, computer science, machine learning, and analytics. The series will publish cutting-edge research, industry applications, and textbooks in data science.

The inclusion of concrete examples, applications, and methods is highly encouraged. The scope of the series includes titles in the areas of machine learning, pattern recognition, predictive analytics, business analytics, Big Data, visualization, programming, software, learning analytics, data wrangling, interactive graphics, and reproducible research.

Recently Published Titles

Big Data Analytics
A Guide to Data Science Practitioners Making the Transition to Big Data
Ulrich Matter

Data Science for Sensory and Consumer Scientists
Thierry Worch, Julien Delarue, Vanessa Rios De Souza and John Ennis

Data Science in Practice
Tom Alby

Introduction to NFL Analytics with R
Bradley J. Congelio

Soccer Analytics
An Introduction Using R
Clive Beggs

Spatial Statistics for Data Science
Theory and Practice with R
Paula Moraga

Research Software Engineering
A Guide to the Open Source Ecosystem
Matthias Bannert

The Data Preparation Journey
Finding Your Way With R
Martin Hugh Monkman

Getting (more out of) Graphics
Practice and Principles of Data Visualisation
Antony Unwin

Introduction to Data Science
Data Wrangling and Visualization with R Second Edition
Rafael A. Irizarry

Data Science
A First Introduction with Python
Tiffany Timbers, Trevor Campbell, Melissa Lee, Joel Ostblom and Lindsey Heagy

Mathematical Engineering of Deep Learning
Benoit Liquet, Sarat Moka, and Yoni Nazarathy

For more information about this series, please visit: https://www.routledge.com/Chapman--HallCRC-Data-Science-Series/book-series/CHDSS

Mathematical Engineering of Deep Learning

Benoit Liquet, Sarat Moka, and Yoni Nazarathy

CRC Press
Taylor & Francis Group
Boca Raton London New York

CRC Press is an imprint of the
Taylor & Francis Group, an **informa** business

A CHAPMAN & HALL BOOK

by CRC Press
2385 Executive Center Drive, Suite 320, Boca Raton, FL 33431, U.S.A.

and by CRC Press
4 Park Square, Milton Park, Abingdon, Oxon, OX14 4RN

CRC Press is an imprint of Taylor & Francis Group, LLC

© 2025 Benoit Liquet, Sarat Moka, and Yoni Nazarathy

ISBN: 978-1-032-28829-1 (hbk)
ISBN: 978-1-032-28828-4 (pbk)
ISBN: 978-1-003-29868-7 (ebk)

DOI: 10.1201/9781003298687

Typeset in LM Roman
by KnowledgeWorks Global Ltd.

Publisher's note: This book has been prepared from camera-ready copy provided by the authors.

A ma maman

Benoit Liquet.

To my mother Mariyamma and my wife Toshali

Sarat Moka.

To Emily, Kayley, and Yarden

Yoni Nazarathy.

Contents

Preface

In the last few years, deep learning has seen explosive growth and even been dubbed as the "new electricity". The field has shown incredible success in automated applications and predictive tasks. Deep learning models are mathematical in nature, and hence to understand deep learning, one needs to understand the mathematical description of the models. This book aims to provide such an understanding via a concise, accessible, and self-contained presentation.

Many deep learning resources focus on programming while making an effort to hide the mathematics. Other resources focus on theoretical results without an attempt to disambiguate the terminology of the field. A third breed of resources puts heavy emphasis on historical progression. Each of these viewpoints is important for a specific purpose, however, we conjecture that for a mathematical audience, these are all suboptimal ways to learn about deep learning. Hence we created this book.

Our focus is on the basic mathematical constructs that make up the field. We call this **mathematical engineering of deep learning**. Using the language of equations and algorithms, deep learning objects interface together to make very powerful models. A reader familiar with mathematical notation and with knowledge of basic calculus, basic probability, and basic linear algebra can go a long way in understanding deep learning quickly. For this, we use simple mathematical notation to outline the mechanisms used for training, execution, and application of deep neural networks.

Deep learning is certainly not solely about mathematics, as it also requires good software, hardware, and data. However, we aim to present the mathematical technology of deep learning without focusing on the implementation aspects that one would consider if trying to use deep learning frameworks in practice. We also aim to focus on the current state of the art as opposed to the historical progression of the field. Finally, we aim to minimize the focus on the human brain and the loose analogies that one can make between deep artificial neural networks and actual biological neurons. All of these aspects that we downplay are important, but we believe that if in an initial exposure to the field one spends too much time on implementation, history, or bio-neurological analogies, then the simplicity of deep learning is missed.

The book is primarily intended for readers from engineering, signal processing, statistics, physics, econometrics, operations research, quantitative management, pure mathematics, bioinformatics, applied machine learning, or even applied deep learning. A reader with a background in one of these domains will be able to get a concentrated and concise description of deep learning. In cases where a mathematical refresher is needed, appendices provide a condensed review, such as a review of key aspects of multivariable calculus.

The book can be read sequentially, or alternatively readers may wish to jump between chapters for quick lookup. It is assumed that readers have had exposure to mathematical notation at the level equivalent to at least three or four university courses. Hence, set notation, matrices, basic probability, and calculus are used without apology. However, no explicit knowledge of machine learning, statistics, optimization, or advanced probability is needed or assumed. We hope that we strike the right balance so that a mathematically equipped non-expert can easily read the book in a self-contained manner.

While the focus of the book is "mathematical engineering", we fully acknowledge the importance of applications and the ability to use software and hardware effectively. For this you may also use the companion website, `https://deeplearningmath.org/`, where additional examples, links, and software usage details are provided.

Outline of the Contents

The book has eight chapters and two appendices. Chapters 1–4 introduce the field, outline key concepts from machine learning, present an overview of optimization concepts needed for deep learning, and focus on fundamental models and concepts. Chapter 5 introduces fully connected deep neural networks. Chapters 6 and 7 deal with the core models and architectures of deep learning, including convolutional networks, recurrent neural networks, and transformers. Chapter 8 covers additional popular domains such as generative models, reinforcement learning, and graph neural networks. Appendices A and B provide mathematical support. Here is a detailed outline of the contents.

Chapter 1 – Introduction: In this chapter we present an overview of deep learning, demonstrate key applications, survey the associated ecosystems of high-performance computing, discuss big and high-dimensional data, and set the tone for the rest of the book. The chapter discusses key terminology including data science, machine learning, and statistical learning, and with this we place these terms in the context of the book. Key popular datasets such as ImageNet and MNIST digits are also presented together with a description of the deep learning culture that emerged.

Chapter 2 – Principles of Machine Learning: Deep learning can be viewed as a sub-discipline of machine learning, and hence this chapter provides an overview of key machine learning concepts and paradigms. The reader is introduced to supervised learning, unsupervised learning, and the general concept of iterative-based optimization for learning. The concepts of training sets, test sets, and the like, together with principles of cross validation and model selection are introduced. A key object explored in the chapter is the linear model which can be trained also via iterative optimization. We introduce the most simple gradient descent algorithm, and it is later refined in Chapter 4. Gradient descent is used for training almost any deep learning model. We also explore basic unsupervised learning algorithms including K-means clustering, principal component analysis (PCA), and the singular value decomposition (SVD).

Chapter 3 – Simple Neural Networks: In this chapter we focus on logistic regression (sigmoid) for binary classification and the related multinomial regression model (softmax) for multi-class problems. These models are the most popular shallow neural networks. The chapter sets the tone for more complex models by introducing principles of deep learning such as the cross entropy loss and other basic terminology. The chapter also presents a simple non-linear autoencoder architecture and with this introduces general ideas of autoencoders.

Chapter 4 – Optimization Algorithms: The training of deep learning models involves optimization over the learned parameters. Hence a solid understanding of optimization algorithms is required, as well as an understanding of specialized optimization techniques that work well for deep learning models such as the ADAM algorithm. In this chapter we focus on such techniques. We also study the details of various forms of automatic differentiation, a tool that has become critical in deep learning for computing gradients. Other optimization techniques, not always popular in contemporary deep learning, are also surveyed. This includes various first-order and second-order methods.

Chapter 5 – Feedforward Deep Networks: This chapter is the heart of the book where the general feedforward deep neural network, also known as the multi-layer perceptron, is defined and introduced. After introducing the basic architecture and exploring the expressive power of deep neural networks, we dive into the details of training by understanding the backpropagation algorithm for gradient evaluation. We also explore other aspects such as weight initialization, batch normalization, and dropout.

Chapter 6 – Convolutional Neural Networks: Convolutional neural networks are natural models for images and similar spatial data formats. In this chapter we explore the convolution concept and then see it used in the context of deep learning models. The concepts of channels and general convolutional neural networks are introduced. We then follow with an exploration of common unique architectures that have made a significant impact and are still in use today. We also explore a few key tasks associated with images such as object localization and face identification.

Chapter 7 – Sequence Models: Sequence models are critical for data such as text with applications in natural language processing, conversational agents, and translation. In this chapter we get a taste for the key deep learning ideas of the field. We explore recurrent neural networks and their generalizations including long short term memory (LSTM) models and gated recurrent unit (GRU) models. We then explore encoder-decoder architectures building up to the concept of attention where we formalize the attention mechanism. This idea then integrates into transformer models which in many ways are the state-of-the-art models used in large language models (LLM).

Chapter 8 – Specialized Architectures and Paradigms: In this final chapter we survey key ideas of specialized architectures and paradigms which are used for various types of tasks. This includes generative models, reinforcement learning, and graph neural networks. In terms of generative models we start by diving into the variational autoencoder architecture, a probabilistic deep learning model. We then extend to Markovian hierarchical variational autoencoders of which diffusion models are a special case. We then study generative adversarial networks (GANs) which were the first class of highly powerful deep learning models for realistic looking image generation. The chapter then moves to study reinforcement learning where we first present an overview of basics of Markov decision processes and then hint on how deep reinforcement learning can be implemented. We close with an introduction of graph neural networks. As such, the multitude of ideas in this chapter encompass several paradigms where deep learning models can be modified or joined together for specialized purposes.

With Thanks

We began this project while undertaking instruction at the 2021 AMSI (Australian Mathematical Sciences Institute) summer school. In that course we taught 60 students from all over Australia for 28 lecture hours. See a link to the course material through the book website `https://deeplearningmath.org/`. We thank students for embarking on the journey with us and further appreciate student comments useful for creating the book. We also mention that without support from our families and loved ones, this book would not be possible. We thank Alan White for supplying the banana for Figure 1.1. We thank various family members for appearing in some of our images.

We especially thank Vishnu Prasath and Ajay Hemanth of Richmond Enterprises Pvt Ltd for working on many illustrations of the book. The TikZ source code for these illustrations is now open-sourced with a link available through the course website. A few of the images in our figures, when mentioned, are taken from other research papers and other sources. We thank the authors for their permission to use these images. We also thank Toshali Banerjee for art design.

We also thank the following people for detailed comments and useful discussions: Teo Nguyen, Thomas Grahm, Vektor Dewanto, and Miriam Redding. In addition, useful comments were received from Marcus Gallagher, Matt Dirks, Adam Bennaceur, Gabriel Bianconi, Kwangsoo Cho, Jerzy Filar, Liam Bluett, Fred Roosta-Khorasani, and Maria Vlasiou. Sarat Moka also thanks Celestien Warnaar-Notschaele and Ole Warnaar for friendship and an extensive accommodation period in Brisbane during the extensive Sydney lockdown of 2021.

We hope that you enjoy the book.

Benoit Liquet, Sarat Moka, and Yoni Nazarathy.
February 2024.

1 Introduction

Many methods and techniques of deep learning have been known for a long time. However, it is only recently that deep learning has become a field of its own. At its core, deep learning is a collection of models, algorithms, and techniques, such that when assembled together, efficient automated detection and decision making can take place. Trained deep learning models can detect, classify, translate, create, and take part in systems that execute simple human-like tasks and beyond. The basic building block of deep learning is the *artificial neural network* or *neural network* for short. Deep learning is all about defining, training, and using such objects, often with multiple layers—hence the term "deep".

While most of this book is about the mathematical engineering of deep learning, this introductory chapter takes a more general viewpoint. It aims to introduce the field and the terminology in broad terms. In Section 1.1 we introduce deep learning by discussing the general nature of the field and key terms involved. We continue in Section 1.2 where we present an overview of some of the most popular tasks and their associated deep learning architectures. These architectures then re-appear in detail in the remainder of the book. In Section 1.3 we discuss key ingredients of the field. It is here where we briefly touch on connections between deep learning and neuroscience, discuss computation power, and discuss the availability of large datasets. In the remainder of the book, since our focus is on mathematical formulation, these topics are seldom considered, yet we consider them here for completeness. In Section 1.4 we introduce common openly available datasets that are often used to train deep learning models, both in practice and industrially. In Section 1.5 we introduce the concept **mathematical engineering of deep learning** which is the title of this book. We close with Section 1.6 where we introduce common mathematical notation used in the book.

1.1 The Age of Deep Learning

Classically, computers can be programmed to do complicated repetitive tasks very well. This includes automating calculations, sorting and filtering data, finding shortest paths between two locations on a map, and even executing extremely complicated weather simulations. For many of these tasks, one may consider specialized, well-defined algorithms whose specification is detailed by a sequence of steps including logical expression evaluation, conditional statements, iteration, and similar constructs.

With such a classical algorithmic approach, software can carry out computational tasks in a faster and more accurate manner than any living creature. No human can sort a large list of numbers faster than a computer, and no dog or other non-human animal can be expected to reliably sort numbers at all. However humans and other animals are able to think! This is something that computers (still) cannot do. On a more elementary level, many living beings can quickly learn to determine types of objects based on their appearance, sound, feel, or smell in a manner that up to **recently** was not possible by a machine.

DOI: 10.1201/9781003298687-1

The key here is **recently**. The second decade of the 21st century has witnessed incredible advances in the ability of computers to carry out (simple) human-like tasks. These include interpretation of images, voice, and text, learning how to execute smart decision making in uncertain environments such as playing games like Go, and highly impressive conversational agents which can communicate with humans in ways that appear human. The main vehicle for this success is **deep learning**.

The general computational area dealing with algorithms for classification, prediction, and decision making based on data is generally called *machine learning*. It has been around since the late 1950s together with the general phrase of *artificial intelligence*. Machine learning often involves programmed statistical techniques for tuning parameters of algorithms based on data, so that later when used on unseen data, the algorithms hopefully work well. There are dozens if not hundreds of well-developed machine learning techniques and models. One class of models involves artificial neural networks that have recently also become known as **deep learning**.

Up to about 2010, deep learning models were generally perceived as just another class of machine learning models, which while being interesting, in many cases were often not the most successful or insightful tools to use. However, together with the advent of large collected datasets and the availability of GPUs (graphical processing units), deep learning has emerged as the norm in the machine learning world rather than the exception. There are now tens of thousands of companies around the world that are deploying and developing automated applications that use deep learning technology. Indeed, the emergence of deep learning has brought artificial intelligence and machine learning to the forefront of technology and it is by now an integral part of almost any application. Many human-like tasks which include image processing, voice analysis, as well as dealing with general complex datasets are handled exceptionally well by deep learning techniques.

The world of *machine learning (ML)*, *artificial intelligence (AI)*, and *deep learning (DL)* is often characterized via multiple competing terms. In addition to ML, AI, and DL which are often used interchangeably, other key terms include *data science* and *statistics*. Related terms that have somewhat stepped out of the spotlight include *statistical learning*, *data mining*, and *big data (analytics)*. When considered in isolation, each of these terms may be broadly defined with a slightly different meaning. However, when considered together, there are significant intersections between the fields and meanings of the terms. One general way to describe the key terms is to say that **DL is a suite of very popular ML techniques that power much of today's AI**. Further, DL is one of the key components in the tool-box of data science and complements more traditional (as well as new and novel) methods from statistics. We generally do not use the ML, AI, and DL acronyms further in the book but rather use the general term deep learning to describe almost all of the methods and activities in the field.

A First Dive Into Deep Learning

A deep learning model is a very versatile function, denoted as $f_\theta(x)$, where the input x is typically high dimensional and the parameter θ is high dimensional as well. The function returns an output $y = f_\theta(x)$, where y is typically of significantly lower dimension than both θ and x. Given large datasets comprised of many x inputs matching desired outputs y, it is often possible to find good parameter values θ such that $f_\theta(\cdot)$ approximately satisfies the desired relationship between x and y. The process of finding such θ is called *training* or *learning.* Once trained, the model $f_\theta(\cdot)$ can be applied to *unseen input data*, x, with a hope of making good predictions, classifications, or decisions.

Depending on the application, the input data x can be in the form of an image, text, a sound waveform, tabular heterogeneous data, or some other variant. The output data y can be the probability of a label indicating the meaning/content of the image, a numerical value, or an object of similar form to x such as translated text in case x is text, a masked image in case x is an image, or similar.

Focusing on vision (images) for our initial example, one very popular source of data is the *ImageNet database* which has been used for the development and benchmarking of many deep learning models. The database has nearly 15 million color images. In many models, a subset of about 1.5 million images are used for training where the basic form of y is given by a label which is one of 1,000 categories indicating the content of the image. More on ImageNet is in Section 1.4 where we outline other popular datasets as well.

To get a feel for a deep learning model $f_\theta(\cdot)$, consider the *VGG19 model*[1] which is one of several popular (now classical) deep learning models for images. For this model, x is a 224×224 color image. It is thus comprised of $3 \times 224 \times 224 = 150,528$ values, with every coordinate of x representing the intensity of a specific pixel color (red, green, or blue). The output y is a 1,000-dimensional vector where each coordinate of the vector corresponds to a different type of object, e.g., car, banana, etc. The numerical value of the coordinate y_i, where say i is the index which matches banana is the model's prediction of the probability that the input x is a picture of a banana.

Ideally when fed with an input image x of a banana, the output $y = f_\theta(x)$ will have y_i as a high probability value e.g., 0.85 for $i =$ banana, while the other y_j for j other than banana will be low. This then allows one to use $f_\theta(\cdot)$ to classify bananas and other objects by choosing the label (banana, car, etc.) with the highest output probability y_i. Such a machine learning task is called *classification*.

At around the year 2014 when VGG19 was introduced, it was fed about 1.5 million ImageNet images, x, each with a corresponding label, e.g., banana, which is essentially a desired output y. The process of training VGG19 then involved finding good parameters θ so that when the model is presented with a new unseen image x, it predicts the label of the image well. Note that in VGG19, θ has a huge number of parameters; 144 million!

So to recap, there were about 1.5×10^6 input data samples each of size of about 1.5×10^5 (pixel values). Hence the training data size has about 2.25×10^{11} values (numbers). This data was then used to learn about 1.44×10^8 parameters. This is a lot of data, and a lot of trained parameters, but the resulting trained VGG19 model, $f_\theta(\cdot)$, works well. At the time when this specific model was introduced, it took days to train and much longer to fine tune. Today such a model may take around 8 hours to train on current state-of-the-art hardware and software. Further, it can take about a fifth of a second to make a prediction with this model, that is to evaluate $f_\theta(x)$. This is not an insignificant duration and can be improved upon by other models.

---

[1]Note that we only chose the VGG19 model here as an illustration. More efficient modern models are discussed later in the book, with image models discussed specifically in Chapter 6.

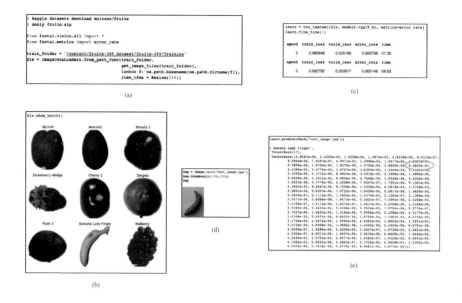

Figure 1.1: Illustrative fast.ai Python code for training (fine tuning) a VGG19 model to recognize fruits. The VGG19 model was originally trained using ImageNet and this pre-trained model is easily downloaded. Retraining this model from scratch using ImageNet would take several hours. Instead, in this example VGG19 is adapted to data from the Fruits 360 dataset. This adaptation takes about 10 minutes using a GPU. We then use the trained model on a single ad hoc image of a banana and it is correctly classified as `Banana Lady Finger`.

As a matter of illustration, Figure 1.1 presents code that uses the VGG19 model which was previously trained on ImageNet. This training of VGG19 from scratch on ImageNet is not something one would typically do in practice. Instead, we use the pre-trained VGG19 model parameters and adapt them based on another dataset called *Fruits 360*[2] which has nearly 100, 000 images of fruits, most of which were taken with repetitions by rotating a single fruit of each type and taking pictures at different angles. Here we use the *fast.ai* library with the *Python* language which also uses *PyTorch* under the hood. However, we could have presented alternatives with other languages (e.g., Julia or R) as well as other deep learning libraries such as *Keras* which uses *TensorFlow*. Indeed, there are many books dealing with deep learning libraries and code with notable ones mentioned in the notes and references at the end of the chapter. This current book is not about using such libraries and the practicalities of deep learning, it is rather about the concepts and key ideas.

In Figure 1.1 (a) we present setup code including downloading Fruits 360 from Kaggle. See also Section 1.4 where we discuss publicly available datasets. In (b) we present a few Fruits 360 images and their labels. In (c) there is the actual code and output needed for training (fine tuning) of the pre-trained VGG19 model (`models.vgg19_bn`) using fast.ai's `fine_tune()` method. This takes about 10 minutes to execute. This practice is called *transfer learning* and is also known as *fine-tuning*. In (d) we present an example banana image not related to the Fruits 360 dataset. Finally in (e) we execute model prediction using fast.ai's `learn.predict()` method where the probability associated with `Banana Lady Finger` is highest among all possible fruits. It is highlighted in index 17. This code takes a fraction of

---

[2]See https://www.kaggle.com/datasets/moltean/fruits.

a second to execute. We note that Figure 1.1 is the first and last example in this book that includes computer code.

Beyond Classification

Deep learning models can be used to perform various forms of *tasks*. The classification task described above is one type of these tasks and is often considered the most basic task. Other tasks include *regression* where y is a numerical value associated with x which needs to be predicted. Chapter 2 presents an introduction to regression and classification tasks in the context of general machine learning, together with an overview of other aspects of machine learning.

There are many more involved tasks as well. One example in the context of image data is *localization* where the goal is to determine the location of an object within an image. This task can also be handled by a variant of VGG19 where the input data is still a $3 \times 224 \times 224$ image but the output data y encodes the location of an object in the image (as well as possibly the type of object). Such a model is no longer $f_\theta(\cdot)$ above, but some other function, say $\tilde{f}_{\tilde{\theta}}(\cdot)$. While the models differ, one of the useful things about deep learning is that the parameters θ for the classification task and the parameters $\tilde{\theta}$ for the localization task are often similar and the bulk of the training effort can be used to learn both θ and $\tilde{\theta}$.

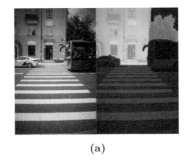

(a)

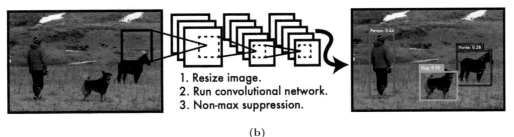

1. Resize image.
2. Run convolutional network.
3. Non-max suppression.

(b)

Figure 1.2: Different types of image tasks.[3] (a) Semantic segmentation of images. (b) Object detection and localization.

In the context of images, other tasks include *semantic segmentation* where all individual pixels of the input image x are marked as belonging to specific categories or classes. In such a case, y is of a similar form to x since it includes an indication of the class of each individual pixel. This case also clearly requires training data where each individual input pixel is considered. See for example Figure 1.2 (a). Further one may wish to identify multiple

---

[3]Image (a) is attributed to B. Palac under the creative commons license and available via Wikimedia Commons. Image (b) is thanks to "You Only Look Once: Unified, Real-Time Object Detection" by J. Redmon, S. Divvala, R. Girshick, and A. Farhadi, [346].

objects in an image as in Figure 1.2 (b). There are other tasks as well when considering text, video, sound, or tabular data and these are discussed in Section 1.2 and throughout the book.

Deep Learning: Where Is It Applied?

If you are reading this book then you probably already know that deep learning is applied in many diverse contexts. By around the time of publication of this book, many automated systems that involve sensing of noisy data already use some variant of deep learning. Many statistical modeling arenas also make use of deep learning. And finally, at the time of publication of this book, some mundane programming or writing tasks are beginning to be replaced by deep learning.

A medical doctor's decision can be viewed as $y = f_\theta(x)$. Here x may be the full medical history of the patient, or more specifically it may be the pixel information of medical imaging. In this context, y may simply be an indication of `benign` vs. `malignant` or may involve more complicated outputs such as `rest for one week and then get checked up again`. On much smaller time scales, the steering wheel actions of a driver are also of the form $y = f_\theta(x)$. Here x can be based on the fusion of multiple sensory data of a car and y is the steering command. Similarly, a farm worker that decides which tomatoes to pick today and which to wait on also repeatedly uses a relationship of the form $y = f_\theta(x)$. The same goes for automated voice recognition systems, for determination of the importance of text messages, and for comprehending handwritten text or handwritten mathematical equations. Even the task of converting a *prompt* such as `please write code for creating a web server for e-commerce for...`, can be carried out via deep learning where the output is a complete set of actions that create the web-server.

In all of these application domains, a trained human makes decisions based on sensory inputs and determines an outcome. It is true that other statistical and machine learning techniques for $y = f_\theta(x)$ may also do the job instead of deep learning. However, experience of the past decade has shown that with ample training data and multiple features, deep learning methods work exceptionally well and in many cases surpass the performance of other machine learning methods such as support vector machines, random forests, or other methods.

Who Are the Personas Involved?

Due to its success, deep learning has been dubbed the "new electricity". If that is the case then who are the "new electricians"? The answer to this question is still evolving because the age of deep learning has just begun. Nevertheless we can try and answer this question based on where deep learning stands as of 2024. As a first go you may use general labels such as *data scientists*, *statisticians*, *computer scientists*, and *machine learning engineers*. People who train deep learning models would probably classify themselves as one of these. If their work involves using modeling to make predictions, gaining business insight from data, or help finding scientific relationships between variables then the "data scientist" label is probably most appropriate. Further, if they use more statistical rigor for model selection and making conclusions, then "a statistician" is probably the right persona classification to use. This is especially true when the work of individuals involves experimental design and analysis of experimental data. Finally, if the work involves training and integrating deep learning solutions in software, then a "machine learning engineer" is probably the correct description.

In terms of research and core development of ideas, many of the developers of deep learning ideas and technology are probably rightly called "computer scientists". Computer science is obviously a broad field which at one extreme deals with discrete mathematical theoretical questions such as $P \stackrel{?}{=} NP$, and at the other extreme includes ideas from artificial intelligence pioneers such as Frank Rosenblatt, the initial creator of the perceptron in the 1950's and many others that worked on neural networks during the second half of the 20th century. In recent years computer scientists responsible for the deep learning revolution include names[4] such as Yann LeCun, Yoshua Bengio, Geoffrey Hinton, and many others. Other important names related both to development of ideas, development of software, and education include Ian Goodfellow, Jeremey Howard, and Andrew Ng. These individuals, together with thousands of others, are responsible for recent advances and education of deep learning that made it what it is today. Most of these researchers would probably feel comfortable with the title "computer scientists", however you will also find "mathematicians", "statisticians", and "neuroscientists" were heavily involved.

In addition to some of the deep learning pioneers mentioned above, many *researchers* from other domains are also now focusing heavily on deep learning. This involves experts from the world of *pure mathematics, probability, statistics, information theory*, and *control theory*. In the years to come we may witness development of theoretical results that describe deep learning; what works, what does not, and why. The focus of this book is not on such theoretical results. Our focus is rather on *mathematical engineering*, a phrase that we define in Section 1.5.

1.2 A Taste of Tasks and Architectures

A reader completing this book is to gain a mathematically aided understanding of deep learning. This includes understanding deep learning model architectures designed for a variety of tasks as well as an understanding of the motivation behind several architectural choices. Further, since deep learning architectures and their training algorithms go hand in hand, the journey also encompasses key methods used to train such models.

To get a feel for the models and methods covered, we now present an overview of the key tasks and architectures. Figure 1.3 presents schematics that include the simplest deep models called *feedforward fully connected neural networks*[5]; *autoencoders* which combine such feedforward networks with an *encoder* and a *decoder*; *convolutional neural networks* useful for image analysis; *recurrent neural networks* useful for sequence data; *transformer models* useful for *large language models* and other applications internally using the *attention mechanism* and also utilizing an encoder and decoder like autoencoders; *diffusion models* that are useful for image generation; *generative adversarial network* architectures also useful for fake data generation; and the paradigm of *reinforcement learning* useful for control of dynamic systems.

---

[4]In general, this book aims to minimize historical notes, yet you may find more historical information in the notes and references at the end of this chapter.

[5]Such models are also known as *fully connected networks, feedforward networks*, or *dense neural networks* where each of these terms may also be augmented with the phrase "deep" as well as the phrase "general". Another name for such models is *multi-layer perceptrons* (MLP).

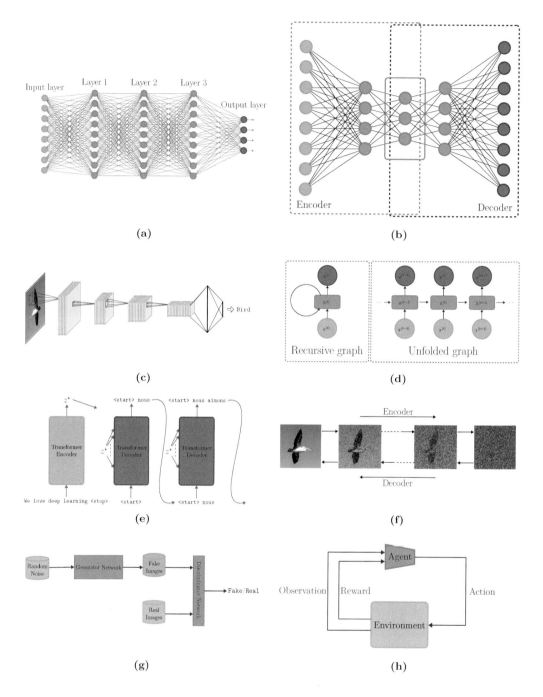

Figure 1.3: Illustrations of some common deep learning architectures and paradigms: (a) Feedforward fully connected neural networks covered in Chapter 5. (b) Autoencoders with shallow versions covered in Chapter 3. (c) Convolutional neural networks covered in Chapter 6. (d) Recurrent neural networks covered in Chapter 7. (e) Transformer architectures with an encoder and repeated application of a decoder covered in Chapter 7. (f) Diffusion models for image generation covered in Chapter 8. (g) Generative adversarial networks covered in Chapter 8. (h) Reinforcement learning covered in Chapter 8.

Feedforward Fully Connected Neural Network

The most basic deep neural network is the *feedforward fully connected neural network*. It is illustrated in Figure 1.3 (a) and covered in detail in Chapter 5. Simple special cases of this network are the *linear model*, analyzed in Chapter 2, as well as *logistic regression* (sigmoid) and *multinomial regression model* (softmax), both covered in Chapter 3.

Mathematically, feedforward fully connected neural networks are simply the combinations of affine (linear) transformations and non-linear activation functions. They constitute a very basic mechanism for enhancing the classical linear model with non-linearities. It turns out that this enhancement gives the model, $f_\theta(\cdot)$, an incredible ability to express complex relationships, $y = f_\theta(x)$, while supporting an algorithmically tractable way of finding θ (training). Classically these models are also called *multi-layer perceptrons* since they are descendants of the first ever neural network model, the *perceptron*, developed by Frank Rosenblatt in the late 1950s.

This architecture is useful for tasks such as classification, regression, or feature extraction. The models provide a highly expressive ability, which is especially useful when the data does not have a specific structure. While the models are heavily parameterized, they often work well for such ad-hoc tasks. Components of these networks, called *fully connected layers*, can also be components of more complex architectures such as convolutional networks, transformer models, and others.

Understanding training of these feedforward fully connected architectures, where gradients are computed via the famous *backpropagation algorithm*, covered in Chapter 5 (after automatic differentiation is introduced in Chapter 4), is key to understanding the essence of deep learning. Fully connected networks are also useful for understanding key ideas such as *dropout*, *batch normalization*, and *weight initialization*, which are all at the heart of deep learning.

Autoencoders

A simple *fully connected deep autoencoder architecture* is illustrated in Figure 1.3 (b). Shallow variants of this paradigm are studied at the end of Chapter 3. In many cases one may wish to use the internal representation of the architecture to extract meaningful features from the inputs. This is called *feature extraction*, and we call the outputs of this process *computed features* or *derived features*. After carrying out feature extraction, the computed features may then be transformed into outputs, used for clustering the data or for other manipulations of data.

Variants of architectures that make use of feature extraction fall under the name of *encoders* since they encode inputs into computed features; *decoders* since they decode computed features to outputs; as well as *autoencoders* since these architectures combine the tasks with an aim of having an output which matches the input. Particularly with an autoencoder, the goal is to find parameters θ, such that $x = f_\theta(x)$ is (approximately) maintained. The application of encoders, decoders, and autoencoders to different tasks is omnipresent in deep learning architecture design. There are hundreds of applications and variations with a few applications outlined at the end of Chapter 3.

Convolutional Neural Networks

The VGG19 model used in the discussion of Section 1.1 is one example of a *convolutional neural network architecture*. This class of models is illustrated in Figure 1.3 (c) and is covered in detail in Chapter 6. These types of models constitute the most famous specialization of fully connected neural networks. Convolutional models are primarily used for image analysis, and it is fair to say that their recent success has shuffled the cards in the broad field of *image processing*. Beyond images, these models can be adapted for other domains such as radiology data or audio. As alluded to in Section 1.1, in the context of images, there are multiple related tasks including classification, semantic segmentation, and localization, and all of this can be handled via convolutional neural networks.

Convolutional neural networks can be viewed as adaptations of fully connected networks, where the action of each layer is not based on the full connections between activations but rather on smaller trainable convolutions. These convolutions maintain a spatially homogenous structure in the network. Such a setup significantly reduces the number of parameters, enables deeper architectures, and most importantly capitalizes on spatial relationships present in the input. As a consequence, for a similar size of θ (e.g., 144 million parameters as in the VGG19 model), one may have a much deeper architecture than would have been possible with a fully connected network. This results in training that is much more efficient and the model is more efficient in production as well.

In addition to the core *trainable convolutions* idea, convolutional networks introduce additional architectural concepts such as the use of *channels* and the use of *pooling*. Huge leaps with convolutional neural networks were made during the first half of the second decade of this century. The incredible success of the so-called *AlexNet* model in the 2012 ImageNet challenge boosted neural networks within the world of machine learning. This in many ways started the deep learning revolution and put deep learning at the forefront of machine learning after the field was on the sidelines for many years.

Recurrent Neural Networks, LSTMs, and GRUs

Figure 1.3 (d) illustrates a recurrent neural network where on the left side of (d) we see the basic architectural components and on the right side we see what is called an unfolded representation of the network, illustrating its recursive operation. While key advances during 2010–2015 were in the convolutional domain focusing on images, the second half of that decade witnessed deep learning becoming an integral part of *natural language processing* (NLP). Today, automatic translation engines, language generation models, and other solutions for tasks associated with text almost always involve deep neural networks or are entirely based on deep learning models. The use of recurrent neural networks is the most rudimentary modeling paradigm for such purposes. These models are covered extensively in Chapter 7.

There are multiple variations of recurrent neural networks and the most basic one is illustrated in Figure 1.3 (d). However, the internal structure can vary and some popular and powerful variations include *long short term memory* (LSTM) models and *gated recurrent unit* (GRU) models. These architectures, also surveyed in Chapter 7, have been very impactful. In addition to NLP, there are many other domains where such models for sequence data is a natural choice. These domains include genomic sequencing, multivariable time series, audio, and even video.

Transformers and the Attention Mechanism

It is probably fair to say that the greatest advances in deep learning at the start of the third decade of the current century are centered around *large language models*. These systems, including the highly popularized *ChatGPT* service, among others, are able to execute many language tasks and are reshaping our view on intelligence at large. To date, the underlying model in most of these systems is the *transformer architecture*. Figure 1.3 (e) is a loose sketch of how such an architecture engages in the task of *machine translation*, i.e., translating from one human language to another. The reader should keep in mind that the illustrated encoder and decoder blocks are each composed of multiple sub-components (not appearing in the figure). Some of these components include feedforward layers as in Figure 1.3 (a), and other components are based on a concept called the *attention mechanism*. The full details of the attention mechanism and transformer models are in Chapter 7.

Large language models that use transformers appear to comprehend and generate human-like text. They engage in natural language understanding, extracting information from textual data, answering questions, and providing contextually relevant responses. Machine translation, our core example activity of Chapter 7, can also be handled by large language models. Beyond machine translation, large language models are versatile as they can handle multiple tasks including summarization of text, facilitating efficient communication across diverse languages, and more. Recent advances also include *multimodal models* which handles text, images, and other formats both as input and output.

Diffusion Models and Other Variational Autoencoders

In Figure 1.3 (f) we see a schematic of a diffusion model which here simply appears as a process of either adding noise to an image in an *encoder* or alternatvely removing noise from an image with a *decoder*. The overarching idea of a diffusion model is to learn not just how to add noise, but also how to create an image out of noise. With this, a trained decoder can generate realistic looking images that are actually random. Diffusion models recently arose as extremely powerful image generation models and are able to generate images that are both realistic looking and highly creative in their style and nature.

Diffusion models and their generalizations are probabilistic in nature. The complete details are in Chapter 8 where we first describe *variational autoencoder* models, then modify them to a class of models called *hierarchical Markovian variational autoencoders* of which diffusion models are a special case.

Generative Adversarial Networks

A *generative adversarial network* (GAN) architecture is illustrated in Figure 1.3 (g) and introduced in Chapter 8. Like diffusion models, GANs are very useful for creating random data that is realistic in nature. The rise and popularity of GANs predated that of diffusion models and today GANs and diffusion models compete for the state of the art in artificial data generation. GANs and diffusion models differ in their architecture and analysis. While diffusion models are probabilistic, the analysis and study of GANs is close to the field of *game theory*.

The key idea of a GAN is simultaneously training two deep neural networks, a *generator* and a *discriminator*. The former generates fake data, while the latter attempts to determine if the data is `fake` or `real`. As the training of both of these networks progresses, the generator is ultimately able to fool the discriminator, and as a consequence, it also creates "real looking" data. Much of the choice of architecture is then with finding measures of the quality of the data in the discriminator as well as with the algorithms for jointly training these networks.

Deep Reinforcement Learning

In Figure 1.3 (h) we illustrate the paradigm of reinforcement learning. Here the basic setup is that a system, or *environment*, is controlled by an *agent*. For example, one may think of the environment as a home, and the agent as a cleaning robot traversing and cleaning the home. As time progresses, the agent makes decisions in the form of *actions*, for example, `move right 5 cm`, and these are interfaced with the environment. The agent in turn receives *reward* from the environment as well as *observations*, where the reward is a mechanism that helps to drive toward better goals, for example, "cleaning in a quick and energy efficient manner", and the observations can include sensory input. The goal of reinforcement learning is to develop meaningful ways for the agent to choose actions.

One of the great leaps of AI during the second decade of this century is in the game of Go; see references at the end of this chapter. This strategic board game was long considered much more difficult "to program" in comparison to other games such as Chess.[6] Yet in 2015 a team from DeepMind through a series of advances and competitions designed a system called AlphaGo which beat the world's best Go players. This highly publicized achievement made the dream of artificial intelligence a bit more concrete by showing the ability of neural networks to solve complicated tasks. The key ideas of this achievement are from the field of *reinforcement learning*. We outline basic ideas in Chapter 8.

Graph Neural Networks

An additional category of neural network models that we explore in Chapter 8 are *graph neural networks*. These models operate on graphical structures, i.e., nodes and edges with attributes. Graph neural networks are suitable for social networks data, the study of chemical compounds, and many other applications where relationships between entities are well described via graph structures. In contrast to other types of neural networks, graph neural network models are often not directly used to try and mimic human-level performance but rather for discovery and insight within data.

1.3 Key Ingredients of Deep Learning

After getting a taste of deep learning, we now discuss the key ingredients that leverage its success. These are notably the availability of large datasets, advances in computer architectures, advances in software, the internet, and the interplay of cognitive science and artificial intelligence research. While the remainder of the book focuses on the mathematical description of deep learning, this section aims to overview the non-mathematical key ingredients which are attributed to the success of the field.

---

[6]Computers have shown their superiority in the game of Chess since the mid-1990s with a notable victory of the Deep Blue Chess playing expert system defeating the champion Garry Kasparov over a six-game match in 1996.

Neural Networks as Artificial Brains?

Brains are composed of (biological) neurons that are interconnected in unstructured ways; see Figure 1.4 where display (a) illustrates a single biological neuron and display (c) illustrates an interconnected network of biological neurons. A human brain has an estimated 85 billion neurons. A single human action such as movement of an arm may induce the firing of around 80 million such neurons, whereas the identification of a visual object may use the bulk of the estimated 150 million neurons that are in the *visual cortex*. Unquestionably, brains are fascinating organs whose scientific understanding, while still at its infancy, will undoubtedly grow in the years to come and have profound effects on human endeavor.

Deep neural network models are neither brains nor attempts to create artificial brains. Nevertheless, the development of these models is highly motivated by the biological structure of the brain. The basic building block of a deep neural network model is the (artificial) neuron abstracting the synapse connection between neurons via a single number called an activation value. See display (b) of Figure 1.4 for a single (artificial) neuron and display (d) which presents a combination of multiple neurons as part of a feedforward (artificial) neural network similar to Figure 1.3 (a). Pioneering and landmark work in AI research was inspired by neuroscience since brains are essentially the only complete proof we have for the existence of what we call "general intelligence". Further, many tasks of deep learning models involve the mimicking of human level (or animal level) tasks such as understanding images or conversational tasks. Thus for example one of the most well-known benchmarks in the world of artificial intelligence is the *Turing test*, originally named the *imitation game* when introduced by Alan Turing in 1950. It is essentially a test to see if a computer can engage in long conversation with a human, without another observing human distinguishing between the computer and the human.

At the time of publishing of this book, the state-of-the-art *large language models* are on the verge of passing the Turing test, and in fact, researchers are seeking alternative more suitable criteria. This is because while the test may appear to be (nearly) achieved, it is still believed that these language models do not constitute general intelligence. In fact, *artificial general intelligence* (AGI) systems are at best at their infancy. Deep learning models generally only achieve *narrow tasks* such as pattern recognition, conversational agents, or playing specific games, as opposed to general intelligence tasks of creative problem solving. Nevertheless, large language models of 2024 and onward appear to be very powerful in multi-modal activities.

With both the fields of neuroscience and AGI still awaiting major breakthroughs, it is natural for researchers to continue to draw parallels between neuroscience and artificial intelligence. On the cognitive sciences side of the spectrum, an increasing number of researchers are making use of artificial neural networks as abstractions for understanding cognitive tasks. However, on the AI side of the spectrum, while in the early years, neuroscientific-motivated models were central, today they have become more of a niche research area. In our context, the mathematical description of deep learning in this book is completely agnostic to biological brains.

Computational Power

It is well known that deep learning models require fast computers with plenty of memory. Training deep learning models can take hours or days using current state-of-the-art hardware

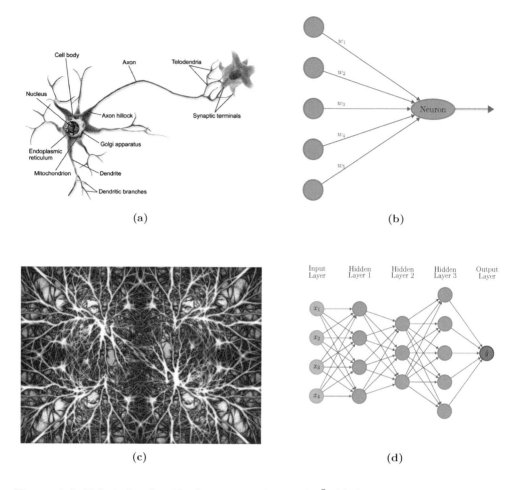

Figure 1.4: Biological and artificial neurons and networks.[7] (a) A single biological neuron. (b) A neuron in an artificial neural network. (c) Connection of multiple biological neurons in a brain. (d) An artificial neural network connecting multiple neurons in a feedforward structured manner.

and would not have been practically possible on machinery of the 1990s or earlier. Similarly, the large-scale application of (trained) deep learning models, such as in self-driving cars, requires massive resources as is attested by the fact that the power consumption for computing in a self driving car can sometimes equate or exceed the power used by the engine. At around the first decade of the 21st century, a technological threshold was passed and the availability of fast computing hardware made earlier deep learning ideas realizable and successful.

A complete description of hardware advances and their relationship with deep learning is beyond our scope and is not the focus of this book. Indeed, practical machine learning engineers often need to consider the computing power and hardware at play to train or implement effective deep learning models. A core component that has made a huge difference to the field is the development and availability of *graphical processing units* (GPUs). In contrast to standard *central processing units* (CPUs) which are optimized for

---

[7]Image (a) is attributed to B. Blaus under the creative commons license and available via Wikimedia Commons. Image (c) is sourced from pixabay.com.

logical operations, branching, and general computations, GPUs are optimized for repetitive large-scale matrix operations and can execute deep learning training or prediction in the order of 20 to 200 times faster than state-of-the-art CPUs. The GPU industry initially grew due to demand from the video gaming market, however, by 2020 their importance in the AI revolution is well understood. Today the needs of deep learning influence the design and development of future-generation GPUs. In summary it is fair to say that without GPUs, deep learning would not be anywhere where it is today.

For example, the aforementioned success of the so-called *AlexNet* model in the 2012 ImageNet challenge, was based on a neural network model specifically designed to be trained on two parallel GPUs which were the state-of-the-art of the time. From a programming perspective, utilizing such GPUs required considerable effort at that time but since then they have become much more accessible with better software. In the past decade, more specialized computing systems, including GPUs, similar *tensor processing units* (TPUs), and dedicated driver software were specifically adapted for deep learning applications. Indeed today, a machine learning engineer engaging in deep learning almost always needs to make use of such tools. Also central in this arena is the availability of *cloud computing*. Today many deep learning applications use dedicated cloud computing services both for training and model application.

For light applications including *scientific machine learning* with small datasets, or for pedagogical purposes, one may often use non-GPU machines (as well as laptops). See the notes and references at the end of the chapter for recommended reading of applied deep learning.

Large Datasets

Deep learning models work exceptionally well in cases where input data has a lot of features, a setting that is loosely called *high dimensional*. In a more statistical context, a proper definition of high dimensionality is that the number of features exceeds the number of datapoints. However, in deep learning, such a distinction is not common. Large datasets with many samples are ubiquitous and often critical to the success of deep learning.

Large enough and annotated datasets were rarely available prior to the turn of the century, yet in recent times humanity has witnessed an explosion in the volume of data stored. As a general guide consider Figure 1.5 showing a projection of the total stored data on earth. While clearly not all of this data is open, available, and suitable for deep learning models, this trend in total data storage is also characteristic with data suitable for deep learning models.

A key example which was pivotal in the success of the field is the aforementioned ImageNet database which contains nearly 15 million images where early deep learning models were trained with 1.5 million annotated images out of the total. Another example is with large-scale text models trained on all of Wikipedia which as of early 2024 includes over 59 million pages with a compressed text size of about 22 gigabytes. Up to the turn of the century, such datasets were much harder to come by, were rarely openly available, and disk sizes of most computer workstations were often too small to accommodate them. However, in recent years the availability of plenty of rich datasets has been pivotal for the success of deep learning. More information on annotations and popular datasets is in Section 1.4.

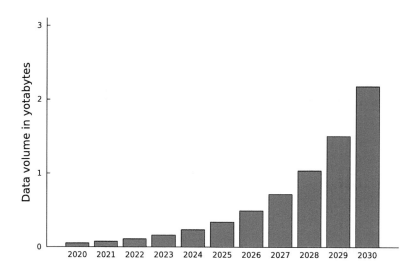

Figure 1.5: Predictions of the world's total data storage during the third decade of the 21st century. The graph is in yotabytes where each yotabyte is 2^{80} bytes or approximately 10^{24} bytes. Note that at the time of publishing of this book, estimates still use the zetabyte unit (a zetabyte is 2^{70} bytes, or $1/1024$ of a yotabyte). Current estimates at the time of publishing of the book are at around 150 zetabytes. The predictions in this plot are speculative and are based on a simple extrapolation that we carried out.[8]

The Internet, Software Practices, and Open Source

In addition to fast computation and the availability of large datasets, the recent success of deep learning was also fueled by new software development practices that evolved around the proliferation of the internet. Up to the last decade of the 20th century, most software developments involved relatively isolated groups working in companies, in research groups, or individually. As such, there was not much sharing of ideas, information, packages, and modules. However, by the end of the first decade of the current century, global collaborative community practices solidified and eventually resulted in (generally) free services such as Stack Overflow, GitHub, GitLab, Kaggle, and many more which today are a natural part of software development and data science culture. A key attribute of these new services is that they incentivize individuals (and groups) to share their ideas and source with the global community. These developments went hand in hand with the growth of the open source ethos that is shared and respected by many.

By the middle of the second decade of the 21st century, as the strength of deep learning became evident, new age collaborative global software practices were already quite mature. The timing of these events greatly helped the deep learning revolution as it allowed thousands of contributors around the globe to develop software, supporting documentation, and examples suited for deep learning applications. As a consequence, within a period of a few years, deep learning software frameworks such as the now popular TensorFlow, PyTorch, Keras,

---

[8]As there is not one credible openly available data source, we crowd sourced estimates via search results for a few years that had estimates searchable via Google. With this, we obtained estimates for the years, 2006, 2007, 2010, 2012, 2018, 2021, and 2022, some of which were for the total data stored and some for the data generated during that year. We then fit an exponential model to the data under a few minor additional assumptions. Details of our extrapolation are in the source code notebook available through https://deeplearningmath.org/.

fast.ai, Flux.jl, and others became available, quickly matured, and are now widely used. By today, new deep learning ideas stemming from research are often published together with open source software. As a consequence, machine learning engineers and other users of deep learning are able to easily and quickly use the state-of-the-art models and methods.

All of these practices are key ingredients not just of deep learning, but of data science and software development at large. However, in terms of the deep learning revolution, the timing of events was just right.

1.4 DATA, Data, data!

Effective training of deep learning models requires large datasets. We now present examples of various forms of data as well as a few popular datasets that are used for educational purposes, for training of real models, or for benchmarking. Our focus here is on *annotated* datasets typically used for *supervised learning*, a concept discussed at greater depth in Chapter 2. These are datasets consisting of *feature vectors*, each denoted by x, and associated *labels*, each denoted by y.

One should keep in mind that single data values are typically either *numerical* or *categorical*. The latter is the case when the values come from small discrete sets. In some cases categorical data has order such as the level of customer satisfaction which may be recorded in the range `unsatisfied, ..., very satisfied`. These are called *ordinal categorical variables* and in certain cases they may be directly converted to numerical variables which encode the order. However, other categorical variables such as `banana`, `car`, etc. have no specific order. These are called *nominal categorical variables* and they are typically treated by expanding the values into unit vectors. More on dealing with such data is in Chapter 2.

Here are a few generic examples for the features x and labels y. In some cases there is a natural *dimension* to the data, such as (mono) audio being one-dimensional, black and white images being two dimensional, and color images being three dimensional. When treating this data mathematically, we often attempt to represent x as a one-dimensional vector of length p, where p is the number of features.

Audio recording: A simple representation of audio is the vector of amplitudes of the recording at regular intervals. For example at $44, 100$ samples per second is a common sampling rate for high-quality audio. The amplitude can be a positive or negative number.[9] Hence for example, an audio recording taking a snapshot of music for 5 seconds will be a vector x with $p = 220, 500$ entries (features). An associated label y, in case it is music audio, may be the genre of the music such as `jazz`, `hip-hop`, etc.

A monochrome image: Here consider an image of p_1 by p_2 pixels with each pixel signifying the intensity e.g., 0 is black and 255 (or some other maximal value) is white. We can consider the image as a $p_1 \times p_2$ matrix and the vector x of length $p = p_1 \cdot p_2$ as a vectorized form of the matrix using either a *column-major* or *row-major* representation of the matrix. As an example consider a 200×100 (portrait) image. In this case the vector x has $p = 20, 000$ features (pixels). A simple example for y may be an indication of the class of the image such as `banana`, `car`, etc. Alternatively y may represent a bounding box inside the image where the object we are looking for is localized.

---

[9]Often such amplitudes are two bytes each, or 16 bits each, and this means that they obtain values from a finite range of $2^{16} = 65, 536$ possibilities.

A color image: Here each pixel is not just an intensity but rather an RGB (Red, Green, Blue) 3-tuple. One way to represent this image is via a 3-tensor which is $3 \times p_1 \times p_2$ dimensional. A vectorized image would be a $p = 3 \cdot p_1 \cdot p_2$ dimensional vector. Note that while color images are typically stored in a compressed format, such as JPEG, for deep learning purposes, images are typically considered in a bit map format as described here.

A text corpus: Text encoded in the ASCII format, the Unicode format, or other means is a sequence of characters. However treated as a datasource for deep learning we may sometimes break up the text into words (or tokens), associate a unique vector with each word, and represent the text as a sequence of vectors. Here y may be the associated text in a different language, or it may be a level of *sentiment* of the text which is a number between -1 and 1 indicating the text is of a negative tone (angry, critical, etc.), or is of a positive tone.

Heterogeneous datasets: In many cases the features are heterogeneous in nature as one would expect for example in the case of individual customer records in a database. For such a case, each customer can be associated with dozens, hundreds, or thousands of features, either numerical or categorical. Here y may be a continuous variable indicating the propensity (probability) of a customer to leave the service.

A Few Popular Datasets Examples

In most cases, datasets are created using a *manual annotation process* where after the x's are collected, standardized, and cleaned, human annotators look at each x and assign the matching value to y. Needless to say, the process of annotating datasets is generally time-consuming and expensive. The deep learning revolution was sparked by the creation of a few large annotated datasets and to date, the process of annotating datasets for new purposes is often a costly barrier for engaging in new deep learning ventures or research.

We now list a few selected notable datasets that are likely to appear in elementary deep learning practical examples, in research papers, and are sometimes also used professionally. See also Figure 1.6. Of the examples which we list here, the ImageNet database is the most industrially applicable dataset since many vision models are trained via this dataset and can then be ported to other models using transfer learning.

Some datasets break up the data a-priori into some random but fixed partition between a *train set* and a *test set*. This, in principle, allows one to benchmark models by training them on only the training set and evaluating performance on the test set. Such practices are discussed in greater detail in Chapter 2.

MNIST[10] **digit images**: One of the most basic machine learning datasets is the *MNIST database*. In this dataset, each sample x is a 28×28 black and white image of a handwritten digit, and when vectorized it is a $p = 784$ long vector. Each label y is an element from 0,1, ..., 9 indicating the digit. The distribution of digits is *balanced*, meaning that there are approximately the same number of images for each digit (class). The dataset is broken into a train set comprised of 60,000 images, and a test set with an additional 10,000 images.

---

[10]Modified National Institute of Standards and Technology.

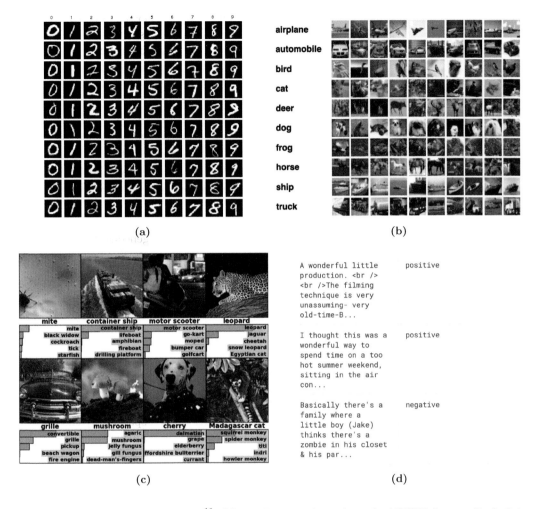

(a)

(b)

(c)

(d)

Figure 1.6: Some popular datasets.[11] (a) Handwritten digits from the MNIST dataset. Each digit is 28 × 28 pixels. (b) The CIFAR-10 dataset comprised of 32 × 32 color images from 10 classes. (c) Some images from the ImageNet dataset together with the top-5 classified labels. The correct label is in pink. (d) An extract from the IMDB movie reviews dataset. Each review is labeled as either **positive** or **negative**.

It is fair to say that this dataset does not have much industrial value, but is rather used for educational and academic purposes. Training a machine learning model to classify MNIST images is one of the most basic practice tasks one can consider. Further, many machine learning papers use MNIST to illustrate ideas and benchmarks.

A similar dataset with the same characteristics, is the *fashion MNIST dataset* where the images and labels are **shirt**, **shoe**, etc. and the dimensions of the data are set exactly like MNIST. The digit MNIST can be trained for over 99.8% accuracy with state of the art models, yet fashion MNIST is slightly more challenging and can be trained to achieve around 96.5% accuracy.

---

[11]Image (b) is sourced from the website https://www.cs.toronto.edu/~kriz/cifar.html. Image (c) is thanks to A. Krizhevsky, I. Sutskever, and G. E. Hinton, sourced from "ImageNet Classification with Deep Convolutional Neural Networks", [239]. Image (d) is from https://www.kaggle.com/lakshmi25npathi/imdb-dataset-of-50k-movie-reviews.

CIFAR[12]-10 small color images: Similarly to MNIST, the *CIFAR-10 dataset* has small images (32×32 pixels in this case) each broken up into 10 classes, namely `airplane`, `frog`, etc. as appearing in Figure 1.6 (b). However, the images are in color and hence $p = 3 \times 32 \times 32 = 3,072$. Here there are $50,000$ train images and $10,000$ test images and like MNIST this is a balanced dataset between classes. This dataset is popular with research papers and is occasionally used as a benchmark dataset. State-of-the-art models achieve around 99% prediction accuracy.

ImageNet: The *ImageNet database* is undoubtedly the most popular dataset for deep learning benchmarking, as well as for industrial use in image analysis. It was created at the end of the first decade of this century and coupled with the associated *ImageNet challenge*[13]. The challenge uses 1,000 non-overlapping image classes. ImageNet has nearly 15 million color images with varying resolutions and an average resolution of around 470×385. The images are labeled very specifically to more than $20,000$ categories where some images are also labeled with bounding boxes around objects.

In 2012 the AlexNet deep learning model achieved a top-5 accuracy of nearly 85%, meaning that in 85% of the tested images the correct label out of the 1,000 labels is one of the top-5 predicted probabilities. See also Figure 1.6 (c) where the top 5 label probabilities are presented for a few example images with the correct label marked in pink. More on top-5 accuracy and other performance measures is in Chapter 2. By today, state-of-the-art performance for top-5 accuracy is over 99% and for top-1 (or absolute accuracy) state-of-the-art is just over 90%.

IMDb[14] movie reviews: This textual dataset has $50,000$ movie reviews. It is split into $25,000$ for training and the remainder for testing. For each review an indication of either `positive` or `negative` indicates the sentiment of the review. The typical task with such a dataset is to predict the sentiment for an unseen review. This dataset is often used as a practice dataset for introductory tasks within *natural language processing* (NLP).

Beyond these datasets there are now thousands of additional quality available open sourced datasets, some of which are only really useful for practice or experimentation, while others, like ImageNet have real industrial value. Deep learning frameworks and specialized libraries often come with example datasets, and Kaggle is considered the most popular general source of openly available datasets since it also holds competitions and educational activities associated with the data.

1.5 Deep Learning as a Mathematical Engineering Discipline

While computing power and the abundance of data are key to the success of deep learning, the importance of mathematics cannot be underplayed. We now explain the term **mathematical engineering** and justify its use in the book's title.

Human engineering of systems has relied on mathematics almost since the dawn of time. The pyramids of Egypt could not have been constructed without some prior geometric calculations and later down the road, the machines that drove the industrial revolution were

---

[12]Canadian Institute For Advanced Research.

[13]More formally known as ILSVRC (ImageNet Large Scale Visual Recognition Challenge).

[14]IMDb is short for Internet Movie Database. The dataset was made publicly available by the authors of [275].

designed with the aid of trigonometry, algebra, calculus, and other mathematical techniques. For such reasons and others, the term **engineering mathematics** is often used to describe the combination of calculus, linear algebra, basic differential equations, Fourier analysis, Laplace transforms, and many other mathematical tools that can interplay to help design and develop physical engineered systems. So why does this book use the permuted term **mathematical engineering**?

Our answer lies in the fact that in the field of deep learning, mathematics is used directly as an engineered component. When one designs an electrical circuit, the flow of electricity is engineered and hence the field is called **electrical engineering**. When one designs a robot arm, the mechanics are directly considered and hence the term used is **mechanical engineering**. What about the design of a deep neural network for deep learning? The specification of deep learning models is a mathematical specification about the flow of data through a combination of affine functions, non-linear functions, and at times other mathematical components. Hence the design of deep learning models is the act of **mathematical engineering**.

To further understand the phrase **mathematical engineering** consider the following display borrowed from Chapter 5. It represents an action that is at the heart of deep learning:

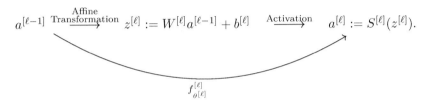

$$a^{[\ell-1]} \xrightarrow{\text{Affine Transformation}} z^{[\ell]} := W^{[\ell]}a^{[\ell-1]} + b^{[\ell]} \xrightarrow{\text{Activation}} a^{[\ell]} := S^{[\ell]}(z^{[\ell]}).$$
$$f^{[\ell]}_{\theta^{[\ell]}}$$

The function $f^{[\ell]}_{\theta^{[\ell]}}$ represents the mathematical action of a single layer of a deep neural network on neurons/activations/inputs $a^{[\ell-1]}$ to obtain neurons/activations/outputs $a^{[\ell]}$. It involves an affine transformation to reach an intermediate value $z^{[\ell]}$ via a matrix multiplication by the (weight) matrix $W^{[\ell]}$ and an addition of a (bias) vector $b^{[\ell]}$. It then involves the application of a non-linear (activation) function $S^{[\ell]}$. The trainable parameters of the layer are $W^{[\ell]}$ and $b^{[\ell]}$ and their combination is denoted via $\theta^{[\ell]}$. A deep neural network $f_\theta(\cdot)$ is then a functional composition of many such layers, say L,

$$y = f_\theta(x) = f^{[L]}_{\theta^{[L]}}(f^{[L-1]}_{\theta^{[L-1]}}(\ldots(f^{[1]}_{\theta^{[1]}}(x))\ldots)),$$

where $x = a^{[0]}$ and $y = a^{[L]}$. The parameters of the whole model, θ are comprised of the individual parameters of the layers $\theta^{[1]}, \ldots, \theta^{[L]}$.

Some aspects of the mathematical engineering of deep learning involve choosing the non-linear functions $S^{[\ell]}(\cdot)$ and other aspects involve defining the dimensions of $W^{[\ell]}$ and $b^{[\ell]}$ as well as any sparsity structures in these. Importantly, deep learning training is essentially the iterative solution of an optimization problem where the decision variables are the many components of θ. For such a problem, in addition to specifying efficient optimization algorithms, one needs to determine the optimization objective which in the language of deep learning is called the loss function. All these activities and many more comprise the **mathematical engineering of deep learning**.

The Mathematics Used

While deep learning is "deep", the application of mathematics in the field is relatively shallow. The mathematical engineering of deep learning is mostly based on linear algebra, multivariate calculus, and basic probability. It thus uses mathematics at a level comparable to the first 2 years of university studies of an engineering degree or a similar field. In fact, even within these fields, one mostly requires only elementary operations and there is not much reliance on theoretical results. For example in terms of linear algebra, matrix multiplication operations are key, but vector spaces, eigenvalues, and matrix decompositions are mostly not essential. Similarly in terms of calculus, derivatives and their multi-variable counterparts are central to the field, but integration, vector fields, manifolds, or properties of real functions are mostly not used. Finally in terms of probability, one simply needs to account for basic probabilities and occasionally evaluate expectations or variances for random variables.

This accessibility of deep learning has motivated pedagogical approaches focusing on a practical coding perspective as in "Deep Learning for Coders with fast.ai and PyTorch" [190] as well as many other resources; see also the notes at the end of this chapter. In contrast to such code-centric approaches, our approach relies on mathematical notation as a means of communicating ideas. With our approach, even though theoretical mathematical results are mostly not essential, mathematical notation plays a central role in conveying ideas concisely. See also Section 1.6, where we introduce basic notation for the remainder of the book as well as a few supporting mathematical results summarized in the appendices.

Development and Investigation of Deep Learning via Advanced Mathematics

It is important to note that while the mathematics used in this book is quite simple, many of the models and techniques that we present have advanced theoretical underpinnings. These advanced counterparts have often played a role in the development of the current simple models. One such case which will become evident in Chapter 3 is the logistic regression model. With this model one can analyze the model as either a simple neural network, or alternatively position it as a statistical model leveraging on the (slightly more advanced) theory of maximum likelihood estimation (MLE). Interestingly, historically logistic regression was first developed using MLE and only later appeared as a basic machine learning model. Similar trends in the machine learning community are also in association with support vector machines (SVM), a topic that we do not cover further in this book. The theory and application of SVMs hinges on beautiful mathematical results from functional analysis and their popularity has certainly inspired multiple ideas in deep learning. Nevertheless, deep learning models can be understood without considering SVMs. Other such examples where more advanced mathematics are lurking behind the scenes is in our consideration of optimization algorithms in Chapter 4. In this context, while most of today's methods are simple first-order optimization methods, the path to realize that (at the moment) these techniques work best has also gone through the study of much more advanced second-order optimization techniques and their variations.

We also mention that there is ongoing research on the theoretical properties of deep neural networks. In this domain, researchers make use of functional analysis, measure theory, information theory, stochastic processes, differential geometry, topology, category theory, and other mathematical fields to describe and understand theoretical properties of deep learning. While such research efforts are very exciting, they are not the topic of this book.

1.6 Notation and Mathematical Background

It is assumed that the reader has an understanding of basic mathematical notation and operations including sets, function notation, basic probability, and basic matrix algebra. Appendix A reviews key results from multi-variable calculus that are used, but other basics are assumed known. We now highlight a few notational elements that we use throughout and you may also consult the notes and references at the end of this chapter for suggestions of a few resources that may help with a review.

Vectors are in general considered as elements of $\mathbb{R}^n$ with n being some positive integer. We can denote a vector $u \in \mathbb{R}^n$ as $u = (u_1, \ldots, u_n)$ where the scalar u_i is the i-th coordinate of the vector. Written in this *tuple* form, we consider the vector as a column vector. That is, we may consider for example a matrix $A \in \mathbb{R}^{m \times n}$ and then have the matrix-vector product $v = Au$ with $v = (v_1, \ldots, v_m) \in \mathbb{R}^m$, and each v_i given via,

$$v_i = \sum_{j=1}^{n} A_{i,j} u_j.$$

An alternative way to represent a vector is using square brackets in which case we take its orientation literally. For example $w = [w_1 \; \cdots \; w_m]$ is a row vector and its transpose, $w^\top$ is a column vector and an element of $\mathbb{R}^m$. In this case, we may, for example consider the vector-matrix product $x = wA$ which yields x, an m-dimensional row vector. Hence in summary, with our vector notation for $u \in \mathbb{R}^n$ we have,

$$u = \begin{bmatrix} u_1 \\ u_2 \\ \vdots \\ u_n \end{bmatrix} = [u_1 \; \cdots \; u_n]^\top \equiv (u_1, \ldots, u_n).$$

Some key vectors we consider are $\mathbf{1} \in \mathbb{R}^n$ which is a vector of all 1s. Similarly, 0 can be taken as a vector of all 0s. However, the same symbol, 0, is clearly also used for scalars, and matrices with the actual meaning clear from context. There are also the unit vectors e_i with dimension taken from context and $i = 1, \ldots, n$ in case they are n dimensional. Each such unit vector has 0 entries everywhere except for 1 in the ith coordinate. So as an example of this notation, see that the identity matrix $I \in \mathbb{R}^{n \times n}$ can be represented as a sum of outer products,

$$I = \sum_{i=1}^{n} e_i e_i^\top.$$

The norm $\| \cdot \|$ is often used and unless stated otherwise it is the vector L_2 norm. That is, for $u \in \mathbb{R}^n$, $\|u\| = \sqrt{u^\top u}$. Similarly $\|u\|^2 = u^\top u$.

Surprisingly, beyond matrix vector arithmetic, norms, and a few basic operations, we only seldom use linear algebra machinery such as eigenvalues, and the singular value decomposition. This is because most of the book is focused on the basic mathematical engineering of deep learning which at its core, does not use very sophisticated mathematics but rather employs basic mathematical concepts to create sophisticated models. We thus leave it to the reader to seek a self-refresher on such topics as they arise. For example, some of our discussion of unsupervised learning in Chapter 2 builds on the understanding of linear algebra. However, these topics are in general tangential to the core material of the book.

In terms of concepts of optimization theory, Chapter 4 covers these in a self-contained manner that does not require further background beyond a few multivariate calculus results, reviewed in Appendix A. Hence for example, convexity is briefly introduced and summarized directly in Chapter 4. At certain points, the reader may want to dig deeper and review topics individually elsewhere.

Finally, in terms of probabilistic concepts, it is assumed the reader is aware of the informal definition of probability $\mathbb{P}(\cdot)$ and basics of random variables. This also includes the expectation $\mathbb{E}[X]$ of a random variable X, the variance, denoted $\mathrm{Var}(\cdot)$, and covariance. Similarly to the linear algebra concepts, we do not make heavy use of probability theory, however certain probabilistic computations occasionally arise.

Notes and References

Since this book is purposefully agnostic both to implementation issues and to the historical account of deep learning, we use an unnumbered section such as this at the end of every chapter to provide keynotes and references. More additional information is also available via the book's website.[15] The website also includes links to articles and references, tutorials, videos, and mathematical background.

In terms of implementation, at the time of writing this book, almost any introductory (or advanced) work in deep learning uses the *Python* language. See for example [342] for a general intermediate guide to Python. Neat examples in Python for general machine learning are in [240], and that resource can also serve as a comprehensive mathematical introduction to the whole field of machine learning and data science. Another language slowly stepping into the spotlight is *Julia*, see [303] for a statistical introduction to the language and [338] to get a taste of scientific machine learning, where Julia is often used. Further, the *R statistical computing system* has a vast user base in statistics and supports tens of thousands of packages as well as multiple interfaces to deep learning frameworks. See for example [245] for a classical introduction to statistics with R and [10] for ways of using R with deep learning. Until around 2015, programming in *C/C++* for deep learning was the norm. However, these days it is much more of a specialty and is only required in very specific cases. See for example [319] for a classical text involving pattern recognition and C/C++. Paid language platforms such as *Mathematica*, *Maple*, and *MATLAB* are attempting to retain some of their user base by incorporating deep learning libraries. However to date, these commercial systems have shown to be much less mature with deep learning in comparison to the open source arena.

Beyond the computer languages in use, a key aspect of deep learning is the usage of deep learning frameworks. These are software packages for training and execution of the model. In the near past, the most popular framework was *TensorFlow* supported with the higher level interface of *Keras*. See for example [133]. However more recently *PyTorch*, see for example [320], has gained much more popularity and it has higher level interfaces such as *fast.ai* or *PyTorch Lightning*. See also *Flux.jl* in the Julia arena. In general, [190] provides a comprehensive applied introduction to deep learning with fast.ai and PyTorch. Further, consider the video lecture series coordinated by Andrew Ng[16]; see also [306]. Our book's website provides more information and suggested resources.

Some quantitative measures of the size of the human brain and other brains is in [175]. As we do not provide a complete historical account of deep learning and its relationship with neuroscience, we recommend the article [364] for a taste of recent work on neuroscience inspired by deep learning. Also see [165] for a comprehensive condensed review of the interface of the fields, both historically and today. An interesting style of neural network architecture which attempts to resemble the brain more closely than standard architectures is the *spiking neural network*; see for example [398]. More relationships with neuroscience as well as other views of deep learning can be found in other deep learning texts. A now classical deep learning book is [142], and a slightly more recent comprehensive textbook is [4]. See also the book [308], as well as a more recent book taking a mathematical approach [69].

A recognized reference for the *Turing test* from 1950 is [404]. The first major work on neural network architectures is the *perceptron* by Frank Rosenblatt [353] in 1958. A major landmark that helped spark general attention in deep learning is the *AlexNet* success in the 2012 ImageNet challenge [239]. The literature and work between Rosenbladtt's early work and the 2012 AlexNet success is too numerous to list here. It involved an evolution of many ideas and experiments over decades. A good starting point for the key references is the nature paper [249]. Further key references are mentioned at the end of the chapters in the sequel.

The creation of the *ImageNet* database by Fei-Fei Li, her team, and collaborators has had a profound affect on deep learning in the second decade of this century. ImageNet, [102], was created with the aid of Amazon's *Mechanical Turk* service and with the *WordNet*, [289], hierarchical word database. The *VGG19* model (see [380] as well as the *VGG16* model) mentioned in this chapter is one of several models that at the time of creation had near-top performance in the ImageNet challenge, associated with the ImageNet database. Many additional models are both more efficient and outperform VGG19 and VGG16. The frontier keeps improving. See for example the empirical analysis [70] to get a feel for the performance metrics at play. Chapter 6 provides more references to landmark convolutional architectures.

---

[15]See https://deeplearningmath.org/.
[16]See https://www.deeplearning.ai/.

Diffusion models gained significant prominence following [183]. The *generative adversarial network* idea and the association to game theory first appeared in Ian Goodfellow's co-authored work in 2015 [143]. For a key reference dealing with *AlphaGo* see [377] from 2016. An article describing the earlier (1999) *deep blue* chess playing system is [164]. Modern approaches to deep reinforcement learning are surveyed extensively in the texts [125] and [393]. The more recent *GPT-3* model is described in [66]. General texts about graph neural networks are [157] and [273]. The subsequent chapters contain more detailed references for each of these subject areas.

Our exposition has completely ignored the many *ethical aspects* associated with machine learning, deep learning, and artificial intelligence. This includes issues of disinformation, bias, and fairness (also in datasets), privacy and surveillance, algorithmic colonialism, and the many negative applications that one can cook up with deep learning and generative models. Ethical considerations are critical for anyone embarking on an applied deep learning project. As this is not the topic of this book, we suggest the resources and content from the fast.ai website[17] as a starting point on ethics in deep learning.

Since the rest of the chapters of this book are more mathematical than this chapter, some readers may wish to strengthen or revive their mathematical background. In the context of *linear algebra* for data science, the introductory book [56] is recommended. Further [391] provides more context with a rich variety of linear algebra topics and their interface with data science. A general mathematical review for machine learning is in [101]. Finally, in addition to exploring the appendices of the current book we also recommend the appendices of [240]. See also further resources in our book's website.

The research world of deep learning is expanding very quickly with multiple important directions that are beyond the scope of the book. One direction is *Bayesian neural networks*, see for example [213]. Another component is *causal modeling*; see for example [325]. Further, in the field of quantum computing, research on *quantum deep learning* is already available; see for example [130]. The applied world of deep learning is currently in a massive growth phase with large language models being one of the most exciting directions. A comprehensive survey relevant for the time of publishing of this book is [449].

---

[17]See https://ethics.fast.ai/.

2 Principles of Machine Learning

At its core, deep learning is a class of machine learning models and methods. Hence, to understand deep learning, one must have at least a basic understanding of machine learning principles. There are dozens of general machine learning methods and models that one can cover and our purpose here is certainly not to present a detailed account of all of these methods. Instead, we take a path that presents a general overview of machine learning, and then focuses mostly on linear models which are the most elementary neural networks out there. In the process we explore gradient-based learning for the first time, a topic that plays a key role in the chapters that follow.

In Section 2.1 we present an overview of the key activities of machine learning including supervised learning, unsupervised learning, and variants of these. In Section 2.2 we explore key elements of supervised learning. In Section 2.3 we introduce linear models. These form the basis for many other models as well as for the deep learning networks of this book. We then explore the basic gradient descent algorithm in Section 2.4. Linear models can often be trained without gradient descent, yet exploring gradient descent in the context of linear models is a useful warmup for the following chapters. In Section 2.5 we discuss generalization ability, overfitting, and introduce techniques of regularization. We further discuss the training process including splitting of the data and cross validation methods. Most of this book deals with supervised learning methods, however understanding basic techniques from unsupervised learning is also important. Hence, in Section 2.6, we take a brief look at unsupervised methods including K-means clustering, principal component analysis (PCA), and also touch the singular value decomposition (SVD).

2.1 Key Activities of Machine Learning

The world of machine learning intersects heavily with both the worlds of statistics and computer science. In statistics data and randomness are key. In contrast, in computer science, algorithms and computation are the focus. Machine learning borrows from both worlds and is about the combination of data and algorithms. It is all about training mathematical models on a computer in order to classify data, predict outcomes, estimate relationships, summarize data, control complex processes, and more.

We now present and characterize a few key activities of machine learning. These activities are carried out when training, calibrating, adjusting, or designing machine learning models and algorithms. Many of these activities are loosely called *learning* while other activities involve prediction or decision making.

Any activity of machine learning can be described as an interaction between the following entities: *data*, *models*, *algorithms*, and the *real world*. By *data* we mean both collected data such as the (x, y), features-label, pairs described in the previous chapter, or data that is generated as output of models or algorithms. By *models* we mean mathematical objects

DOI: 10.1201/9781003298687-2

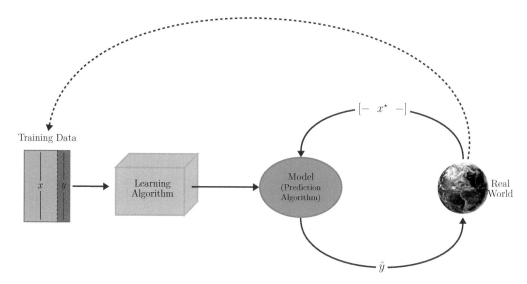

Figure 2.1: Supervised learning. Activities include training a model and prediction in production.

stored and implemented on a computer, together with the parameters that specify these objects. By *algorithms* we mean the procedures for creating models, procedures for creating output datasets, as well as procedures for using the models themselves for prediction or related tasks. Finally, by the *real world* we refer to scenarios that support generation of data, annotation of data, as well as usage of output data for decision making, and control.

Machine learning activities are often dichotomized into two broad categories, *supervised learning*, and *unsupervised learning*. With supervised learning, data is assumed to be available as (x, y) pairs where each feature vector x is labeled via y. In the case of unsupervised learning, one only observes data points x and tries to find relationships between the various elements, variables, or coordinates of x. To understand this terminology consider the learning of babies or toddlers, which only involves the exploration of input sensory data without any indication of what is what. This is unsupervised learning since toddlers are typically not told explicitly "this does this" and "that does that". Then later on, for example during school, they engage in supervised learning since language and text are used to present the learners with examples x and their outcomes y.

A key activity in supervised learning is the usage of data to learn/train models for *prediction*. See Figure 2.1. This prediction is called *classification* in case the labels y are from a finite discrete set and it is called *regression* in case the labels y are continuous variables. There are also other cases of prediction where the labels y are vectors, images, or similar. A related activity is obviously to use the trained models for prediction when presented with unlabeled data from the real world; this is illustrated on the right of Figure 2.1. Both the training of models and usage of models for prediction involves the execution of algorithms. Sometimes the trained model is called an "algorithm" as well since it may be integrated in part of bigger systems that use it. Most of this book focuses on supervised learning and we begin in the next section, Section 2.2, by overviewing key concepts of supervised learning.

With unsupervised learning there are other activities beyond prediction, regression, and classification. See Figure 2.2. One important activity is *clustering*, which focuses on finding groups of similar data samples. The output of algorithms that perform clustering are typically not considered as models but are rather modified datasets that incorporate the clustering information. Another key unsupervised learning activity is to carry out *data reduction* where high-dimensional vectors are transformed into lower-dimensional vectors that still encode some of the key relationships between variables. While unsupervised learning is not the focus of this book, several unsupervised learning algorithms are overviewed in Section 2.6.

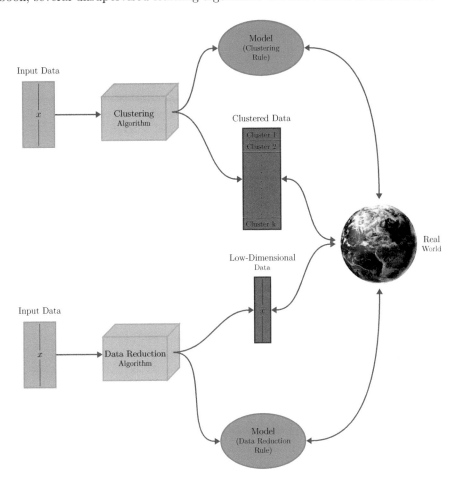

Figure 2.2: Some activities of unsupervised learning: Clustering (top) partitions the data. Dimension reduction (bottom) reduces the size of the features.

Beyond the dichotomy of machine learning into supervised and unsupervised, there are also additional popular activities that are not directly categorized as such. One popular class of activities is *reinforcement learning* introduced in Chapter 8. Here a temporal component is key and an *agent* is trained to carry out tasks in a dynamic environment. An additional class of activities is *generative modeling* also introduced in Chapter 8 in the context of variational autoencoders, diffusion models, and generative adversarial networks. Here models are trained to create artificial datasets with characteristics (or a distribution) similar to the input dataset. An additional suite of activities is *transfer learning*. Transfer learning is all about taking models that have been trained for one domain and adapting them to other domains with new data. Related is *active learning* where the learning process is not static but is rather informed by the performance of the model on unseen data. This is very closely related to *semi-supervised learning* where like supervised learning, there are both feature vectors x and labels y, yet only a subset of the feature vectors have accompanying labels. The learning process tries to use all of the available data. Finally, *self-supervised learning*, briefly discussed in the context of deep learning natural language processing in Chapter 7, creates models where sequences of data are used to self-predict the future or missing elements of the sequences. This is useful for language models and related tasks.

Data: Seen, Unseen, Training, and Test

Data is a central part of machine learning. In considering data it is important to distinguish between *seen data* and *unseen data*. Seen data is the data available for learning, namely for training of models, model selection, parameter tuning, and testing of models. Unseen data is essentially unlimited since it is all data from the real world that is not available while learning takes place but is later available when the model is used. This can be data from the future, or data that was not collected or labeled with the seen data.

Needless to say, for machine learning to work well, the nature of the seen data should be similar to that of the unseen data. The underlying assumption of machine learning is that the seen data used to create models is generated by underlying processes of the real world that are similar to the processes generating unseen data. Practically, one needs to carry out data collection and labeling so that this resemblance between the seen and unseen data is maintained.

A common practice in the world of machine learning is to split the seen data into *training data* and *testing data*. These are sometimes called the *training set* and *test set*; an additional name for the test set is the *hold out set*. The key idea with such a split is to use the training data for learning and to use the testing data for mimicking a scenario of unseen data. As described in Section 1.4 some popular example datasets come with such a predefined split. In other cases, it is up to the machine learning engineer to split the data randomly according to some predetermined proportions. Examples follow in this chapter.

Since the purpose of the train-test data split is to mimic the unseen data with the test set, one should not recalibrate, adjust, or tune models on the training set while testing repeatedly on the test set. Carrying out such a repetitive use of the test set would invalidate its resemblance of unseen data. For this reason one sometimes performs an additional split of the training data by removing a chunk out of the training data and calling it the *validation set*. More on this practice and other alternatives such as *K-fold cross validation* is in Section 2.5.

The practice of splitting data into the train set and test set is very popular in machine learning so long as a large number of samples is available. However, in some cases, there is not enough data to be able to separate a test set and thus one wishes to use all available data for model fitting. This is sometimes the case when working with experimental data and biomedical data. In such cases, statistical inference approaches for evaluating the quality of the model fit make heavier use of the model at hand. Some approaches for comparing models are *likelihood* based and include performance measures such as the *Akaike information criterion* (AIC) or *Bayesian information criterion* (BIC). The world of model fitting in a statistical content is vast and we do not focus on such methods further in this book.

Data Preprocessing

Raw data often requires *preprocessing* before it can be used for training, prediction, or as input to other machine learning models. Although a full description of data processing steps and practical aspects of data processing is beyond our scope, one important activity that we cover is *standardization of the data*, also sometimes called *normalization of the data*. This involves subtraction of the mean of each feature and division by the standard deviation of the feature.

Assume the values for some feature i are $x_i^{(1)}, \ldots, x_i^{(n)}$ where n is the number of data samples. The *sample mean* and *sample variance*[1] of the feature are respectively computed as,

$$\overline{x}_i = \frac{1}{n} \sum_{j=1}^{n} x_i^{(j)}, \qquad s_i^2 = \frac{1}{n} \sum_{j=1}^{n} (x_i^{(j)} - \overline{x}_i)^2. \tag{2.1}$$

Further, the *sample standard deviation* is the square root of the sample variance and is denoted via s_i. With these basic descriptive statistics of the feature available we may standardize the data samples of each feature $i = 1, \ldots, p$ to obtain *standardized samples*,

$$z_i^{(j)} = \frac{x_i^{(j)} - \overline{x}_i}{s_i} \qquad \text{for} \qquad j = 1, \ldots, n. \tag{2.2}$$

Now the standardized data for feature i, $z_i^{(1)}, \ldots, z_i^{(n)}$, has a sample mean of exactly 0 and a sample standard deviation of exactly 1. In the case the data samples of the feature are distributed according to a normal distribution,[2] then most standardized samples would lie in the range $[-3, 3]$. Even if the data is not normally distributed, the standardized samples will still lie in the vicinity of this range and are centered about 0.

Such standardization is useful as it places the dynamic range of the model inputs on a uniform scale and thus improves the numerical stability of algorithms. It also allows us to use similar models for different datasets that may, without standardization, have completely different dynamic ranges. In Section 2.4 we discuss how such standardization can also help optimization performance.

Learning $\approx$ Optimization

Almost any form of a learning or model training activity involves optimization either explicitly or implicitly. This is because learning is the process of seeking model parameters that are "best" for some given task. In fact, all of Chapter 4 is devoted to optimization techniques in the context of machine learning and deep learning, and a few of the sections of this current chapter contain aspects of optimization as well.

In some cases optimization is carried out directly on some performance measure that quantifies how good the model at hand performs. This is, for example, the case when one considers the *mean square error* criterion for regression problems, a concept which we study in detail in this chapter. However in other cases, a *loss function* is engineered for the problem at hand in a way that minimization of the loss function is a proxy for minimization of

---

[1]In a statistical context one often uses $n - 1$ in the denominator of the sample variance instead of n. For non-small n this distinction is insignificant.

[2]A few attributes of the normal distribution, also known as the Gaussian distribution, are in Appendix B.

the actual performance measures that are of interest. This is, for example, the case when considering classification problems and aiming to get the most accurate classifier. In such a case, optimization is typically not carried out directly on the accuracy measure but rather on a loss function such as the mean square error, or cross entropy defined in Chapter 3. In any case, the design of loss functions as part of the learning procedure is central to machine learning and deep learning and appears throughout this book.

Note that in some cases, machine learning algorithms such as *decision trees* do not directly specify an optimization procedure but rather execute a predefined algorithm for fitting a model. However, even in such cases, there is typically an inherent hidden optimization problem associated with the procedure. Hence in general we can think of "learning" as the process of carrying out some sort of "optimization".

2.2 Supervised Learning

We now focus on supervised learning and outline key concepts, practices, and terminology. Supervised learning is about predicting an outcome $\hat{y}$ for y, where the prediction is based on a vector of input features x. When y is from a finite discrete set, then the task is called *classification* and when y is a continuous variable, then the task is called *regression*.

We begin with overviewing basic regression in the context of linear models and feature engineering. We then discuss aspects of binary classification which is the case when y only attains one of two possibilities. The more general *multi-class classification* case in which y takes on multiple possibilities is presented as part of specific examples in Section 2.3. We close this section with a high-level overview of several methods and general approaches to supervised learning.

Regression and Feature Engineering

We begin by considering a very simple *univariate* example where the scalar ($p = 1$) feature x is the average number of rooms per dwelling and y is the median value of owner-occupied homes in thousands of dollars. These variables, respectively denoted rm and medv, represent data from the well-known Boston housing dataset.[3] A regression model $y = f_\theta(x)$ attempts to predict the median house price as a function of the average number of rooms.

To illustrate this concept, consider the well-known *simple linear regression model* where $f_\theta(x) = \beta_0 + \beta_1 x$ and the parameter vector is $\theta = (\beta_0, \beta_1)$. Notice that in this case the dimension of the parameter vector is $d = 2$. This model can also be described statistically via,

$$y = \beta_0 + \beta_1 x + \epsilon, \tag{2.3}$$

where ϵ represents the *error* or noise term as it models the gap between y and $f_\theta(x)$. In statistical theory and practice, assumptions about the probability distribution of ϵ go a long way as they support inference outputs such as *confidence bands, hypothesis tests*, and more. However in practical machine learning culture, one often ignores ϵ and such statistical assumptions.

---

[3]This dataset originally published in [161], has 506 observations where each observation is associated with a suburb or town in the Boston Massachusetts area. Of these observations, 16 are capped at rm = 50 and we remove these to stay with $n = 490$ observation.

Provided feature data $x^{(1)}, x^{(2)}, \ldots, x^{(n)}$ ($\mathtt{rm}$ – average number of rooms per house in a geographical area), and corresponding label data $y^{(1)}, y^{(2)}, \ldots, y^{(n)}$ ($\mathtt{medv}$ – median house prices in a geographical area), the training process involves finding a suitable or best $\hat{\theta} = (\hat{\beta}_0, \hat{\beta}_1)$. In Section 2.3 we study the process for finding $\hat{\theta}$ via minimization of a loss function, yet at this point let us just consider the *model parameters*, also known as *parameter estimates* $\hat{\theta}$, as an outcome of training.

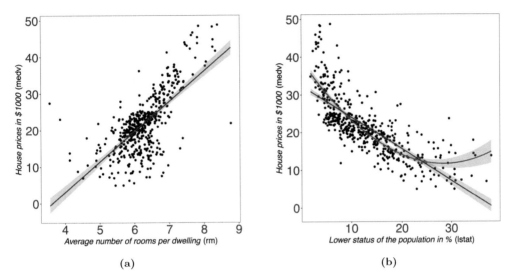

(a)   **(b)**

Figure 2.3: Examples of elementary linear models. (a) Median house prices per locality ($\mathtt{medv}$) as a function of the average number of rooms per dwelling ($\mathtt{rm}$) is described via a simple linear (affine) relationship. (b) House prices as a function of lower status of the population in % ($\mathtt{lstat}$) is not described well with a linear relationship (red), but by introducing an additional quadratic engineered feature it is described well via a three-parameter linear model resulting in a quadratic fit (blue).

Figure 2.3 (a) presents a scatter of the $(x^{(i)}, y^{(i)})$ pairs. The parameters estimated for this model are $\hat{\beta}_0 = -30.01$ and $\hat{\beta}_1 = 8.27$. The figure also includes a plot of the fit or estimated model $f_{\hat{\theta}}(\cdot)$ as a red line. Clearly, with such a model, any unseen (new) observation $x^\star$ can be used to make a prediction $\hat{y} = f_{\hat{\theta}}(x^\star)$. If one is willing to make statistical assumptions about the error ϵ and probabilities of error then an extra benefit of the model is the confidence bands, presented as a gray shaded area around the red line. Most of the modeling described in this book uses very complex models that do not take such a statistical approach and hence built-in inference outputs such as these confidence bands are often not available.

We also mention that the estimated parameters in such a model have an *interpretation*. For example, $\hat{\beta}_1 = 8.27$ indicates that increasing the average number of rooms by one room implies an average rise in median price of $\$8.27K$. Many types of statistical models have the benefit of interpretable parameters, yet in the world of machine learning where models are often very complex, parameter interpretation is an exception rather than the rule.

As a follow up example arising from the same dataset, consider the relationship between the variable x taken as the percentage of the population that is of a low social economic status ($\mathtt{lstat}$) and the variable y taken again as $\mathtt{medv}$. A simple linear model fit to this will yield parameter estimates $\hat{\theta}$ since in almost any case one can fit any model to data. However the model may not always be suitable for the data or process at hand. For example, in Figure 2.3 (b), a scatter plot of the data is presented and it is apparent that the downward sloping linear model fit in red does not do a good job in describing the relationship between x and y. Observe also that confidence bands for this simple linear model may look deceptively appealing (tight gray bands around the red line) especially if one was only to look at these and not the actual scatter plot of the data. The pitfall is that these confidence bands are computed under the assumption that the model fits the underlying process and data well, a case that does not hold here. Such a phenomena is often loosely called *model misspecification* and is one of the risks that one undertakes (and needs to mitigate) when using statistical inference techniques.

An alternative to the simple linear model is to seek a richer relationship such as,

$$y = \beta_0 + \beta_1 x + \beta_2 x^2 + \epsilon. \tag{2.4}$$

One way of describing this relationship is via the function $f_\theta(\cdot)$ defined for $d = 3$-dimensional θ via, $f_\theta(x) = \beta_0 + \beta_1 x + \beta_2 x^2$ where x is still a scalar ($p = 1$). An alternative description of the same model is to consider the squaring of the `lstat` variable as a new *engineered feature* and thus now consider x as a $p = 2$-dimensional vector with x_1 being the original feature and $x_2 = x_1^2$ the new engineered feature. In this case the model function is linear (affine) $f_\theta(x_1, x_2) = \beta_0 + \beta_1 x_1 + \beta_2 x_2$, and the fact that x_2 is the square of x_1 is considered a feature engineering aspect and not a model function aspect. In practice, in this case, both approaches are identical and yield the same θ. Figure 2.3 (b) presents the fit and corresponding error bands of this quadratic model in blue.

We mention that linear models for regression, which are the workhorse of classical statistics, can be extended in many ways. The notes at the end of this chapter point at some extensions such as *Generalized Linear Models* (GLM), *mixed models*, and more. Note also that non-linear relationships in a regression context could be explored using smoothing techniques. Popular techniques in this framework are the *Generalized Additive Model* (GAM), the *Locally Estimated Scatterplot Smoothing* (LOESS) method, as well as *Nadaraya-Watson kernel regression*. We also mention that generally when one considers feature engineering, one very important aspect is dealing with *interaction terms*. This means creating new engineered features based on the products of other features.

The world of machine learning has adopted these models often removing statistical assumptions (e.g., about the noise ϵ) while introducing additional non-linearities and mechanisms that yield very expressive models. The deep neural networks we cover in this book include one such rich class of examples. We also mention that while the numerical house price context that we presented here appears to be simple and low-dimensional, Regression problems can often involve extremely high-dimensional input feature vectors. For example, any regression problem where the input data is an image is of this nature. A concrete example of such a case is using images of a human face to predict the age of the person.

Binary Classification

Moving on from regression problems where y is continuous, we now consider binary classification where y attains one of two values, which are sometimes referred to as `positive` or `negative`. There are dozens of machine learning methods for binary classification and our purpose here is not to explore how these methods work. Instead we wish to illustrate how their performance is quantified.

Our exposition relies on a *logistic regression* based classifier where `positive` samples are encoded via $y = 1$ and `negative` samples are encoded via $y = 0$. Note that, in other scenarios positive and negative samples are sometimes encoded via $y = +1$ and $y = -1$, respectively. Logistic regression is explored in depth in Sections 3.1 and 3.2 of Chapter 3. For now, we can treat such a classifier as being based on a function $f_\theta(x)$ where x is the feature vector. The output of $f_\theta(x)$ is a number in the continuous range $[0, 1]$ indicating the probability that x matches a `positive` label. Hence, the higher the value of $f_\theta(x)$, the more likely it is that the label associated with x is $y = 1$.

With the model $f_\theta(\cdot)$ at hand, a classifier can be constructed via a *decision rule* based on a threshold τ, with the predicted output being,

$$\widehat{y} = \begin{cases} 0 \text{ (negative)}, & \text{if} \quad \hat{y} \le \tau, \\ 1 \text{ (positive)}, & \text{if} \quad \hat{y} > \tau, \end{cases} \qquad \text{where} \qquad \hat{y} = f_\theta(x). \qquad (2.5)$$

In many cases, one selects the threshold at $\tau = 0.5$. However, as we see below, τ can often be adjusted. Also note the notational difference between $\widehat{y}$ and $\hat{y}$. Throughout the book, we use the former to signify the actual predicted label for classification problems and latter to denote the output of the model which is usually (continuous) numerical in nature.

As an example we consider breast cancer prediction where the label, or outcome variable y has $y = 1$ in case of malignant lumps and $y = 0$ in case of benign lumps. A popular dataset in this context is called the *Wisconsin Breast Cancer Dataset* and is based on clinical data released in the early 1990s.[4] We make further use of this dataset in Section 3.1 where we dive into logistic regression. The feature vector x is of dimension $p = 30$ and is composed of continuous variables such as `radius_mean`, `texture_mean`, etc., each potentially affecting the probability of malignancy. The data has $n = 569$ observations and here we use the *80-20 splitting strategy* where we split it randomly between training data and testing data to have $n_{\text{train}} = 456 \approx 0.8 \times n$ training observations and $n_{\text{test}} = 113 \approx 0.2 \times n$ testing observations. Note that in some of the sections below, we simply use n to denote n_{train}, when not considering the test set explicitly.

We may train different forms of logistic regression for this data where we take x to be some subset or transformation of the original feature vector. At one extreme we can take x to be a single variable, and at another extreme x can be the full feature vector or even have additional engineered features similar to the home pricing example above.

Here for simplicity, we consider two logistic regression models. For the first model we use only a single feature in x which is `smoothness_worst` where the actual physical meaning of this variable is not critical for our understanding at this point. This model is denoted $f_{\theta;\, p=1}(\cdot)$. In the second model we consider all 30 features in the dataset. This model is denoted $f_{\theta;\, p=30}(\cdot)$. For each of these models, we use the n_{train} observations to obtain an estimated parameter vector where in the case of $p = 1$, the estimated parameters $\hat{\theta}$ is of dimension $d = 2$ and in the case of $p = 30$ the estimated parameters $\hat{\theta}$ are of dimension $d = 31$. Details on the actual meaning of the models and parameters of logistic regression are in Chapter 3. With these models at hand, at first let's fix τ and consider several performance measures of the classifiers defined via (2.5).

---

[4]See https://archive.ics.uci.edu/ml/datasets/breast+cancer+wisconsin+(Diagnostic) for more information and relevant references.

The standard way to compute binary a classification performance measure is to evaluate the classifier on the test set. This then allows us to consider the predictions $\widehat{y}^{(i)}$ and compare them to the test set labels $y^{(i)}$, for $i = 1, \ldots, n_{\text{test}}$. Observations where $\widehat{y}^{(i)} = y^{(i)}$ are counted as either *True Positive* (TP) or *true negative* (TN), depending on the value of $y^{(i)}$ being 1 or 0, respectively. Similarly, observations where $\widehat{y}^{(i)} \neq y^{(i)}$ are counted as either *False Positive* (FP) or *False Negative* (FN). These four counts, TP, TN, FP, and FN total up to n_{test} and are customarily summarized in the *2 by 2 confusion matrix*:

		Decision		
		Decide 0	Decide 1	
Reality	Label 0	True Negative (TN)	False Positive (FP)	Specificity $\frac{TN}{TN+FP}$
	Label 1	False Negative (FN)	True Positive (TP)	Sensitivity/Recall $\frac{TP}{TP+FN}$
		Negative Predictive Value (NPV) $\frac{TN}{TN+FN}$	Precision/Positive Predictive Value (PPV) $\frac{TP}{TP+FP}$	

Observe the ratios at the margins of the matrix that present various performance indicators, namely *sensitivity* (also known as *recall*), *specificity*, *precision* (also known as *positive predictive value*), and *negative predictive value*. In general, we would like all of these ratios to be as close to 1 as far as possible, however as we describe below, there are often tradeoffs.

An additional natural performance measure is the *accuracy*. It is simply defined as the proportion of correctly classified samples, namely,

$$\text{Accuracy} = \frac{\text{TP} + \text{TN}}{n_{\text{test}}}. \tag{2.6}$$

This is often the first performance measure one considers[5], yet for unbalanced data it can be an extremely misleading measure. Indeed, a degenerate classifier that always predicts the most abundant class will have an accuracy equal to the proportion of that class. For example, in binary classification if the `positive` class constitutes only 5% of the samples, then the degenerate classifier which always predicts $\widehat{y} = 0$ will have an accuracy of 95%. Then, even when considering a non-degenerate classifier, one observes an accuracy that is typically not worse than 95% and if for example the accuracy is 99%, it is still not an indication that the classifier works well.

A more robust analysis of performance considers the competing objectives of *sensitivity* and *specificity*, where a high sensitivity (or recall) value indicates a good ability of the classifier to detect `positive` samples and a high specificity value indicates a good ability to detect `negative` samples. In the case of a threshold-based classifier as in (2.5), varying τ alters the confusion matrix and thus the sensitivity and specificity values are modified as well.

A parametric curve based on τ known as the *Receiver Operating Characteristic (ROC) curve* is often used to visualize this tradeoff between sensitivity and specificity where the x-axis

---

[5]Note that the formula here is for binary classification, yet this performance measure is also used for multi-class classification where the numerator TP+TN needs to be replaced by the total number of correctly classified samples.

plots one minus the specificity also known as the *false positive rate*. Observe from (2.5) and the formulas at the bottom margin of the confusion matrix, that as $\tau \to 0$ the number of false negatives vanishes and hence the sensitivity approaches 1. Similarly, as $\tau \to 1$ the number of false positives vanishes and hence the specificity approaches 1. More generally, as we vary τ between 0 and 1 a tradeoff emerges as captured in the ROC curve. This allows us to tune the threshold τ for balancing sensitivity and specificity, depending on the problem at hand. Figure 2.4 presents (smoothed) ROC curves for the breast cancer example where we compare the ROC curves for the $f_{\theta;\, p=1}(\cdot)$ and $f_{\theta;\, p=30}(\cdot)$ models, as well as a "coin flip" model (chance line) and the "ideal" model.

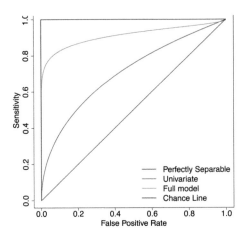

Figure 2.4: Receiver operating characteristic (ROC) curves for the breast cancer data. One model is a univariate model and the other is a full model. A chance line (guessing model) and a perfectly separable line (ideal model) are also plotted. For each model, the ROC captures the tradeoff between the sensitivity and the false positive rate (one minus specificity).

Receiver operating characteristic curves allow us to assess the quality of models taking all possible threshold parameters into account. A related measure that tries to quantify the quality of a curve into a single number is the *area under the curve* (AUC) measure. For a classifier with an ROC curves that achieves perfect sensitivity under any level of specificity, this measure is at 1 and corresponds to the perfectly separable green curve in Figure 2.4. However, for classifiers that just choose a random class, this measure is at 0.5 corresponding to the chance line red line in Figure 2.4. In the case of the breast cancer data, we see that on the test set the AUC for the $f_{\theta;\, p=1}(\cdot)$ model is 0.70 and for the $f_{\theta;\, p=30}(\cdot)$ model it is at 0.92. This may give an indication that the additional features in the richer model help obtain a better predictor.

Let us now fix the threshold at $\tau = 0.5$ and compare a few more performance measures. The test accuracy in this case is 0.73 for the $f_{\theta;\, p=1}(\cdot)$ model and 0.89 for the $f_{\theta;\, p=30}(\cdot)$ model. However, since the number of `positive` samples in the test set is 40 (out of $n_{\text{test}} = 113$), we see that this dataset is somewhat unbalanced, and hence accuracy is not a good measure of performance. In such cases, machine learning practice typically focuses on both precision and recall (sensitivity). Note that this could have alternatively been a focus on specificity and sensitivity (recall), but in machine learning the precision–recall pair is more popular. Precision, similar to specificity, approaches 1 as the number of false positives FP approaches 0. However precision is based on the true positives number (TP), while specificity is based on the true negatives (TN) value.

For the $f_{\theta;\,p=1}(\cdot)$ model with $\tau = 0.5$ we have

$$\text{Precision} = \frac{\text{TP}}{\text{TP} + \text{FP}} = 0.70 \qquad \text{and} \qquad \text{Recall} = \frac{\text{TP}}{\text{TP} + \text{FN}} = 0.4,$$

and for the $f_{\theta;\,p=30}(\cdot)$ model with $\tau = 0.5$ we have Precision $= 0.82$ and Recall $= 0.9$.

A popular way to consider both precision and recall is by averaging them using the harmonic mean of the two values. This is called the F_1 *score* and is computed as follows:

$$F_1 = \frac{2}{\frac{1}{\text{Precision}} + \frac{1}{\text{Recall}}} = 2\frac{\text{Precision} \times \text{Recall}}{\text{Precision} + \text{Recall}}. \tag{2.7}$$

In our example, for the $f_{\theta;\,p=1}(\cdot)$ model with $\tau = 0.5$ we have $F_1 = 50.8\%$ and for the $f_{\theta;\,p=30}(\cdot)$ model with $\tau = 0.5$ we have $F_1 = 85.7\%$. Note that one may also use F_1 scores to calibrate the threshold τ. Sometimes one uses a generalization of F_1 called the F_β where β determines how much more important recall is in comparison to precision. However, in general, if there is not a clear reason to price false positives and false negatives differently, then using the F_1 score as a single measure of performance is sensible.

In general, cases of unbalanced data should be treated with caution not just in terms of performance measures and threshold calibration, but also in terms of inference. There are multiple techniques for handling unbalanced data, some of which include over-sampling or under-sampling to balance the data. One of the more popular techniques is called *synthetic minority oversampling technique* (SMOTE). See the notes and references for further details.

Approaches and Algorithms for Supervised Learning

We cannot cover all aspects of supervised learning in a single section, a single chapter, or even in a single book. Yet now, after getting a taste of supervised learning in the context of regression and classification, let us discuss a few general approaches for supervised learning. Toward that end we first distinguish between *discriminative models* and *generative models*. Most of the models in this book are discriminative. This means that when viewed through a probabilistic lens (even though we mostly do not do that), they are based on learning aspects of the distribution $\mathbb{P}(y \mid x)$, i.e., the conditional distribution of the label y given the feature vector x. This is the case for linear models, logistic regression, general neural networks, and multiple additional models and algorithms that are not covered in this book.

In contrast, generative models involve learning the joint distribution $\mathbb{P}(x, y)$. As a byproduct knowledge of $\mathbb{P}(x, y)$ also means knowing the marginal distributions $\mathbb{P}(x)$ and $\mathbb{P}(y)$ as well as the conditional distributions $\mathbb{P}(y \mid x)$ and $\mathbb{P}(x \mid y)$. Hence generative models consider all of the data relationships as being learned, not just $\mathbb{P}(y \mid x)$. In this book, the most prominent appearance of generative models is in the context of variational autoencoders, diffusion models, and generative adversarial networks, appearing in Chapter 8. Another elementary generative type of model that we do not cover is the famous *naive Bayes classifier*, most notably known for early success of e-mail spam detection applications. One more common type of generative model is *linear discriminant analysis* (LDA) used in many experimental statistical contexts.

Naive Bayes classifiers are based on certain independence assumptions for $\mathbb{P}(x, y)$. We assume that given the label y, all features $x_1, \ldots, x_p$, are mutually independent. This (naive) independence assumption then allows us to represent the likelihood function[6] of the sample easily and in turn this enables efficient generative learning. LDA-based classifiers are also generative models (even though the name "discriminant" might be misleading). These classifiers fit a multivariate normal model to the data to carry out classification based on linear boundaries. Further details of these classifier models and algorithms are beyond our scope.

In terms of discriminative models, the linear models, logistic regression models, and more general deep learning models in this book are very common examples. Linear models and logistic regression models are simple deep neural networks. Linear models are studied in detail in this chapter. Logistic regression models and generalizations are the focus of Chapter 3, general fully connected deep learning models are the focus of Chapter 5, and other specialized deep learning models are in Chapters 6 and 7. Beyond these deep learning models, other types of popular machine learning models that we do not cover in the book include *support vector machines (SVM)*, *decision trees*, and their generalizations, which include *random forests* as well as *gradient boosted trees*. There are also additional elementary models often used for instruction of machine learning such as the class of *K-nearest neighbors* classification models. Indeed, the world of machine learning is vast with ideas and algorithms for creating both discriminative and generative models. The notes and references at the end of this chapter point at key resources.

2.3 Linear Models at Our Core

In this section, we focus on linear models which are the basis for many other models including deep neural networks. This is the first model in the book where we explicitly use a loss function for learning. The basic principles of the linear model and the associated loss function alternatives extend to more advanced models covered in the sequel. Similarly, other concepts that we cover here in the context of the linear model, such as the treatment of categorical variables and aspects of multi-class classification, are also relevant for more advanced models.

For the linear model, let us consider a feature vector $x = (x_1, \ldots, x_p) \in \mathbb{R}^p$ and the output/response variable $y \in \mathbb{R}$. The linear model links the output y to the features x through

$$y = b + w^\top x + \epsilon \tag{2.8}$$

where the scalar parameter b is called the *intercept* or *bias*, the vector parameter w is called the *regression parameter* or *weight vector*, and the ϵ term represents the noise or error. This is a generalization of the simple linear regression model (2.3) allowing x to be a vector and we now denote β_0 via b. To facilitate the presentation of key concepts of linear models, we often use a more compact representation of the model

$$y = \begin{bmatrix} b & w_1 & \cdots & w_p \end{bmatrix} \begin{bmatrix} 1 \\ x_1 \\ \vdots \\ x_p \end{bmatrix} + \epsilon = \theta^\top \tilde{x} + \epsilon, \tag{2.9}$$

---

[6]The likelihood function is a basic statistical concept that we survey in Chapter 3 in the context of logistic regression.

where $\theta = (b, w_1, \ldots, w_p)$ encapsulates both b and w and the feature vector x is extended to $\tilde{x}$ via a constant unit in its first position. The dimension of θ is the number of parameters and in this case it is $d = p + 1$.

In order to use the linear model for prediction of unseen data we have to *learn* the model. This means finding appropriate values for the parameters in θ based on training data such that the model performs well in prediction. Such a suitable learned parameter is further denoted via $\hat{\theta}$ and with such an estimate at hand, a prediction for a new data point $x^\star \in \mathbb{R}^p$ is given by,

$$\hat{y}(x^\star) = \hat{b} + \hat{w}_1 x_1^\star + \ldots + \hat{w}_p x_p^\star = \hat{\theta}^\top \tilde{x}^\star.$$

Learning the Linear Model

Consider a training dataset $\mathcal{D} = \{(x^{(1)}, y^{(1)}), \ldots, (x^{(n)}, y^{(n)})\}$ composed of a collection of n samples. For such data it is convenient to define the $n \times d$ dimensional *design matrix* X for the features, and the corresponding output response vector y, via,

$$X = \begin{bmatrix} | & | & & | \\ 1 & x_{(1)} & \cdots & x_{(p)} \\ | & | & & | \end{bmatrix} \quad \text{with} \quad x_{(i)} = \begin{bmatrix} x_i^{(1)} \\ \vdots \\ x_i^{(n)} \end{bmatrix}, \quad \text{and} \quad y = \begin{bmatrix} y^{(1)} \\ \vdots \\ y^{(n)} \end{bmatrix}. \tag{2.10}$$

Using this notation we can express the linear model for all the samples of the training set via

$$y = X\theta + \epsilon,$$

with $\epsilon = (\epsilon_1, \ldots, \epsilon_n)$ representing a vector of noise. From this representation, given a learned parameter vector $\hat{\theta}$, we can further define the predicted output vector of the model for the input training data via,

$$\hat{y} = X\hat{\theta}, \qquad \text{where} \qquad \hat{y} = \begin{bmatrix} \hat{y}^{(1)} \\ \vdots \\ \hat{y}^{(n)} \end{bmatrix}.$$

A suitable value for $\hat{\theta}$ will yield $\hat{y} \approx y$. This closeness is captured via a *loss function*,

$$C(\theta \, ; \, \mathcal{D}) = \frac{1}{n} \sum_{i=1}^{n} C_i(\theta), \tag{2.11}$$

where $C_i(\theta) := C_i(\theta \, ; \, y^{(i)}, \hat{y}^{(i)})$ is the loss for the i-th data sample. Specifically $\hat{\theta}$ is typically chosen so that the loss function is minimal at the point $\theta = \hat{\theta}$.

For the linear model, the most popular loss function is the *square loss* function also called *quadratic loss* where the loss for each data sample is

$$C_i(\theta) = (y^{(i)} - \hat{y}^{(i)})^2. \tag{2.12}$$

This loss penalizes each element $e^{(i)} := y^{(i)} - \hat{y}^{(i)}$, also known as the *error* or *residual* for sample i, quadratically. In this case, the loss for the entire training data can be represented

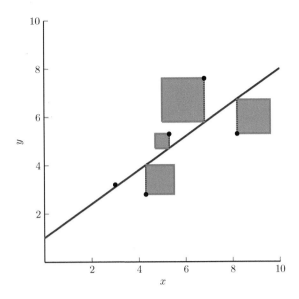

Figure 2.5: Squared loss visualization for one input feature $p = 1$. The sum of the area of the squares is the loss.

in terms of the L_2 norm $\| \cdot \|$ of the corresponding error vector,

$$C(\theta \, ; \, \mathcal{D}) = \frac{1}{n} \sum_{i=1}^{n} (y^{(i)} - \hat{y}^{(i)})^2 = \frac{1}{n} \|y - \hat{y}\|^2 = \frac{1}{n} \|e\|^2.$$

With this notation, by treating the learning of θ as an optimization problem and observing that the objective can be manipulated via monotonic transformations, we can now represent the learned parameter vector as

$$\hat{\theta} = \operatorname*{argmin}_{\theta \in \mathbb{R}^d} \frac{1}{n} \|y - X\theta\|^2 = \operatorname*{argmin}_{\theta \in \mathbb{R}^d} \|y - X\theta\|^2. \tag{2.13}$$

The search for θ that optimizes (2.13) is known as the *least squares problem*. Figure 2.5 presents a visual representation of the squared loss in the case of a single input feature ($p = 1$ and $d = 2$). In this case we seek a line specified by b and w_1 such that the sum of the squares (total area of blue boxes in the figure) is minimized.

The least squares solution can be easily derived by first computing the gradient of $\|X\theta - y\|^2$ with respect to θ using vector and matrix differentiation rules (see Appendix A) as

$$
\begin{aligned}
\frac{\partial \|y - X\theta\|^2}{\partial \theta} &= \frac{\partial (y - X\theta)^\top (y - X\theta)}{\partial \theta} \\
&= \frac{\partial \left(y^\top y - 2 y^\top X\theta + \theta^\top X^\top X\theta \right)}{\partial \theta} \\
&= -2X^\top y + 2X^\top X\theta.
\end{aligned}
\tag{2.14}
$$

Then, by setting the gradient to 0, we get the *normal equations*,

$$X^\top X\theta = X^\top y, \tag{2.15}$$

which describe vectors θ that obtain a zero gradient, and, in this case, it can also be shown that they are global minima of the objective (see further discussion in Chapter 4 about global and local minima).

The normal equations have a unique solution when the $d \times d$ matrix $X^\top X$, also known as the *Gram matrix* of X, is invertible. In this case we can represent the estimator as $\hat{\theta} = (X^\top X)^{-1}X^\top y$, or, by setting $X^\dagger = (X^\top X)^{-1}X^\top$, we have,

$$\hat{\theta} = X^\dagger y, \tag{2.16}$$

where $X^\dagger$ is called the *Moore-Penrose pseudo inverse* of X.

In fact, the Moore-Penrose pseudo-inverse, $X^\dagger$, can be represented in different ways. An alternative form to $(X^\top X)^{-1}X^\top$ is based on the *singular value decomposition*[7] (SVD) of X. Here, $X = U\Delta V^\top$ where U is an $n \times n$ orthogonal matrix, Δ is an $n \times d$ matrix with non-zero elements only on the main diagonal, and V is a $d \times d$ orthogonal matrix. Using the SVD we can represent the Moore-Penrose pseudo-inverse as

$$X^\dagger = V\Delta^+ U^\top, \tag{2.17}$$

where Δ^+ contains the reciprocals of the non-zero (main diagonal) elements of Δ, and has 0 values elsewhere. This SVD-based representation holds both if $X^\top X$ is singular or not. Hence (2.17) can be viewed as the more general representation of the pseudo-inverse. Note that $X^\top X$ is non-singular if and only if the matrix X is a full column rank matrix (i.e., the columns of X are linearly independent).

Note that, if X is not full column rank (i.e., $X^\top X$ is singular), then there is not a unique solution to the normal equations (2.15). However, the solution given via (2.16), using the SVD form (2.17), has a minimal norm for θ out of all possible solutions.

In the context of high-dimensional data when the number of features p is greater than the number of samples n, the design matrix X is never full column rank. This issue also appears when some of the features are linear combinations of the others. Even if X is mathematically full column rank, in some situations there is (strong) *multicollinearity* among the features, meaning that some of the features are approximately linear combinations of the others. This yields an $X^\top X$ matrix that is *ill-conditioned* and difficult to invert. In all these cases, the SVD-based representation (2.17) can be used in (2.16) to obtain a solution for (2.15).

Other Loss Functions

A first appealing result for the choice of the squared error loss function (2.12) is the closed-form solution (2.16) for (2.15). Also, when it is assumed that y is measured with uncorrelated Gaussian noise ϵ, using the squared error loss function (2.12) is equivalent to using a solution derived by the *maximum likelihood estimation method*. This is a technique widely used in

---

[7]Note that the form of SVD presented here is sometimes called the *full SVD*. A different form called the *reduced SVD* is used in Section 2.6 in the context of PCA.

statistics for parameter estimation which is further discussed in Section 3.1 in the context of logistic regression.

A second popular loss function for the linear model is the *absolute error loss*,[8]

$$C_i(\theta) = \left| y^{(i)} - \hat{y}^{(i)} \right|. \tag{2.18}$$

It is known to be more robust to outliers, however, even in the case of the linear model, there is no closed-form solution for estimating the parameters. From a statistical point of view, minimizing the *absolute error loss* is equivalent to maximum likelihood estimation when assuming a Laplace distribution[9] for the error noise ϵ. The Laplace distribution has heavier tails than the Gaussian distribution, see Figure 2.6 (b). With these tails, large noise values, i.e., outliers, are more probable than with the Gaussian noise, and hence the loss (2.18) is typically more robust to extreme values.

A third alternative, which is a hybrid between the absolute error loss and squared error loss, is the *Huber error loss*. It is parameterized by a hyper-parameter δ and represented via,

$$C_i(\theta) = \begin{cases} \frac{1}{2}(y^{(i)} - \hat{y}^{(i)})^2, & \text{if } \left| y^{(i)} - \hat{y}^{(i)} \right| < \delta, \\ \delta \left| y^{(i)} - \hat{y}^{(i)} \right| - \frac{1}{2}\delta^2, & \text{otherwise.} \end{cases} \tag{2.19}$$

This loss function penalizes small errors quadratically and deals with outliers by penalizing larger errors similarly to the absolute error loss. Figure 2.6 (a) provides a visual representation of these three loss functions. Even if it appears to be an appealing tradeoff between the absolute error loss and squared error loss, the Huber loss has the disadvantage of having to calibrate the arbitrary extra hyper-parameter δ and, like the absolute error loss, it does not have a closed-form solution.

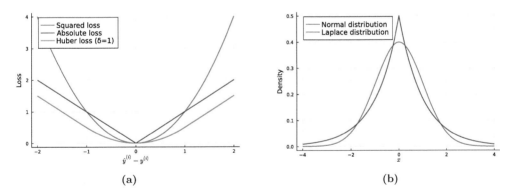

(a)

(b)

Figure 2.6: Loss function and error distribution alternatives. (a) Squared, absolute, and Huber loss functions. (b) Gaussian (normal) and Laplace error distributions.

---

[8]Note that the use of the absolute error loss (2.18) in (2.11) is sometimes called the L_1 loss as it is related to the L_1 norm. Similarly the use of the square loss (2.12) is sometimes called the L_2 loss.

[9]This is a probability distribution over $\mathbb{R}$ with density function in the variable u, proportional to $e^{-\frac{|u-\mu|}{b}}$, where $\mu \in \mathbb{R}$ is the mean and $b > 0$ is a scaling parameter.

Categorical Input Features

We have so far considered numerical input features. We now describe methods for dealing with categorical input features. The methods we present are useful for linear models as well as almost any machine learning and deep learning model.

Before describing the general method of using *one-hot encoding*, let us highlight two cases that sometimes receive special treatment. One such case is when the categorical feature is binary. For example, assume a feature that only takes on two values `red` or `blue`. In such a case, it can be encoded via 0 and 1 respectively and used as a numerical variable. A second special case is when the categorical feature is an ordinal variable, which may be interpreted as or converted to a numerical value. For example if the feature is a "user satisfaction rating" with values `low`, `medium`, and `high`, it may be transformed to numerical values 0, 1, and 2. Note however that this practice is sometimes problematic since different spacings between the assigned numerical values would yield different interpretations of the features and in general yield different models. For example, assigning numerical values of 0, 1, and 4 would indicate a bigger gap between `medium` and `high` than between `low` and `medium`.

Moving on to the general case of non-binary, nominal, categorical features one can use one-hot encoding, a method that we present now. Denote the number of possible values that the feature attains via L and here for simplicity, assume the feature values are 1, ..., L. The idea is to create L binary features in place of the categorical feature where, if the categorical feature is z, we construct an L-dimensional vector $\tilde{z} = (\mathbf{1}\{z=1\}, \ldots, \mathbf{1}\{z=\mathsf{L}\})$ with $\mathbf{1}\{\cdot\}$ denoting the indicator function taking on 0 or 1. That is, $\tilde{z} = e_z$ where e_z is the unit vector with 1 in the position z and 0 elsewhere; see the unit vector notation defined in Section 1.6.

Thus, each one-hot encoded categorical feature is expanded into such a vector and, with this encoding, the new transformed total number of features is,

$$\tilde{p} = p_{\text{num}} + \sum_{i \text{ categorical}} L_i,$$

where p_{num} is the number of numerical features and L_i is the number of possible values for categorical feature i.

To see how this one-hot encoding affects the design matrix, assume for simplicity that in addition to the p_{num} numerical features there is a single categorical feature with L levels. In this case the $n \times (1 + p_{\text{num}} + L)$ dimensional design matrix is

$$X = \begin{bmatrix} | & | & & | & | & & | \\ 1 & x_{(1)} & \cdots & x_{(p_{\text{num}})} & \tilde{z}_{(1)} & \cdots & \tilde{z}_{(L)} \\ | & | & & | & | & & | \end{bmatrix} \quad \text{with} \quad \tilde{z}_{(j)} = \begin{bmatrix} \tilde{z}_j^{(1)} \\ \vdots \\ \tilde{z}_j^{(n)} \end{bmatrix}.$$

Here $\tilde{z}_{(j)}$ for $j = 1, \ldots, L$ is the vector of indicator variables that marks which observations are at level j for the categorical feature. In statistics, the new L columns $\tilde{z}_{(j)}$ are called *indicator* or *dummy* variables. However, in statistics the practice is to include only $L-1$ dummy variables instead of L. One reason for this is that when using L dummy variables the design matrix X will never be a full column rank matrix since the sum of the L dummy columns is equal to the first column of 1s. Thus, traditional statistical practice only includes $L-1$ dummy variables in the model and the remainder is considered as the *reference level*.

An alternative is to remove the bias term from the model (meaning drop the first column of X) and then keep all L dummy variables.

Multi-Class Classification

Now that we understand the linear model as well as ways of dealing with categorical variables, let us consider an application of the linear model for multi-class classification. We note that linear models are generally far from the state-of-the-art when it comes to their application for classification problems, yet seeing the linear model applied to classification is instructive.

In classification problems each of the labels y takes on one of a finite number of values. The number of possible values is denoted via K. When $K = 2$ it is a binary classification problem but generally for $K > 2$, it is a *multi-class classification problem*. Notationally it is convenient to denote the set of label indices as $\{1, \ldots, K\}$ and consider some bijection between these indices and the actual label values e.g., banana, dog, etc. As a concrete example we consider the MNIST digits dataset where the label values are the digits 0,1,...,9 and notationally we use the label indices $\{1, \ldots, K = 10\}$. Here the obvious bijection shifts by 1.

A general scheme for multi-class classification is introduced in Section 3.3 in the context of multinomial (softmax) regression, and is further employed with other deep learning models in the following chapters. An alternative, which we introduce now in the context of linear models, is based on the fusion of multiple binary classification models into a multi-class classifier. For this we introduce two general methods, namely *one vs. rest* (also known as *one vs. all*) and *one vs. one*.

Both of these methods assume we have trained binary classifiers for sub-problems. With the one vs. rest strategy, we assume the availability of models for binary classification $f_{\theta_i}(\cdot)$ for $i = 1, \ldots, K$ where the ith model can discriminate between the label index i treated as positive and otherwise if the label index is not i then negative. With the one vs. one strategy we have $K(K - 1)$ binary classifiers[10] denoted $f_{\theta_{i,j}}(x)$ for all $i, j = 1, \ldots, K$ such that $i \neq j$. Here the $(i,j)th$ classifier discriminates between the label being of index i (positive) or index j (negative).

The output range obtained by $f_{\theta_i}(\cdot)$ or $f_{\theta_{i,j}}(\cdot)$ is generally a value on the real number line $\mathbb{R}$. Positive outputs indicate positive while negative outputs indicate negative. The farther the model output is from 0 the stronger the confidence of the classification decision. A cutoff in similar nature to (2.5) is to apply the $\text{sign}(\cdot)$ function[11] to the model output and conclude either positive in case of $+1$ or negative in case of -1.

Now the one vs. rest or one vs. one strategies carry out prediction via,

$$\hat{y} = \begin{cases} \text{argmax}_{i=1,\ldots,K} \, f_{\theta_i}(x) & \text{in case of one vs. rest,} \\ \text{argmax}_{i=1,\ldots,K} \sum_{j \neq i} \text{sign}\big(f_{\theta_{i,j}}(x)\big) & \text{in case of one vs. one.} \end{cases}$$

The idea of the one vs. rest strategy is to pick the label index which is most probable among the K classification models where each model focuses on a different label index. The idea of

---

[10]In practice only half of this number of classifiers is needed because the classifier for (i, j) can be reverted to the classifier (j, i).

[11]In case the classifiers were trained with the 1, 0 encoding as in the case of logistic regression it is easy to transform them.

the one vs. one classifier is to pick the label i that when compared to the other $K - 1$ labels, was chosen most often. This is achieved via comparison of a summation of $\text{sign}(f_{\theta_{i,j}}(x))$ for all other labels j. In both cases, one needs to supply rules for handling ties in the argmax in the final decision, yet these details are generally insignificant.

We proceed with an example of using both strategies for the MNIST dataset using a linear model. The crux in creating the supporting binary classifiers is to set the label vector y used in (2.16) to have values of $+1$ for samples that are `positive` and values of -1 for samples that are `negative`.

For example, when learning the $f_{\theta_3}(\cdot)$ classifier we consider all digit images in the original dataset with the label value 2 as having[12] $y = +1$ and otherwise -1. Out of the $60,000$ MNIST training samples $5,958$ training samples that satisfy $y = +1$ and then for $y = -1$ there are $54,042$ samples. Obtaining the parameters for this classifier using (2.16) uses the design matrix X as in (2.10) which is of dimension $60,000 \times 785$ where $785 = 1 + 28 \times 28$. Each row of X corresponds to a different image and each of the columns 2 to 785 corresponds to a different pixel.

Similarly, when learning the $f_{\theta_{3,8}}(\cdot)$ classifier (this compares the digit 2 and the digit 7) used in one vs. one, we set $+1$ for all $5,958$ training samples that have 2 and set -1 for all $6,265$ samples that have a label value of 7. Here the design matrix X is of dimension $12,223 \times 785$ since $5,958 + 6,265 = 12,223$.

To obtain predictors $\hat{\theta}_i$ or $\hat{\theta}_{i,j}$ using (2.16) we compute the pseudo inverse of the respective design matrices using (for example) numerical procedures for (2.17). Note that the $\hat{\theta}_i$ classifiers require only a single pseudo inverse for all i while $\hat{\theta}_{i,j}$ has a different design matrix for each (i, j) pair and hence requires its own pseudo inverse.

Now after training on the $60,000$ MNIST training samples and evaluating performance on the $10,000$ testing samples, we obtain an accuracy of 0.8603 using one vs. rest and an accuracy of 0.9297 using one vs. one. As MNIST is generally a balanced dataset, the use of accuracy to evaluate performance is sensible, and the level of accuracy obtained is impressive since the linear model is very simple and training these models using the pseudo-inverse computation only takes a few seconds at most. However, to get industrial grade performance one requires more advanced models such as the convolutional neural networks of Chapter 6. As mentioned in the previous chapter state-of-the-art performance for MNIST is at an accuracy of over 99.8%.

When evaluating multi-class classifiers it is common to use a *confusion matrix* similar to the 2×2 confusion matrix presented in Section 2.2 for the case of binary classification. Table 2.1 presents the confusion matrices for both one vs. rest and one vs. one. It is insightful to pick out the entries where non-negligible misclassification occurs. For example with the one vs. rest classifier multiple real digits of 7 were classified as 9. This occurred 77 times. Similarly in the one vs. one case, there were 36 misclassifications of the digit 5 as the digit 8.

---

[12]Here 2 is the label value that matches label index 3 when using label indexing $\{1, \ldots, K = 10\}$.

Table 2.1: Confusion matrices for the MNIST digit test set using linear classifiers trained on the training set. (a) one vs. rest achieves an accuracy of 0.8603. (b) One. vs. one achieves an accuracy of 0.9297.

		Decision									
		0	1	2	3	4	5	6	7	8	9
Reality	0	944	0	18	4	0	23	18	5	14	15
	1	0	1107	54	17	22	18	10	40	46	11
	2	1	2	813	23	6	3	9	16	11	2
	3	2	2	26	880	1	72	0	6	30	17
	4	2	3	15	5	881	24	22	26	27	80
	5	7	1	0	17	5	659	17	0	40	1
	6	14	5	42	9	10	23	875	1	15	1
	7	2	1	22	21	2	14	0	884	12	77
	8	7	14	37	22	11	39	7	0	759	4
	9	1	0	5	12	44	17	0	50	20	801

(a)

		Decision									
		0	1	2	3	4	5	6	7	8	9
Reality	0	961	0	9	9	2	7	6	1	7	6
	1	0	1120	18	1	4	5	5	16	17	5
	2	1	3	936	18	6	3	12	17	8	1
	3	1	3	12	926	1	30	0	3	23	11
	4	0	1	10	2	931	8	5	11	10	30
	5	6	1	5	20	1	800	19	1	36	12
	6	8	4	10	1	7	17	908	0	10	0
	7	3	1	10	7	4	2	1	955	10	21
	8	0	2	22	21	3	15	2	1	840	3
	9	0	0	0	5	23	5	0	23	13	920

(b)

2.4 Iterative Optimization-Based Learning

Linear models coupled with quadratic loss (2.12) are gifted with a closed-form solution for the parameter estimate as appearing in (2.16). This solution, which is based on the pseudo-inverse (2.17), is well-studied in the field of numerical linear algebra and is typically suited for efficient numerical evaluation, a topic that we do not cover here any further. For example, in the section above, the pseudo inverse computation of the $60,000 \times 785$ design matrix X used for MNIST digit classification using the one vs. rest method can be evaluated in about a second on a modern laptop. Nevertheless, there are multiple scenarios where this closed-form solution may not be used. This is the case when we use an alternative loss function to the quadratic loss, such as (2.18) or (2.19). Problems with using the explicit optimal quadratic solution may also arise when the number of features p is very large. In such cases and others, more generic methods for optimization are needed.

We now present a general iterative optimization method for obtaining this solution. It can be used in such scenarios where the pseudo-inverse-based computation does not work, yet more importantly it serves to illustrate how gradient-based optimization interplays with machine learning. Indeed the more complex deep learning models on which this book focuses use this type of method, and its generalizations, almost exclusively. We note that other methods are used as well and Chapter 4 focuses entirely on optimization. Our purpose here is simply to introduce the essence of the most basic technique, namely *gradient descent*.

Algorithm 2.1: Gradient descent

      Input: Dataset $\mathcal{D} = \{(x^{(1)}, y^{(1)}), \ldots, (x^{(n)}, y^{(n)})\}$,
             objective function $C(\cdot) = C(\cdot \, ; \mathcal{D})$, and
             initial parameter vector θ_{init}
      Output: Approximately *optimal* θ

1 $\theta \leftarrow \theta_{\text{init}}$
2 **repeat**
3   Compute the gradient $\nabla C(\theta)$
4      $\theta \leftarrow \theta - \alpha \nabla C(\theta)$
5 **until** θ satisfies a *termination condition*
6 **return** θ

This gradient descent procedure, presented in Algorithm 2.1, executes over iterations indexed by $t = 0, 1, 2, \ldots$ and works by taking small steps in the direction opposite to the gradient. That is, it traverses "downhill" each time trying to descend in the steepest direction. In its simplest form, steps sizes are controlled by a fixed $\alpha > 0$ called the *learning rate*. After some initialization with the vector $\theta^{(0)} = \theta_{\text{init}}$, in each iteration t, the next vector $\theta^{(t+1)}$ is obtained via,

$$\theta^{(t+1)} = \theta^{(t)} - \alpha \nabla C(\theta^{(t)}). \tag{2.20}$$

The algorithm repeats (2.20) where the key object that requires computation at each iteration t is the gradient of the loss function $\nabla C(\theta^{(t)})$. For complex deep learning models, this computation is one of the core components of a deep learning framework and is carried out using the backpropagation algorithm studied in Chapter 5. However, for simpler models such as the linear model or logistic regression type models of Chapter 3, we have explicit expressions for the gradient.

In the case of the linear model with quadratic loss, we have already computed the gradient in (2.14) and we can represent it as

$$\nabla C(\theta) = \frac{2}{n} X^\top (X\theta - y). \tag{2.21}$$

In general the algorithm iterates over t until $\theta^{(t+1)}$ and $\theta^{(t)}$ are close as measured by some stopping criteria, such as

$$\|\theta^{(t)} - \theta^{(t+1)}\| < \varepsilon, \tag{2.22}$$

with some fixed $\varepsilon > 0$. Other stopping criteria and variants of this method are studied in detail in Chapter 4. The final $\theta^{(t+1)}$ is used as $\hat{\theta}$ and if α and ε are well chosen, the algorithm output may closely approximate the optimal θ.

One of the main difficulties with the application of gradient descent is choosing the learning rate α. As an illustration, Figure 2.7 presents a contour plot of the squared error loss function associated with simple linear regression similar in nature to the Boston housing price data example of Figure 2.3 (a) where we optimize to find $\hat{\theta} = (\hat{\beta}_0, \hat{\beta}_1)$. When running with $\varepsilon = 10^{-5}$ and starting θ_{init} at the origin, we see that if $\alpha = 0.01$, the algorithm terminates near the optimal parameters in 3,109 iterations. If $\alpha = 0.0235$, the algorithm terminates

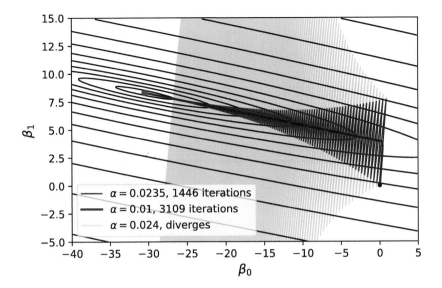

Figure 2.7: A contour plot of the loss function for a simple linear regression problem. The optimal point in green is reached when starting gradient descent at the origin (black point) with $\alpha = 0.01$ or $\alpha = 0.0235$. However with a learning rate slightly higher at $\alpha = 0.024$ gradient descent does not converge.

near the optimal parameters quicker with 1,446 iterations yet follows a more jagged path. Finally, when running with a learning rate that is just slightly higher at $\alpha = 0.024$, the algorithm diverges and does not terminate at all. The plot in this divergent case is only for the first 300 iterations where the growing oscillations are still in the vicinity of the optimum. With further iterations, the values quickly diverge.

This simple example illustrates that the value of the learning rate α is crucial. Different values imply drastically different behaviors. A more thorough investigation of gradient descent and its generalizations is in Chapter 4. Interestingly, for linear models more explicit results are available, as we present now.

Learning Rate Analysis for Linear Models

In general there is not a simple analytical way to determine a suitable learning rate α. Chapter 4 presents adaptive generalizations of gradient descent. Yet, universally there are no closed-form recipes. Nevertheless, when it comes to the special case of the linear model, analysis of the dynamics of gradient descent is analytically attractive and we may explicitly describe the range of α for which convergence takes place. This description is not necessarily of direct practical use. However, it gives insight into the nature of gradient descent.

Consider the linear model with design matrix X as in (2.10) and denote $\lambda_{\max}$ as the maximal eigenvalue of the Gram matrix $X^\top X$. We can now show that gradient descent converges[13]

---

[13]Here "convergence" formally means that for any $\varepsilon_2 > 0$ there is an $\varepsilon > 0$ of (2.22) where the algorithm terminates in a finite number of iterations with $\|\theta^* - \theta^{(t+1)}\| < \varepsilon_2$ and θ^* is a minimizer of the optimization problem.

to a solution of the normal equations (2.15) as long as

$$\alpha < \frac{n}{\lambda_{\max}}. \tag{2.23}$$

Note that for the data used in Figure 2.7, $\lambda_{\max} = 20{,}670.33$ and $n = 490$. Hence in this case the algorithm converges for α in the range $(0, 0.02371)$ and this bound is in agreement with the examples of Figure 2.7 where the first two paths have α in this range and the third path with $\alpha = 0.024$ diverges.

To see (2.23) use the gradient expression (2.21) and the gradient update rule of (2.20) to obtain the recursion,

$$\theta^{(t+1)} = \theta^{(t)} - \alpha \frac{2}{n} X^\top (X\theta^{(t)} - y)$$
$$= \underbrace{\left(I - \alpha \frac{2}{n} X^\top X\right)}_{A} \theta^{(t)} + \underbrace{\alpha \frac{2}{n} X^\top y}_{c},$$

or in short $\theta^{(t+1)} = A\theta^{(t)} + c$. Such a recursion is known as an *affine discrete time linear dynamical system* and equilibrium points of such a system, denoted θ^* satisfy, $\theta^* = A\theta^* + c$ or $(I - A)\theta^* = c$. In our case, using A and c, it is evident that such equilibrium points are solutions of the normal equations (2.15). For simplicity let us assume here that X is full column rank and hence there is a unique θ^*.

It follows from linear systems theory that the spectral radius[14] of the matrix A determines the convergence or non-convergence of $\theta^{(t)}$ to θ^*. Specifically, if the spectral radius of the matrix A is less than unity, then $\theta^{(t)}$ converges to θ^* for any initial $\theta^{(0)}$ and, if the spectral radius is greater than unity, then for any initial $\theta^{(0)}$ the sequences diverge (the border case of the spectral radius being 1 is indeterminate). Hence, putting aside the border case of a spectral radius of 1, we see that convergence occurs if and only if the eigenvalues of A are in $(-1, 1)$.

Now, since since $X^\top X$ is a symmetric matrix, the eigenvalues of $X^\top X$ are real and since $X^\top X$ is positive semidefinite (this is a property of any Gram matrix), the eigenvalues lie in the range $(0, \lambda_{\max}]$ with $\lambda_{\max} > 0$. Further, the eigenvalues of $-2\alpha n^{-1} X^\top X$ lie in the range $[-2\alpha n^{-1}\lambda_{\max}, 0)$ and the eigenvalues of $A = I - 2\alpha n^{-1} X^\top X$ lie in the range $[1 - 2\alpha n^{-1}\lambda_{\max}, 1)$. Thus the critical inequality that ensures that the spectral radius of A is less than 1 is

$$-1 < 1 - 2\alpha \frac{1}{n}\lambda_{\max},$$

which is equivalent to (2.23).

The Loss Landscape and Standardization of Inputs

In Section 2.1 we presented the standardization of inputs via (2.2) where the sample mean and sample variance are computed via (2.1). We now show that such standardization also affects the *loss landscape*, often yielding improvement in the execution of gradient descent. As above, our analysis is in the realm of linear models where explicit analysis is possible.

---

[14]The spectral radius of a square matrix is the largest of all magnitudes of the eigenvalues.

To illustrate this concept, we consider a simple case of a linear model with two input features x_1 and x_2, where

$$y = w_1 x_1 + w_2 x_2 + \epsilon.$$

This is similar to the simple regression models of Section 2.2 yet is without an intercept term. Here, with n data samples, the $n \times 2$ design matrix X has columns composed of the vectors $(x_i^{(1)}, \ldots, x_i^{(n)})$ for $i = 1, 2$; the labels vector is $y = (y^{(1)}, \ldots, y^{(n)})$; and the model parameters are $\theta = (w_1, w_2)$. The square loss function where for simplicity we omit the $1/n$ term in (2.11) is

$$C(\theta ; X, y) := \|y - X\theta\|^2 = (y - X\theta)^\top (y - X\theta)$$
$$= \theta^\top X^\top X\theta - 2y^\top X\theta + y^\top y.$$

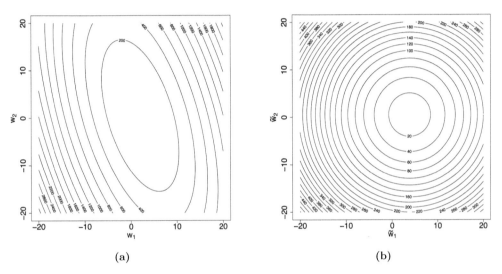

(a)   (b)

Figure 2.8: Contour levels of a loss function for an example with $p = 2$ parameters. (a) The loss as a function of (w_1, w_2) for the original data has accentuated elliptical contours. (b) The loss as a function of the parameters, $(\tilde{w}_1, \tilde{w}_2)$, associated with standardized data, yields much less accentuated contours.

A contour plot for some arbitrary (not standardized) data is illustrated in Figure 2.8 (a). It is evident that the contour levels of the loss function are ellipsoids. For some given loss level C, it can be shown that the lengths of the principal axes of the ellipse are $\sqrt{\frac{k}{\lambda_i}}$ for $i = 1, 2$, where λ_1 and λ_2 are the eigenvalues of the Gram matrix $X^\top X$ and $k = C - y^\top y$. With this, the elongation of the ellipse, which is the ratio of these lengths, is,

$$R_X = \frac{\sqrt{\frac{k}{\lambda_1}}}{\sqrt{\frac{k}{\lambda_2}}} = \sqrt{\frac{\lambda_2}{\lambda_1}}.$$

Now, when standardizing the inputs, each of the columns of X is transformed separately via (2.2). In such a case, the loss function associated with the standardized data is $C(\theta ; Z, y)$ where we denote the design matrix of the standardized data by Z. It can now be shown that

$$Z^\top Z = n \begin{pmatrix} 1 & \rho \\ \rho & 1 \end{pmatrix}, \qquad \text{where} \qquad \rho = \frac{1}{n} \sum_{j=1}^{n} z_1^{(j)} z_2^{(j)}$$

is also known as the *sample correlation* between the two features. Note that $\rho \in [-1, 1]$.

With such a normalization and an explicit form for the Gram matrix $Z^\top Z$, we can compute the eigenvalues of $Z^\top Z$ to be $\lambda_1 = 1 + |\rho|$ and $\lambda_2 = 1 - |\rho|$. Thus, the elongation of the ellipsoid is

$$R_Z = \sqrt{\frac{\lambda_2}{\lambda_1}} = \sqrt{\frac{1 + |\rho|}{1 - |\rho|}}.$$

It is evident that in cases where the correlation between the features is low, then $R_Z \approx 1$ which implies that the loss landscape of the standardized data is much more similar to Figure 2.8 (b).

Now, in terms of gradient descent, taking steps in a loss landscape such as Figure 2.8 (b) is generally more efficient than using the loss landscape in Figure 2.8 (a). Hence it is expected that as long as ρ is not close to 1 or -1, carrying out standardization will help gradient descent converge faster. Note that, while this analysis is carried out on a simplistic $p = 2$, $d = 2$ linear model with quadratic loss, the principle often applies to more complicated loss landscapes as well. Further, note that with standardization of the features or any other transformation one also has to encode the standardization transformation as part of the deployed model, since the model is now for the standardized features z instead of x.

2.5 Generalization, Regularization, and Validation

The data available while learning and calibrating a model is called the *seen data* and future data is called *unseen data*. These concepts were introduced in Section 2.1. Our purpose of fitting a model based on seen data is that it will ultimately work well for unseen data, a property known as *generalization ability*. With this view, when seeking models that generalize well, there are two competing negative attributes that one needs to balance, *underfitting* and *overfitting*. Underfitting is a case when the model is too simple and fails to capture the complexity of the underlying data. On the other hand, overfitting is a case where the model is so specialized to the training data such that unseen examples that slightly differ from the training data do not perform well. The theme of *model selection* in machine learning and statistics deals with the calibration of underfitting and overfitting to yield models that generalize well.

Model selection, or the quest for optimal generalization ability, is one of the hardest problems in machine learning primarily because the unseen data is not available. For this, one needs to judiciously budget the seen data by splitting it into the training set, the test set, and also carry out validation in one of several ways that we outline in this section. In quantifying generalization ability, there are several plots and measures that one can use. These include the quantification of *model bias*, *model variance*, and the *bias and variance tradeoff*. We present these in this section.

Some classes of models are by construction designed to enable calibration of underfitting, overfitting, or the bias-variance tradeoff. One general technique for this is called *regularization* which in one common form, includes the introduction of additional terms to the model's loss function. We present a taste of regularization techniques here and then in Section 5.7 we focus on regularization in the context of deep neural network models.

In terms of notation, throughout this book, we use $\mathcal{D}$ to denote data with n samples. This sometimes means only the training set and in other cases as all of the seen data. When we focus on training specific types of models such as in Section 2.3, the symbol $\mathcal{D}$ is treated as the data is allocated specifically for training and hence we assume there are n training samples for training and potentially other samples for testing and/or validation that we do not account for. In other cases, $\mathcal{D}$ is treated as all of the available seen data, part of which may be used for testing via a testing or hold-out set which we denote via $\mathcal{D}_{\text{test}}$ with n_{test} samples.

Performance on Unseen Data

We have already introduced several examples of *performance metrics* in Section 2.2. These include accuracy (2.6), the F_1 score (2.7), mean square error in the case of regression, and others. In some cases one wishes to maximize the performance metric whereas in other cases one wishes to minimize it. Note that the loss function used in model training is in some instances directly related or equal to the performance metric, and, in other instances, it is different.

It is notationally convenient to relate a *performance function* to the performance metric. We denote the performance function via $\mathcal{P}(\cdot, \cdot)$ and it penalizes differences between a single predicted label $\hat{y}$ and the actual label y. For example, when the mean square error performance metric is used the performance function is $\mathcal{P}(\hat{y}, y) = (\hat{y} - y)^2$. As another example, if the accuracy performance metric is used in classification then the performance function is $\mathcal{P}(\widehat{\mathcal{Y}}, y) = \mathbf{1}\{\widehat{\mathcal{Y}} \neq y\}$, where $\mathbf{1}\{\cdot\}$ is the indicator function and y is taken as the actual label. Note that we construct the performance function such that small values are desirable.[15]

When we train a model and create a predictor either for regression or classification, we use the data $\mathcal{D}$ and based on the model obtain a predictor denoted by $\hat{y}(\cdot\,; \mathcal{D})$. Now, for some data pair (x, y), the value $\hat{y}(x\,; \mathcal{D})$ is the prediction of y and the performance function evaluated for the prediction of this data pair is $\mathcal{P}\big(\hat{y}(x\,; \mathcal{D}), y\big)$.

As outlined in Section 2.1, we ensure that the nature of the seen data is similar to that of unseen data and with this, the underlying modeling assumption is that both seen and unseen data are generated by the same underlying processes. Hence, for both theoretical and empirical analysis, unless we know otherwise, we assume that the probability distribution of each data sample $(x^{(i)}, y^{(i)})$ is the same for all $i = 1, \ldots, n$ and is further the same as the distribution of each unseen data sample $(x^\star, y^\star)$. That is, we assume there is an underlying probability space for the observations and we vaguely denote the joint probabilities of the features and label via $\mathbb{P}(x, y)$. Our usage of probabilistic statements here is only via expected values where we denote the expectation operator via $\mathbb{E}[\cdot]$ and often use a subscript for the expectation to denote the objects that are treated as random.

With this notation, the expected value of the performance of the trained model for unseen data points $(x^\star, y^\star)$ is denoted via,

$$E_{\text{unseen}} = \mathbb{E}_{(x^\star, y^\star)}\big[\mathcal{P}\big(\hat{y}(x^\star; \mathcal{D}_{\text{train}}), y^\star\big)\big], \tag{2.24}$$

---

[15] In this section, to avoid notational confusion between $\hat{y}$ and $\widehat{\mathcal{Y}}$, we use the notation $\hat{y}$ for both cases.

where $\mathcal{D}_{\text{train}}$ is the training data. This quantity is called the *generalization performance* or *generalization error*. It may be viewed as an average over all possible unseen data points and hence E_{unseen} evaluates how well the predictor or model generalizes. With a given training dataset $\mathcal{D}_{\text{train}}$, our aim is to build a model that yields the smallest possible E_{unseen}.

Unfortunately, since it is based on unseen data, E_{unseen} is a theoretical construct and since we do not know the probability law $\mathbb{P}(x, y)$ exactly, we cannot compute E_{unseen}. However, as a first attempt, we can approximate the expectation by averaging over available training data. That is,

$$E_{\text{train}} = \frac{1}{n_{\text{train}}} \sum_{(x,y) \in \mathcal{D}_{\text{train}}} \mathcal{P}\big(\widehat{y}(x\,;\,\mathcal{D}_{\text{train}}), y\big), \tag{2.25}$$

where n_{train} is the number of observations in $\mathcal{D}_{\text{train}}$. It turns out that E_{train} is typically a poor estimator of E_{unseen} because the same training observations that were used to create the predictor are also used to evaluate the predictor performance. That is, the learned parameters of the model $\hat{\theta}$ that are used to construct $\widehat{y}(\cdot)$ depend on $\mathcal{D}_{\text{train}}$. Hence while E_{train} does present us with some insight about the ability of our model to reproduce the data that has been learned, it lacks the ability to estimate performance on unseen data.

In order to get a better estimate of E_{unseen}, it is preferable to average over data that has not been used for training the model, namely over the test set. In an ideal situation where we use the test set only once and do not calibrate and adjust the model based on the test set, the test set observations are completely independent of the model. In such a case, the estimator,

$$E_{\text{test}} = \frac{1}{n_{\text{test}}} \sum_{(x,y) \in \mathcal{D}_{\text{test}}} \mathcal{P}\big(\widehat{y}(x; \mathcal{D}_{\text{train}}), y\big), \tag{2.26}$$

is a good estimator of E_{unseen}, especially for significantly large n_{test}. Specifically under the assumption that the unseen data and the test set have the same distribution, the expected value of E_{test} is exactly E_{unseen} making it a statistically *unbiased estimator* of performance. Further it is statistically *consistent* in the sense that if we are able to allocate more testing data and $n_{\text{test}} \to \infty$ then $E_{\text{test}} \to E_{\text{unseen}}$. This is simply a consequence of the law of large numbers.[16] Note that these desirable statistical properties are only for a fixed $\mathcal{D}_{\text{train}}$.

The straightforward statistical properties of unbiasedness and consistency enjoyed by E_{test} make the practice of holding out a test set for performance evaluation attractive. However, setting aside a test set is costly as we effectively "throw away" n_{test} observations and do not use them for improving the model. For this reason, it is often tempting in practice to iteratively evaluate (2.26) while adjusting model settings or hyper-parameters. This frowned upon practice breaks the independence between $\mathcal{D}_{\text{test}}$ and the model at which point the desirable statistical properties of E_{test} are lost. Hence, as an alternative, we use a validation set or some other method as described below.

In addition to using an independent test set as in (2.26), other alternatives for estimation of performance also exist, which include using K-fold cross validation. This is a topic we describe below in the context of validation and hyper-parameter optimization, yet it may also be used for purposes of performance evaluation.

---

[16]Formally the convergence $E_{\text{test}} \to E_{\text{unseen}}$ may be seen as convergence in probability in one form or almost sure convergence in a different form. We do not focus on such subtleties here.

Model Choice, Underfitting, and Overfitting

The generalization performance in (2.24) is specific to a fixed single training dataset $\mathcal{D}_{\text{train}}$. However, for a given problem, when considering which type of model to use and what hyper-parameters to choose, it is often useful to think about the expectation over all possible training datasets. For this we define the *expected generalization performance*,

$$\widetilde{E}_{\text{unseen}} = \mathbb{E}_{\mathcal{D}_{\text{train}}}[E_{\text{unseen}}] = \mathbb{E}_{\mathcal{D}_{\text{train}}}\big[\mathbb{E}_{(x^\star, y^\star)}[\mathcal{P}(\widehat{y}(x^\star; \mathcal{D}_{\text{train}}), y^\star)]\big]. \tag{2.27}$$

It represents the average of E_{unseen} over all possible datasets $\mathcal{D}_{\text{train}}$ of a given size from the same probability law $\mathbb{P}(x, y)$ where we keep in mind that each dataset potentially yields a different representation of the model.

A similar quantity for the training set is

$$\widetilde{E}_{\text{train}} = \mathbb{E}_{\mathcal{D}_{\text{train}}}[E_{\text{train}}] = \frac{1}{n_{\text{train}}}\mathbb{E}_{\mathcal{D}_{\text{train}}}\bigg[\sum_{(x,y)\in\mathcal{D}_{\text{train}}} \mathcal{P}(\widehat{y}(x; \mathcal{D}_{\text{train}}), y)\bigg]. \tag{2.28}$$

Keep in mind that $\widetilde{E}_{\text{unseen}}$ and $\widetilde{E}_{\text{train}}$ are functions of the type of model used, the hyper-parameters, and the training dataset size. With such relationships present, the machine learning engineer can in principle ponder about the theoretical shape of $\widetilde{E}_{\text{unseen}}$ and $\widetilde{E}_{\text{train}}$ and seek a model that appears best. In this respect the *generalization gap* defined as $\widetilde{\Delta} = \widetilde{E}_{\text{unseen}} - \widetilde{E}_{\text{train}}$ is also important.

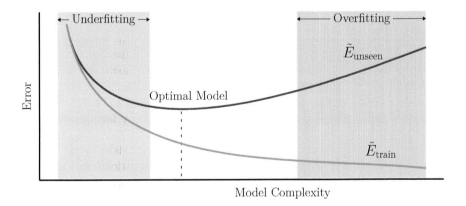

Figure 2.9: Typical behavior of expected generalization performance and expected training performance as a function of model complexity.

The combination of $\widetilde{E}_{\text{unseen}}$, $\widetilde{E}_{\text{train}}$, and the generalization gap $\widetilde{\Delta}$ based on estimates, allows one to seek a balance between underfitting and overfitting. There are multiple suggestions on "best practice" for using the available data to estimate $\widetilde{E}_{\text{unseen}}$, $\widetilde{E}_{\text{train}}$, $\widetilde{\Delta}$, and to select the best model. A thorough discussion of such best practices is beyond our scope. The notes and references at the end of this chapter link to further reading. Instead, let us consider the schematic Figure 2.9, which presents typical behavior of $\widetilde{E}_{\text{unseen}}$ and $\widetilde{E}_{\text{train}}$ as a function of model complexity.

Generally, as model complexity increases, expected training performance, $\widetilde{E}_{\text{train}}$, improves (decreases) since complex structured models can explain the training data better. At high extremes this is overfitting. Similarly an opposite phenomenon is that models with low complexity are not able to describe the data well. The tradeoff between these two regimes is obtained at the minimum of $\widetilde{E}_{\text{unseen}}$ marked by the vertical dashed line.

In practice, unless presented with an infinite pool of data, one is not able to evaluate $\widetilde{E}_{\text{unseen}}$ and $\widetilde{E}_{\text{train}}$ directly and one is certainly not able to evaluate these quantities over all possibilities of models, hyper-parameters, and sample sizes. Nevertheless, much of the practice of model selection revolves around getting a feel for the dependence of $\widetilde{E}_{\text{unseen}}$ and $\widetilde{E}_{\text{train}}$ on model choice, hyper-parameters, and sample size. This is typically done using very limited measurements from one or several training and validation executions.

Typical practice is to monitor empirical estimates of these quantities as a function of model complexity, hyper-parameter choice, or sample size. The most basic practice is evaluation of E_{train} of (2.25) together with a validation performance measurement that is of similar nature to E_{test} of (2.26) such as the validation performance or K-fold cross validation performance which are defined in the sequel. See (2.37) and (2.38).

As one simple illustrative example capturing the tradeoffs of model complexity, let us consider linear models with polynomial features applied to synthetic univariate ($p = 1$) data. The model

$$y = \beta_0 + \beta_1 x + \beta_2 x^2 + \ldots + \beta_k x^k + \epsilon \tag{2.29}$$

is denoted by $\mathcal{M}_k$ where k is the order of the polynomial. Hence $\mathcal{M}_0$ is the constant model, $\mathcal{M}_1$ is the simple linear model, $\mathcal{M}_2$ is the quadratic model, and so on. A quadratic model of this nature was used in (2.4) of Section 2.2. In this framework, model complexity corresponds to the degree of the polynomial model.

Now taking one possible realization of $\mathcal{D}_{\text{train}}$, in Figure 2.10 (a) we use this family of models to fit data of size $n_{\text{train}} = 10$. With this single realization we clearly see underfitting behavior for models $\mathcal{M}_0$ and $\mathcal{M}_1$. In contrast, model $\mathcal{M}_9$ appears to overfit the observed data. Between these two extremes, model $\mathcal{M}_3$ looks like an appropriate representation of the observed data.

In Figure 2.10 (b) the red curve presents E_{train} for this dataset. It is obvious that as k increases training fit improves. Further, in this hypothetical example since we know the underlying process with probability law $\mathbb{P}(x, y)$ used for purposes of simulation of synthetic data, we may sample as many $(x^\star, y^\star)$ pairs as we wish, to obtain a reliable estimate of E_{unseen}. This curve is plotted in black where in this case we use $10,000$ repetitions for each k, each time with the fixed model based on our single available dataset, $\mathcal{D}_{\text{train}}$. This Monte Carlo simulation makes it clear that when $k = 9$ or $k = 8$, there is overfitting and when $k = 0, 1, 2$, there is underfitting. In practice plots exactly like Figure 2.10 (b) cannot be produced because we do not know $\mathbb{P}(x, y)$. Instead one can resort to estimates based on cross validation to obtain curves similar to the black curve in Figure 2.10 (b).

We also mention that, while we stated that key elements that affect expected performance are the model type, hyper-parameters, and sample size, in the world of deep learning there is also an additional major factor, *training time*. For deep learning models, since the number of parameters in the model is often huge, letting the model train for longer is similar to using a more complex model as presented in Figure 2.9.

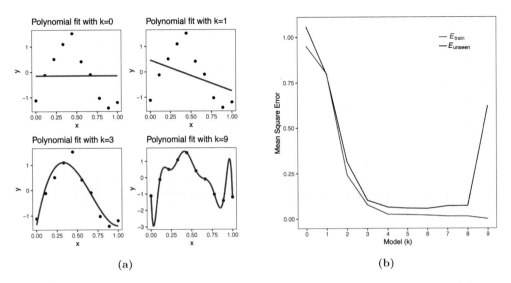

Figure 2.10: Increasing model complexity illustrated via linear models with polynomial features where k, the order of the polynomial, captures the complexity. (a) Fitting several models to a single realization with $n = 10$ data-points. (b) The training performance is in red and simulation estimates of the generalization performance is in black.

Bias and Variance Decomposition

A related view to the analysis of expected generalization performance and the generalization gap is the so-called bias and variance decomposition. It focuses on the expected generalization performance in production, $\widetilde{E}_{\text{unseen}}$, and decomposes it into a sum of terms related to model bias, model variance, and the noise magnitude. With this decomposition, underfitting is said to be a situation with high model bias and overfitting is said to be a situation with high model variance. Using this terminology, balancing model bias and model variance is equivalent to balancing underfitting and overfitting respectively. This is known as the *bias and variance tradeoff*.

The bias and variance decomposition is mathematically elegant in the special case of the square error performance function $\mathcal{P}(\hat{y}, y) = (\hat{y} - y)^2$ and a specifically assumed underlying random reality

$$y = f(x) + \epsilon, \quad \text{with} \quad \mathbb{E}[\epsilon] = 0, \quad \text{and } \epsilon \text{ is independent of } x. \tag{2.30}$$

Here x a vector of features and y a scalar real valued label. Further $\mathbb{E}[\epsilon^2]$ is the variance of the noise term and is called the *inherent noise*. In this setting, for some unseen feature vector $x^\star$, the predictor trained on data $\mathcal{D}$ is $\hat{y}(x^\star\,;\mathcal{D})$, which we also denote via $\hat{f}(x^\star\,;\mathcal{D})$ since it estimates $f(x^\star)$. Hence the expected generalization performance of (2.27) becomes

$$\widetilde{E}_{\text{unseen}} = \mathbb{E}_{\mathcal{D},x^\star,\epsilon}\left[\left(\hat{f}(x^\star;\mathcal{D}) - (f(x^\star) + \epsilon)\right)^2\right]. \tag{2.31}$$

Now, a standard algebraic manipulation common in statistics is to add and subtract $\mathbb{E}_{\mathcal{D},x^\star}[\hat{f}(x^\star;\mathcal{D})]$ inside (2.31), expand the expression, apply the external expectation operator, and then cancel out terms that have zero expectation (resulting from $\mathbb{E}[\epsilon] = 0$ and the fact that ϵ and $x^\star$ are independent). This manipulation transforms (2.31) to the *bias-variance-noise decomposition equation,*

$$\widetilde{E}_{\text{unseen}} = \underbrace{\left(\mathbb{E}[\hat{f}(x^\star\,;\,\mathcal{D})] - \mathbb{E}[f(x^\star)]\right)^2}_{\text{Bias squared of } \hat{f}(\cdot)} + \underbrace{\mathbb{V}\text{ar}\left(\hat{f}(x^\star\,;\,\mathcal{D})\right)}_{\text{Variance of } \hat{f}(\cdot)} + \underbrace{\mathbb{E}[\epsilon^2]}_{\text{Inherent Noise}} . \qquad (2.32)$$

Here the first term is the square of the bias, the second term is the variance taking into consideration variability both from $\mathcal{D}$ used for training and $x^\star$, and the third term is the inherent noise. The expectations and variances in the bias and variance terms are with respect to the training dataset $\mathcal{D}$ and the arbitrary unseen feature vector $x^\star$. The main takeaway from (2.32) is that if we ignore the inherent noise, the loss of the model has two key components, *model bias* (technically it is the model bias squared), and *model variance*.

The model bias is a measure of how a typical (expectation over all possible data samples) model $\hat{f}(\cdot\,;\,\mathcal{D})$ misspecifies the correct relationship $f(\cdot)$. Model classes with high bias, have that $\hat{f}(\cdot\,;\,\mathcal{D})$ does not accurately predict $f(\cdot)$. That is, high bias generally implies *underfitting*. Similarly, model classes with low model bias are detailed descriptions of reality since the expected difference in the bias term is near zero.

The model variance is a measure of the variability of the model class $\hat{f}(\cdot\,;\,\mathcal{D})$ with respect to the random sample $\mathcal{D}$ and the distribution of $x^\star$ as implicitly implied by the probability law of the data $\mathbb{P}(x, y)$. Model classes with high model variance are often *overfit* (to the training data) and do not *generalize* (to unseen data) well. Similarly, model classes with low model variance are much more robust to the training data and generalize to the unseen data much better.

Similar analysis to the derivation that leads to (2.32) can also be attempted for other performance functions other than square error, and model structures other than (2.30). With such other settings, the mathematical elegance of (2.32) is often lost. Nevertheless, the concepts of model bias, model variance, and the bias and variance tradeoff still persist. For example, in a classification setting we may compare the accuracy obtained on the training set to that obtained on a validation set. If there is a high discrepancy where the training accuracy is much higher than the validation accuracy, then there is probably a variance problem indicating that the model is overfitting.

Addition of Regularization Terms

One natural way to control model variance is to induce or force model parameters to remain within some confined subset of the parameter space. This is called *regularization*. At the extreme case where all model parameters are 0, the model variance vanishes as well. In less extreme cases where there is only some constraint on model parameters, model variance is still controlled. Such decreases in model variance may imply an increase in model bias. Nevertheless, the ultimate goal of optimizing the expected performance loss typically merits such adjustments.

A common way to keep model parameters at bay is to augment the optimization objective $\min_\theta C(\theta\,;\,\mathcal{D})$ with an additional *regularization term* $R_\lambda(\theta)$. The revised objective is then,

$$\min_\theta\ C(\theta\,;\,\mathcal{D}) + R_\lambda(\theta). \qquad (2.33)$$

The regularization term $R_\lambda(\theta)$ depends on a *regularization parameter* λ, which is often a scalar in the range $[0, \infty)$ but also sometimes a vector. This *hyper-parameter* allows us to optimize the bias and variance tradeoff.

A common general regularization technique called *elastic net* has regularization parameter $\lambda = (\lambda_1, \lambda_2)$ and,

$$R_\lambda(\theta) = \lambda_1 \|\theta\|_1 + \lambda_2 \|\theta\|^2 \quad \text{with} \quad \|\theta\|_1 = \sum_{i=1}^{d} |\theta_i| \text{ and } \|\theta\|^2 = \sum_{i=1}^{d} \theta_i^2, \tag{2.34}$$

where d is the dimension of the parameter space.[17] Hence the values of λ_1 and λ_2 determine what kind of penalty the objective function will pay for high values of θ_i.

Clearly, with $\lambda_1, \lambda_2 = 0$ the original objective is unmodified. In contrast, as $\lambda_1 \to \infty$ or $\lambda_2 \to \infty$ the estimates $\theta_i \to 0$ and any information in the data $\mathcal{D}$ is fully ignored. Indeed, as λ_1 or λ_2 grow, the model bias grows while model variance is decreased and overfitting is mitigated. With regularization there is often a magical "sweet spot" for λ where the objective (2.33) does a good job of fitting the model.

Particular cases of elastic net are the classic *ridge regression*, also called *Tikhonov regularization*, and *LASSO* standing for *least absolute shrinkage and selection operator*. In the former $\lambda_1 = 0$ and only λ_2 is used, and in the latter $\lambda_2 = 0$ and only λ_1 is used. One of the benefits of LASSO, also present in the more general elastic net case, is that the $\|\theta\|_1$ loss allows the algorithm to remove variables from the model by "zeroing out" their θ_i values completely. Hence LASSO is very useful as a model selection technique.

The case of ridge regression is slightly simpler to analyze than LASSO and it fits well within the framework of linear models presented in Section 2.3. We thus present the details now. For ridge regression the data fitting problem can be represented as

$$\min_{\theta \in \mathbb{R}^d} \|y - X\theta\|^2 + \lambda \|\theta\|^2, \tag{2.35}$$

where the design matrix X is as in (2.10) and we now consider λ as a scalar (previously in (2.34) it was denoted as λ_2) in the range $[0, \infty)$. Compare (2.35) with the original least squares objective (2.13). Now by manipulating the $\|\cdot\|^2$ expressions, the problem can be recast as[18]

$$\min_{\theta \in \mathbb{R}^d} \left\| \begin{bmatrix} y \\ 0 \end{bmatrix} - \tilde{X}_\lambda \, \theta \right\|^2 \quad \text{with} \quad \tilde{X}_\lambda = \begin{bmatrix} X \\ \sqrt{\lambda} I \end{bmatrix},$$

where I is the $d \times d$ identity matrix and 0 is the zero vector in $\mathbb{R}^d$. The pseudo-inverse associated with $\tilde{X}_\lambda$ is $\tilde{X}_\lambda^\dagger = (X^\top X + \lambda I)^{-1} [X^\top \ \lambda I]$. Hence, returning to (2.16), the parameter estimate for ridge regression is

$$\hat{\theta} = (X^\top X + \lambda I)^{-1} X^\top y. \tag{2.36}$$

---

[17] Note that in cases such as linear regression or deep neural networks where there is a constant term (β_0 for example), the parameters for the constant term are typically not regularized and hence the norms are taken only on the other parameters.

[18] In practice we often do not regularize the intercept term and this requires adjusting the identity matrix in $\tilde{X}_\lambda$.

As an aside, note that for any $\lambda > 0$ the matrix $X^\top X + \lambda I$ is not singular even if $X^\top X$ is singular. Also as $\lambda \to 0$ it can be shown that the pseudo-inverse $\tilde{X}_\lambda^\dagger$ converges to the SVD-based pseudo-inverse (2.17) associated with X.

We note that while for linear models, the results are very elegant, in other cases closed-form solutions such as (2.36) do not exist. Still, many machine learning loss functions can be augmented with a regularization term. We revisit these concepts in Section 5.7 in the context of deep learning where other regularization methods are presented as well. Also note that the regularization parameter λ is a first class example of a hyper-parameter that one would like to calibrate during learning. This specific parameter serves as a good lever for optimizing the bias and variance tradeoff. We now discuss the general topic of hyper-parameter optimization.

Hyper-Parameter Calibration and Cross Validation

As alluded to above, calibrating the model choice and the hyper-parameters while reusing the test set for performance evaluation is bad practice since it pollutes the test set performance estimator E_{test} of (2.26). For this reason it is common to further split the training data $\mathcal{D}_{\text{train}}$ using one of several ways while experimenting with model configurations and hyper-parameters. With such an approach $\mathcal{D}_{\text{test}}$ is reserved only for final performance evaluation before rolling out the model to production. Such use of the training data where some parts of the data are used for training parameters and the other parts are used for checking performance and tuning hyper-parameters is generally called *cross validation*.

There are multiple common cross validation techniques with many variants used in practice. Here we present only two main approaches, the *train-validate split* approach and *K-fold cross validation*. The train-validate split approach is common in situations where the total number of datapoints n is large. The K-fold cross validation approach is useful when data is limited.

The train-validate split approach simply implies that the original data with n samples is first split into training and testing as before and then the training data is further split into two subsets where the first is (confusingly) again called the *training set* and the latter is the *validation set*. Hence considering all of the available data $\mathcal{D}$, with this approach,

$$\mathcal{D} = \mathcal{D}_{\text{train}} \cup \mathcal{D}_{\text{validate}} \cup \mathcal{D}_{\text{test}}, \quad \text{where the unions are of disjoint sets.}$$

When considering all of the data, this approach is also called the *train-validate-test split* approach. If, for example, we use a 80-20 rule for both splits and assume that divisibility holds, then $n_{\text{test}} = 0.2 \times n$, $n_{\text{train}} = 0.64 \times n$ and $n_{\text{validation}} = 0.16 \times n$.

As an example, assume the model is fixed yet regularized with elastic net as presented above. Hence the hyper-parameters in question are $\lambda = (\lambda_1, \lambda_2)$ and the choice of these needs to be tuned. The approach is then to evaluate the estimator on $\mathcal{D}_{\text{train}}$ over a grid of such hyper-parameters, retraining from scratch for every λ. We then choose $\lambda^* = \text{argmax}_\lambda \ E_{\text{validation}}(\lambda)$ with

$$E_{\text{validation}}(\lambda) = \frac{1}{n_{\text{validation}}} \sum_{(x,y) \in \mathcal{D}_{\text{validation}}} \mathcal{P}\big(\widehat{y}(x \,;\, \mathcal{D}_{\text{train}}, \lambda), y\big), \qquad (2.37)$$

where we can see that the predictor $\widehat{y}$ depends on the hyper-parameter. With the optimal λ^* pair selected, the model with this λ^* is evaluated on the test set once via (2.26) before being rolled out to production. Note that this approach has many variants used in practice.

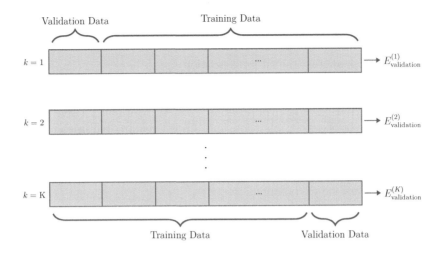

Figure 2.11: K-fold cross validation. For each $k = 1, \ldots, K$ the data is split into training data and validation data differently. This yields K estimates for performance and these estimates can be averaged.

In case of limited observations, a train-validate-test split may be too wasteful of data and an alternative approach is K-fold cross validation as illustrated in Figure 2.11. This approach may be used on all of the data $\mathcal{D}$ or only on the training data after a train-test split is performed. Here for simplicity, we apply it to some dataset $\mathcal{D}$. The approach is useful both for model selection, hyper-parameter optimization, and performance evaluation.

The value K of this approach which determines the number of data chunks or repetitions is a static configuration parameter with a typical value being $K = 5$ or $K = 10$. The approach is to split $\mathcal{D}$ into K equally sized data chunks each denoted $\mathcal{D}^k$ with,

$$\mathcal{D} = \mathcal{D}^1 \cup \mathcal{D}^2 \cup \ldots \cup \mathcal{D}^K, \quad \text{where again the unions are of disjoint sets.}$$

Then for each $k = 1, \ldots, K$ we fix a training set to be composed of all of the observations except for $\mathcal{D}^k$ and the validation set (may also be called a test set) to be $\mathcal{D}^k$. That is, denoting set difference with '\' we set,

$$\mathcal{D}_{\text{train}}^{(k)} = \mathcal{D} \setminus \mathcal{D}^k, \quad \text{and} \quad \mathcal{D}_{\text{validation}}^{(k)} = \mathcal{D}^k. \quad \text{for} \quad k = 1, \ldots, K.$$

We may now retrain and evaluate the model separately for each data chunk k where each time we use $\mathcal{D}_{\text{train}}^{(k)}$ as the training data and $\mathcal{D}_{\text{validation}}^{(k)}$ as the validation (or testing) data. That is, if for example $K = 10$ and originally $\mathcal{D}$ has n observations, then for each k we have $n_{\text{train}} = 0.9 \times n$ and $n_{\text{validation}} = 0.1 \times n$ (again assuming n is properly divisible).

With the model trained separately for each repetition k, we can now estimate performance via,

$$E_{\text{cv}} = \frac{1}{K} \sum_{k=1}^{K} E_{\text{validation}}^{(k)}, \tag{2.38}$$

with

$$E^{(k)}_{\text{validation}} = \frac{1}{n_{\text{validation}}} \sum_{(x,y) \in \mathcal{D}^{(k)}_{\text{validation}}} \mathcal{P}\big(\widehat{y}^{(k)}(x \,;\, \mathcal{D}^{(k)}_{\text{train}}), y\big),$$

where $\widehat{y}^{(k)}$ is the predictor trained for repetition k.

Once again, if needed, hyper-parameter optimization may take place by treating E_{cv} as a function of the hyper-parameter in question. Also, as mentioned above in situations where the total number of observations is low and if not tweaking parameters then K-fold cross validation may serve as an alternative approach to general performance evaluation using a train-test split. Again as with the train-validate-test split approach, there are multiple variations for K-fold cross validation with the exact method used in practice often depending on the specific situation encountered.

2.6 A Taste of Unsupervised Learning

Now that we have explored key aspects of supervised learning in the sections above, let us get a taste for the basics of unsupervised learning. In this context the data is unlabeled and is denoted via $\mathcal{D} = \{x^{(1)}, \ldots, x^{(n)}\}$. Here we assume that each sample or observation $x^{(j)}$ is a p-dimensional vector of features in Euclidean space. Observe that there are no labels $y^{(j)}$.

We briefly introduce two popular unsupervised learning methods, one for clustering and one for data reduction. These are respectively the *K-means* algorithm and the framework of *principal component analysis* (PCA). In exploring PCA we also take a slightly deeper look at the linear algebraic concept of singular value decomposition (SVD) already used in Section 2.3 in the context of pseudo-inverse representation. Here we see how SVD has applications to data compression, a notion sometimes used in more complex deep learning models. Further reference to more advanced supervised learning methods are in the notes and references at the end of the chapter.

K-Means Clustering

The machine learning activity of *clustering* allows us to identify meaningful groups, or clusters, among the data points and find representative centers of these clusters. The aim is that the samples within each cluster are more closely related to one another than samples from different clusters.

Formally, for the dataset $\mathcal{D}$, clustering is the act of associating a cluster ℓ with each observation, where ℓ comes from a small finite set, $\{1, \ldots, K\}$. That is, a clustering algorithm works on the data $\mathcal{D}$ and outputs a function $c(\cdot)$ which maps individual data points to the label values $\{1, \ldots, K\}$. The clustered data (algorithm output) is then a collection of clusters denoted via

$$C_\ell = \big\{x^{(j)} \mid c(x^{(j)}) = \ell, \; j \in \{1, \ldots, n\} \big\}, \quad \text{for} \quad \ell = 1, \ldots, K.$$

A clustering algorithm attempts to choose the clusters such that the elements of each C_ℓ are as homogenous as possible.

The K-means algorithm is one very basic, yet powerful heuristic algorithm. With K-means, as with several other types of clustering algorithms, we pre-specify a number K, determining the number of clusters that we wish to find. Hence K may be treated as a hyper-parameter. As the algorithm seeks the function $c(\cdot)$, or alternatively the partition $C_1, \ldots, C_K$, it also seeks representative *centers* (also known as *centroids*) of the clusters, denoted by $J_1, \ldots, J_K$, each an element of $\mathbb{R}^p$.

One may view the ideal aim of K-means as minimization of

$$\text{Clustering loss} = \sum_{\ell=1}^{K} \sum_{x \in C_\ell} \|x - J_\ell\|^2. \tag{2.39}$$

Such a minimization is generally computationally intractable since it requires considering all possible partitions of $\mathcal{D}$ into clusters. Yet it can be approximately minimized via the K-means algorithm using a classic iterative approach. The K-means algorithm does this by separating the problem into two sub-problems or sub-tasks called *mean computation* and *labeling*. We define these now.

Mean computation: Given $c(\cdot)$, or a clustering $C_1, \ldots, C_K$ where $|C_\ell|$ denotes the number of elements in cluster ℓ, find $J_1, \ldots, J_K$ that minimizes (2.39) via,

$$J_\ell = \frac{1}{|C_\ell|} \sum_{x \in C_\ell} x, \qquad \text{for} \qquad \ell = 1, \ldots, K. \tag{2.40}$$

Here each J_ℓ is the vector obtained via the element-wise average over all the vectors in C_ℓ where each of the p coordinates is averaged separately.

Labeling: Given, $J_1, \ldots, J_K$ and assuming these values are fixed, find $c(\cdot)$ that minimizes (2.39) for every $x \in \mathcal{D}$. This is done by setting,

$$c(x) = \operatorname*{argmin}_{\ell \in \{1,\ldots,K\}} \|x - J_\ell\|. \tag{2.41}$$

That is, the label of each element is determined by the closest center in Euclidean space.

The K-means algorithm starts with randomly initialized[19] centers $J_1, \ldots, J_K$. A labeling step is then executed and these initial random centers are then used to determine initial labels according to (2.41). The algorithm then iterates over the mean computation step (2.40), followed by the labeling step (2.41) and repeats the two steps one after the other. This is done until no more changes are made to the labels and the means. Such an iteration generally does not find the absolute minimum of the objective (2.39), however the approximation found is often satisfactory.

In Figure 2.12 we illustrate the workflow of the algorithm on synthetic data where we choose $K = 3$. The top left plot is the initialization with three random means represented by the black circle, the red triangle, and the green cross. Then each row of Figure 2.12 represents one iteration of the algorithm where the plots on the left column show the output of mean computation (except for the first row which is initialization), and the plots on the right column show the output of labeling.

Note that in practice when presented with data $\mathcal{D}$, one typically first standardizes the data as presented in Section 2.1. Then the process of selecting the hyper-parameter K which is

---

[19]These may be randomly selected elements of $\mathcal{D}$ or some other random set of vectors.

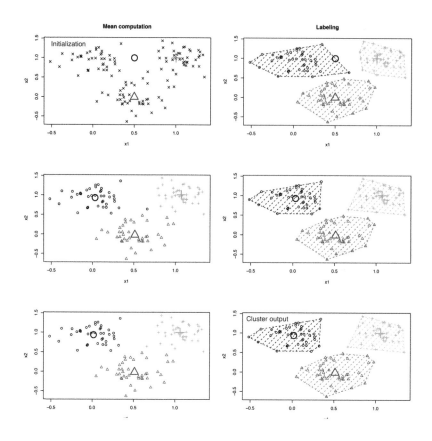

Figure 2.12: Workflow of the K-means algorithm on synthetic data with $K = 3$. The left column is the mean computation step and the right column is the labeling step. Each row is an iteration and the algorithm converges after three iterations. Initialization is presented in the top left corner and after three iterations the algorithm converges with the output presented in the bottom right corner.

external to the K-means algorithm is carried out. One way to do so is to run K-means for increasing values of K and seek a *knee point* or *elbow* when plotting (2.39) as a function of K. As K increases the objective (2.39) generally decreases, however beyond a certain K the value of adding further clusters quickly diminishes. In some cases, such as the visual pixel segmentation we present below, visual subjective measures can be used to find the most appropriate K.

Image Segmentation with K-means

We have already briefly discussed image segmentation in Section 1.1; see Figure 1.2 in that section. As discussed, the goal of image segmentation is to label each pixel of an image with a unique class from a finite number of classes. In Chapter 6 we briefly describe a supervised approach called semantic image segmentation which uses labeled data, namely class masks in addition to the image for training. Nevertheless, in the absence of such information, one may still carry out unsupervised image segmentation. One way to carry out this task is to use the K-means clustering algorithm where each pixel of the image is considered a point in $\mathcal{D}$ and the dimension of each point is typically $p = 3$ (red, green, and blue) for color images.

This can produce impressive image segmentation without any other information except for the image.

Figure 2.13 presents the segmentation of a color image where K-means is used for grouping the pixels into K different clusters. This color image in Figure 2.13 (a) is a $n = 640 \times 640 = 409,600$ pixel color image ($p = 3$). The segmentation consists of running the K-means algorithm which groups similar pixels based on their attributes and assigns the attributes of the corresponding cluster center to the pixel in the image. Figure 2.13 (b) presents the result of the segmentation using $K = 6$ and Figure 2.13 (c) does so with $K = 2$.

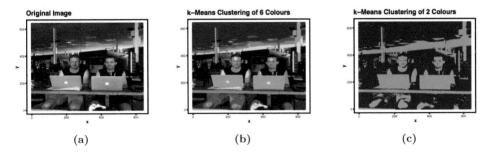

(a)   (b)   (c)

Figure 2.13: Unsupervised image segmentation using K-means. (a) Original image. (b) $K = 6$. (c) K=2.

Matrices in Unsupervised Learning

We often organize the data $\mathcal{D} = \{x^{(1)}, \ldots, x^{(n)}\}$ in the *data matrix* $X_{\mathcal{D}}$, similar to the design matrix (2.10) by stacking each observation vector $x^{(j)}$ in a separate row. The difference between the design matrix and $X_{\mathcal{D}}$ is that the latter does not have a first column of 1s. Thus $X_{\mathcal{D}}$ is an $n \times p$ matrix where the ith column has the data samples for feature i.

It is useful to *de-mean* the data by defining the *centered data matrix*,

$$X = X_{\mathcal{D}} - \mathbf{1}\overline{x}^{\top}, \tag{2.42}$$

where we (re)use the notation X for the matrix previously used for the design matrix and where $\mathbf{1}$ is a column vector of 1s of length p. In the centering process, the p-dimensional vector $\overline{x}$ has coordinates $\overline{x}_i$ which are sample means of the features as defined in (2.1). That is, for each column (feature) in $X_{\mathcal{D}}$ we subtract the mean of the feature. Thus the new $n \times p$ matrix X has features that each have a sample mean of 0.

An important matrix for such data is the $p \times p$ *sample covariance matrix*[20]

$$S = \frac{1}{n} X^{\top} X. \tag{2.43}$$

Written in scalar form, the (i, j)th element of the symmetric matrix S is

$$S_{i,j} = \frac{1}{n} \sum_{k=1}^{n} (x_i^{(k)} - \overline{x}_i)(x_j^{(k)} - \overline{x}_j).$$

---

[20]In a statistical context one often uses $n - 1$ in the denominator instead of n. For non-small n this distinction is insignificant. See a similar comment in relation to (2.1)

and it estimates the *covariance* between feature i and feature j. On the diagonal of S where $i = j$, $S_{i,i}$ equals the sample variance s_i^2 of (2.1). The off diagonal entries account for the measure and direction of linear dependence between features.

We note that the sample covariance matrix can be further normalized to a *sample correlation matrix* by dividing each (i,j)th entry by the product of the sample standard deviations, $s_i s_j$. Sample correlation matrices are important for multivariate descriptive statistics, yet we do not use them explicitly now. Our focus is rather on PCA which we introduce in terms of the de-meaned data matrix X and the sample covariance matrix S.

Principal Component Analysis

It is often the case that not all p dimensions of the data are equally useful. This is especially the case in the presence of high-dimensional data (large p). Moreover, many features may be either completely redundant or uninformative and these cases are referred to as *correlated features* or *noise features*, respectively. In such cases and others, *PCA* is often employed. It is a well-known and widely used dimensionality reduction technique applicable to a wide variety of applications such as data compression, feature extraction, and visualization.

The basic idea of PCA is to project each point of $\mathcal{D}$ which has many correlated coordinates onto fewer coordinates, called *principal components*, which are uncorrelated. This is done while still retaining most of the variability present in the data. In this setting, PCA offers a low-dimensional representation of the features that attempt to capture the most important information from the data. The principal components found via PCA are a new reduced set of features, indexed by $i = 1, \ldots, m$ where $m < p$ is some specified lower dimension. For visualization we often take $m = 2$ or $m = 3$. In other applications, for example the integration of PCA as part of other machine learning procedures, m is often calibrated as a hyper-parameter.

As input, PCA uses the de-meaned data from the centered data matrix X of (2.42) where we denote by $x_{(i)}$ the ith column of X (corresponding to a vector of feature i for all n observations). PCA uses a linear combination of these columns to arrive at the vectors of the new features $\tilde{x}_{(1)}, \ldots, \tilde{x}_{(m)}$. This can simply represented as

$$\tilde{x}_{(i)} = v_{i,1} \begin{bmatrix} | \\ x_{(1)} \\ | \end{bmatrix} + v_{i,2} \begin{bmatrix} | \\ x_{(2)} \\ | \end{bmatrix} + \ldots + v_{i,p} \begin{bmatrix} | \\ x_{(p)} \\ | \end{bmatrix} \quad \text{for} \quad i = 1, \ldots, m,$$

where each new n-dimensional vector, $\tilde{x}_{(i)}$, is a linear combination of the original features. The coefficients of this linear combination can be organized in the vector $v_i = (v_{i,1}, \ldots, v_{i,p})$ which is called the *loading vector* for i. Thus $\tilde{x}_{(i)} = X v_i$. This can also be represented for all the reduced features and loading vectors together via,

$$\underbrace{\begin{bmatrix} | & & | \\ \tilde{x}_{(1)} & \cdots & \tilde{x}_{(m)} \\ | & & | \end{bmatrix}}_{\substack{\widetilde{X}_{n \times m} \\ \text{Reduced data}}} = \underbrace{\begin{bmatrix} | & & & | \\ x_{(1)} & \cdots & \cdots & x_{(p)} \\ | & & & | \end{bmatrix}}_{\substack{X_{n \times p} \\ \text{Original de-meaned data}}} \times \underbrace{\begin{bmatrix} | & & | \\ v_1 & \cdots & v_m \\ | & & | \end{bmatrix}}_{\substack{\widetilde{V}_{p \times m} \\ \text{Matrix of loading vectors}}} . \qquad (2.44)$$

It turns out that a very useful way to represent the loading vectors $v_1, \ldots, v_m$ is by normed eigenvectors associated with eigenvalues of the sample covariance matrix S as in (2.43). Specifically, since S is symmetric and positive semidefinite, the eigenvalues of S are real and non-negative, a fact which allows us to order them via $\lambda_1 \geq \lambda_2 \geq \ldots \geq \lambda_p \geq 0$. We then pick the loading vector v_i to be a normed eigenvector associated with λ_i, namely,

$$Sv_i = \lambda_i v_i, \tag{2.45}$$

while keeping in mind that the first loading vector is associated with the highest eigenvalue; the second is associated with the second highest eigenvalue; and so forth. The symmetry of S also means that its eigenvectors are orthogonal and hence $\tilde{V}$ is a matrix with orthonormal columns. In this setting we assume that at least the first m eigenvalues are strictly positive, namely $\lambda_m > 0$.

In the subsection below we derive the main result to show why this choice of loading vectors based on eigenvectors is attractive. At this point let us consider a numerical example.

We return to the Wisconsin breast cancer data used in Section 2.2 where $p = 30$ and $n = 569$. To visualize this data using PCA, we set $m = 2$ and compute the first two loading vectors from the 30×2 matrix $\tilde{V}$ using standard numerical procedures for eigenvalues and eigenvectors. Then by multiplying the 569×30 demeaned data matrix X by $\tilde{V}$ as in (2.44) we get the 569×2 matrix $\tilde{X}$ of principal components. We then plot each row which is two-dimensional in Figure 2.14 (a).

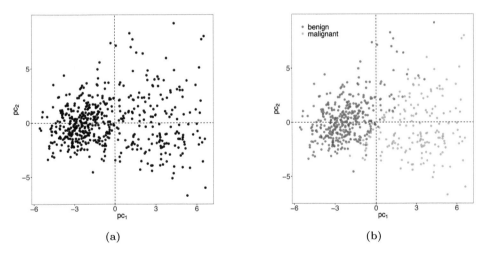

(a)	(b)

Figure 2.14: Breast tumor data samples projected on the two first principal components from the PCA. (a) Unlabeled data. (b) Once adding the label to each sample a pattern and separation between **benign** vs. **malignant** appears.

On their own, the two-dimensional points in Figure 2.14 (a) may not be insightful. After all the principal components coordinates `pc1` and `pc2` do not have any physical meaning in this context. Nevertheless, if we consider Figure 2.14 (b) where we frame this as a supervised learning problem and color the points based on the labels **benign** vs. **malignant**, a useful pattern emerges. There is quite a clear separation between the two classes and hence there is potential to classify points by separating the region in the principal components plane. We do not discuss concrete examples of constructing a classifier in this case. We rather point out that the data following a PCA transformation with reduced dimension $m < p$ can often be used as input to a supervised learning algorithm.

Derivation of PCA

The PCA framework tries to project the data in the directions with maximum variance. Returning to (2.44), since $\tilde{x}_{(i)} = Xv_i$ we can formulate this by maximizing the sample variance of the components of $\tilde{x}_{(i)}$. Keeping in mind that $\tilde{x}_{(i)}$ is a 0 mean vector, its sample variance using (2.1) is simply $\tilde{x}_{(i)}^\top \tilde{x}_{(i)}/n$. Hence substituting $\tilde{x}_{(i)} = Xv_i$, we have,

$$\text{Sample variance of component } i = \frac{1}{n} v_i^\top X^\top X v_i = v_i^\top S v_i,$$

where S is the sample covariance of the data as in (2.43). Thus in searching for the first loading vector v_1 we have the optimization problem,

$$\max_{v \in \mathbb{R}^p} v^\top S v, \quad \text{subject to} \quad \|v\| = 1. \tag{2.46}$$

Note the constraint which seeks a normalized direction v with $\|v\| = 1$ which is equivalent to $v^\top v = 1$. This representation allows us to use Lagrange multiplier techniques where we convert this constrained problem to an unconstrained quadratic problem. The objective is then,

$$\max_{v, \lambda} v^\top S v + \lambda \left(1 - v^\top v\right) \quad \text{or} \quad \max_{v, \lambda} v^\top S v - v^\top \lambda v + \lambda, \tag{2.47}$$

with the Lagrange multiplier λ and the constraint $1 - v^\top v = 0$. Now taking derivatives with respect to the vector v (see also Appendix A) we obtain $Sv - \lambda v$. Thus the first-order conditions in terms of v reduce to the eigenvalue problem $Sv = \lambda v$. This means that any eigenvector v of S adheres to the first-order conditions for the optimization problem (2.46).

By multiplying the eigenvalue equation by $v^\top$ we get $v^\top Sv - v^\top \lambda v = 0$. As apparent from the representation on the right hand side of (2.47), this means that any eigenvector yields a maximization objective that is equal to the corresponding eigenvalue λ. Hence the optimization problem is solved by choosing the maximal eigenvalue λ_1 with v being an associated normalized eigenvector, v_1 as in (2.45).

The subsequent directions v_i for $i = 2, \ldots, m$ are chosen by maximizing the variance of new linear combinations that are orthogonal to previous ones. That is, the directions capture the part of variance that has not been previously captured. It can be shown that a normalized eigenvector that matches the second eigenvalue, v_2 maximizes the variance once the direction of v_1 is removed. This then continues for $i = 3, \ldots, m$ and in summary principal components are determined via (2.45).

PCA Through SVD

We have already used the *singular value decomposition* (SVD) in Section 2.3 in the context of the Moore-Penrose pseudo-inverse. We now further revisit the construction of SVD from linear algebra and see the relationship between SVD and PCA.

Any $n \times p$ dimensional matrix X of rank r can be represented as

$$X = U\Delta V^\top = \sum_{i=1}^{r} \delta_i \, u_i \, v_i^\top, \quad \text{with} \quad \Delta = \text{diag}(\delta_1, \ldots, \delta_r), \quad \text{and} \quad \delta_i > 0. \tag{2.48}$$

Here the $n \times r$ matrix U and the $p \times r$ matrix V are both with orthonormal columns denoted u_i and v_i, respectively for $i = 1, \ldots, r$. These columns are called the left and right *singular vectors* respectively. The values δ_i in the $r \times r$ diagonal matrix Δ are called *singular values* and are ordered as $\delta_1 \geq \delta_2 \geq \cdots \geq \delta_r > 0$. Note that this representation of the SVD differs from the one employed near (2.17) in Section 2.3 where U and V were taken as square matrices and Δ was not necessarily square. The form used in Section 2.3 is sometimes called the *full SVD* and the form we present here is called the *reduced SVD*.

Consider now X again as the de-meaned data matrix (2.42). Now using its SVD representation in the sample covariance (2.43) we obtain

$$S = \frac{1}{n} \underbrace{V\Delta^\top U^\top}_{X^\top} \underbrace{U\Delta V^\top}_{X} = \frac{1}{n} V\Delta^2 V^\top, \quad \text{with} \quad \Delta^2 = \text{diag}(\delta_1^2, \ldots, \delta_r^2).$$

Here the fact that U has orthonormal columns implies $U^\top U$ is the $r \times r$ identity matrix and hence it cancels out. Hence,

$$S = \sum_{i=1}^{r} \frac{\delta_i^2}{n} \, v_i \, v_i^\top. \tag{2.49}$$

We can now compare to the eigenvector-based representation of PCA where $\widetilde{V}$ is the matrix of PCA loading vectors as in (2.44). Take $m = r$ and denote by Λ the diagonal matrix with diagonal entries as the eigenvalues of S in decreasing order $\lambda_1 \geq \ldots \geq \lambda_r > 0$. Now using (2.45) we have the *spectral decomposition* of S,

$$S = \widetilde{V}^\top \Lambda \widetilde{V} = \sum_{i=1}^{r} \lambda_i \, v_i \, v_i^\top. \tag{2.50}$$

We now compare (2.49) and (2.50) and see that with $\lambda_i = \delta_i^2 / n$ the loading vectors in (2.50) are the right singular vectors in (2.49). That is, $\widetilde{V} = V$.

Further, to obtain the data matrix of principal components, $\widetilde{X}$ of (2.44) we set $\widetilde{X} = XV$. Now using the SVD representation of X (2.48) and assuming $m = r$, PCA can be represented as,

$$\widetilde{X} = \underbrace{U\Delta V^\top}_{X} V = U\Delta = \begin{bmatrix} | & | & & | \\ \delta_1 u_1 & \delta_2 u_2 & \cdots & \delta_r u_r \\ | & | & & | \end{bmatrix}. \tag{2.51}$$

That is, each column of the reduced data matrix $\widetilde{X}$ is a left singular vector u_i stretched by the singular value δ_i. Further, for $m < r$ we only take the first m columns.

With these relationships between PCA and SVD, numerical methods for computing the SVD decomposition of X can be used for PCA. Indeed in practice, efficient and numerically robust computational methods for SVD are employed for PCA.

SVD for Compression

The singular value decomposition can also be viewed as a means for compressing any matrix X. Specifically, consider the SVD representation in (2.48) with $\delta_1 \geq \delta_2 \geq \ldots \geq \delta_r$. Then a rank $m < r$ approximation of X is

$$\widehat{X} = \sum_{i=1}^{m} \delta_i\, u_i\, v_i^{\top} \approx X, \qquad \text{where} \qquad X - \widehat{X} = \sum_{i=m+1}^{r} \delta_i\, u_i\, v_i^{\top}. \tag{2.52}$$

The rank of $\widehat{X}$ is m and since one often uses m significantly smaller than r, this is called a *low-rank approximation*. For small enough δ_{m+1} the approximation error is negligible since the summation of rank one matrices $\delta_i\, u_i\, v_i^{\top}$ for $i = m+1, \ldots, r$ is small. Observe that the number of values used in this representation of $\widehat{X}$ is $m \times (1 + n + p)$ and for small m this number is generally much smaller than $n \times p$ which is the number of values in X. Hence this may viewed as a compression method.

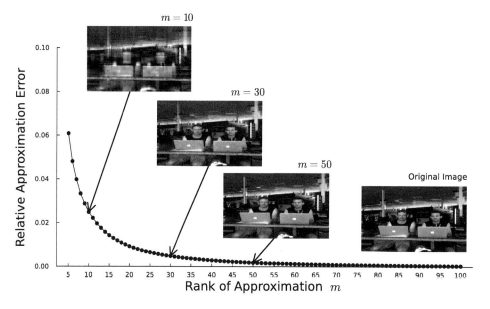

Figure 2.15: SVD for data compression: The original image is presented based on compression with $m = 10$ singular values, $m = 30$ singular values, and $m = 50$ singular values. The images are presented in terms of the relative approximation error based on the Frobenious norm.

The usefulness of such low-rank approximations is validated by a theoretical result called the *Eckart-Young-Mirsky theorem*. Here we consider the approximation-error matrix $X - \widehat{X}$ and we seek to have the best rank m approximation in terms of minimization of $\|X - \widehat{X}\|$. The theorem works for several types of matrix norms, yet here let us focus on the Frobenious norm[21] denoted $\|A\|_F$ for any matrix A. We now have for the Frobenious norm,

$$\min_{\widehat{X} \text{ of rank } m} \left\| X - \widehat{X} \right\|_F^2 = \left\| X - \sum_{i=1}^{m} \delta_i\, u_i\, v_i^{\top} \right\|_F^2 = \sum_{i=m+1}^{r} \delta_i^2. \tag{2.53}$$

---

[21]This is the square root of the sum of the squared elements of the matrix.

Singular value decomposition-based matrix approximations such as (2.52) are useful in multiple domains including improvement of neural network model size. We do not discuss these topics specifically in this book. Instead, consider a simple visual example with a 353×469 monochrome (grayscale) image appearing at the bottom right of Figure 2.15; this is X. Then the other images in Figure 2.15 are $\widehat{X}$ with $m = 10$, $m = 30$, and $m = 50$. As is evident, the $m = 50$ approximation appears close to the original image. The main plot in the figure is the relative approximation error as given by the right hand side of (2.53) divided by the sum of all singular values squared.

Note that the original image uses $353 \times 469 = 165,557$ values while the $m = 50$ approximation only uses $50 \times (1 + 353 + 469) = 41,150$ values. That is the approximation yields $\widehat{X}$ which is compressed to about 25% of the size of X and looks very similar.

Note that variants of the types of plots as in Figure 2.15 are also common when carrying out PCA. In that context the plot is called a *scree plot* and it presents the percentage of variance explained by the principal components.

Notes and References

One does not need to master all other branches of machine learning to understand deep learning, nevertheless getting a taste for key elements of the field is useful. Beyond the basics that we presented in this chapter, one may consult several general machine learning texts. We recommend [240] for a comprehensive mathematical account of practical machine learning and the more classic [39] as an additional resource. Further, the book [299] provides a probabilistic approach. Focusing on linear algebra, the introductory book [56] is a good introduction to foundations such as K-means, least squares, and ridge regression. Further, [391] provides a richer context covering PCA, SVD, and many aspects of matrix algebra appearing in machine learning. Finally for a short read which provides an overview of many practical aspects of machine learning, see [68]. An additional recommended reference is [263].

The worlds of machine learning and statistical inference are intertwined and methods developed in one field are often used in the other field and vice versa. For those with expertise in one or both of the fields it is quite easy to spot the differences between the approaches, however for those entering these worlds afresh it may be helpful to read the survey paper "Statistical modeling: The two cultures" by Leo Breiman, [61]. On that note, to get a feel for many statistical aspects of *linear regression*, see, e.g., [296] or one of many other statistical books. Note that [296] is also a good reference for understanding *interaction terms*, a concept that we mentioned in the chapter and did not cover. A general text that integrates methodology and algorithms with statistical inference and machine learning together with speculations of future directions is [115].

Throughout this chapter we have made reference to several aspects of statistics or machine learning that are not studied further in this book. Here are some references for each. In general, a good reference for likelihood based inference is [31]. Specifically *Akaike information criterion* (AIC), introduced in [7], and the *Bayesian information criterion* (BIC) are surveyed in [432]. A general class of models also appearing in the next chapter is *generalized linear models* (GLMs); these first appeared in [304] and a good contemporary applied reference is [121]. Other models are *general additive models* (GAMs) which extend generalized linear models in which some predictor variables are modeled by smooth functions; see [168]. In terms of non-linear regression the *LOESS* method is a generalization of moving average and polynomial regression, see [88]. Further, *Nadaraya-Watson kernel regression* is a non-parametric regression method in which a kernel function is exploited; see [378].

We have covered the basics of decision theory via binary classification however there are many more studies for these aspects. See the comprehensive survey [118] on metrics for binary classification as well as [158]. For a discussion on different uses of receiver operating curves and different approaches see [58] and [315]. The origins of the F_1 score can be attributed to Cornelis Joost van Rijsbergen who introduced the *effectiveness function* of which F_1 score is a special case; see [408]. The SMOTE method for dealing with unbalanced data is from [75]. See also the surveys [153], [211], and [348].

We briefly mentioned the differences between discriminative and generative learning. More on the topic is in chapter 9 of [299] together with a treatment of the *naive Bayes classifier* and *linear discriminant analysis* (LDA). The area of *support vector machines* became extremely popular in the world of machine learning with their height of popularity during the 1990s and the decade that followed. A complete treatment of these methods is in [240] together with associated ideas of *kernel methods*. Specific to this area is the concept of *VC dimension* (standing for Vapnik–Chervonenkis) which we did not cover here; see [409].

Decision trees are also very popular machine learning techniques; see chapter 8 of [240] for an overview. Within the study of machine learning, generic methods of *boosting* and *bagging* are prominent in the context of decision trees. Specifically see [366] and [59]. The *random forest* algorithm is one such method that has been hailed the most usable ad-hoc generic method when there is no further information about the problem; see [60]. *Gradient boosting* has become very popular due to a software package called *XGBoost*; see [77]. The *K-nearest neighbors* classification algorithm that we mention is often used as an introductory example. See for example Section 2.3 of [166].

The origins of *least squares* fitting are from the turn of the 19th century, initially with applications to astronomy. The first least squares publication is typically attributed to an 1809 paper by Gauss [132] although an earlier 1805 publication by Legendre publicized the concept first. An interesting historical investigation into "who invented least squares" is in [389]. Since then, least squares methods have become some of the most prominent tools in applied mathematics. The *Moore-Penrose pseudo inverse* was independently described in [297], [40], and [328]. *Singular value decomposition* (SVD) has origins in differential geometry with the first linear algebra publication typically associated with the 1936 Eckart Young paper [114]. To the best our knowledge the first association between SVD and least squares is in [139]. The survey [105] may also be of interest as it contrasts different numerical methods for least squares.

Aspects of *multi-collinearity* are treated in many statistical contexts, see for example [281]. Using regression with other methods such as *absolute error loss* (robustness) is covered in [302] and for a reference on the *Huber error loss* see [197]. See also [160] for a discussion on dealing with categorical input features by conversion to numbers.

The origins of gradient descent are attributed to Cauchy from 1847 with [72], way before the invention of any digital electronic computer; see [253] for a historical account. The analysis of the loss landscape in machine learning has been studied multiple times, see for example [279] for a survey, and [284] for theoretical result in a high-dimensional context.[22]

We have only touched the tip of the iceberg in terms of *model selection*. See [390] for a survey as well as the book [87]. There are also recent developments such as [26] dealing with other approaches for balancing underfitting and overfitting or bias and variance. In general, model selection is still a very active open avenue of research. Our discussion surrounding the generalization gap follows similar lines to [263]. Our example in Figure 2.10 is inspired by a similar example in [39].

For an excellent reference dealing with the *LASSO* method see [167]. The original *Tikhonov regularization* (*ridge regression*) technique appeared in 1943 in [401] yet it is believed to have been invented in parallel in other contexts as well. *Elastic net* models are more of a specialty; see [453] and also relations to support vector machines in [452]. Generalizations, variants, and discussion of issues arising with *K-fold cross validation* are in [424] and [232]. See also [215] and [62] for recent developments as well as [443] for stratified cross-validation and variants.

The accepted first reference for *K-means* clustering is [277] from 1967 although the method was known prior. An applied book to understand the main concept and algorithms for cluster analysis is [226]. A recent comprehensive survey about clustering adding to what we presented here is in [120]. Further clustering approaches include *hierarchical clustering*, see [300]. See also an older general survey in [208]. Principal component analysis was first proposed by Pearson in [326] with initial ideas also attributed to Hotelling in [189]. A substantial book on the topic is [212]. Relationships to SVD are well explained in [391] and the first appearance of the *Eckart-Young-Mirsky theorem* for SVD was in [114] by Eckart and Young. It was further independently extended by Mirsky in [292]. One popular efficient method for numerical computation of SVD is the so-called *Golub-Reinsch algorithm* first introduced in [140]. A further overview of additional SVD algorithms is in [89].

---

[22]See also `https://losslandscape.com/` for a visual presentation of different loss landscapes.

3 Simple Neural Networks

In the previous chapter we explored machine learning in general and also focused on linear models trained via gradient-based optimization. In this chapter we move up a notch and consider logistic regression and multinomial (softmax) regression models used for regression and classification. These models often play a key role in statistical modeling and are also very important from a deep learning perspective. Logistic regression and multinomial regression models are shallow neural networks, but also involve non-linear activation functions and hence understanding their structure and means of training is a gateway to understanding general deep learning networks.

In exploring logistic and multinomial regression, we are introduced to some of the basic mathematical engineering elements of deep learning. These include the sigmoid activation function, the cross-entropy loss, the softmax function, and the convexity properties that these models enjoy. We also use the opportunity to consider simple but non-trivial autoencoders that generalize PCA, introduced in the previous chapter.

In Section 3.1 we consider the statistical viewpoint of the logistic regression model. In the process we see how binary cross entropy loss results from maximum likelihood estimation. In Section 3.2 we consider the same model, only this time as a shallow neural network trained via gradient descent. In Section 3.3 we adapt the model to multi-class classification. Here we introduce the softmax function as well as the categorical cross entropy loss. In Section 3.4 we investigate feature engineering for such models and see how additional features extend the resulting classifiers to have non-linear decision boundaries. In Section 3.5 we move onto unsupervised learning and consider simple non-linear autoencoder models. In that section we also discuss general autoencoder concepts and describe a few applications of autoencoders. Note that a reader returning to Section 3.5 after reading Chapter 5, may also envision how the simple shallow autoencoders that we present in Section 3.5, can be generalized to deep autoencoders.

3.1 Logistic Regression in Statistics

Logistic regression can be viewed as the simplest non-linear neural network model. However, outside of the context of deep learning, logistic regression is a very popular statistical model. In this section we present logistic regression via a statistical viewpoint. We show how to estimate parameters via maximum likelihood estimation and in the process are introduced to the (binary) cross entropy loss function common in deep learning models including logistic regression, but also beyond.

Note that the statistical view of logistic regression also incorporates parameter uncertainty intervals, hypothesis tests, and other statistical inference aspects. Literature for such statistical inference aspects of logistic regression is provided at the end of the chapter.

DOI: 10.1201/9781003298687-3

The Model

Logistic regression is a model for predicting the probability that a binary response is 1. It is suitable for classification tasks, a concept described in Section 2.2, as well as for prediction of proportions or probabilities. From a statistical perspective, it is defined by assuming that the distribution of the binary response variable, y, given the features, x, follows a Bernoulli distribution[1] with success probability $\phi(x)$ that is defined below. That is, if we represent the random variables of the feature vector and the (scalar) response via X and Y respectively, then

$$\mathbb{P}(Y = 1 \mid X = x) = \phi(x) \quad \text{and} \quad \mathbb{P}(Y = 0 \mid X = x) = 1 - \phi(x). \tag{3.1}$$

To capture the relationship between x and $\phi(x)$, the model uses the *odds* of $\phi(x)$, namely $\phi(x)/\big(1 - \phi(x)\big)$, via the *log odds* represented as the *logit* function

$$\text{Logit}(u) = \log \left(\frac{u}{1 - u} \right). \tag{3.2}$$

Using this notation, the logistic regression model assumes a linear (affine) relationship between the feature vector x and the log odds of $\phi(x)$, namely,

$$\text{Logit}\big(\phi(x)\big) = b + w^\top x. \tag{3.3}$$

Here for feature vector $x \in \mathbb{R}^p$, the parameter space is $\Theta = \mathbb{R} \times \mathbb{R}^p$ and the parameters $\theta \in \Theta$ are denoted via $\theta = (b, w)$. In this case, the number of parameters is $d = p + 1$ similarly to the linear regression models of Section 2.3. Like the linear regression model introduced in Section 2.3, the scalar parameter b is called the *intercept* or *bias* and the vector parameter w is called the *regression parameter* or *weight vector*.

The fact that (3.3) presents $\text{Logit}(\cdot)$ on the left hand side in contrast to simply the expected response y as one would have in linear regression, sets logistic regression as a non-trivial form of a *generalized linear model* (GLM). We do not discuss general GLM further, nevertheless it is good to note that using GLM terminology, $\text{Logit}(\cdot)$ plays the role of a *link function*. See notes at the end of this chapter for details and references.

A closely related function to $\text{Logit}(\cdot)$, central to logistic regression and deep learning, is the *sigmoid function*, also called the *logistic function*. It is denoted $\sigma_{\text{Sig}}(\cdot)$ and is also discussed in Section 5.3 where it is plotted in Figure 5.6. Its expression is

$$\sigma_{\text{Sig}}(u) = \frac{1}{1 + e^{-u}}. \tag{3.4}$$

It has domain $\mathbb{R}$, range $(0, 1)$, and is monotonically increasing. Note that the logit function and the sigmoid function are inverses. That is, $\sigma_{\text{Sig}}\big(\text{Logit}(u)\big) = \text{Logit}\big(\sigma_{\text{Sig}}(u)\big) = u$.

Applying $\sigma_{\text{Sig}}(\cdot)$ to the representation of the model in (3.3) we obtain the common representation of the logistic regression model,

$$\phi(x) = \mathbb{P}\Big(Y = 1 \mid X = x\,;\, \theta = (b, w)\Big) = \sigma_{\text{Sig}}(b + w^\top x) = \frac{1}{1 + e^{-(b + w^\top x)}}. \tag{3.5}$$

---

[1]This is a probability distribution of a random variable having only two outcomes, 0 or 1. It is a special case of the *binomial distribution* where the parameter for the "number of trials" is set at 1.

Side Note: The Logistic Distribution

When considering the statistical logistic regression model, it is also interesting to represent the relationship between X and Y via a continuous *latent variable*, denoted via Z. A latent variable is an unobserved quantity. In the case of logistic regression, Z allows us to use the linear model representation,

$$Z = b + w^\top X + \epsilon, \qquad \text{with} \qquad \begin{cases} Y = 1, & \text{if } Z \geq 0, \\ Y = 0, & \text{if } Z < 0. \end{cases} \qquad (3.6)$$

Here the noise component ϵ follows a (standard) *logistic distribution* whose cumulative distribution function, $F_\epsilon(u) = \mathbb{P}(\epsilon \leq u)$, is given by $F_\epsilon(u) = \sigma_{\text{Sig}}(u)$. Such a representation agrees with (3.5) since,

$$\begin{aligned} \phi(x) = \mathbb{P}(Y = 1 \mid X = x\,; \theta) &= \mathbb{P}(Z \geq 0 \mid X = x\,; \theta) \\ &= \mathbb{P}(b + w^\top x + \epsilon \geq 0) \\ &= \mathbb{P}(b + w^\top x - \epsilon \geq 0) \\ &= \mathbb{P}(\epsilon \leq b + w^\top x) = \sigma_{\text{Sig}}(b + w^\top x). \end{aligned}$$

Note that in the step between the second line and the third line, we use the fact that the (standard) logistic distribution of ϵ is symmetric about 0, and hence ϵ and $-\epsilon$ are equal in distribution.

We do not use the latent representation (3.6) any further. Yet it may be interesting to note that if one chooses to use a Gaussian distribution for ϵ in (3.6) in place of the logistic distribution, the model turns out to be a *probit regression model* instead of (3.3). Hence such a latent representation of the model is generally interesting.

Estimation Using the Maximum Likelihood Principle

When statistical assumptions are imposed, *maximum likelihood estimation* (MLE) is a very common statistical method for estimating parameters. While most deep learning models in this book do not use MLE or other statistical point estimation techniques directly, exploring how MLE works for logistic regression is insightful.

Central to MLE is the *likelihood function $L : \Theta \to \mathbb{R}$*. It is a function of the parameters $\theta \in \Theta$ which is obtained by the probability (or probability density) of the data for any given parameter value θ. The idea of MLE is to choose values for θ that maximize the likelihood (function). We see this in action for the logistic regression model now.

Consider the training observations $\mathcal{D} = \left\{ (x^{(1)}, y^{(1)}), \ldots, (x^{(n)}, y^{(n)}) \right\}$ denoted in a similar way to the way $\mathcal{D}$ is defined for the linear model Section 2.3. Here each label $y^{(j)}$ is encoded via either 0 or 1, and the feature vector $x^{(j)}$ is a p-dimensional vector in $\mathbb{R}^p$.

A common assumption in statistics and machine learning is that all features-label pairs $(x^{(j)}, y^{(j)})$ are identically distributed and mutually independent; this is often denoted *i.i.d.*. With this assumption, we aim to estimate the parameters $\theta = (b, w)$. The likelihood function is

$$L(\theta\,;\,\mathcal{D}) = \prod_{i=1}^{n} \mathbb{P}(Y = y^{(i)} \mid X = x^{(i)}\,;\,\theta), \tag{3.7}$$

where $L(\theta\,;\,\mathcal{D})$ is viewed as a function of θ given the observed sample $\mathcal{D}$. The product is due to the independence assumption arising as part of the i.i.d. assumption. In the context of the underlying logistic model, using (3.1) and (3.5), since the labels are either 0 or 1, the likelihood evaluates as

$$L(\theta\,;\,\mathcal{D}) = \prod_{i=1}^{n} \phi(x^{(i)})^{y^{(i)}} \left(1 - \phi(x^{(i)})\right)^{1-y^{(i)}} \text{ where } \phi(x) = \sigma_{\text{Sig}}(b + w^{\top}x). \tag{3.8}$$

The maximum likelihood estimate is defined as a value of the parameter θ which maximizes the likelihood function. That is, MLE is the value which maximizes the probability of the observations $\mathcal{D}$ assuming the logistic model.

With (3.8) available, one may proceed to optimize the likelihood directly. However, in this case of logistic regression, and in many other cases where MLE is applied to i.i.d. data, it is more convenient to maximize the logarithm of the likelihood called the *log-likelihood* denoted via $\ell(\theta\,;\,\mathcal{D}) = \log\left(L(\theta\,;\,\mathcal{D})\right)$. This is equivalent to maximizing the likelihood since the log function is a monotonic increasing function. For logistic regression, the log-likelihood expression is,

$$\ell(\theta;\mathcal{D}) = \sum_{i=1}^{n} \left[y^{(i)} \log\left(\phi(x^{(i)})\right) + (1 - y^{(i)}) \log\left(1 - \phi(x^{(i)})\right) \right], \tag{3.9}$$

and the maximum likelihood estimate (MLE) is represented via,

$$\hat{\theta}_{MLE} := \underset{\theta \in \mathbb{R}^1 \times \mathbb{R}^p}{\operatorname{argmax}} \ \ell(\theta\,;\,\mathcal{D}) = \underset{\theta \in \mathbb{R}^1 \times \mathbb{R}^p}{\operatorname{argmin}} \ -\frac{1}{n}\ell(\theta\,;\,\mathcal{D}), \tag{3.10}$$

since maximizing $\ell(\theta\,;\,\mathcal{D})$ is the same as minimizing $-\ell(\theta\,;\,\mathcal{D})$ and the positive factor $1/n$ does not change the optimization, yet is useful in the presentation that follows.

One can show that the function $-\frac{1}{n}\ell(\cdot\,;\,\mathcal{D})$ is convex and hence a global minimum exists. Aspects of optimization and convexity are also further overviewed in Section 4.1. However in contrast to the linear model where an explicit analytic solution as in (2.16) exists, in the case of logistic regression there is no analytic solution for the minimizer, and hence optimization algorithms are needed to find the MLE (3.10).

The Binary Cross-Entropy Loss

We have already claimed throughout Chapter 2 that learning almost always involves optimization. The MLE based paradigm for logistic regression certainly reinforces this claim. We now reposition logistic regression MLE in terms of minimization of a loss function, following similar lines to the learning of the linear model of Section 2.3. This setup continues straight into more involved deep learning models that follow.

Recall the general loss function formulation which we first presented for linear models in (2.11) where the loss is $C(\theta\,;\,\mathcal{D}) = \frac{1}{n}\sum_{i=1}^{n} C_i(\theta)$. In that context of linear models, the individual loss is $C_i(\theta) = C_i(\theta\,;\,y^{(i)}, \hat{y}^{(i)}) = (y^{(i)} - \hat{y}^{(i)})^2$. Logistic regression learning follows similar lines except that $C_i(\theta)$ is not the quadratic loss. To see this, revisit the minimization form of (3.10) together with the log-likelihood expression (3.9). The scaled negative likelihood that needs to be minimized can then be represented as a loss function via

$$C(\theta\,;\,\mathcal{D}) = \frac{1}{n}\sum_{i=1}^{n}\left[-\left(y^{(i)}\log\left(\hat{y}^{(i)}\right) + (1-y^{(i)})\log\left(1-\hat{y}^{(i)}\right)\right)\right], \qquad (3.11)$$

where $\hat{y}^{(i)} = \phi(x^{(i)}) = \sigma_{\mathrm{Sig}}(b + w^{\top}x^{(i)})$. This then implies that for logistic regression the loss for each data sample is $C_i(\theta) = \mathrm{CE}_{\mathrm{binary}}(y^{(i)}, \hat{y}^{(i)})$, where

$$\mathrm{CE}_{\mathrm{binary}}(y, \hat{y}) = -\left(y\log\left(\hat{y}\right) + (1-y)\log\left(1-\hat{y}\right)\right) \qquad (3.12)$$

is called the *binary cross entropy* (estimate) applied to observation y and prediction $\hat{y}$.

In general the phrase "cross entropy" and specifically "entropy" is rooted in information theory. Appendix B outlines relationships between cross entropy and related quantities. However, these relationships are not critical for understanding of deep learning. In terms of relating (3.11) to the probabilistic meaning of cross entropy, see first the definition of the cross entropy for two probability distributions in (B.2) and then see the specialization to distributions with binary outcomes in (B.5).

We also mention that while we arrived to the binary cross entropy here as a bi-product of maximum likelihood estimation for logistic regression, in general, binary cross entropy has become the default loss function for more complex binary classification deep learning models, as surveyed in Chapter 5 and onward.

Predicted Probabilities and Parameter Interpretability

For any observed or postulated feature vector $x^{\star} \in \mathbb{R}^p$, the output of logistic regression is $\hat{y} = \phi(x^{\star})$. This is a probability which can be interpreted as in the left hand side of (3.1). Hence at its core, the logistic regression model (3.5) yields probabilities as outputs. Below we see how these probabilities can be used for classification, yet first let us consider prediction of probabilities or proportions.

Recall the breast cancer example presented in Section 2.2 where we overviewed concepts of binary classification. In that example, based on the Wisconsin breast cancer dataset, we used a training set to create two logistic regression models. The first model used the `smoothness_worst` feature as the single coordinate of x and the second model used all possible 30 features. We now continue to use this dataset, this time using all $n = 569$ observations for statistical inference and setting a single feature model based on the `area_mean` feature, and a two feature model based on `area_mean` and `texture_mean`.

With the estimated two feature model based on the estimated parameters $\hat{\theta} = (\hat{b}, \hat{w}_1, \hat{w}_2)$, the predicted probability for an observation $x^{\star} \in \mathbb{R}^2$ is

$$\hat{y} = \sigma_{\mathrm{Sig}}(\hat{b} + \hat{w}_1 x^{\star}_{\texttt{area\_mean}} + \hat{w}_2 x^{\star}_{\texttt{texture\_mean}}),$$

where for clarity we subscript coordinates of the features vector $x^\star$ with the feature name. Using (3.3) this can also be represented as,

$$\log \left(\frac{\hat{y}}{1 - \hat{y}} \right) = \hat{b} + \hat{w}_1 x^\star_{\text{area\_mean}} + \hat{w}_2 x^\star_{\text{texture\_mean}}. \tag{3.13}$$

The representation in (3.13) is appealing since it endows the estimated parameters $\hat{b}$, $\hat{w}_1$, and $\hat{w}_2$ with a concrete *interpretation*. First observe that the log odds is linearly described by the features `area_mean` and `texture_mean`. This means that it is estimated that a unit increase in `area_mean` will see the log odds increase by $\hat{w}_1$, a unit increase in `texture_mean` will see the log odds increase by $\hat{w}_2$, and similarly when both features are at zero, the log odds is at the value $\hat{b}$. This interpretation of parameters is clearly not limited to a model with two features as it would work for any number of features.

In practice, it is more convenient to interpret the odds given by

$$\frac{\hat{y}}{1 - \hat{y}} = e^{\hat{b} + \hat{w}_1 x^\star_{\text{area\_mean}} + \hat{w}_2 x^\star_{\text{texture\_mean}}}.$$

Specifically, an increase of `area_mean` by one unit implies a multiplicative increase for the odds by a factor of $e^{\hat{w}_1}$, and similarly for `texture_mean` by a factor of $e^{\hat{w}_2}$. With this view, it is typically common to consider the *odds ratio* where the meaning of the multiplicative factor $e^{\hat{w}_i}$ is the ratio between the odds of a model where the feature is at some level x_i between the odds at some level $x_i + 1$. This is especially useful where features are binary but also in general. Further, note that when $e^{\hat{w}_i} > 1$ we say the feature has an increasing effect on the probability of the outcome being `positive`, and in the opposite direction a feature with $e^{\hat{w}_i} < 1$ yields a decrease in this probability when the feature is increased.

Such parameter interpretation transcends from linear models to logistic regression models as well as to other statistical models. This interpretability property often makes models extremely attractive in biostatistics and similar fields. It means that estimated parameters of the model are not only useful for their predictive ability, but also for reasoning about the relationships between variables.

As we progress beyond this chapter to more complicated deep neural networks, direct interpretability of parameter values is often lost. In such *non-interpretable* cases, the parameter estimate $\hat{\theta}$ is not a useful learning outcome on its own, but is rather only useful for the tasks of the model. Nevertheless we mention that an active area of machine learning and deep learning research is to seek *interpretable models*.

In Figure 3.1 we present the actual fit of the single feature model and the two features model for the Wisconsin breast cancer observations. The response $\hat{y}$ is the probability of malignant lumps whereas the data involves observations with $y = 1$ (`positive`, i.e., malignant) and $y = 0$ (`negative`, i.e., benign). In (a) we see the sigmoid function applied to $\hat{b} + \hat{w}_1 x_1$ for the estimated single feature model directly and also plot all the observations with a slightly random jitter on the y axis so that they can be visualized easily, where we cut off high outliers. In (b) the sigmoid function is applied to $\hat{b} + \hat{w}_1 x_1 + \hat{w}_2 x_2$ yielding a surface where we omit plotting the actual observations. In both the univariate and multivariate cases, it is evident that the models present a monotonic relationship between each of the variables and the predicted probability. This property holds for any number of features.

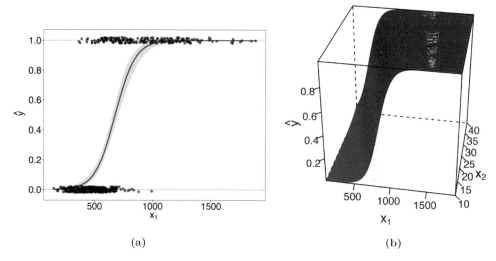

Figure 3.1: Logistic regression models for probability prediction fit to the Wisconsin breast cancer dataset. (a) A $p = 1$ model with the feature `area_mean` (x_1)and a confidence band. (b) A $p = 2$ model based on the features `area_mean` (x_1) and `texture_mean` (x_2).

Logistic Regression for Classification Is a Linear Classifier

In addition to the application of probability prediction as described above, logistic regression models can also be naturally used for binary classification. We have already seen how to convert such models into binary classifiers. This was first seen in Section 2.2 where we introduced the threshold-based classifier in (2.5) with the label prediction $\widehat{\mathcal{Y}}$ set to 1 (`positive`) if $\hat{y} > \tau$ and 0 (`negative`) otherwise.

As an illustration, consider the two features breast cancer prediction model with a response surface as in Figure 3.1 (b). By setting $\tau = 0.5$ for this model, we get a classifier as illustrated in Figure 3.2. The red region corresponds to potential feature vectors that would be classified as `negative` (benign) and the blue region corresponds to `positive` (malignant). It is seen that many, but not all, of the training samples are correctly classified.

Interestingly the classification boundary appears like a straight line, or more precisely a hyperplane in the feature space. One may then wonder if this is a property of logistic regression. We now show that in fact, logistic regression-based binary classifiers are *linear classifiers*. Such linear classifiers separate the feature space via a single hyperplane, $\mathcal{H}(x) = \breve{b} + \breve{w}^\top x$, where $\breve{b} \in \mathbb{R}$ and $\breve{w} \in \mathbb{R}^p$ are the parameters of the hyperplane. Since every hyperplane cuts Euclidean space into two half-spaces, a natural way to use a hyperplane for classification of a feature vector $x^\star \in \mathbb{R}^p$ is

$$\widehat{\mathcal{Y}} = \begin{cases} 0 \ (\texttt{negative}), & \text{if} \quad \mathcal{H}(x^\star) \leq 0, \\ 1 \ (\texttt{positive}), & \text{if} \quad \mathcal{H}(x^\star) > 0. \end{cases} \tag{3.14}$$

This means that if $x^\star$ falls in one of the half spaces the classification is `negative` and in the other it is `positive`. The distinction on the boundary is arbitrary.

To see that logistic regression is a linear classifier consider estimated model parameters $\hat{\theta} = (\hat{b}, \hat{w})$ and probability prediction $\hat{y}$. The `positive` classification then occurs if $\hat{y} > \tau$,

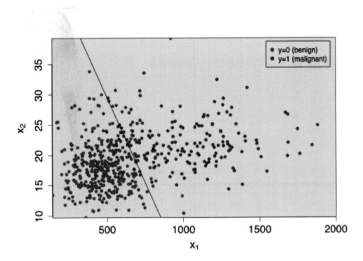

Figure 3.2: Binary classification of malignant lumps (blue dots) and benign lumps (red dots) using a logistic model with two features, `area_mean` (x_1) and `texture_mean` (x_2).

i.e., $\sigma_{\mathrm{Sig}}(\hat{b} + \hat{w}^\top x^\star) > \tau$. We can then apply the Logit($\cdot$) function (3.2) to this inequality. Since Logit($\cdot$) is a monotonic increasing function and is the inverse of the sigmoid function, we obtain,

$$\hat{b} + \hat{w}^\top x^\star > \log\left(\frac{\tau}{1 - \tau}\right), \qquad \text{or} \qquad \hat{b} + \log(\tau^{-1} - 1) + \hat{w}^\top x^\star > 0.$$

We thus see that the hyperplane parameters associated with logistic regression classification are $\breve{b} = \hat{b} + \log(\tau^{-1} - 1)$ and $\breve{w} = \hat{w}$. This shows that logistic regression with a threshold rule yields a linear classifier. Note in fact that with the same argument, any model of the form $\hat{y} = \sigma(\hat{b} + \hat{w}^\top x^\star)$ where $\sigma(\cdot)$ is some bijection (invertible function) from $\mathbb{R}$ to $\mathbb{R}$ yields a linear classifier. This naturally includes the linear model-based binary classification outlined in Section 2.3. It also includes the probit model mentioned earlier in this section, as well as any shallow neural network for binary classification that has a strictly monotonic activation function, a concept introduced in the coming section. Other very common linear classifiers, including support vector machine models are not surveyed in this book.

Note also that one often transforms classifiers with linear decision boundaries into more expressive classifiers via feature engineering. In Section 3.4 we explore how feature engineering-based transformations of the features may yield non-linear decision boundaries for the models studied in this chapter.

3.2 Logistic Regression as a Shallow Neural Network

We now position the logistic regression model as a deep learning model. However, as we shortly explain, it is not really "deep" but is rather "shallow" since it does not have hidden layers. This is the first instance in this book where we explicitly consider deep learning models with some mathematical detail. The general (fully connected) deep learning model is outlined in Chapter 5 and our presentation here serves as a shallow introduction. Note that the linear model of Section 2.3 can also be positioned as a deep learning model; we shed light on this too. Further, the multinomial regression model of the next section is a close relative of logistic regression, and as we show in that section, it is also a shallow neural network.

Logistic Regression Is an Artificial Neuron

Let us first represent the logistic regression model (3.5) as

$$\hat{y} = \sigma \underbrace{\left(\overbrace{b + w^\top x}^{z} \right)}_{a}. \qquad (3.15)$$

Observe that in (3.15), we omit the subscript from $\sigma_{\text{Sig}}(\cdot)$ used in (3.5). We call $\sigma(\cdot)$ a scalar *activation function* which in the case of logistic regression needs to be $\sigma_{\text{Sig}}(\cdot)$, but in other cases can be a different function. Section 5.3 is devoted to specific forms of such activation functions beyond the sigmoid function. At this point, let us just mention a trivial alternative, the *identity activation function* $\sigma(z) = z$. With this identity activation function, the model in (3.15) is clearly just the linear model $\hat{y} = b + w^\top x$.

The form of (3.15) represents what we may call a *single layer* of a deep learning model, a *shallow neural network*, or simply an *artificial neuron*. In this case the vector inputs $x \in \mathbb{R}^p$ are transformed to a scalar output $\hat{y} \in \mathbb{R}$. However, in general, deep learning models (as well as shallow neural networks) allow for vector outputs. The next section presents such a case.

Observe that the artificial neuron is composed of an affine transformation $z = b + w^\top x$ followed by a (generally) non-linear transformation $a = \sigma(z)$. This notation of using z for the result of the affine transformation and a for the result of the non-linear transformation is common in deep learning models and heavily used in Chapter 5. Note that the actual definition of an artificial neuron is sometimes taken as the combination of z and a, sometimes just as the non-linear result a, and sometimes as the computation mechanism defined by the right hand side of (3.15). Figure 3.3 summarizes the components of the artificial neuron focusing on the specific $\sigma(\cdot) = \sigma_{\text{Sig}}(\cdot)$ activation function.

A "deeper" deep learning model would have "hidden layers" based on the composition of constructs similar to (3.15). This would involve multiple z and a values computed along the way. Thus when there is a single layer as in (3.15), the neural network is called *shallow* and otherwise it is called *deep*. Logistic regression is the simplest non-linear scalar output shallow neural network that one can consider.

Training Logistic Regression

We have already seen in (3.10) how maximum likelihood estimation positions parameter fitting of logistic regression as an optimization problem. We then saw that maximization of the likelihood is identical to minimization of the loss $C(\theta\,;\,\mathcal{D}) = \frac{1}{n}\sum_{i=1}^{n} C_i(\theta)$, with $C_i(\theta)$ defined via the cross entropy cost (3.12). Importantly, when considered as a deep learning model or a machine learning model, one sometimes ignores the maximum likelihood interpretation of logistic regression and starts off with the cross entropy loss as an engineered loss function which requires minimization.

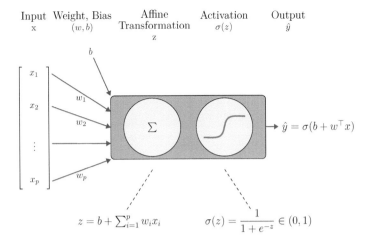

Figure 3.3: Logistic regression represented with neural network terminology as a shallow neural network. The gray box represents an artificial neuron composed of an affine transformation to create z and an activation $\sigma(z)$.

When statistical software packages are used to estimate parameters of logistic regression, this optimization is typically tackled using second-order methods; see Section 4.6 for an overview of such techniques. In general, such methods make use of the *Hessian matrix* of the loss function or approximations of it; see Appendix A for a review.

In contrast to statistical practices, in deep learning and machine learning culture, one often considers the number of features p to be very large in which case standard second-order optimization methods tend to struggle computationally. Thus when treated as a deep learning model, the default technique used for logistic regression training is gradient descent. Gradient descent was already introduced in Section 2.4 in the context of the linear model, and variants of gradient-based learning are studied in detail in the next chapter as well; see sections 4.2 and 4.3.

General deep learning models do not have explicit expressions for the gradient of $C(\theta\,;\mathcal{D})$ and certainly not for the Hessian matrix. Hence learning such models requires computational techniques for gradient evaluation. The most common technique is the backpropagation algorithm, described in Section 5.4, which is a form of automatic differentiation, overviewed in Section 4.4. However, in the case of logistic regression, like the linear model, there are explicit expressions both for the gradient and the Hessian. We see these now.

In the case of logistic regression, the gradient of $C(\theta\,;\mathcal{D}) = \frac{1}{n}\sum_{i=1}^{n} C_i(\theta)$ with respect to $\theta = (b, w) \in \mathbb{R} \times \mathbb{R}^p$ is a vector in $\mathbb{R}^d$ with $d = p + 1$. It is denoted as $\nabla C(\theta)$ and can be represented as the average of the gradients of each $C_i(\theta)$, namely,

$$\nabla C(\theta) = \frac{1}{n}\sum_{i=1}^{n} \nabla C_i(\theta). \tag{3.16}$$

This general relationship between the gradient of the total loss function $\nabla C(\theta)$ and the gradients of the loss for each observation $\nabla C_i(\theta)$ is common throughout deep learning. In the case of logistic regression we have that $C_i(\theta) = \mathrm{CE}_{\text{binary}}(y^{(i)}, \hat{y}^{(i)})$ with $\mathrm{CE}_{\text{binary}}(\cdot, \cdot)$

from (3.12). We thus require expressions for the gradient of this binary cross entropy loss with observation label $y^{(i)}$ and predicted value $\hat{y}^{(i)} = \sigma_{\mathrm{Sig}}(b + w^\top x^{(i)})$. That is,

$$\nabla C_i(\theta) = -\nabla \left(y^{(i)} \log \sigma_{\mathrm{Sig}}(b + w^\top x^{(i)}) + (1 - y^{(i)}) \log \left(1 - \sigma_{\mathrm{Sig}}(b + w^\top x^{(i)})\right)\right),$$

where the gradient is with respect to the vector $\theta = (b, w)$. Now using basic differentiation rules and the structure of $\sigma_{\mathrm{Sig}}(\cdot)$, we obtain,

$$\frac{\partial C_i}{\partial b} = \sigma_{\mathrm{Sig}}(b + w^\top x^{(i)}) - y^{(i)},$$

$$\frac{\partial C_i}{\partial w_j} = \left(\sigma_{\mathrm{Sig}}(b + w^\top x^{(i)}) - y^{(i)}\right) x_j^{(i)} \qquad \text{for} \qquad j = 1, \ldots, p.$$

Thus, by combining these components we can represent the $d = p + 1$ dimensional gradient vector as

$$\nabla C_i(\theta) = \left(\sigma_{\mathrm{Sig}}(w^\top x^{(i)} + b) - y^{(i)}\right) \begin{bmatrix} 1 \\ x^{(i)} \end{bmatrix}, \tag{3.17}$$

where the first scalar expression on the right hand side of (3.17) is the prediction difference for observation i and the second expression is the vector of features for observation i including the constant 1 feature.

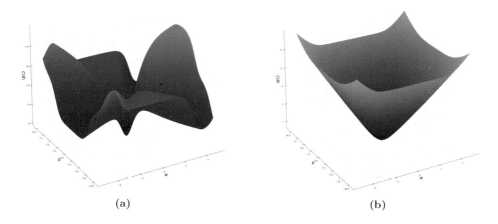

       (a)   (b)

Figure 3.4: The loss landscape of logistic regression for a synthetic dataset. (a) Using the squared loss $C_i(\theta) = (y^{(i)} - \hat{y}^{(i)})^2$. (b) Using the binary cross entropy loss $C_i(\theta) = \mathrm{CE}_{\mathrm{binary}}(y^{(i)}, \hat{y}^{(i)})$.

Some Benefits of Cross Entropy Loss

If one considers the problem of logistic regression training purely based on an optimization approach and not based on a maximum likelihood approach, then the cross entropy loss can potentially be replaced by other loss functions such as for example quadratic cost. However, it turns out that using the cross entropy loss, generally yields desirable loss landscapes.

As an illustration consider Figure 3.4 based on logistic regression with synthetic data of a single feature ($p = 1$ and $d = 2$). The parameters to be optimized are the scalars b and w. In (a) we use the squared error loss and in (b) we use the cross entropy loss. It is clear from this image, that at least in this case, the cross entropy loss landscape is more manageable to navigate for an optimization algorithm like gradient descent. In fact, in the case of logistic regression, cross entropy always yields a convex loss landscape (further details are in the next chapter), while other losses such as the squared error loss generally yield non-convex loss landscapes often presenting multiple local minima as well as saddle points.

Importantly, when considering classification (or probability prediction problems), deep learning models that are more complex than logistic regression are still often easier to optimize using the cross entropy loss than the squared error loss or other losses. In such more complex models, the neat mathematical convexity property that logistic regression enjoys with cross entropy is lost, and multiple local minima can exit. Nevertheless, computational and research experience has shown that in general using cross entropy is preferable.

As an illustration of gradient descent applied to a concrete example we return to the Wisconsin breast cancer dataset, now splitting the observations as we did in Chapter 2 into $n_{\text{train}} = 456$ and $n_{\text{test}} = 113$. We use a model with $p = 10$ features ($d = 11$) and learn the parameters via gradient descent with some arbitrary initialization. In doing so, we obtain trajectories as in Figure 3.5.

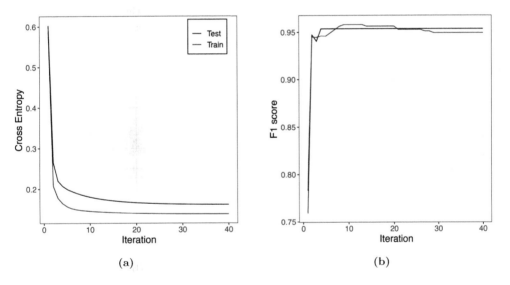

Figure 3.5: Training logistic regression via gradient descent for the Wisconsin breast cancer dataset with an 80-20 train-test split (we can treat the test set as a validation set). (a) Loss over iterations. (b) Performance over iterations using the F_1 score when $\tau = 0.5$.

3.3 Multi-Class Problems with Softmax

The *multinomial regression model* as it is known in statistics, or *softmax regression* as it is known in machine learning[2] is the generalization of logistic regression from the binary response case to the case of $K > 2$ classes. Now the response random variable Y takes on values $1, 2, \ldots, K$. The feature vector remains X just like in logistic regression.

---

[2]In some machine learning fields, this is also called *softmax logistic regression, multinomial logistic regression,* or *multi-class logistic regression.*

We denote the training observations via $\mathcal{D} = \{(x^{(1)}, y^{(1)}), \ldots, (x^{(n)}, y^{(n)})\}$, where each label $y^{(j)}$ is one of K class values $\{1, \ldots, K\}$. The purpose of the model is to predict class probability vectors, or if used for classification, to predict a class in a multi-class setting.

The Model

Just like logistic regression predicts two classes and uses $\phi(x)$ for the probability of the positive response, in multinomial regression the predicted response is the probability vector

$$\phi(x) = \big(\phi_1(x), \ldots, \phi_K(x)\big), \tag{3.18}$$

where,

$$\phi_k(x) = \mathbb{P}(Y = k \mid X = x) \quad \text{for} \quad k = 1, \ldots, K. \tag{3.19}$$

When comparing (3.19) with (3.1), observe that (3.1) is like (3.19) with $K = 2$ where $\phi(x) = \phi_1(x)$ and $1 - \phi(x) = \phi_2(x)$.

The name "multinomial regression" stems from the *multinomial distribution* which is a probability distribution over vectors of length K consisting of non-negative integers that sum up to some specified integer N. Specifically a random vector $(U_1, \ldots, U_K)$ follows a multinomial distribution with parameters N (positive integer) and $\phi = (\phi_1, \ldots, \phi_k)$ (probability vector) if,

$$\mathbb{P}\,(U_1 = u_1, \ldots, U_K = u_K) = \frac{N!}{u_1! \cdot \ldots \cdot u_K!} \prod_{k=1}^{K} \phi_k^{u_k} \qquad \text{for} \qquad \sum_{k=1}^{K} u_k = N,$$

with the probability at 0 when $\sum_{i=k}^{K} u_k \neq N$. A specific case of the multinomial distribution with $N = 1$ is called a *categorical distribution*. In this case the random vector $(U_1, \ldots, U_K)$ is like a one-hot encoded vector since exactly one coordinate is 1 and the rest are 0s.

A statistical assumption in multinomial regression is that the one-hot encoded response, given the features vector $X = x$, follows a categorical distribution. The random vector of the one-hot encoded response of Y is denoted $(Y_1, \ldots, Y_k)$ where $Y_k = \mathbf{1}\{Y = k\}$, i.e., the kth coordinate equals 1 if $Y = k$ and equals 0 otherwise. Now the categorical distribution assumption is

$$\mathbb{P}\,(Y_1 = u_1, \ldots, Y_K = u_K \mid X = x) = \prod_{k=1}^{K} \phi_k(x)^{u_k}. \tag{3.20}$$

The key model assumption in multinomial regression[3] is the way in which the probability vector $\phi(x)$ depends on the features vector x. There are two possible parameterizations, which we refer to as the statistical parameterization and the machine learning parameterization. The former presents a unique (*identifiable*) model while the latter is slightly simpler but has some redundant parameters.

Starting with the statistical parameterization, it is assumed that,

$$\phi_k(x) = \frac{e^{b_k + w_{(k)}^\top x}}{1 + \sum_{j=1}^{K-1} e^{b_j + w_{(j)}^\top x}} \qquad \text{for} \qquad k = 1, \ldots, K-1, \tag{3.21}$$

---

[3]Like logistic regression, this assumption is rooted in the theory of generalized linear models (GLM), a topic we defer to the notes and references at the end of the chapter.

and further,

$$\phi_K(x) = 1 - \sum_{j=1}^{K-1} \phi_j(x). \tag{3.22}$$

This directly generalizes the sigmoidal relationship of logistic regression (3.5) from $K = 2$ to $K \geq 2$. With this statistical parameterization, the parameters of multinomial regression are

$$\theta = (b_1, \ldots, b_{K-1}, w_{(1)}, \ldots, w_{(K-1)}) \in \underbrace{\mathbb{R} \times \ldots \times \mathbb{R}}_{K-1 \text{ times}} \times \underbrace{\mathbb{R}^p \times \ldots \times \mathbb{R}^p}_{K-1 \text{ times}} := \Theta, \tag{3.23}$$

and hence the number of parameters is $d = (K-1)(p+1)$. The scalar parameters $b_1, \ldots, b_{K-1}$ are bias (intercept) parameters and each of the vector parameters $w_{(1)}, \ldots, w_{(K-1)}$ is a weight vector (regression parameter). With this parameterization, we may also view the final bias b_K and final weight vector $w_{(K)}$ as the scalar 0 and vector 0, respectively. Such a restriction on b_K and $w_{(K)}$ allows us to combine (3.21) and (3.22) into

$$\phi_k(x) = \frac{e^{b_k + w_{(k)}^\top x}}{\sum_{j=1}^{K} e^{b_j + w_{(j)}^\top x}} \qquad \text{for} \qquad k = 1, \ldots, K. \tag{3.24}$$

Moving onto the machine learning parameterization, the last class also has free parameters b_K and $w_{(K)}$ like the other classes. In this case the parameter space is $\Theta = \mathbb{R}^K \times (\mathbb{R}^p)^K$ and thus $d = K(p+1)$. Now again the representation (3.24) is valid, yet unlike the statistical parameterization, the last term in the summation in the denominator is not restricted to be 1. In this sense the machine learning parameterization is simpler, however when estimating θ with maximum likelihood estimation, there is never a unique θ that maximizes the likelihood (or minimizes the loss).

The Softmax Function and Multinomial Regression as a Shallow Neural Network

A common function in deep learning, especially when considering classification problems, is the $\mathbb{R}^K \to \mathbb{R}^K$ *softmax activation function*. For $z = (z_1, \ldots, z_K) \in \mathbb{R}^K$, it is defined as,

$$S_{\text{Softmax}}(z) = \frac{1}{\sum_{i=1}^{K} e^{z_i}} \begin{bmatrix} e^{z_1} \\ \vdots \\ e^{z_K} \end{bmatrix}. \tag{3.25}$$

Softmax and its derivative expressions are further discussed in Section 5.3. At this point let us just point out that for any vector $z \in \mathbb{R}^K$, the result of $S_{\text{Softmax}}(z)$ is a probability vector.

Now with $S_{\text{Softmax}}(\cdot)$ defined we can revisit the multinomial regression model (3.24) and represent it as the softmax of a vector of length K that has $b_k + w_{(k)}^\top x$ at the kth coordinate. More succinctly we have

$$\hat{y} = \underbrace{S_{\text{Softmax}}\big(\overbrace{b + Wx}^{z \in \mathbb{R}^K}\big)}_{a \in \mathbb{R}^K}, \tag{3.26}$$

where $\hat{y}$ is the prediction of $\phi(x)$ as in (3.18), and the K-dimensional vector b and the $K \times p$ dimensional matrix W are

$$b = \begin{bmatrix} b_1 \\ \vdots \\ b_K \end{bmatrix} \quad \text{and} \quad W = \begin{bmatrix} \underline{\qquad} & w_{(1)}^\top & \underline{\qquad} \\ \underline{\qquad} & w_{(2)}^\top & \underline{\qquad} \\ & \vdots & \\ \underline{\qquad} & w_{(K)}^\top & \underline{\qquad} \end{bmatrix}, \text{ respectively.} \quad (3.27)$$

Compare (3.26) with (3.15) of logistic regression. In the logistic regression case, z is a scalar and so is a. In contrast, in the multinomial regression case the affine transformation converts a vector $x \in \mathbb{R}^p$ into a vector $z \in \mathbb{R}^K$ and then the softmax activation retains the same dimension to arrive at a.

Note also that (3.26) is valid both in the statistical parameterization and the machine learning parameterization. In the former, the last bias b_K and the last row $w_{(K)}^\top$ of (3.27) are zeros, where in the latter these are free variables.

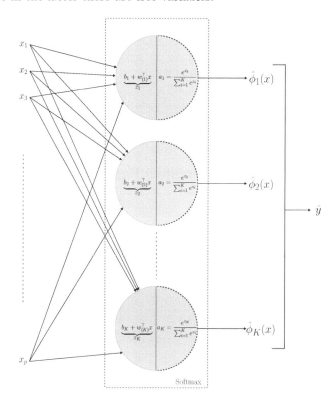

Figure 3.6: Multinomial regression as a neural network model with output $\hat{y}$ which is a probability vector. Note that each a_i is a function of all of $z_1, \ldots, z_K$ due to the softmax operation.

This representation positions multinomial regression as a shallow neural network similarly to the way logistic regression is a shallow neural network. The difference is that multinomial regression has vector outputs while logistic regression has a scalar output. Figure 3.6 illustrates the multinomial regression model as a neural network. Here, each circle can again be viewed as an "artificial neuron", however note that the softmax activation affects all neurons together via the normalization in the denominator of (3.25). Hence the activation value of each neuron is not independent of the activation values of the other neurons.

Likelihood and Cross Entropy

Now that we understand the multinomial regression model, both from a statistical perspective using the probabilistic interpretation (3.20) and as a deep learning model using (3.26), let us consider maximum likelihood estimation, and equivalently loss function minimization.

To obtain an estimate $\hat{\theta}$ using maximum likelihood estimation, we follow a procedure analogous to the one used for logistic regression. Specifically, the likelihood of the model for the data $\mathcal{D}$ is defined as in (3.7). Now using the probability law of the categorical distribution, (3.20), we obtain the likelihood,

$$L(\theta \,;\, \mathcal{D}) = \prod_{i=1}^{n} \prod_{k=1}^{K} \phi_k(x^{(i)})^{\mathbf{1}\{y^{(i)}=k\}} \qquad \text{for} \qquad \theta \in \Theta,$$

where we use the parameter space Θ as in the statistical parameterization[4] (3.23). Now considering the log-likelihood by applying $\log(\cdot)$ to $L(\theta \,;\, \mathcal{D})$ we obtain,

$$
\begin{aligned}
\ell(\theta \,;\, \mathcal{D}) &= \sum_{i=1}^{n} \sum_{k=1}^{K} \mathbf{1}\{y^{(i)} = k\} \log \phi_k(x^{(i)}) \\
&= \sum_{i=1}^{n} \log \phi_{y^{(i)}}(x^{(i)}) \\
&= \sum_{i=1}^{n} \left(b_{y^{(i)}} + w_{(y^{(i)})}^{\top} x^{(i)} - \log \sum_{j=1}^{K} e^{b_j + w_{(j)}^{\top} x^{(i)}} \right).
\end{aligned}
\tag{3.28}
$$

In the second step, we use the subscript $y^{(i)}$ because except for the index k where $y^{(i)} = k$, all other summands of the internal sum of the first row are zero. The third step follows from (3.24).

Now similarly to our development of logistic regression MLE in (3.10), we can represent the problem as a minimization problem and define the estimator via,

$$\widehat{\theta}_{MLE} = \operatorname*{argmin}_{\theta \in \Theta} \; -\frac{1}{n} \ell(\theta \,;\, \mathcal{D}).$$

Further, we can also view this problem as a loss minimization problem, where as before the loss is of the form $C(\theta \,;\, \mathcal{D}) = \frac{1}{n} \sum_{i=1}^{n} C_i(\theta)$, and $C_i(\theta)$ is the loss for an individual observation. Based on the expressions of the log-likelihood (3.28) and using $\hat{y}^{(i)}$ for the vector $\phi(x^{(i)})$, the loss for an individual observation is,

$$C_i(\theta) = -\sum_{k=1}^{K} \mathbf{1}\{y^{(i)} = k\} \log \hat{y}_k^{(i)} = -\log \hat{y}_{y^{(i)}}^{(i)}. \tag{3.29}$$

---

[4]Alternatively one may opt for the machine learning parameterization with the bigger parameter space. In this case the MLE is never unique.

3 Simple Neural Networks

Here, the first and second equalities arise from the first and the second lines of (3.28), respectively. The expressions on the right hand side of (3.29) are in fact called the *categorical cross entropy*. More precisely, for a label $y \in \{1, \ldots, K\}$ and a probability vector $\hat{y}$ of length K, we define the categorical cross entropy as

$$\text{CE}_{\text{categorical}}(y, \hat{y}) = -\sum_{k=1}^{K} \mathbf{1}\{y = k\} \log \hat{y}_k = -\log \hat{y}_y, \qquad (3.30)$$

where in the final expression $\hat{y}_y$ means taking the element at index y from the vector $\hat{y}$. We thus see that

$$C_i(\theta) = \text{CE}_{\text{categorical}}(y^{(i)}, \hat{y}^{(i)}). \qquad (3.31)$$

We have already seen the binary cross entropy in (3.12). Similar to this, the categorical cross entropy (3.30) is the empirical estimate of the cross entropy of two probability distributions appearing in (B.2) of Appendix B. We note here that in many contexts of deep learning, one just uses the phrase "cross entropy" without the prefix "binary" or "categorical".[5] Then the distinction between $\text{CE}_{\text{binary}}(\cdot, \cdot)$ and $\text{CE}_{\text{categorical}}(\cdot, \cdot)$ is based on context and the type of arguments y and $\hat{y}$. In the former y is a binary outcome in $\{0, 1\}$ and $\hat{y}$ is a scalar probability in $[0, 1]$. In the latter, y is a multi-class choice in $\{1, \ldots, K\}$ and $\hat{y}$ is a probability vector of length K in $[0, 1]^K$. The binary cross entropy is a special case of the categorical cross entropy and $\text{CE}_{\text{binary}}(y, \hat{y})$ with $y \in \{0, 1\}$ and $\hat{y} \in [0, 1]$ can be represented as $\text{CE}_{\text{categorical}}(2 - y, [\hat{y}, 1 - \hat{y}]^\top)$.

It may be useful to also see other notational forms for the loss of an individual observation $C_i(\theta)$. Making use of (3.26) we obtain

$$C_i(\theta) = \text{CE}_{\text{categorical}}\left(y^{(i)}, S_{\text{Softmax}}(b + W x^{(i)})\right), \qquad (3.32)$$

where we parameterize θ to be composed of the vector b and the matrix W. Namely, $\Theta = \mathbb{R}^K \times \mathbb{R}^{K \times p}$. Alternatively,

$$C_i(\theta) = -(b_{y^{(i)}} + w_{(y^{(i)})}^\top x^{(i)}) + \log \sum_{j=1}^{K} e^{b_j + w_{(j)}^\top x^{(i)}}. \qquad (3.33)$$

Compare (3.33) with the last line of (3.28).

Derivatives and Learning

Like logistic regression, inference for multinomial regression can be efficiently carried out using second-order methods. Nevertheless, when multinomial regression is viewed as a deep learning model with very large p, one often uses gradient descent (first-order optimization) in place of second-order methods. We now present the derivative expressions of $C_i(\theta)$. These expressions allow one to use gradient descent just as in the case of logistic regression.

Using either the statistical parameterization with $d = (K - 1)(p + 1)$ parameters or the machine learning parameterization with $d = K(p + 1)$ parameters, we have that $\theta = (b, W)$ consists of both a vector b and a matrix W. Thus in dealing with the gradient of $C_i(\theta)$ with respect to θ, denoted as $\nabla C_i(\theta)$, one needs to either vectorize θ into a vector of length d, or

---

[5]In practice, when deep learning systems implement the binary and categorical cross entropy, $\log(u)$ in (3.12) or (3.30) is replaced with $\log(u + \varepsilon)$ where ε is a small fixed parameter. This allows to seamlessly include the probability $u = 0$ as an input.

alternatively, to make use of the derivative of a real valued function with respect to a matrix, as defined in Appendix A, (A.7). Our presentation uses both variants, and it is a prelude for further derivative expressions involving matrix valued parameters, used in Chapter 5 and the chapters that follow.

Let us first consider the derivative of (3.33) with respect to the individual scalar elements of $\theta = (b, W)$. Specifically, for $k = 1, \ldots, K$ we obtain

$$\frac{\partial C_i}{\partial b_k} = \frac{e^{b_k + w_{(k)}^\top x^{(i)}}}{\sum_{j=1}^K e^{b_j + w_{(j)}^\top x^{(i)}}} - \mathbf{1}\{y^{(i)} = k\},$$

$$\frac{\partial C_i}{\partial w_{k,\ell}} = \left(\frac{e^{b_k + w_{(k)}^\top x^{(i)}}}{\sum_{j=1}^K e^{b_j + w_{(j)}^\top x^{(i)}}} - \mathbf{1}\{y^{(i)} = k\} \right) x_\ell^{(i)}, \qquad \text{for} \qquad \ell = 1, \ldots, p.$$

These derivatives can then be placed in a d-dimensional vector[6] to form $\nabla C_i(\theta)$.

Now observe that the expressions above involve the softmax function where the ratio of the exponent with the sum of exponents in each expression is the kth coordinate of $S_{\text{Softmax}}(b + W x^{(i)})$. This hints at a simpler expressions and indeed we may use unit vector notation (e_j) in place of the indicator functions $\mathbf{1}\{\cdot\}$ to obtain,

$$\frac{\partial C_i}{\partial b} = S_{\text{Softmax}}(b + W x^{(i)}) - e_{y^{(i)}},$$

$$\frac{\partial C_i}{\partial W} = \left(S_{\text{Softmax}}(b + W x^{(i)}) - e_{y^{(i)}} \right) x^{(i)\top}.$$

With such a representation, the first expression which is a derivative with respect to a vector is in fact composed of the coordinates associated with the bias vector b of the gradient $\nabla C_i(\theta)$. Further, the second expression is a derivative of a real-valued function with respect to a matrix, which we represent using the notation of (A.7) in Appendix A to represent the matrix with (k, ℓ)th element denoting $\frac{\partial C_i}{\partial w_{k,\ell}}$.

In general, when one implements gradient-based optimization as in Step 3 of Algorithm 2.1 of Chapter 2, or using one of the multiple variants presented in Chapter 4, these derivatives need to be used with the right shape dimensions taken into account.

We also mention that it is sometimes common to subsume the bias term expressions within the weight parameter expressions by augmenting the feature vector to be of length $p+1$ where the first feature is the constant 1. This practice was already used in the linear regression treatment of Section 2.3, and could have also been employed in the logistic regression treatment of Section 3.1. However, we mention that from a deep learning perspective[7] the weight parameters in W often receive a different treatment than the bias parameters in b and thus keeping b separate from W notationally is instructive.

---

[6]In fact, when one seeks a Hessian expression for $C_i(\theta)$, using such a d-dimensional vector layout is the standard approach. We skip presentation of such a Hessian expression here, but note that (in contrast to more complicated deep learning models) it is tractable, and is often used in second-order methods for multinomial regression.

[7]This is made clearer in Chapters 5 and 6, for example, in the contexts of dropout and convolutional neural networks.

Note that like logistic regression, the loss minimization problem of multinomial regression is convex and thus with proper hyper-parameter choices (e.g., learning rate), a global minimum can in principle always be reached. We establish the convexity of this optimization problem in the next chapter.

Classification with Multinomial Regression Yields Convex Polytope Decision Regions

Recall that with multi-class classification, our goal is to provide a prediction $\widehat{\mathcal{Y}}$ for the label $\{1, \ldots, K\}$ associated with the input feature vector. In Section 2.2 we introduced concepts of binary classification and then via the example of the linear model in Section 2.3 we saw strategies for adapting binary classifiers to a multi-class classifier. This was via the one vs. rest and one vs. one approaches. In contrast to these approaches, the multinomial model provides a direct solution for multi-class classification since the output $\hat{y}$ is already a probability vector over $\{1, \ldots, K\}$. This approach is also common in more complicated deep learning models presented in the chapters that follow.

In general, an output vector $\hat{y}$, which is a probability vector, can be used to create a classifier by choosing the class that has the highest probability (and breaking ties arbitrarily). Namely,

$$\widehat{\mathcal{Y}} = \operatorname*{argmax}_{k \in \{1, \ldots, K\}} \hat{y}_k. \tag{3.34}$$

This approach is called a *maximum a posteriori probability* (MAP) decision rule since it simply chooses $\widehat{\mathcal{Y}}$ as the class that is most probable. It is the most common decision rule when using deep learning models for classification. As a side note, for binary classification, a MAP decision agrees with binary classification threshold prediction (2.5) with a threshold of $\tau = 0.5$, and similarly to the fine-tuning of the parameter τ, in the multi-class case we may also deviate from MAP decisions and adjust (3.34) if needed.

As an example, consider Figure 3.7 that represents a MAP-based rule applied to the output of a multinomial model trained on $n = 17$ synthetic data observations with $p = 2$ features

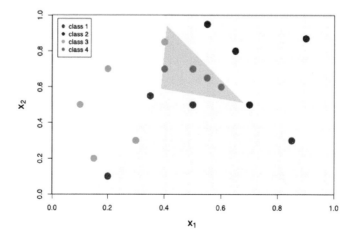

Figure 3.7: Multinomial regression for multi-class classification applied to synthetic data. In this example, the training accuracy is 15/17.

and $K = 4$ classes. Like the binary classification example of Figure 3.2, this multi-class classification example illustrates the output of $\widehat{\mathcal{Y}}(x^\star)$ for any $x^\star \in \mathbb{R}^p$. Similarly to the logistic regression case, we observe that the decision boundaries are straight lines. This is not a coincidence but rather a property of multinomial regression.

Denote by $\mathcal{C}_k$ the set of input features vectors $x \in \mathbb{R}^p$ such that $\widehat{\mathcal{Y}}(x) = k$. We now see that $\mathcal{C}_k$ is an intersection of half spaces, i.e., it is a convex polytope. To see this, consider some arbitrary point $x^\star \in \mathbb{R}^p$ and fix some class label $k \in \{1, \ldots, K\}$. We can now compare $\hat{y}_k(x^\star)$ and $\hat{y}_j(x^\star)$ for all other labels j. Specifically, since the denominator of the softmax expression (3.24) is independent of the index, $\hat{y}_k(x^\star) \geq \hat{y}_j(x^\star)$ if and only if

$$e^{b_k + w_{(k)}^\top x^\star} \geq e^{b_j + w_{(j)}^\top x^\star}, \quad \text{or} \quad (b_k - b_j) + (w_{(k)} - w_{(j)})^\top x^\star \geq 0.$$

Thus by defining a hyperplane $\mathcal{H}_{kj}$ with $\mathcal{H}_{kj}(x) = (b_k - b_j) + (w_{(k)} - w_{(j)})^\top x$, we see that $\hat{y}_k(x^\star) \geq \hat{y}_j(x^\star)$ holds if and only if $\mathcal{H}_{kj}(x^\star) \geq 0$ for all other classes j. Now $\widehat{\mathcal{Y}}(x^\star) = k$ only if $\hat{y}_k(x^\star) \geq \hat{y}_j(x^\star)$ for all other j. Hence $\mathcal{C}_k$ is an intersection of $K - 1$ hyperplanes.

As a concrete example we return to MNIST digit classification explored in Section 2.3. In that section we used the one vs. rest and one vs. one strategies to build classifiers based on linear models. See Table 2.1 where we presented confusion matrices for these classifiers when trained on the $60,000$ MNIST train images and tested on the $10,000$ test images. We now report the performance of multinomial regression on this dataset as well as the application of one vs. rest and one vs. one with logistic regression.[8] A summary is in Table 3.1. Our purpose with this comparison is not to claim which method is best since in practice all these methods are significantly beat by convolutional neural networks as presented in Chapter 6. We rather aim to highlight that a direct multi-class strategy such as multinomial regression requires training and application of a single model while the other approaches require multiple models, and are not significantly superior.

Table 3.1: Different approaches for creating an MNIST digit classifier. It is evident that in general one vs. one classifiers outperform one vs. rest classifiers, yet have many more parameters. Further, on the same type of classification scheme, logistic regression generally outperforms linear regression. Finally observe that multinomial regression with only a single model and a low number of parameters, generally performs almost as good as the top scheme (logistic regression with one vs. one).

Strategy and model type	Number of models to train	Total number of parameters	Test accuracy
Linear regression one vs. rest	10	7,850	0.8603
Linear regression one vs. one	45	35,325	0.9297
Logistic regression one vs. rest	10	7,850	0.9174
Logistic regression one vs. one	45	35,325	0.9321
Multinomial regression	1	7,850 or 7,065	0.9221

---

[8]The linear regression-based classifiers use the pseudo-inverse and hence the accuracy on the test set of size 10,000 is exact (up to numerical error). The logistic and multinomial classifiers were trained with gradient-based learning with an ADAM algorithm with a learning rate of 0.02, mini-batches of size 2,000, and 200 epochs; see Chapter 4. With this gradient-based learning, there is room for error with the accuracy as it depends on the optimization run. Hence the best achievable accuracy with the non-linear classifiers can potentially be slightly better.

3.4 Beyond Linear Decision Boundaries

Recall Figures 3.1, 3.2, and 3.7. Logistic regression naturally yields a sigmoidal relationship between the input features and the response probability. In a classification setting, this translates to linear (affine) decision boundaries. Similarly, multinomial regression used for classification also yields straight line boundaries via convex polytope decision boundaries. One may then ask if these shallow neural network models can create other forms of response curves or decision boundaries? We now show that this is indeed possible via feature engineering which was introduced in Section 2.2.

Enhancing the Sigmoidal Response

Recall first the linear regression example illustrated in Figure 2.3 of Chapter 2. In display (b) of that figure we saw how linear regression with an engineered quadratic feature yields a curve that better fits the data. A similar type of idea may also be applied in the case of logistic regression.

As an example consider a dataset $\mathcal{D}$ where each observation is for a different geographic location and feature vector has a single coordinate which is the level of precipitation at that location (in millimeters/month). For each observation, the associated $y^{(i)} \in \{0,1\}$ determines the absence (0) or presence (1) of a certain species. Observations from such a dataset[9] are presented in Figure 3.8. At locations i that are not too dry or not too wet, the species tends to be present, whereas when the precipitation is very low or very high the species tends to be absent.

Such a relationship between the precipitation and the probability of presence of the species cannot be captured by a sigmoidal curve since $\sigma_{\text{Sig}}(b + w^\top x)$ is a monotonic function of x. In fact, when fitting a sigmoidal curve to this data, the response, plotted via the red curve of Figure 3.8, tends to be almost flat in the region of the observations because in this case $\hat{w}_1$ is close to zero.

Nevertheless, if we wish to use logistic regression for this problem, we may do so via feature engineering. Similarly to the housing price example of Section 2.2 we introduce a new feature $x_2 = x_1^2$. With the addition of this new feature, the model (as a function of the single feature $x = x_1$) becomes

$$\phi(x) = \frac{1}{1 + e^{-(b+w_1 x + w_2 x^2)}}.$$

Following from basic properties of the parabola $b + w_1 x + w_2 x^2$, if $w_2 < 0$, then we have that $\phi(x) \to 0$ as $x \to -\infty$ or $x \to \infty$ and further $\phi(x)$ is maximized at $x = -w_1/w_2$ which is the maximal point of the parabola. Similarly, if $w_2 > 0$, the shape of $\phi(x)$ is reversed and $\phi(x)$ has a minimum point at the minimum point of the parabola. In both cases, $\phi(x)$ is symmetric about $x = -w_1/w_2$.

Fitting logistic regression to this feature engineered model yields a negative value for $\hat{w}_2$ and this results in the blue curve of Figure 3.8. We thus see that feature engineering in the context of logistic regression allows us to extend the form of the response beyond the sigmoid function.

---

[9]This is a synthetic dataset motivated by ecological applications.

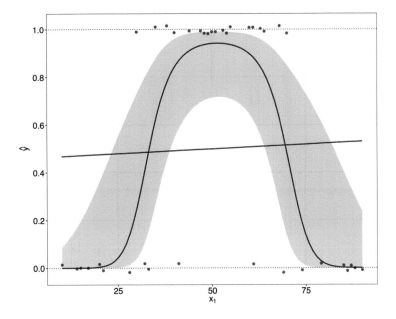

Figure 3.8: Logistic regression fit with prediction $\hat{y}$ for a feature engineered model with the feature $x_2 = x_1^2$. The response curve is plotted in blue together with confidence bounds. The red curve is a sigmoidal function fit to the single feature $x = x_1$.

The General Setup of Polynomial Feature Engineering

As in the example above, in general it is quite common to use powers for feature engineering and this makes the linear combination of the engineered features a polynomial. In examples with a small number of features such as the $p = 1$ example above, we can tweak feature engineering visually. However, when p is large, other performance-based methods are needed.

When there are initially p input features, we can automate the creation of more features by choosing each new feature as a *power product* or *monomial* form $x_1^{k_1} x_2^{k_2} \cdot \ldots \cdot x_p^{k_p}$ for some non-negative integers $k_1, k_2, \ldots, k_p$. With such a process, it is common to limit the *degree* by a constant r via

$$k_1 + k_2 + \ldots + k_p \leq r.$$

For example when $r = 2$, the set of engineered features is

$$\tilde{x} = (x_1, x_2 \ldots, x_p, x_1^2, x_2^2, \ldots, x_p^2, x_1 x_2, x_1 x_3, \ldots, x_{p-1} x_p).$$

In this case $\tilde{x} \in \mathbb{R}^{p(p+3)/2}$ and thus $d = 1 + p(p + 3)/2$ for logistic regression.[10] So if for example there are initially $p = 1{,}000$ features, then there are about half a million engineered features with $d = 501{,}501$. One may of course cull the number of engineered features to reduce the number of learned parameters. This can sometimes be done via regularization introduced in Section 2.5.

---

[10]This also holds for any model where the number of parameters is one plus the number of features such as for example linear regression.

Let us also go beyond $r = 2$ to higher degrees of the monomial features. From basic combinatorics, we have that the number of non-negative integer solutions to $k_1 + \ldots + k_p = \ell$ is $(\ell + p - 1)!/((p - 1)!\,\ell!)$ and thus when using a polynomial feature scheme with degree up to r, we have for logistic regression

$$d = \sum_{\ell=0}^{r} \frac{(\ell + p - 1)!}{(p - 1)!\,\ell!},$$

and for multinomial regression, d is K or $K - 1$ times this number with the machine learning or statistical parameterizations respectively.

Using *Stirling's approximation* of factorials we have that when ℓ is significantly smaller than p then $d \approx p^r/r!$ for logistic regression. As an example with $p = 1{,}000$ input features and setting $r = 4$, we have approximately 4.16×10^{10} parameters or and exact number of $42{,}084{,}793{,}751$. This is over 40 trillion parameters to learn!

Having 40 trillion parameters in a model is indeed borderline astronomical and as of today still infeasible. Thus for non-small p, going beyond $r = 2$ is rarely used in practice. In fact, in Section 5.2 of Chapter 5 we argue that with deeper neural networks one may sometimes get more expressivity without creating such a large number of parameters.

Versatile Classification Boundaries

To further explore feature engineering let us now consider classification problems. We have seen above that both logistic regression and multinomial regression yield decision boundaries defined by hyperplanes which we may generally denote via $\mathcal{H}(x)$. Each such hyperplane has parameters $\breve{b} \in \mathbb{R}$ and $\breve{w} \in \mathbb{R}^p$ which depend on the estimated model parameters in a simple manner. Decision rules for classification with given input $x^\star \in \mathbb{R}^p$ are then made based on if $\mathcal{H}(x^\star) \geq 0$ or not (where the case of multinomial regression involves multiple such hyperplanes and comparisons).

Now in a feature engineered scenario, the hyperplane parameters are adapted to be of larger dimension and in place of $\mathcal{H}(x^\star)$ we evaluate $\tilde{\mathcal{H}}(\tilde{x}^\star)$ with parameters $\tilde{b}$ and $\tilde{w}$. As an example consider a scenario with $p = 2$ features where we carry out polynomial feature engineering as above with $r = 2$. Then the number of parameters is now $d = 6$ and the hyperplane comparison is then

$$\breve{b} + \breve{w}_1 x_1 + \breve{w}_2 x_2 + \breve{w}_3 x_1^2 + \breve{w}_4 x_2^2 + \breve{w}_5 x_1 x_2 \geq 0. \tag{3.35}$$

Now here the set of input feature vectors $(x_1, x_2) \in \mathbb{R}^2$ that satisfies the above inequality is no longer a half space but rather represents a curved subset of $\mathbb{R}^2$. We thus see that such feature engineering enables non-linear decision boundaries.

As an example, consider Figure 3.9 based on synthetic data that requires binary classification with two input features. In (a) we see that for this type of data, logistic regression without feature engineering is not suitable. The linear decision boundary is not able to capture the pattern in the data. In (b) we expand the set of features and get a decision boundary of the form (3.35). It appears that going from $d = 3$ learned parameters to $d = 6$ learned parameters is worthwhile since the pattern in the data is well captured in (b).

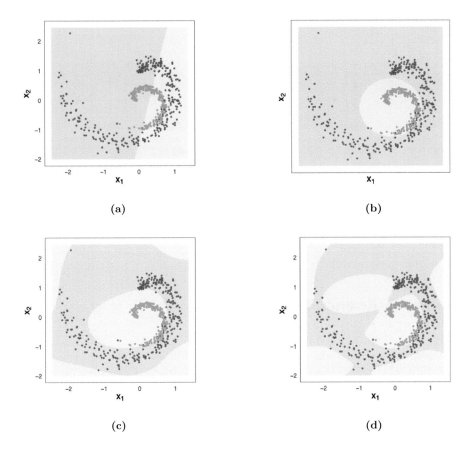

Figure 3.9: Decision boundaries for binary classification with an expanded set of features for synthetic data with $p = 2$ input features. (a) No feature engineering ($r = 1$, $d = 3$). (b) Quadratic $r = 2$, $d = 6$. (c) Quartic $r = 4$, $d = 15$. (d) $r = 8$, $d = 45$.

We may continue with higher orders in an attempt to gain more expressivity. However, as we see in (c) and (d), for this data, the higher order models yield obscure classification decision boundaries. In this example, these higher orders certainly appear like an overfit. Such overfitting would probably lead to high generalization error (recall the discussion in Section 2.5). Moreover, the new set of features could become highly correlated and can cause difficulty in inference of parameters when taking a statistical approach.

As another visual example, return to the synthetic multi-class classification example illustrated in Figure 3.7. We expand the set of features for this example and plot the decision regions in Figure 3.10. As this is just a synthetic example, our purpose is to show that by increasing the number of engineered features we can get more curvature in the decision boundaries. In (a) we consider quadratic features and in (b) we consider an extreme case of $r = 8$ which has 180 parameters with the machine learning parameterization.

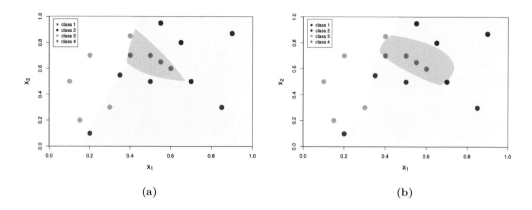

Figure 3.10: Decision boundaries with an expanded set of features for the multinomial regression model ($K = 4$). (a) $r = 2$, $d = 24$ (b) $r = 8$, $d = 180$.

3.5 Shallow Autoencoders

So far in this chapter we explored logistic and multinomial regression. Both of these models are shallow neural networks, which do not involve hidden layers, for supervised learning where the data is labeled. They are special cases of "deep learning models". The same goes for the linear regression model of Section 2.3 (it was made evident that linear regression is also a deep learning model in Section 3.2). Thus, all the key supervised learning models that we discussed up to this point are simple neural networks that fall under the umbrella of deep learning.

We now devote this last section of this chapter to simple neural networks that are used for unsupervised learning where the unlabeled data is of the form $\mathcal{D} = \{x^{(1)}, \ldots, x^{(n)}\}$. Now we introduce *autoencoder* architectures and focus primarily on shallow versions of these architectures. Recall that in Section 2.1 we presented an overview of the concept of unsupervised learning in the context of machine learning activities, and in Section 2.6 we surveyed basic unsupervised learning techniques including principal components analysis (PCA). As we see in this current section, PCA can be cast as a special case of an autoencoder and this positions PCA as a form of (simple) neural network-based learning as well.

Autoencoder Principles

Before we explore several varied applications of *autoencoders*, let us focus on their basic architecture. Consider Figure 3.11 which presents a schematic of an autoencdoer with a single *hidden layer*. The input $x \in \mathbb{R}^p$ is transformed into a *bottleneck*, also called the *code*, which is some $\tilde{x} \in \mathbb{R}^m$ and is the hidden layer of the model. Then the bottleneck is further transformed into the output $\hat{x} \in \mathbb{R}^p$. The part of the autoencoder that transforms the input into the bottleneck is called the *encoder* and the part of the autoencoder that transforms the bottleneck to the output is called the *decoder*. Both the encoder and the decoder have parameters that are to be learned.

Note that many other deep learning models in this book will also have hidden layers. In fact, representing and computing gradients for the parameters of such layers is the focus of

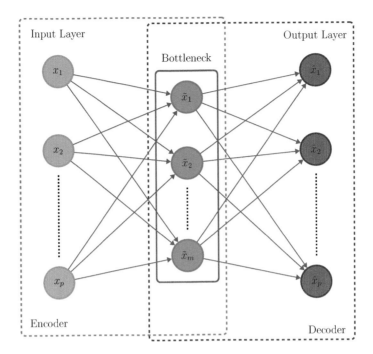

Figure 3.11: An autoencoder architecture with an encoder, decoder, and a bottleneck in between which is the single hidden layer of this neural network.

Chapter 5 and stands at the heart of deep learning. In this autoencoder example, the single hidden layer is also the bottleneck of the autoencoder and thus we informally consider this shallow autoencoder as "simple". Other "deeper" autoencoders may have multiple hidden layers of which one should be treated as the bottleneck or code.

Interestingly for input x, once parameters are trained, we generally expect the autoencoder to generate output $\hat{x}$ that is as similar to the input x as possible. This property of autoencoders, after which they are named, may at first seem obscure since it means that the autoencoder is a form of an identity function. However, this architecture is actually very useful for a variety of applications, some of which are detailed in this section and others appearing as parts of more complicated models such as sequence models which we discuss in Chapter 7.

For now, as the most basic application, consider the activity of data reduction already surveyed in Section 2.6 in the context of PCA and SVD (singular value decomposition). For this, assume that the dimension of the bottleneck m is significantly smaller than the input and output dimension p (in other non-data reduction applications we may also have $m \geq p$). For example, return to the case of MNIST digits (initially introduced in Section 1.4) where $p = 784$. For our example here, assume we have an autoencoder with $m = 30$.

If indeed a trained autoencoder yields $x \approx \hat{x}$, then it means that we have an immediate data reduction method. With the trained encoder we can to convert digit images, each of size $28 \times 28 = 784$, into much smaller vectors, each of size 30. With the trained decoder, we can to convert back and get an approximation of the original image. This choice of m implies a rather remarkable compression factor of about 26.

MNIST test set

Reconstruction with PCA

Reconstruction with shallow autoencoder

Reconstruction with deep autoencoder

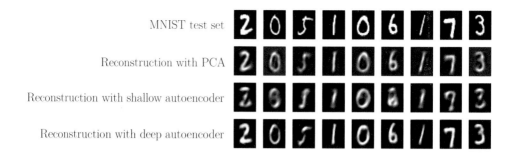

Figure 3.12: Reconstruction of the test set of MNIST (first row) using various types of autoencoders.

Figure 3.12 presents the compression/decompression effect of several types of autoencoders with $m = 30$. The first row presents untouched MNIST images. The second row presents the effect of reducing the images via PCA (a shallow linear autoencoder) and the reverting back to the image. The third row presents the effect of a shallow non-linear autoencoder. Finally the last row presents the effect with a richer autoencoder that has several hidden layers (a deep autoencoder).

There are multiple other applications for autoencoders which we soon survey. However, let us first formulate these types of models more precisely.

Single-Layer Autoencoders

As already mentioned, we may view an autoencoder as a function $f_\theta : \mathbb{R}^p \to \mathbb{R}^p$, where θ are the trainable parameters of this function. These parameters θ ideally influence the function's operation such that $f_\theta(x^\star) \approx x^\star$ where $x^\star$ is an arbitrary observation from either the seen or unseen data. The approximate equality, "$\approx$", can be considered informally as closeness of two vectors.

In practice when faced with training data $\mathcal{D} = \{x^{(1)}, \ldots, x^{(n)}\}$, we train the autoencoder (learn the parameters θ) such that $C(\theta ; \mathcal{D}) = \frac{1}{n} \sum_{i=1}^{n} C_i(\theta)$ is minimized. This is a similar loss function setup to that used in supervised contexts such as linear regression, logistic regression, multinomial regression, or the deep learning models that are discussed in the following chapters. For autoencoders, we construct the loss for an individual observation, $C_i(\theta)$, as some distance penalty measure between the input observation $x^{(i)}$ and the output $\hat{x}^{(i)} = f_\theta(x^{(i)})$. Contrast this with supervised learning where we compare the observed label and the predicted label. That is, with autoencoders the target of the model is the input in contrast to a label $y^{(i)}$ in supervised learning.

The most straightforward choice for the distance penalty in $C_i(\theta)$ is the square of the Euclidean distance, namely,

$$C_i(\theta) = \|x^{(i)} - f_\theta(x^{(i)})\|^2 \quad \text{and thus} \quad C(\theta\,;\,\mathcal{D}) = \frac{1}{n}\sum_{i=1}^{n}\|x^{(i)} - f_\theta(x^{(i)})\|^2. \tag{3.36}$$

With this cost structure, learning the parameters, θ, of an autoencoder based on data $\mathcal{D}$ is the process of minimizing $C(\theta\,;\,\mathcal{D})$.

In general, autoencoders may have architectures similar to the fully connected deep neural networks that we study in Chapter 5 as well as to extensions in the chapters that follow. This may include multiple hidden layers, convolutional layers, and other constructs. At this point, to understand the key concepts, let us revert to Figure 3.11 and consider autoencoders composed of the same components of that figure.

Specifically, we decompose $f_\theta(\cdot)$ to be a composition of the encoder function denoted via $f^{[1]}_{\theta^{[1]}}(\cdot)$ and the decoder function denoted via $f^{[2]}_{\theta^{[2]}}(\cdot)$, where $\theta^{[1]}$ are the parameters of the encoder and $\theta^{[2]}$ are the parameters of the decoder.[11] That is,

$$\hat{x} = f_\theta(x) = \big(f^{[2]}_{\theta^{[2]}} \circ f^{[1]}_{\theta^{[1]}}\big)(x) = f^{[2]}_{\theta^{[2]}}\big(f^{[1]}_{\theta^{[1]}}(x)\big),$$

where the notation using the square bracketed superscripts for the functions and parameters is in agreement with the notation we use for layers of deep neural networks in Chapter 5 and onward.

In general, one may construct all kinds of structures for the encoder, decoder, and their parameters. In our case, we consider a specific single-layer neural network structure. We define

$$\begin{aligned}
f^{[1]}_{\theta^{[1]}}(u) &= S^{[1]}(b^{[1]} + W^{[1]}u) \quad \text{for} \quad u \in \mathbb{R}^p \quad \text{(Encoder) and} \\
f^{[2]}_{\theta^{[2]}}(u) &= S^{[2]}(b^{[2]} + W^{[2]}u) \quad \text{for} \quad u \in \mathbb{R}^m \quad \text{(Decoder),}
\end{aligned} \tag{3.37}$$

where the notation is somewhat similar to (3.26). Specifically for $\ell = 1, 2$, $b^{[\ell]}$ are vectors, $W^{[\ell]}$ are matrices, and $S^{[\ell]}(\cdot)$ are vector activation functions with $S^{[1]} : \mathbb{R}^m \to \mathbb{R}^m$ and $S^{[2]} : \mathbb{R}^p \to \mathbb{R}^p$. Before describing the exact details of these functions, let us focus on the autoencoder parameters.

The encoder parameters $\theta^{[1]}$ are composed of the bias $b^{[1]} \in \mathbb{R}^m$ and weight matrix $W^{[1]} \in \mathbb{R}^{m\times p}$, and the decoder parameters $\theta^{[2]}$ are composed of the bias $b^{[2]} \in \mathbb{R}^p$ and weight matrix $W^{[2]} \in \mathbb{R}^{p\times m}$. Thus the complete list of parameters for the autoencoder is

$$\theta = (b^{[1]}, W^{[1]}, b^{[2]}, W^{[2]}). \tag{3.38}$$

Returning to the vector activation functions $S^{[1]}(\cdot)$ and $S^{[2]}(\cdot)$, we construct these based on scalar activation functions $\sigma^{[\ell]} : \mathbb{R} \to \mathbb{R}$ for $\ell = 1, 2$ such as the sigmoid function (3.4), the identity function $\sigma(u) = u$, or one of many other variants (see also Section 5.3). Specifically, we set $S^{[\ell]}(z)$ to be the element-wise application of $\sigma^{[\ell]}(\cdot)$ on each of the coordinates of z.

---

[11] An alternative way to denote these parameters would have been using ϕ (in place of $\theta^{[1]}$) for encoder and θ (in place of $\theta^{[2]}$) for the decoder. This is the notation used in variational autoencoders in Chapter 8.

Namely,

$$S^{[\ell]}(z) = \begin{bmatrix} \sigma^{[\ell]}(z_1) \\ \vdots \\ \sigma^{[\ell]}(z_r) \end{bmatrix}. \tag{3.39}$$

The choice of the type of scalar activation function may depend on the domain of the input x since the output of the model aims to reconstruct the input. For example, the use of sigmoid function for $\sigma^{[2]}(\cdot)$ restricts the output to be in the range $[0, 1]$ and this will clearly be unsuitable in cases where the input is not limited to this range.

With this notation in place, it may also be useful to see the individual representation of the bottleneck units $\tilde{x}_1, \ldots, \tilde{x}_m$, and the outputs $\hat{x}_1, \ldots, \hat{x}_p$. Specifically, with $a_k = \tilde{x}_k$,

$$\tilde{x}_i = \sigma^{[1]}\left(b_i^{[1]} + \sum_{k=1}^{p} w_{i,k}^{[1]} x_k\right), \quad \text{for} \quad i = 1, \ldots, m,$$

$$\hat{x}_j = \sigma^{[2]}\left(b_j^{[2]} + \sum_{k=1}^{m} w_{j,k}^{[2]} a_k\right). \quad \text{for} \quad j = 1, \ldots, p.$$

This set of equations is the first non-shallow (single hidden layer) neural network which we see in the book. Note also that we often use the notation $a_k = \tilde{x}_k$ as it is an 'activation' of a *unit* or neuron within the encoder.

Also, it may be useful to see the loss function representation as,

$$C(\theta\,;\mathcal{D}) = \frac{1}{n}\sum_{i=1}^{n} \|x^{(i)} - \underbrace{S^{[2]}(W^{[2]}S^{[1]}(W^{[1]}x^{(i)} + b^{[1]}) + b^{[2]})}_{f_\theta(x^{(i)})}\|^2. \tag{3.40}$$

With this loss function, for given data $\mathcal{D}$, the learned autoencoder parameters $\hat{\theta}$ are given by a solution to the optimization problem $\min_\theta C(\theta\,;\mathcal{D})$.

PCA Is an Autoencoder

It turns out that autoencoders generalize principal component analysis (PCA), introduced in Section 2.6. We overview the details by seeing that PCA is essentially a shallow autoencoder with identity activation functions $\sigma^{[\ell]}(u) = u$ for $\ell = 1, 2$, also known as a *linear autoencoder*. Specifically, we now summarize how PCA yields one possible solution to the learning optimization problem for linear autoencoders. Note that some of the mathematical details below may be skipped on a first reading without loss of continuity. The key outcome of this subsection is that PCA and linear autoencoders are essentially the same.

Consider the loss (3.40) with identity activation functions. In this case the vector activation functions $S^{[\ell]}(\cdot)$ of (5.3) are each vector identity functions and (3.40) reduces to,

$$C(\theta;\mathcal{D}) = \frac{1}{n}\sum_{i=1}^{n} \|x^{(i)} - W^{[2]}W^{[1]}x^{(i)} - W^{[2]}b^{[1]} - b^{[2]}\|^2. \tag{3.41}$$

Now it can be shown[12] that by considering the de-meaned data, we can formulate the objective without the bias vectors $b^{[1]}$ and $b^{[2]}$ in (3.41) and focus on optimizing the loss function

$$C(W^{[1]}, W^{[2]}; \mathcal{D}_{\text{de-meaned}}) = \frac{1}{n}\sum_{i=1}^{n} \|x^{(i)} - W^{[2]}W^{[1]}x^{(i)}\|^2. \qquad (3.42)$$

Here when the data is denoted $\mathcal{D}_{\text{de-meaned}}$, we reuse the notation of the data samples $x^{(i)}$, taking them now as de-meaned feature vectors (see also (2.42) in Chapter 2). It can be shown that optimization of this new loss function (3.42) is equivalent to optimization of (3.41). Specifically, if minimizing (3.42) and obtaining minimizers $\hat{W}^{[1]}$ and $\hat{W}^{[2]}$, then $\hat{b}^{[1]}$ can be set to any value and $\hat{b}^{[2]}$ has a specific expression. With these, together with $\hat{W}^{[1]}$ and $\hat{W}^{[2]}$ minimize (3.41).

Now it is possible to go one step further in reducing the parameter space. In fact, it can be shown that for the optimum of (3.42) we have $\hat{W}^{[1]} = \hat{W}^{[2]\dagger}$. This is the pseudoinverse as introduced in Section 2.3. Specifically when $\hat{W}^{[2]}$ is full column rank, the pseudoinverse can be represented as

$$\hat{W}^{[2]\dagger} = (\hat{W}^{[2]\top}\hat{W}^{[2]})^{-1}\hat{W}^{[2]\top},$$

and thus the matrix in (3.42) is,

$$\hat{W}^{[2]}\hat{W}^{[1]} = \hat{W}^{[2]}\hat{W}^{[2]\dagger} = \hat{W}^{[2]}(\hat{W}^{[2]\top}\hat{W}^{[2]})^{-1}\hat{W}^{[2]\top}.$$

This is the $p \times p$ projection matrix which projects vectors of length p onto the m-dimensional column space of $W^{[2]}$. Now using the QR decomposition[13] this projection matrix may be represented as $VV^\top$ where the $p \times m$ matrix V has orthonormal columns. This means that an equivalent optimization problem to the problem of minimizing (3.42) is

$$\min_{V \in \mathbb{R}^{p\times m}, \ V^\top V = I_m,} \frac{1}{n}\sum_{i=1}^{n}\|x^{(i)} - VV^\top x^{(i)}\|^2, \quad \text{where} \quad x^{(i)} \in \mathcal{D}_{\text{de-meaned}}, \qquad (3.43)$$

with I_m denoting the $m \times m$ identity matrix. This *constrained optimization* problem limits the search space to the space of $p \times m$ matrices that have orthonormal columns. Any minimizer V^* of (3.43) can then be used as a minimizer of the losses (3.41) or (3.42) by setting,

$$\hat{W}^{[1]} = V^{*\top}, \quad \text{and} \quad \hat{W}^{[2]} = V^*.$$

Now let us see the connection to PCA. Recall (2.51) representing the encoded lower-dimensional PCA data as $\widetilde{X} = XV$ where the $n \times p$ matrix X is a (de-meaned) data matrix as in Section 2.6, the $n \times m$ matrix $\widetilde{X}$ is the reduced data matrix, and the matrix V has columns that are singular vectors from the SVD decomposition of X. Hence the projection of each data point $x^{(i)}$ into this lower-dimensional space is given by

$$\tilde{x}^{(i)} = V^\top x^{(i)} \qquad (3.44)$$

with V the $p \times m$ matrix of columns $v_1, \ldots, v_m$ that are an orthonormal basis of a reduced subspace of $\mathbb{R}^p$. Further, with PCA, the matrix V can also be used to reconstruct the data

---

[12]This is shown by considering the derivative with respect to $b^{[2]}$ as a first step and then reorganizing the objective (3.41) with the expression of $b^{[2]}$ that sets the objective to 0. See also the notes and references at the end of the chapter

[13]See the notes and references of Chapter 1 for suggested linear algebra background reading.

as a decoder. Specifically, the decoded data points in the original space can be obtained via

$$\hat{x}^{(i)} = V\tilde{x}^{(i)}. \tag{3.45}$$

Now piecing together the encoder in (3.44) and the decoder in (3.45), the reconstruction error is

$$\frac{1}{n}\sum_{i=1}^{n}\|x^{(i)} - \hat{x}^{(i)}\|^2 = \frac{1}{n}\sum_{i=1}^{n}\|x^{(i)} - VV^{\top}x^{(i)}\|^2. \tag{3.46}$$

We see that (3.46) agrees with the objective in (3.43) and further the matrix V of PCA agrees with the constraint $V^{\top}V = I_m$. Now from the *Eckart-Young-Mirsky theorem* appearing in (2.53) of Section 2.6 as well as the relationship between SVD and PCA captured in (2.51), we have that V of PCA is indeed one of the optimal solutions[14] of (3.43). Hence in summary, PCA is a (shallow) linear autoencoder.

In practice, one would not use gradient-based optimization to learn the parameters of linear autoencoders, but rather employ algorithms from numerical linear algebra. Further, the specific basis vectors obtained via PCA are insightful since they order vectors according to their variance contributions. In contrast, optimizing (3.43) without considering PCA, yields arbitrary V (with orthonormal columns). Nevertheless, as we see now, our positioning of PCA as a special case of the (non-linear) autoencoder is insightful.

Autoencoders as a Form of Non-Linear PCA

As we discussed above, encoding and decoding with PCA projects the p-dimensional feature vector x onto an m-dimensional subspace. When $p = 2$ and $m = 1$, this can be viewed as a projection of points from the plane onto a line and when $p = 3$ and $m = 2$, this is a projection of points from three-dimensional space onto some plane, and similarly in more realistic higher dimensions of p, we project onto a linear subspace of dimension m. As such, the bottleneck of the linear encoder (PCA) encodes the location of the points on this projected space.

Linear subspace projection is sometimes a sufficient data reduction technique and at other times is not. In such cases, there are other multiple forms of *non-linear PCA* where points are projected onto manifolds that are generally curved. Since autoencoders generalize PCA, they present us with one such rich class of non-linear PCA models.

As an illustration consider Figure 3.13 where we consider synthetic data with $p = 2$ which we wish to encode with $m = 1$. This means that the bottleneck layer, or the code, represents a value on the real line for each data point. If we use PCA (red), then this encoding translates to a location on a linear subspace of $\mathbb{R}^2$. However, if we modify the identity activation functions in the autoencoder to be non-linear (blue), then the projection is on a manifold which is generally curved. In this example the non-linear activation functions are taken as tanh functions; see also Section 5.3.

As a further example, consider using an autoencoder on MNIST where $p = 28 \times 28 = 784$ and we use $m = 2$. We encode this via PCA, a shallow non-linear autoencoder, and a deep autoencoder that has hidden layers. In Figure 3.14 we present scatter plots of the codes for various types of autoencoders for both the training and test sets. That is, the autoencoders

---

[14]There is an infinite number of solutions.

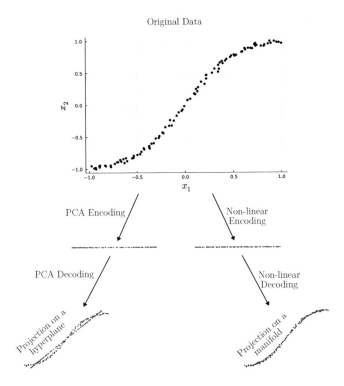

Figure 3.13: Encoding and decoding of synthetic $p = 2$ data with a linear autoencoder (PCA) in red and a non-linear shallow autoencoder (blue). The reconstruction of PCA falls on a hyperplane while the non-linear autoencoder projects onto a manifold that is not an hyperplane.

are trained on the training set and the codes presented are both for the training set and test set data. While the training and testing do not directly involve the labels (the digits 0–9), in our visualization we color the code points based on the labels. This allows us to see how different labels are generally encoded onto different regions of the code space. Recall also Figure 2.14 (b) which is of a similar nature.

Keeping in mind that one application of such data reduction is to help separate the data, it is evident that as model complexity increases (moving right along the displays of the figure), somewhat better separation occurs in the data. In particular, compare (d) based on the test set using PCA, and (f) based on the test set using the deep autoencoder. Refer also to Figure 3.12 which illustrates the reconstruction effect for various types of autoencoders (here $m = 30$). In terms of reconstruction, it is also evident in this case that more complex models exhibit better reconstruction ability.

Applications and Architectures

We have already seen the archetypical autoencoder application, namely data reduction. Yet there are many more applications and associated architectures of autoencoders. We now discuss some of these. We first consider various ways in which data reduction can be employed to help with additional machine learning activities. We then discuss *de-noising* which is a different application to data reduction. We close the chapter with ways in which interpolation in the *latent space* of the bottleneck are useful. The discussion here is just a

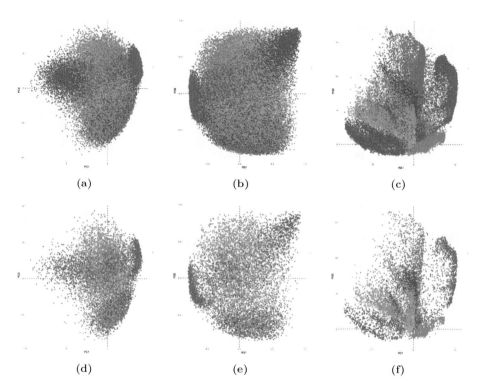

Figure 3.14: Autoencoders applied to MNIST with each of the 10 digits color coded. (a)–(c) are for the training set and (d)–(f) are for the test set. In the first column with (a) and (d) we use PCA. In the middle column with (b) and (e) we use a shallow non-linear autoencoder. In the last column with (c) and (f) we use a deep autoencoder.

brief summary and more information is provided in the notes and references at the end of the chapter.

In terms of uses of data reduction, let us consider both supervised and unsupervised learning. In the supervised setting, whenever a dataset $\mathcal{D} = \{(x^{(1)}, y^{(1)}), \ldots, (x^{(n)}, y^{(n)})\}$ is used where the dimension of $x^{(i)}$ is p, one may attempt to reduce the dimension of the input features to $m < p$ and obtain a dataset $\widetilde{\mathcal{D}} = \{(\tilde{x}^{(1)}, y^{(1)}), \ldots, (\tilde{x}^{(n)}, y^{(n)})\}$. With such a dataset, training a supervised learning model to predict y in terms of $\tilde{x} \in \mathbb{R}^m$ may sometimes yield much better performance than using the original $x \in \mathbb{R}^p$. This process requires training an autoencoder as well as training of the model. Then in production with unseen data, for any $x^\star$ we first use the trained encoder to obtain to obtain $\tilde{x}^\star$. We then use the trained model on $\tilde{x}^\star$ to obtain $\hat{y}$. Note that with such an activity we do not use the decoder per-se.

To take this application of data reduction even further, consider a situation with $y^{(i)} \in \mathbb{R}^q$ where q is not small (high-dimensional labels such as segmentation maps on images for example). In this case one may encode both the feature vector and the label, each with their own autoencoder, to obtain encoded data of the form $\widetilde{\mathcal{D}} = \{(\tilde{x}^{(1)}, \tilde{y}^{(1)}), \ldots, (\tilde{x}^{(n)}, \tilde{y}^{(n)})\}$, say with $\tilde{x}^{(i)} \in \mathbb{R}^{m_p}$, $\tilde{y}^{(i)} \in \mathbb{R}^{m_q}$, $m_p < p$, and $m_q < q$. One may then train a supervised model using this $\widetilde{\mathcal{D}}$. Such a model predicts $\hat{\tilde{y}}^\star \in \mathbb{R}^{m_q}$ for each encoded feature vector $\tilde{x}^\star \in \mathbb{R}^{m_p}$. With this setup, when the model is used in production, one first observes a feature vector $x^\star$, then uses the feature vector encoder to create $\tilde{x}^\star$, then uses the model to predict $\hat{\tilde{y}}^\star$, and finally uses the label decoder to obtain $\hat{y}^{\star}$. Hence with such an activity, in production the encoder of the feature vector and the decoder of the label are used.

In terms of unsupervised learning, there are many secondary ways to use the applications of autoencoder-based data reduction as well. As one example, assume that we wish to cluster images. A naive approach may be to treat each image as a vector and then use an algorithm such as K-means (see Section 2.6) to cluster the vectors. This approach has many drawbacks since the distance between images is based on the exact locations (indices) of pixels within an image. For example two images of the same object with slightly different camera locations would be generally "far" when comparing the Euclidean distance between the associated vectors. A more suitable approach is to use an autoencoder for data reduction of the image to a low dimension and then to perform clustering on the reduced dimensional codes. See for example Figure 3.14 (f). Here $m = 2$ and with such encoding, clustering the vectors on the two-dimensional code space will generally work well.

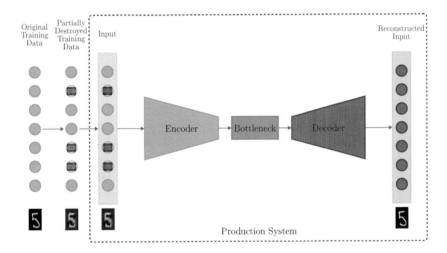

Figure 3.15: A denoising autoencoder where during training partially destroyed (noised) data is fed into the input. The production system is the trained autoencoder.

We now move away from data reduction and consider an architecture called the *denoising autoencoder* which is illustrated in Figure 3.15. This model learns to remove noise during the reconstruction step for noisy input data. It takes in partially corrupted input and learns to recover the original denoised input. It relies on the hypothesis that high-level representations are relatively stable and robust to entry corruption and that the model is able to extract characteristics that are useful for the representation of the input distribution.

In terms of architectures, denoising autoencoders exhibits a similar architecture to the usual autoencoder as they involve an encoder $f_{\theta^{[1]}}^{[1]}(\cdot)$ and decoder $f_{\theta^{[2]}}^{[2]}(\cdot)$. However a key difference is that during the learning process it is trained on noisy samples and the loss function is modified. First, the initial input x is corrupted into $\check{x}$ via some predefined stochastic mapping performing the partial destruction. In practice the main corruption mappings include adding Gaussian noise, masking noise (a fraction of the input chosen at random for each example is forced to 0), or salt-and-pepper noise (a fraction of the input chosen at random for each example is set to its minimum or maximum value with uniform probability).

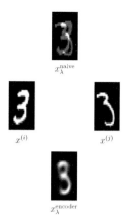

Figure 3.16: Interpolation of images with $\lambda = 1/2$. The left and right images of $x^{(i)}$ and $x^{(j)}$ are raw images. The top image is the naive interpolation and the bottom image is obtained via interpolation using an autoencoder.

Now with the corrupted input $\breve{x}$, the autoencoder encodes and decodes $\breve{x}$ in the usual way yielding a reconstructed $\hat{\breve{x}}$ via

$$f_\theta(\breve{x}) = f^{[2]}_{\theta^{[2]}} \circ f^{[1]}_{\theta^{[1]}}(\breve{x}) \longrightarrow \hat{\breve{x}}.$$

However, instead of using a loss like (3.36) which would compare $\breve{x}$ and $\hat{\breve{x}}$, we modify the loss function for each observation i to be $C_i(\theta) = \|x^{(i)} - \hat{\breve{x}}^{(i)}\|^2$. Hence during training, we seek parameters θ that try to remove the noise as much as possible. When used in production on an unseen data sample $x^\star$, the data corruption step is clearly not employed. Instead, at this point the autoencoder output $f_\theta(x^\star)$ is conditioned to generate outputs $\hat{x}^\star$ that are denoised.

As a third general application of autoencoders, let us consider *interpolation on the latent space*. Take $x^{(i)}$ and $x^{(j)}$ from $\mathcal{D} = \{x^{(1)}, \ldots, x^{(n)}\}$ and consider the convex combination

$$x^{\text{naive}}_\lambda = \lambda x^{(i)} + (1 - \lambda)x^{(j)},$$

for some $\lambda \in [0, 1]$. Ideally such an x^{naive}_λ may serve as a weighted average between the two observations where λ clearly captures which of the observations has more weight. However with most data, such arithmetic on the associated feature vectors is too naive and often meaningless for similar reasons to those discussed above in the context of K-means clustering of images. For example, in the top image of Figure 3.16 we see such interpolation with $\lambda = 1/2$.

When considering the latent space representation of the images, it is often possible to create a much more meaningful interpolation between the images. An example is in the bottom image of Figure 3.16. To carry out this interpolation we train an autoencoder and then encode $x^{(i)}$ and $x^{(j)}$ to obtain $\tilde{x}^{(i)}$ and $\tilde{x}^{(j)}$. We then interpolate on the codes, and finally decode $\tilde{x}_\lambda$ to obtain an interpolated image. That is, using the notation of (3.37) and omitting parameter subscripts we have

$$x^{\text{encoder}}_\lambda = f^{[2]}\Big(\lambda f^{[1]}(x^{(i)}) + (1 - \lambda)f^{[1]}(x^{(j)})\Big).$$

This property of latent space representations and the ability to interpolate with these representations also plays a role in the context of generative models discussed in Chapter 8. At this point let us mention that one potential application of such interpolation is for design purposes, say in art or architecture, where one chooses two samples as a starting point and then uses interpolation to see other samples lying "in between".

Notes and References

A comprehensive book on applied logistic and multinomial regression in the context of statistics is [187] where one can find out about *confidence intervals, hypothesis tests,* and other aspects of inference. The statistical origins of logistic regression are most probably due to Chester Ittner Bliss who developed the related *probit regression model* in the 1930s; see [42]. Probit was initially used for bioassay studies. See also [95] for an historical account where the development of the logistic function (the sigmoid function) as a solution to logistic differential equations is presented. In the context of machine learning, logistic regression may be viewed as one example of a probabilistic generative model, see for example Section 4.2 of [39].

These days, in the context of deep learning, logistic regression is treated as the simplest general non-linear (shallow) neural network. However, the original simplest (non-linear) neural network is not actually logistic regression but rather the *perceptron*; see [353]. That model, created by Rosenblatt in 1958, differs from logistic regression in that the activation function is the *Heaviside step function*; see (5.16) in Chapter 5. With this activation function, gradient-based optimization is not possible. Yet Rosenblatt introduced the *perceptron learning algorithm* which finds a classifier in finite time when the data is *linearly separable*; see also [314].

The phrase "softmax" for (3.25) started to be used in the machine learning community at around 1990; see [63]. The presentation of both logistic regression and multinomial regression as specific cases of generalized linear models is described in [106]. We also mention several variants of these models. These include *Dirichlet regression* which is used to analyze continuous proportions and compositional response data; see [108], [149], and [178]. Also *ordinal regression* is relevant for predicting an ordinal response variable. Examples of ordinal models are the ordered logit and ordered probit models. A comprehensive reference for analyzing categorical variables is [5] and for ordinal variables see [6]. See also chapters 10 and 12 of [179]. Related terms from the world of machine learning are *ranking learning,* also known as *learning to rank.* This field builds on *ordinal regression;* see for example [374]. See [91] as a general reference for inference using maximum likelihood estimation in the context of biostatistics as well as the classic theoretical statistics book [94]. Further, see [44] for additional computational aspects including optimization of multinomial regression using second-order methods.

Our focus on autoencoders in this chapter is mostly on the most popular architecture where the hidden layer presents a lower number of units than the inputs. This specific architecture is called *undercomplete* as opposed to the *overcomplete* case presenting more hidden units than the input. We mention that autoencoders in general, and specifically overcomplete architectures, go hand in hand with regularization, achieving sparse representations of the code; see for example chapter 14 of [142]. A general overview of autoencoder applications and architectures is in [74]. Specifically, different variants of autoencoders have recently emerged including *sparse autoencoders, contractive autoencoders, robust autoencoders,* and *adversarial autoencoders.* One particular class explored in Chapter 8 is *variational autoencoders.* For relationships between PCA and autoencoders see [332] as well as the more classic works [21] and [53].

4 Optimization Algorithms

The field of optimization is an important sub-field of applied mathematics. In the context of deep learning it is just as important. Any form of training process for a deep learning model requires optimization to minimize the loss. The decision variables are the learned model parameters, and the data is typically considered fixed from the point of view of optimization. In this chapter we explore general optimization methods and results, focusing on the essential tools for optimization in the process of training deep learning models.

In Section 4.1 we formulate optimization problems in a general setup, discuss their forms, and review local extrema, global extrema, and convexity. In Section 4.2, we focus on optimization of learned parameters for deep learning. We discuss the nature of such problems, and common techniques including stochastic gradient descent, tracking performance measures, and early stopping. In Section 4.3 we discuss various forms of first-order methods which extend the basic gradient descent algorithm. The focus is on the popular ADAM optimizer which has grown to become very popular for deep learning. In Section 4.4 we introduce automatic differentiation, a computational paradigm for efficient evaluation of derivatives. Here we present both forward and backward mode automatic differentiation in a general context, and later in the next chapter we specialize that to deep learning. We continue with Section 4.5 where additional first-order methods are presented beyond gradient descent and ADAM. In Section 4.6 we introduce and present an overview of second-order methods. These powerful methods are sometimes used in deep learning, and when applicable they can perform very well. Note that on a first reading of the book, one may focus on Sections 4.1, 4.2, 4.3, and 4.4 before continuing to understand the deep learning models of Chapter 5.

4.1 Formulation of Optimization

Optimization algorithms, such as gradient descent described in Algorithm 2.1 of Chapter 2, primarily aim to minimize an objective. We have already seen in previous chapters that "learning" $\approx$ "optimization". Hence to do well in deep learning, it is important to understand the implementation of optimization algorithms as they play a crucial role in minimizing the training loss.

The General Setup

Consider a multivariate real-valued function $C : \mathbb{R}^d \to \mathbb{R}$. An optimization problem can be written as

$$
\begin{aligned}
&\underset{\theta}{\text{minimize}}\ C(\theta) \\
&\text{subject to } \theta \in \Theta,
\end{aligned}
\tag{4.1}
$$

DOI: 10.1201/9781003298687-4

where $\Theta \subseteq \mathbb{R}^d$ is called the *feasible set* or the *constraint set*. In the context of learning, Θ represents the parameter space and the *objective function* $C(\theta)$ is the loss function.

In other words, the optimization problem aims to find a point $\theta \in \Theta$, where $C(\theta)$ is minimum. Such a point, denoted θ^*, is called a *solution* or *minimizer* of the optimization problem (4.1). That is,

$$C(\theta^*) \leq C(\theta), \quad \text{for all } \theta \in \Theta. \tag{4.2}$$

There can be multiple solutions to this problem and the set of all the solutions θ^* is usually denoted by

$$\underset{\theta \in \Theta}{\text{argmin}} \; C(\theta).$$

The formulation (4.1) is general, in the sense that any optimization problem can be written in this form. Particularly, the maximization problem $\text{maximize}_{\theta} \, C(\theta)$ subject to $\theta \in \Theta$ can be stated in the form (4.1) by replacing $C(\theta)$ in (4.1) with $-C(\theta)$. We say that the optimization problem is *unconstrained* if $\Theta = \mathbb{R}^d$; otherwise, it is a *constrained* optimization problem.

Our focus is primarily on unconstrained optimization and hence the "subject to" component of (4.1) may be omitted. This is the common case in the context of deep learning, and in particular, it is the case for the optimization problems that we already encountered in the context of linear regression (Section 2.3), logistic regression (Sections 3.1 and 3.2), multinomial regression (Section 3.3), and most cases of autoencoders (Section 3.5).

General optimization theory and practice comes in many shapes and forms where different methods can be used for different sub-types of the general problem (4.1). In the context of deep learning, we typically make use of multivariate calculus[1] and assume (or construct) the objective $C(\cdot)$ to have desirable properties, such as continuity, differentiability (almost everywhere), and other smoothness properties. With this type of optimization, the gradient ∇C and Hessian $\nabla^2 C$ are typically used. The former is the key component of first-order methods as described in Sections 4.3 and 4.5. The latter sometimes plays a role in second-order methods as described in Section 4.6. Throughout this chapter when we make use of the gradient or Hessian, we are implicitly assuming that these objects exist and are well defined for the problem at hand.

Local and Global Minima

A solution of an unconstrained minimization problem is also known as a *global minimum* because of (4.2). However, there can be other points that are not solutions to the optimization problem but exhibit similar properties within their vicinity. Such points are known as local minima. In particular, a point $\theta \in \mathbb{R}^d$ is called a *local minimum* of a continuous function C if there is an $r > 0$ such that $C(\theta) \leq C(\phi)$ for all $\phi \in B(\theta, r)$, where $B(\theta, r)$ is an open ball with center θ and radius $r > 0$ defined by

$$B(\theta, r) = \{\phi \in \mathbb{R}^d : \|\theta - \phi\| < r\}.$$

---

[1] Refer to Appendix A for a review.

In other words, if θ is a local minimum of C, we cannot decrease the value of C by moving an infinitesimal distance from θ in any direction; see Figure 4.1 for illustrations of global and local minima.

Suppose that the objective function C is differentiable at a local minimum θ. Then, it is well known that the gradient $\nabla C(\theta) = 0$. This is a necessary condition for a point to be a local minimum, but not sufficient. For example, the condition $\nabla C(\theta) = 0$ also holds if θ is a *local maximum* of C, that is, θ is a local minimum of $-C$. Further, the condition $\nabla C(\theta) = 0$ can hold even at a point θ that is neither a local minimum nor a local maximum, and such a point is called a *saddle point* where the objective function slopes up in one direction and slopes down in another direction (for smooth functions we define it below using the Hessian matrix). See Figure 4.1 for examples.

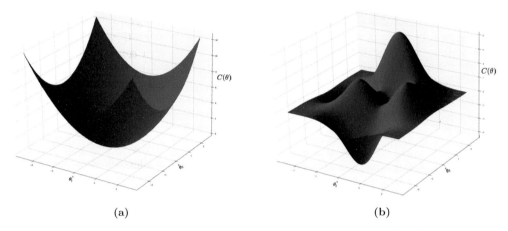

(a)	(b)

Figure 4.1: Two-dimensional smooth functions. (a) The function $C(\theta) = \theta_1^2 + \theta_2^2$ is convex with a unique global minimum. (b) A well-known function called the *peaks function* is an example non-convex function with several local minima, local maxima, and saddle points. Around each local minimum, the function is locally convex. On the figure, two saddle points are marked with red dots.

Convexity and Saddle Points

There are classes of optimization problems for which a local minimum is also guaranteed to be a global minimum. One such popular class is the family of convex problems. While most deep learning-based optimization problems are not convex, understanding convexity helps to better understand optimization concepts.

We can establish stronger necessary conditions for a point to be a local minimum via *convexity*. A set $\Theta_C \subseteq \mathbb{R}^d$ is said to be a *convex set* if for any $\theta, \phi \in \Theta_C$ and $\lambda \in [0, 1]$,

$$\lambda\theta + (1 - \lambda)\phi \in \Theta_C.$$

That is, for a convex set, the line segment connecting any two points in the set, remains in the set. An example of a convex set is an open ball $B(\theta, r)$ for any $\theta \in \mathbb{R}^d$ and $r > 0$.

We are now ready to define *convex functions*. Suppose that $\Theta_C \subseteq \mathbb{R}^d$ is a convex set. Then, a function $C : \mathbb{R}^d \to \mathbb{R}$ is called a *convex function* on Θ_C if for every $\theta, \phi \in \Theta_C$ and $\lambda \in [0, 1]$, we have

$$C\big(\lambda\theta + (1-\lambda)\phi\big) \leq \lambda C(\theta) + (1-\lambda)C(\phi). \tag{4.3}$$

Furthermore, we say that the function C is *strictly convex* on Θ_C if the inequality in (4.3) holds with strict inequality for all $\lambda \in (0,1)$ and for every choice of $\theta, \phi \in \Theta_C$ with $\theta \neq \phi$.

If the function C is convex on the entire space $\mathbb{R}^d$, then all the local minima are global minima. Furthermore, if C is strictly convex on $\mathbb{R}^d$, then there is a unique global minimum. See Figure 4.1 (a) for an example of a convex function.

If we further assume that C is twice differentiable, then we can define convexity and strict convexity in terms of the Hessian matrix. In particular, C is convex on Θ_C if and only if Hessian $\nabla^2 C(\theta)$ is positive semidefinite[2] at every interior point on Θ_C. Furthermore, C is strictly convex on Θ_C if and only if Hessian $\nabla^2 C(\theta)$ is positive definite at every interior point on Θ_C.

Note that if a point θ is a local minimum of a twice differentiable function C, then there exists some $r > 0$ such that C is convex on the open ball $B(\theta, r)$, and hence we sometimes say that C is *locally convex* around θ. See Figure 4.1 (b) for an example of a function with several local minima with local convexity around each minimum point.

For twice differentiable functions, it is easy to define *saddle points*. A point θ with $\nabla C(\theta) = 0$ is called a saddle point if the Hessian $\nabla^2 C(\theta)$ at θ is neither positive semidefinite nor negative semidefinite.

Objective Functions in Deep Learning

In the context of deep learning, the typical form of the objective function is

$$C(\theta) := C(\theta\,;\mathcal{D}) = \frac{1}{n}\sum_{i=1}^{n} C_i(\theta), \tag{4.4}$$

where n is the number of data samples in a given dataset $\mathcal{D}$. We have already seen this form in (2.11) where $C_i(\theta)$ is the loss computed for the i-th data sample. Specific forms for this i-th data sample loss include the square loss, cross entropy loss, and other forms that have already appeared in Chapters 2 and 3. In this chapter the data samples play a more minor role, and hence from now onward we suppress $\mathcal{D}$ from the notation and use $C(\theta)$.

The optimization problem corresponding to the objective function (4.4) is often called the *finite sum problem*. For such problems, it is relatively easy to study the properties of $C(\theta)$ by studying the corresponding properties of each $C_i(\theta)$, its gradient $\nabla C_i(\theta)$ and Hessian $\nabla^2 C_i(\theta)$. Most importantly, the form (4.4) plays an important role in finding the descent direction for optimization methods such as stochastic gradient descent, described in Section 4.2. When each $C_i(\theta)$ is twice differentiable, the gradient and Hessian of the objective $C(\theta)$ are respectively given by

$$\nabla C(\theta) = \frac{1}{n}\sum_{i=1}^{n}\nabla C_i(\theta), \quad \text{and} \quad \nabla^2 C(\theta) = \frac{1}{n}\sum_{i=1}^{n}\nabla^2 C_i(\theta). \tag{4.5}$$

---

[2]See (A.12) in Appendix A for definition of positive semidefinite and positive definite matrices.

A local minimum of $C(\cdot)$ at θ can be characterized via $\nabla^2 C(\theta)$ being positive semidefinite in addition to $\nabla C(\theta) = 0$. It is useful to note that $\nabla^2 C(\theta)$ is positive semidefinite at θ if each $\nabla^2 C_i(\theta)$ is positive semidefinite at θ, because for any $\phi \in \mathbb{R}^d$,

$$\phi^\top \nabla^2 C(\theta) \phi = \frac{1}{n} \sum_{i=1}^n \left(\phi^\top \nabla^2 C_i(\theta) \phi \right). \tag{4.6}$$

Convexity of Certain Shallow Neural Networks

While most deep learning optimization problems are not convex, the loss functions of the logistic regression and multinomial regression models of Chapter 3 are convex. The same is true for the linear regression model of Chapter 2. For completeness, we now establish convexity properties of these simple models. The details of this subsection may be skipped on a first reading.

We begin with the linear regression model of Section 2.3 with the standard quadratic loss. In this case, the associated optimization problem is (2.13) and we may compute the Hessian for the objective $C(\theta) = \|y - X\theta\|^2$ to be

$$\nabla^2 C(\theta) = 2X^\top X,$$

where we use the notation of the design matrix X from (2.10). The Hessian is thus a Gram matrix which implies it is positive semidefinite.[3] Hence the optimization problem is convex. Further if X has linearly independent columns then $\nabla^2 C(\theta)$ is positive definite and the problem is strictly convex and has a unique optimal point.

We now continue to logistic regression of Sections 3.1 and 3.2. An expression for the gradient is in (3.17) and we may follow-up with an expression for the Hessian. Specifically, the $d \times d$ Hessian matrix of each $C_i(\theta)$ is

$$\nabla^2 C_i(\theta) = \sigma_{\text{Sig}}(b + w^\top x^{(i)}) \left(1 - \sigma_{\text{Sig}}(b + w^\top x^{(i)})\right) \begin{bmatrix} 1 \\ x^{(i)} \end{bmatrix} \begin{bmatrix} 1 & x^{(i)\top} \end{bmatrix}, \tag{4.7}$$

which is evidently a product of a strictly positive scalar and a Gram matrix. Thus, it is always positive semidefinite following the same argument used for linear regression with quadratic loss.[4] Now using the argument at (4.6), the matrix $\nabla^2 C(\theta)$ is also positive semidefinite and thus the total loss function $C(\theta)$ is convex. Further, in cases where the vectors $(1, x^{(i)})$ for $i = 1, \ldots, n$ span $\mathbb{R}^d$ (this can only hold when $d \leq n$), we cannot find a non-zero vector $\phi \in \mathbb{R}^d$ orthogonal to all the vectors $(1, x^{(i)})$. Thus, for any vector $\phi \neq 0$, there exists at least one i such that $\phi^\top \nabla^2 C_i(\theta) \phi > 0$, and hence $\nabla^2 C(\theta)$ is positive definite, making the total loss function $C(\theta)$ strictly convex.

We now move on to multinomial regression as well as linear regression with other loss functions introduced in Section 2.3. The optimization problems of these models also enjoy convexity properties, yet expressions of the Hessian are not easy to work with. We thus use other means to argue for convexity.

---

[3]For any matrix A, the associated Gram matrix is $A^\top A$ and thus $\phi^\top A^\top A \phi = \|A\phi\|^2 \geq 0$ for any vector ϕ. Further with linearly independent columns $A\phi \neq 0$ for any $\phi \neq 0$ and thus $\|A\phi\|^2 > 0$ implying that A is positive definite.

[4]Note however that $\nabla^2 C_i(\theta)$ is never (strictly) positive definite because $\nabla^2 C_i(\theta)$ is a rank one matrix.

For multinomial regression recall $C_i(\theta)$ from (3.33) which we can represent as

$$C_i(\theta) = \log \sum_{j=1}^{K} e^{b_j + w_{(j)}^\top x^{(i)}} - (b_k + w_{(k)}^\top x^{(i)}), \qquad (4.8)$$

when $y^{(i)} = k$. Recall here that θ includes both bias terms (scalars) $b_1, \ldots, b_K$ and weight vectors $w_{(1)}, \ldots, w_{(K)}$; see (3.23).

As a building block we first use the *log-sum-exp* function, $h : \mathbb{R}^K \to \mathbb{R}$ of the form

$$h(v) = \log \sum_{j=1}^{K} e^{v_j}.$$

We can show that $h(v)$ is convex by explicitly calculating its Hessian expression and seeing that it is positive semidefinite[5] for all $v \in \mathbb{R}^K$. Further, using first principles of the definition of convexity, (4.3), we can show that for any convex function $h : \mathbb{R}^K \to \mathbb{R}$, $h(\theta^\top u_1, \ldots, \theta^\top u_K)$ is convex in $\theta \in \mathbb{R}^d$ for any fixed $u_1, \ldots, u_K \in \mathbb{R}^d$. Since θ is the vector consisting of all the weights and the biases, by using multiple zeros in each $u_j \in \mathbb{R}^d$, we can construct u_j from the p-dimensional vector $x^{(i)}$ such that

$$v_j = \theta^\top u_j = b_j + w_{(j)}^\top x^{(i)}.$$

As a consequence, the first term of (4.8) is convex in θ. Further, the second term is also convex as it is an affine function of θ. Hence (4.8) is a sum of convex functions and is thus convex.

We now again consider linear regression, but this time with the absolute error loss (2.18) and Huber loss (2.19). In both of these cases, like the standard quadratic loss, we have,

$$C_i(\theta) = g(y^{(i)} - \theta^\top x^{(i)}),$$

where $g : \mathbb{R} \to \mathbb{R}$ is a convex function; see Figure 2.6 (a) in Chapter 2. Here again, we use the fact that composition of a convex function and an affine function is convex and this establishes convexity of $C_i(\theta)$ and thus of $C(\theta)$ for all these loss function alternatives.

Before we depart from this discussion of convexity we suggest that the reader consider the plots of Figure 3.4 of Chapter 3. In that figure, it is evident that logistic regression with "the wrong" loss function, namely squared loss in this case, may yield a non-convex problem. We highlight this here to emphasize that the convexity properties discussed above are more of an exception than the rule. Indeed, most deep learning models of this book have non-convex objectives.

General Approach of Descent Direction Methods

We have seen that the loss functions of logistic regression and multinomial regression are convex, and thus they have global minima. Unfortunately, even for these shallow neural

---

[5]The gradient of the log-sum-exp function is the softmax function for which we present the Jacobian in (5.19), and the Jacobian of the gradient of a function is equal to the Hessian of that function. It also turns out that the matrix in (5.19) is the covariance matrix of a categorical distribution (multinomial distribution with one trial) implying that it is positive semidefinite.

networks, despite being convex, there is no known analytical solution for the minimizer. Further, in more general deep neural networks, we do not have such convexity properties and the objective functions typically exhibit numerous local minima. Thus in general, it is impossible to think of an analytical solution for the optimization problems in deep learning, and thus we have to rely on more advanced numerical optimization methods.

All the popular optimization methods used in deep learning can be put under one general framework called the *descent direction method*. It is an iterative method that attempts to reduce the objective in each iteration. This might seem obvious, however, the reader should keep in mind that in general optimization methods beyond those covered in this book, steps may sometimes increase the objective before further reduction. In contrast, with the descent direction method, while increases may occur in certain iterations, the overall goal is to have a reduction of the objective in each iteration. This method is presented in Algorithm 4.1.

Algorithm 4.1: The descent direction method

> **Input:** Dataset $\mathcal{D} = \{(x^{(1)}, y^{(1)}), \ldots, (x^{(n)}, y^{(n)})\}$,
> objective function $C(\cdot) = C(\cdot\,;\mathcal{D})$, and
> initial parameter vector θ_{init}
>
> **Output:** Approximately *optimal* θ

1   $\theta \leftarrow \theta_{\mathsf{init}}$
2   **repeat**
3 | Determine the *descent step* θ_{s}
4 | $\theta \leftarrow \theta + \theta_{\mathsf{s}}$
5   **until** *termination condition* is satisfied
6   **return** θ

As is evident, Algorithm 4.1 relies on the specific way in which we compute the descent step and the termination condition. For each iteration $t = 0, 1, 2, \ldots$ of Algorithm 4.1, denote the values of the parameter vector θ and the descent step θ_{s} by $\theta^{(t)}$ and $\theta_{\mathsf{s}}^{(t)}$, respectively. With this notation, the sequence of points $\theta^{(1)}, \theta^{(2)}, \ldots$, evolves via

$$\theta^{(t+1)} = \theta^{(t)} + \theta_{\mathsf{s}}^{(t)}, \qquad \text{with} \qquad \theta^{(0)} = \theta_{\mathsf{init}},$$

until $\theta^{(t)}$ satisfies the specified termination condition. Note that the magnitude of the descent step $\|\theta_{\mathsf{s}}^{(t)}\|$ denotes the *step size* in the t-th iteration.

The descent step $\theta_{\mathsf{s}}^{(t)}$ at the t-th iteration is determined by information available in or until the iteration t, where the exact manner in which it is computed depends on the variant of the algorithm used. One specific example that we have already seen is the gradient descent method, Algorithm 2.1 of Section 2.4, which is a special case of Algorithm 4.1. With gradient descent, the descent step is set at $\theta_{\mathsf{s}} = -\alpha \nabla C(\theta)$ where α is the learning rate and since $\alpha > 0$, at each step the gradient descent method moves along the steepest descent direction.[6] Other variants encountered in the sequel are stochastic gradient descent (SGD) presented in Section 4.2, ADAM presented in Section 4.3, second-order methods presented in Section 4.6, and other variants.

---

[6]Note that some authors use the term *steepest descent* to denote a more general class of algorithms of which gradient descent can be taken as a special case. However, in this book "steepest descent" and "gradient descent" are considered synonymous.

When the goal is to minimize the objective function C, ideally, we would like to terminate the algorithm only after reaching a minimum point. This convergence is difficult to guarantee in general as the updates of the algorithm can slow down or oscillate in the vicinity of the minimum. Therefore, in practice, several termination conditions allow some tolerance so that the algorithm can output a point in the vicinity of the minimum. We now list a few traditionally popular termination criteria. However, note that the next section presents optimization in the context of deep learning where other criteria are often used.

Most termination conditions use a small tolerance value, $\varepsilon > 0$. One standard termination condition tracks *absolute improvement*, where we terminate when the change in the function value is smaller than the tolerance, namely,

$$\left| C(\theta^{(t)}) - C(\theta^{(t+1)}) \right| < \varepsilon.$$

An alternative is to use the *gradient magnitude* or step size where we terminate respectively if

$$\| \nabla C(\theta^{(t+1)}) \| < \varepsilon, \qquad \text{or} \qquad \| \theta_{\mathsf{s}}^{(t)} \| < \varepsilon.$$

When the region around the (local) minimum is shallow, it is sometimes useful to use the *relative improvement criterion* where we terminate based on the relative change in the function value. Specifically, we terminate if

$$\frac{\left| C(\theta^{(t)}) - C(\theta^{(t+1)}) \right|}{\left| C(\theta^{(t)}) \right|} < \varepsilon.$$

In addition to these criteria which track the state of the optimization procedure, an alternative is to simply run for a fixed number of iterations $t = 0, \ldots, T$, or similarly to limit the running (clock) time of the algorithm to a desired predefined value.

Similarly, the *termination condition* at Step 5 of Algorithm 4.1 will vary for different methods. The *optimal* output point obtained by the algorithm may not necessarily minimize the objective function $C(\cdot)$. Rather it depends on the termination condition. For instance, as we see in Section 4.2, the final θ returned by Algorithm 4.1 could be a minimizer of the validation set accuracy, but not a minimizer of the objective function $C(\cdot)$.

4.2 Optimization in the Context of Deep Learning

In the previous section we saw that certain optimization problems are tractable in the sense that a local minimum is also the global minimum. Specifically, the loss functions of linear regression, logistic regression, and multinomial regression are convex. However, in most deep learning models, starting with the models that we cover in the next chapter, the loss functions are not as "blessed". For such deep learning models, the number of parameters is huge and the problems are non-convex, often having a large number of local minima.

Nevertheless, basic optimization techniques are very useful for deep learning with the compromise of not finding the global minimum of the loss, but rather finding a point close to some local minimum with satisfactory performance. That is, the obtained point has performance which is in general not far off from the performance at the global minimum. Indeed, in deep learning, we use optimization algorithms trying to minimize the training loss function $C(\theta)$ while our ultimate goal is finding θ that has good performance on the unseen data; see Section 2.5 for a discussion of performance on unseen data. That is, the goal is to actually minimize expectations or averages of performance functions[7] $\mathcal{P}(\theta)$, such as the validation set error rate. We then hope that this minimization resembles performance on unseen data. Hence, as the reader studies the optimization methods of this chapter, they should keep in mind that minimization of the loss $C(\theta)$ is often a proxy for minimization of $\mathcal{P}(\theta)$.

Challenges Posed by Basic Gradient Descent

Almost all cases where optimization is used for deep learning involve some variant of gradient descent; see Algorithm 2.1 of Section 2.4 for an introduction to gradient descent. However, basic gradient descent poses a few important challenges that can make the algorithm impractical in the deep learning context.

The primary challenge is that in each iteration, the gradient descent algorithm uses the entire dataset to compute the gradient $\nabla C(\theta)$. In the context of deep learning, the sample size n of the training data can be large or the dimension d of the parameter θ can be huge, or both. As a consequence, the operation of computing the gradient using the entire dataset for every update can be very slow, or even intractable in terms of memory requirements.

Further, as already mentioned, loss functions of deep neural networks tend to be highly non-convex, with multiple local minima. For such functions, the apriori selection of a good learning rate α is very difficult. A low learning rate can lead to an increased computational cost as a result of slow convergence of the algorithm, and it can lead to the algorithm getting stuck at local minima or flat sections of the loss landscape, eliminating the chance to explore other local minima. On the other hand, a large learning rate can result in the objective function values and decision variables fluctuating in a haphazard manner.

Even for simple convex models, as we saw in Figure 2.7 of Section 2.4, the choice of the learning rate affects learning performance. As a second elementary example with $d = 2$, assume hypothetically that the loss landscape takes the form of the following Rosenbrock function

$$C(\theta) = (1 - \theta_1)^2 + 100(\theta_2 - \theta_1^2)^2, \tag{4.9}$$

which is non-convex yet has a unique global minimum at $\theta = (1, 1)$.

Figure 4.2 illustrates the performance of gradient descent on (4.9). Here we also introduce the option of varying the learning rate as optimization iterations progress—based on a *predefined learning rate schedule* in this case. Specifically, at each iteration t, instead of using the fixed learning rate α we use $\alpha^{(t)}$ which is indexed by iteration t and parameterized by the original α. This strategy is sometimes used in practice, yet as we see in this example, it does not work well for the parameters that we choose, and in none of the four trajectories starting at an initial point θ_{init}, do we find the minimum. Note that in this specific example it is not hard to fine-tune the learning rate α and/or the *exponential decay parameter* of $\alpha^{(t)}$, which is set at 0.99. However, this is a simple, hypothetical, $d = 2$ dimensional loss landscape, whereas in actual problems with $d > 2$, tuning $\alpha^{(t)}$ to achieve efficient learning (with basic gradient descent) is often very difficult or impossible.

---

[7]Note that in Section 2.5 we denoted the performance measure function as a function of $\hat{y}$ and y for each observation in $\mathcal{D}$ whereas here we are taking it as a function of the parameters θ.

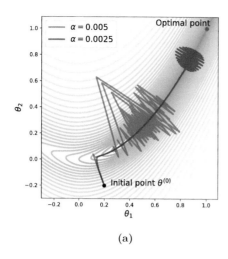 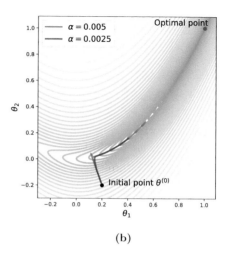

(a)   (b)

Figure 4.2: Attempting minimization of a Rosenbrock function with a minimum at $\theta = (1,1)$ via basic gradient descent for two values of α with 10^5 iterations and $\theta_{\text{init}} = (0.2, -0.2)$. (a) Fixed learning rate: $\alpha^{(t)} = \alpha$. (b) Exponentially decaying learning rate: $\alpha^{(t)} = \alpha\, 0.99^t$.

Instead of adjusting the learning rate according to a predefined schedule, other variants try to adjust the learning rate according to other criteria such as the objective function crossing some threshold values. Further, in some implementations, $\alpha^{(t)}$ is optimized in each iteration, so that the function $C(\theta)$ decreases maximally in the direction of the step size; see for example line search methods described in Section 4.5. However, such strategies are again not fool-proof on their own, or are often not practical for deep learning.

Furthermore, even if a magical ("optimal") learning rate α can be suggested for each iteration, the basic gradient descent method applies the same learning rate to all parameter updates. It turns out that it is better to learn the parameters $\theta_1, \ldots, \theta_d$ at different rates in an adaptive manner because larger rates on rarely changing parameter coordinates may result in faster convergence of the algorithm. To handle such cases, it is common practice to use the ADAM algorithm. This variant of gradient descent, together with its building blocks, are the focus of Section 4.3.

Prior to exploring ADAM and related gradient descent adaptations, let us return to the key issue mentioned above which is the computational complexity of obtaining exact gradients $\nabla C(\theta)$. This issue is typically handled by computing approximate gradients either via stochastic gradient descent, which we describe now, or more practically with the use of mini-batches described in the sequel. Importantly, both of these approaches yield fast computable approximate gradients. These approaches also yield parameter trajectories that manage to escape local minima, flat regions, or saddle points by introducing randomness or noise into the optimization process.

Stochastic Gradient Descent

Recall that the loss function $C(\theta)$ for deep neural networks is typically a finite sum of the form (4.4) with $C_i(\theta)$ being the loss computed for the i-th data sample. The method of *stochastic gradient descent* exploits this structure by computing a noisy gradient using only one randomly selected training sample. That is, it evaluates the gradient for a single $C_i(\theta)$ instead of the gradient for all of $C(\theta)$.

More precisely, the algorithm operates similarly to Algorithm 2.1 (or variations of it), yet in the t-th iteration of stochastic gradient descent, an index variable I_t is randomly selected from the set $\{1, \ldots, n\}$ and the gradient $\nabla C_{I_t}(\theta)$ is computed only for the I_t-th data sample in place of Step 3 of Algorithm 2.1. The update rule for the decision variable is then,

$$\theta^{(t+1)} = \theta^{(t)} - \alpha \, \nabla C_{I_t}(\theta^{(t)}). \tag{4.10}$$

This stochastic algorithm exhibits some interesting properties. Most importantly, the execution of each iteration can be much faster than that of basic gradient descent because only one sample is used at a time (the gain is of order n). Further, if the index variable I_t is selected uniformly over $\{1, \ldots, n\}$, the parameter update (4.10) is *unbiased* in the sense that the expected descent step for each iteration t is the same as that of gradient descent. That is, in each iteration, we have

$$\mathbb{E}\left[\nabla C_{I_t}(\theta)\right] = \frac{1}{n} \sum_{i=1}^{n} \nabla C_i(\theta) = \nabla C(\theta), \qquad \text{for a fixed } \theta. \tag{4.11}$$

and hence the stochastic step (4.10) is "on average correct".[8]

While being unbiased, the stochastic nature of (4.10) introduces fluctuations in the descent direction. Such noisy trajectories may initially appear like a drawback, yet one advantage is that they enable the algorithm to escape flat regions, saddle points, and local minima. Thus, while the fluctuations make it difficult to guarantee convergence, in the context of deep learning their effect is often considered desirable due to highly non-convex loss landscapes.

Importantly, stochastic gradient descent is generally able to direct the parameters θ toward the region where minima of $C_i(\theta)$ are located. To get a feel for this attribute of stochastic gradient descent, we consider another hypothetical example where for $i = 1, \ldots, n$,

$$C_i(\theta) = a_i(\theta_1 - u_{i,1})^2 + (\theta_2 - u_{i,2})^2 \tag{4.12}$$

for some constant $a_i > 0$ and $u_i = (u_{i,1}, u_{i,2}) \in \mathbb{R}^2$. Here u_i is the unique minimizer of $C_i(\theta)$. These individual loss functions for each observation are convex, and hence the total loss function $C(\theta)$ of (4.4) is also convex.[9] Let $\mathcal{S}$ be the *convex hull* of $u_1, \ldots, u_n$, that is, $\mathcal{S}$ is the smallest convex set that contains $\{u_1, \ldots, u_n\}$. It is now obvious that the global minima of C is also in $\mathcal{S}$.

In Figure 4.3 we plot the contours of such a function when $n = 5$ where in Figure 4.3 (a) the points $u_1, \ldots, u_5$ are distinct and in (b) these points are identical. The convex hull is marked by the region bounded in green. The figure also plots the evolution of gradient descent and stochastic gradient descent in each of the cases. As is evident, while the stochastic gradient descent trajectory is noisier, outside of the convex hull $\mathcal{S}$, its trajectory is similar to that of gradient descent.

---

[8]The reader might be tempted to conclude that $\mathbb{E}[\theta_{\text{SGD}}^{(t)}] = \theta_{\text{GD}}^{(t)}$ for all time t, where the subscript SGD is for stochastic gradient descent starting at the same initial condition as GD ("Gradient Descent"). However, in general this is not correct when $\nabla C(\theta)$ is non-linear in θ.

[9]This example does not resemble a general case with multiple local minima since it is convex. Nevertheless, it is useful for understanding some of the behavior of stochastic gradient descent.

Interestingly, in Figure 4.3 (a), once the trajectory hits $\mathcal{S}$ it mostly stays within $\mathcal{S}$ but with a lot of fluctuations. The reason for this behavior is that at any point θ not in $\mathcal{S}$, the descent directions for both gradient descent and stochastic gradient descent are toward the set $\mathcal{S}$ as it contains the minima of $C(\theta)$ and every $C_i(\theta)$. To make this a concrete argument, note that the descent step in the t-th iteration of the stochastic gradient descent is

$$-\alpha \nabla C_{I_t}(\theta^{(t)}) = -\alpha \sum_{i=1}^{n} \mathbf{1}\{I_t = i\} \nabla C_i(\theta^{(t)}).$$

Since every $C_i(\theta)$ has its minimum in the set $\mathcal{S}$, if $\theta^{(t)} \notin \mathcal{S}$, then every $-\nabla C_i(\theta^{(t)})$, the steepest direction of $C_i(\theta)$ at $\theta^{(t)}$, guides the algorithm toward the set $\mathcal{S}$. Therefore, irrespective of the choice of I_t, the update moves $\theta^{(t+1)}$ toward $\mathcal{S}$.

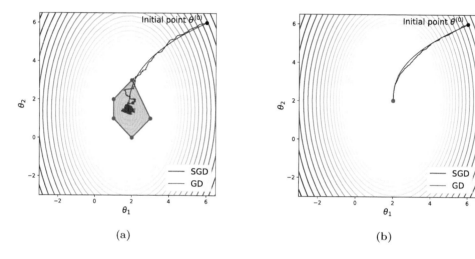

(a)

(b)

Figure 4.3: A hypothetical example where $C(\theta)$ is composed of individual $C_i(\theta)$ as in (4.12). Comparison of gradient descent (GD) and stochastic gradient descent (SGD). (a) A case where the minimizers $u_1, \ldots, u_n$ of $C_i(\theta)$ are different and are each marked by green dots in the plot. (b) A case where the minimizers $u_1, \ldots, u_n$ are the same.

Working with Mini-batches and the Concept of Epochs

On one extreme, computing the full gradient for basic gradient descent is computationally challenging but the obtained gradients are noiseless. On the other extreme, stochastic gradient descent reduces the computational complexity of each update but the updates are noisy. One middle ground approach is to work with mini-batches. Almost any practical deep learning training uses some variant of this approach.

A *mini-batch* is simply a small subset of the data points, of size n_b. Mini-batch sizes are typically quite small in comparison to the size of the dataset and are often chosen based on computational capabilities and constraints. In practice, a mini-batch size n_b is often selected so the computation of gradients associated with n_b observations can fit on GPU memory. This generally makes computation more efficient in comparison to stochastic gradient descent which cannot take full advantage of parallel computing or GPUs since the gradient evaluation for a single observation index I_t is not easy to parallelize.

When using mini-batches, for each iteration t, a mini-batch with indices $\mathcal{I}_t \subseteq \{1, \ldots, n\}$ is used to approximate the gradient via

$$\frac{1}{n_b} \sum_{i \in \mathcal{I}_t} \nabla C_i(\theta^{(t)}). \tag{4.13}$$

This *estimated gradient* is then used as part of gradient descent or one of its variants (such as ADAM which we present in the sequel); for example, it is used in place of Step 3 of Algorithm 2.1. One common practical approach with mini-batches is to shuffle the training data apriori and then use the shuffled indices sequentially. With this process, there are n/n_b mini-batches[10] in the training dataset and each mini-batch is a distinct random subset of the indices. In such a case, when using some variant of gradient descent, every pass on the full dataset has n/n_b iterations, with one iteration per mini-batch. Such a pass on all the data via n/n_b iterations is typically called an *epoch*. The training process then involves multiple epochs. It is quite common to diagnose and track the training process in terms of epochs. Note that one may also randomly shuffle the data (assign new mini-batches) after every epoch to reduce the correlation between the epochs and reduce bias in the optimization process.

Similarly to (4.11) when using the mini-batch approach, the estimated gradient is unbiased in the sense that

$$\mathbb{E}\left[\frac{1}{n_b} \sum_{j=1}^{n_b} \nabla C_{I_{t,j}}(\theta)\right] = \frac{1}{n_b} \sum_{j=1}^{n_b} \frac{1}{n} \sum_{i=1}^{n} \nabla C_i(\theta) = \nabla C(\theta), \qquad \text{for a fixed } \theta.$$

Here $I_{t,j}$ denotes the j-th element in $\mathcal{I}_t$ and the first equality holds because shuffling makes each $I_{t,j}$ to be uniform over $\{1, 2, \ldots, n\}$. To get a feel for the performance of learning with mini-batches, a mathematical simplification is to consider a variant where $\mathcal{I}_t = (I_{t,1}, \ldots, I_{t,n_b})$ is a collection of randomly selected indices on $\{1, \ldots, n\}$ with replacement.[11] As a consequence, each $I_{t,j}$ is not only uniform over $\{1, 2, \ldots, n\}$, but also independent of all the rest. This manner of computing gradients is an approximation to the above mentioned mini-batches with shuffling the data after every epoch.

Importantly, in this setup, we notice that the noise of the gradients decreases as the mini-batch size n_b increases. To see this, recall that for each t, the index variables $I_{t,1}, \ldots, I_{t,n_b}$ are chosen independently. Thus, the variance of the gradient estimate is

$$\mathrm{Var}\left(\frac{1}{n_b} \sum_{j=1}^{n_b} \nabla C_{I_{t,j}}(\theta^{(t)})\right) = \frac{1}{n_b} \mathrm{Var}\left(\nabla C_{I_{t,1}}(\theta^{(t)})\right).$$

Since $\mathrm{Var}\left(\nabla C_{I_{t,1}}(\theta^{(t)})\right)$ is the variance of the gradient for an arbitrary data sample, we see that mini-batch gradient descent has n_b times smaller variance than that of stochastic gradient descent.

---

[10]We assume here that n_b divides n. If it is not the case, then there are $\lceil n/n_b \rceil$ mini-batches with the last one having less than n_b samples.

[11]Some authors refer to this as *mini-batch gradient descent*.

Loss Minimization is a Proxy for Optimal Performance

Recall the discussion in Section 2.5 where we saw that the ultimate focus of learning is to optimize the expected performance on unseen data. This is typically quantified in terms of some error metric which we wish to be minimal. In this section we simply refer to the estimated expected performance function as $\mathcal{P}(\theta)$, keeping in mind that it is typically an average over the validation set. When training, it is often more tractable to minimize the loss $C(\theta)$, however, the true goal is minimization of $\mathcal{P}(\theta)$.

In deep learning, the model choice and training process typically involves choosing a model with a loss function $C(\theta)$, such that loss minimization is computationally easy to carry out and yields low $\mathcal{P}(\theta)$. In that sense, the loss function $C(\theta)$ is often much more amenable to derivative computations and gradient-based learning, in comparison to trying to use the performance function $\mathcal{P}(\theta)$ directly. Further, by tracking a performance function[12] on a validation dataset not used for training, we are often able to make sure that overtraining (also called overfitting) is avoided. This interplay of the loss function $C(\theta)$ and the performance function $\mathcal{P}(\theta)$ implies that deep learning optimization is not traditional optimization where one seeks θ^* that truly optimizes the loss, but is rather a means of using optimization of $C(\theta)$ as a surrogate to good performance on $\mathcal{P}(\theta)$.

We now explore this setting of deep learning optimization in two ways. In one direction we discuss a notion of stopping criteria that differs from the termination conditions presented at the end of Section 4.1. In another direction we highlight that the actual model choice affects the loss function $C(\theta)$, and some loss functions are more amenable to optimization in comparison to others.

First, in terms of stopping criteria, as mentioned above, for training a deep learning model, standard practice is to use mini-batches and epochs. This means that some form of the general Algorithm 4.1 is executed where multiple iterations are grouped into an epoch. As training progresses (this is sometimes a process of hours, days, or more), we track the evolution of the loss,[13] $C(\theta^{(t)})$, as well as one or more performance functions $\mathcal{P}(\theta^{(t)})$, typically focusing on iterations t that are at the end of an epoch. In that sense, a common stopping criterion is to stop at the epoch where $\mathcal{P}(\theta^{(t)})$ is small when evaluated over the validation set.

A popular strategy for this approach is called *early stopping*. With this strategy, training often continues for multiple epochs beyond the epoch where $\mathcal{P}(\theta^{(t)})$ is minimized since at that epoch it is still not evident that the minimum was reached. However, as the training loop progresses, the model parameters for the best epoch are stored such that once training is complete the "best model" in terms of $\mathcal{P}(\theta)$ on the validation set can be used. A typical trajectory of the loss and performance metric is displayed in Figure 4.4. With such a typical trajectory, increasing the number of epochs of training is somewhat similar to the increase of model complexity discussed in Section 2.5; see Figure 2.9 of that section. Running the model beyond the minimum point of $\mathcal{P}(\theta)$ is called *over-training*, similar to *overfitting* discussed in Section 2.5. The early stopping strategy is then a method to use validation performance to try and avoid such over-training. Importantly, the actual minimum point, say θ^*, of $C(\cdot)$ is not a desirable point for this purpose since it is typically a point of over-training and $\mathcal{P}(\theta^*)$ is sub-optimal in terms of the performance function.

---

[12] In practice we may often track multiple performance functions.
[13] In this context it is often called the *training loss* as it is the loss over the training set.

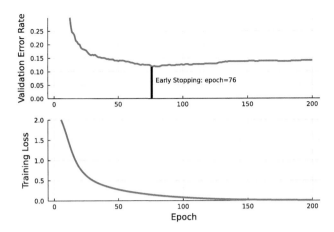

Figure 4.4: Training on a small subset of the MNIST dataset. The validation error rate is tracked together with the training loss. The early stopping strategy uses the model from epoch 76 where the validation performance is best.

As further details for Figure 4.4, we note that it is based on a deep fully connected network (see Chapter 5) with 5 layers and hundreds of thousands of parameters. We train the model using only 1,000 MNIST images and use 500 additional images for the validation set. The performance function used on the validation set is one minus the accuracy, and as is evident, an accuracy of about 90% is obtained. Mini-batch sizes are taken at $n_b = 10$. As we can see, the model is trained for 200 epochs but retrospectively, the early stopping strategy pulls the parameters θ associated with epoch 76. Note that our chosen model is far from a realistic model that one would use for such a low-data scenario, yet we present this example here to illustrate the idea of early stopping.

The second way in which deep learning optimization differs from general optimization is that the model's loss function itself is often specifically engineered for efficient optimization. In classical applications of optimization appearing in applied mathematics, operations research, and other fields, one often needs to optimize a given objective without the ability to modify the objective. In contrast, deep learning models are versatile enough so that the actual model, and subsequently the loss function can be chosen with the goal of facilitating efficient optimization. Discussions of this nature appear in the sequel where for example the choice of activation functions within a deep neural network significantly affects the nature of learning; see for example the discussion of vanishing gradients in Chapter 5.

On this matter, we have already seen in Figure 3.4 of Section 3.2 how the cross entropy loss for logistic regression presents a much better loss landscape for optimization in comparison with the squared loss. In the case of logistic regression, cross entropy loss even implies having a convex loss function, whereas that property is lost with the squared loss function. Further, in more complex models using cross entropy is also beneficial in terms of the loss landscape even if it does not guarantee convexity per-se. We now present one such illustrative example.

Similarly to the synthetic data used for the logistic regression example yielding Figure 3.4, let us use the same synthetic data on the model,

$$f(x) = \sigma_{\text{Sig}}\big(w_2\,\sigma_{\text{Sig}}(w_1\,x)\big) = \frac{1}{1 + e^{-\frac{w_2}{1 + e^{-w_1 x}}}}.$$

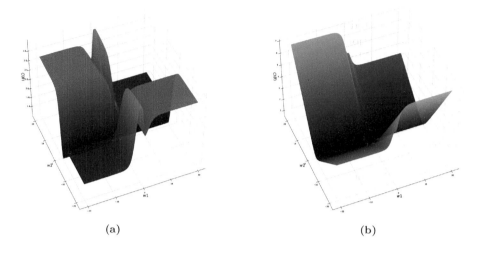

(a)   (b)

Figure 4.5: The loss landscape of a very simple two-layer neural network. (a) Using the squared loss $C_i(\theta) = (y^{(i)} - \hat{y}^{(i)})^2$. (b) Using the binary cross entropy loss $C_i(\theta) = \mathrm{CE}_{\text{binary}}(y^{(i)}, \hat{y}^{(i)})$.

In the context of deep neural networks presented in Chapter 5, this is a two-layer network without bias terms, a single neuron per layer, and sigmoid activation functions. It is not a realistic model one would use in practice, yet for our expository purposes it has the benefit of having only two parameters, and hence we may visualize the loss surface as a function of w_1 and w_2.

Interestingly, as we see in Figure 4.5, using the cross entropy loss presents a much smoother loss landscape in comparison to the squared error loss. Hence using gradient-based optimization with cross entropy would be much more efficient and have the algorithm less likely to get stuck in local minima or saddle points. Further, in much more realistic models with thousands or millions of parameters, while not possible to fully visualize, using cross entropy often yields more sensible loss landscapes as well. Hence in summary we see that when handling optimization in the context of deep learning, there are often other considerations at play, beyond standard optimization.

4.3 Adaptive Optimization with ADAM

Essentially, all deep learning optimization techniques can be cast as some form of the descent direction method outlined in Algorithm 4.1. Basic gradient descent (Algorithm 2.1) is one specific form of this descent direction method, and as we outlined in the previous section, it suffers from multiple drawbacks. While some of these drawbacks are alleviated when using noisy gradients via stochastic gradient descent or mini-batches, there is still much more room for improvement.

In this section we present one of the most popular improvements over basic gradient descent called the *ADAM algorithm* which is short for *adaptive moment estimation*. This algorithm has become the default practical choice for training deep learning models. Like basic gradient descent, ADAM is considered as a *first-order method* since it uses gradients (first derivatives) to determine the descent step used in Algorithm 4.1. For this, ADAM combines the ideas of *momentum*, and *adaptive learning rates per coordinate* into a simple mechanism of computing the descent step. Hence to understand ADAM, we study these two concepts, keeping in mind that the associated algorithms can also be viewed as independent methods in their own right. Other variations of these concepts and additional first-order techniques that are not employed in ADAM are further summarized in Section 4.5.

Adaptive Optimization and Exponential Smoothing

A basic theme of ADAM and many other first-order methods has to do with adapting to the local nature of the loss landscape as iterations progress. For simple $d = 2$ examples as visualized in Figures 4.2 and 4.5, we can quite easily get a taste of the nature of the loss landscape, and it is generally not difficult to choose descent steps that yield good overall performance. However, for practical examples with larger values of d, at any iteration t of the general Algorithm 4.1, the only information available at our disposal is the history of $\theta^{(0)}, \ldots, \theta^{(t-1)}$ and the associated gradients. It thus makes sense to try and "adapt" the descent step based on this history, perhaps with more emphasis on iterations from the near-history. One may loosely call such an approach *adaptive optimization.*

One simple principle used in adaptive optimization is *exponential smoothing*. Beyond optimization, exponential smoothing is popular in time series analysis and many other branches of data science and applied mathematics. In general, exponential smoothing operates on a sequence of vectors $u^{(0)}, u^{(1)}, u^{(2)} \ldots$. In the context of this section, the sequence may be a sequence of gradients or a sequence of the squares of gradients.

The exponentially smoothed sequence is denoted as $\bar{u}^{(0)}, \bar{u}^{(1)}, \ldots$ and is computed via,

$$\bar{u}^{(t+1)} = \beta \bar{u}^{(t)} + (1 - \beta)u^{(t)}, \qquad \text{with} \qquad \bar{u}^{(0)} = 0, \tag{4.14}$$

and $\beta \in [0, 1)$. Thus, for $t \geq 0$, each smoothed vector $\bar{u}^{(t+1)}$ is a convex combination of the previous smoothed vector and the most recent (non-smoothed vector) $u^{(t)}$. When $\beta = 0$, no smoothing takes place,[14] and when β is a high value near 1, the new vector $u^{(t)}$ only has a small effect on the new smoothed vector $\bar{u}^{(t+1)}$.

It is simple to use the update formula (4.14) to represent the smoothed vector $\bar{u}^{(t)}$ as a convex combination of the complete history. Specifically, by iterating (4.14) we have,

$$\bar{u}^{(t+1)} = (1 - \beta) \sum_{\tau=0}^{t} \beta^{t-\tau} u^{(\tau)}, \quad \text{for} \quad t = 0, 1, 2, \ldots, \tag{4.15}$$

and thus each $u^{(0)}, \ldots, u^{(t)}$ contributes to $\bar{u}^{(t+1)}$, with exponentially decaying weights. As we see now, within the context of adaptive optimization, such straightforward exponential smoothing can go a long way.

Momentum

Recall that the descent step of basic gradient descent is α times the negative gradient. This implies that in flat regions of the loss landscape, gradient descent essentially stops. Further, at saddle points or local minima, gradient descent stops completely. Such drawbacks may be alleviated with the use of *momentum*.

---

[14]Note that with our convention of indices, when $\beta = 0$ the exponentially smoothed sequence is actually a time shift of the original sequence.

To get an intuitive feel for the use of momentum in optimization, consider an analogy of rolling a ball downhill on the loss landscape. It gains momentum as it rolls downhill, becoming increasingly faster until it reaches the bottom of a valley. Clearly, the ball keeps memory of the past forces in the form of acceleration. This motivates us to use exponential smoothing on the gradients to obtain an extra step called the *momentum update*. It is then used as follows:

$$v^{(t+1)} = \beta v^{(t)} + (1 - \beta)\nabla C(\theta^{(t)}) \qquad \text{(Momentum Update)}, \qquad (4.16)$$

$$\theta^{(t+1)} = \theta^{(t)} - \alpha v^{(t+1)} \qquad\qquad \text{(Parameter Update)}, \qquad (4.17)$$

for scalar *momentum parameter* $\beta \in [0, 1)$ and *learning rate* $\alpha > 0$, starting with $v^{(0)} = 0$.

Similar to the general exponential smoothing of (4.14), the momentum update[15] is an exponential smoothing of the gradient vectors. When $\beta = 0$, we have the basic gradient descent method and for larger β, information of previous gradients plays a more significant role. In practice, for deep neural networks, the gradient $\nabla C(\theta^{(t)})$ is replaced with a noisy gradient-based on stochastic gradient descent or mini-batches.

The vector $v^{(t)}$ in (4.16) is called the *momentum*[16] at the t-th update. Using (4.15), we have

$$v^{(t+1)} = (1 - \beta)\sum_{\tau=0}^{t} \beta^{t-\tau} \nabla C(\theta^{(\tau)}), \quad \text{for} \quad t = 0, 1, 2, \dots. \qquad (4.18)$$

That is, the momentum accumulates all the past (weighted) gradients, providing acceleration to the parameter θ updates on downward surfaces. Therefore, for $\beta > 0$, the next step taken via (4.17) is not necessarily taken in the steepest descent, instead the direction is dictated by the (exponential smoothing) average of all the gradients up to the current iteration.

Figure 4.6 compares the performance of the gradient descent method with momentum for different values of β on the Rosenbrock function (4.9). We observe that for large β, the momentum method accelerates as it takes downward steps.

Adaptive Learning Rates per Coordinate

Observe that the updates of the gradient descent method, with or without momentum, use the same learning rate α for all components of θ. However, it is often better to learn the parameters $\theta_1, \dots, \theta_d$ with potentially a specific learning rate for each coordinate θ_i.

To get a feel for this issue, let us assume a situation where specific parameters are associated with specific features in the data.[17] For instance, in applications of deep neural networks for natural language processing (see also Chapter 7), there are typically words less frequent than

---

[15]One may alternatively parameterize the momentum update (4.16) as $v^{(t+1)} = \beta v^{(t)} + \nabla C(\theta^{(t)})$. This alternative parameterization has the benefit of allowing one to adjust the momentum parameter β without having significant effects on the step size exhibited in (4.17). Our choice of the parameterization as in (4.16) is for consistency with ADAM presented later in this section.

[16]In physics, momentum is defined as the product of the mass of an object and its velocity vector. The object's momentum points in the direction of an object's movement. In contrast, here the momentum vector $v^{(t)}$ points in the direction opposite to the step taken.

[17]In some simple models such as linear regression, logistic regression, or multinomial regression, this is exactly the case for categorical one-hot encoded variables. In more advanced models, there is a more complicated implicit mapping between parameters and features.

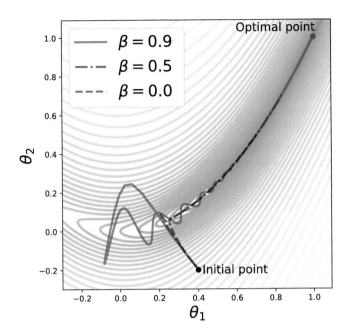

Figure 4.6: Application of momentum to the Rosenbrock function (4.9) for three different values of β with learning rate $\alpha = 0.001/(1 - \beta)$ and a fixed number of iterations. Note that $\beta = 0$ is basic gradient descent.

others. For example in most texts on trees, the word `chaulmoogra` most likely appears less frequently than the word `coconut`, even though both are names of tree species. Now assume that there are associated parameters $\theta_{\text{chaulmoogra}}$ and θ_{coconut}, and consider how these would be updated during learning.

What is likely to occur is that each mini-batch would have on average much fewer occurrences of `chaulmoogra` in comparison to `coconut`. However, if we update all the parameters at the same learning rate, then θ_{coconut} would be updated more than $\theta_{\text{chaulmoogra}}$. Hence with fixed learning rate α, the θ_{coconut} parameter is likely to converge quickly, while the $\theta_{\text{chaulmoogra}}$ parameter will be slow to respond.

A general approach for overcoming this issue is to scale each coordinate of the descent step with a different factor. That is, instead of considering a basic gradient descent step, $\theta_s = -\alpha \nabla C(\theta)$, consider descent steps of the form

$$\theta_s = -\alpha\, r \odot \nabla C(\theta), \tag{4.19}$$

where r is some vector of positive entries which recalibrates the descent steps and $\odot$ is the element-wise product of two vectors.[18] That is, for a parameter such as $\theta_{\text{chaulmoogra}}$ we would expect the coordinate $r_{\text{chaulmoogra}}$ to be high in comparison to the coordinate r_{coconut} which is associated with θ_{coconut}. Such recalibration of the descent steps would result in

---

[18]The element-wise product is also called the *Hadamard product* or *Schur product*.

large steps for $\theta_{\text{chaulmoogra}}$ whenever `chaulmoogra` appears in a mini-batch, and smaller steps for θ_{coconut}.

However, in general, finding a fixed good vector r for (4.19) is difficult, especially when the number of parameters is huge and the actual relationship between parameters and features is not clear. Instead, we use adaptive methods that adjust the descent step during the optimization process.

We now introduce two such approaches called *adaptive subgradient* (or *Adagrad*) and *root mean square propagation* (or *RMSprop*). In both of these approaches the update of coordinate i in iteration t can be represented as

$$\theta_i^{(t+1)} = \theta_i^{(t)} - \frac{\alpha}{\sqrt{s_i^{(t+1)}} + \varepsilon} \frac{\partial C(\theta^{(t)})}{\partial \theta_i}, \tag{4.20}$$

where $s_i^{(0)}, s_i^{(1)}, s_i^{(2)}, \ldots$ is a sequence of non-negative values that are updated adaptively and ε taken to be a small value of order 1×10^{-8} to avoid division by zero. That is, with this representation, the descent step at time t represented in terms of (4.19) has $r_i = \left(\sqrt{s_i^{(t+1)}} + \varepsilon\right)^{-1}$. Generally we aim for s_i to be some smoothed representation of the square of the derivative $\partial C(\theta^{(t)})/\partial \theta_i$. Hence, r_i is roughly the inverse of the magnitude of the derivative, and r_i is low when the changes for coordinate i are steep and vise-versa.

Vector-wise, the parameter update (4.20) can be represented as

$$\theta^{(t+1)} = \theta^{(t)} - \frac{\alpha}{\sqrt{s^{(t+1)}} + \varepsilon} \odot \nabla C(\theta^{(t)}), \tag{4.21}$$

where $s^{(t)} = (s_1^{(t)}, \ldots, s_d^{(t)})$, ε is considered a vector, and the addition, division, and square-root operations are all element-wise operations.

Using the representation (4.20) or (4.21), let us now define how the sequence of $\{s_i^{(t)}\}$ is computed both for Adagrad and RMSprop. Specifically,

$$s_i^{(t+1)} = \begin{cases} \sum_{\tau=0}^{t} \left(\frac{\partial C(\theta^{(\tau)})}{\partial \theta_i}\right)^2, & \text{for Adagrad,} \\ \gamma s_i^{(t)} + (1-\gamma)\left(\frac{\partial C(\theta^{(t)})}{\partial \theta_i}\right)^2, & \text{for RMSprop,} \end{cases} \tag{4.22}$$

where the recursive relationship for RMSprop has the initial value at $s_i^{(0)} = 0$. RMSprop is also parameterized by a *decay parameter* $\gamma \in [0, 1)$ with typical values at $\gamma = 0.999$ implying that for RMSprop the sequence $\{s_i^{(t)}\}$ is a relatively slow exponential smoothing of the square of the derivative.

Adagrad may appear simpler than RMSprop since it does not involve any recursive step or any hyper-parameter like γ and it is simply an accumulation of the squares of the derivatives. However, the crucial drawback of Adagrad is that for each i, the sequence $\{s_i^{(t)}\}$ is strictly non-decreasing. As a consequence, the effective learning rate decreases during training, often making it infinitesimally small before convergence. RMSprop was introduced in the context of deep learning after Adagrad, and overcomes this problem via exponential smoothing of the squares of the partial derivatives.

It is also useful to use (4.15) to see that for RMSprop, the explicit (all-time) representation of the vector $s^{(t+1)}$ is

$$s^{(t+1)} = (1 - \gamma) \sum_{\tau=0}^{t} \gamma^{t-\tau} \left(\nabla C(\theta^{(\tau)}) \odot \nabla C(\theta^{(\tau)}) \right), \qquad (4.23)$$

while for Adagrad a similar representation holds without the $(1 - \gamma)$ and $\gamma^{t-\tau}$ elements in the formula.

Bias Correction for Exponential Smoothing

We have seen that both momentum and RMSprop involve exponential smoothing of the gradients and squares of the gradients respectively. In both cases, we set zero initial conditions at iteration $t = 0$. Specifically, for momentum, we have $v^{(0)} = 0$ and for RMSprop, we have $s^{(0)} = 0$. In the absence of any better initial conditions, this is a sensible choice, yet such initial conditions may introduce some initial (short-term) bias. This bias would eventually "wash away" as t grows but it can play a significant role for short term t. The effect of such bias is especially pronounced when the respective exponential smoothing parameters β or γ are close to 1.

To mitigate such temporal bias, a common technique is using *bias correction*. To see this, first observe from (4.14) that since the initial value $\bar{u}^{(0)} = 0$, the first element of the exponential smoothing is

$$\bar{u}^{(1)} = (1 - \beta)u^{(0)}.$$

Since β is usually set near 1, $\bar{u}^{(1)}$ is close to zero even when $u^{(0)}$ is not close to 0. Consequently, $\bar{u}^{(2)}$ stays close to zero even if $u^{(1)}$ is away from 0 because

$$\bar{u}^{(2)} = \beta \bar{u}^{(1)} + (1 - \beta)u^{(1)}.$$

Thus, initial values of the exponential smoothing sequence $\bar{u}^{(0)}, \bar{u}^{(1)}, \ldots$ remain close to 0 even when the elements of original sequence $u^{(0)}, u^{(1)}, \ldots$ are far from 0.

One way to handle such bias is to consider the special case where the input vectors in (4.15) are fixed at $u^{(t)} = u$. In such a case, via simple evaluation of a geometric sum we have

$$\bar{u}^{(t+1)} = (1 - \beta)\left(\sum_{\tau=0}^{t} \beta^{t-\tau} \right)u = (1 - \beta)\left(\sum_{\tau=0}^{t} \beta^{\tau} \right)u = (1 - \beta^{t+1})u.$$

Hence if we were to divide the exponentially smoothed value $\bar{u}^{(t)}$ by $1 - \beta^{t+1}$, we would get constant vectors when $u^{(t)} = u$ (constant). Further, in (more realistic) cases where the underlying input sequence $u^{(0)}, u^{(1)}, u^{(2)} \ldots$ is not constant but close to a constant, the bias corrected sequence $\{\bar{u}^{(t)}/(1 - \beta^t)\}$ may still do a better job at mitigating initial effects. Clearly as t grows, $\beta^t \to 0$, and the effect of this bias correction disappears.

Now we may apply such bias correction to momentum or RMSprop where we have the specific forms (4.18) or (4.23), respectively. In these cases, the bias corrected updates for v

and s are,

$$\hat{v}^{(t+1)} = \frac{1}{1-\beta^{t+1}}v^{(t+1)} \quad\text{and}\quad \hat{s}^{(t+1)} = \frac{1}{1-\gamma^{t+1}}s^{(t+1)}, \quad t = 0,1,2,\dots. \tag{4.24}$$

Putting the Pieces Together: ADAM

Now that we understand ideas of momentum, RMSprop, and bias correction, we can piece these together into a single algorithm, namely the *adaptive moment estimation* method, or simply *ADAM*. Metaphorically, if the execution of the momentum method is a ball rolling down a slope, the execution of ADAM can be seen as a heavy ball with friction rolling down the slope.

The key update formula for ADAM is

$$\theta^{(t+1)} = \theta^{(t)} - \alpha\frac{1}{\sqrt{\hat{s}^{(t+1)}}+\varepsilon}\hat{v}^{(t+1)}, \tag{4.25}$$

where all vector operations (division, square root, and addition of ε) are interpreted element-wise. Here $\hat{v}^{(t+1)}$ and $\hat{s}^{(t+1)}$ are bias corrected exponential smoothing of the gradient and the squared gradients as given in (4.24).

Algorithm 4.2: ADAM

Input: Dataset $\mathcal{D} = \{(x^{(1)}, y^{(1)}), \dots, (x^{(n)}, y^{(n)})\}$,
objective function $C(\cdot) = C(\cdot\,;\mathcal{D})$, and
initial parameter vector θ_{init}
Output: Approximately *optimal* θ

1 $t \leftarrow 0$ (Initialize iteration counter)
2 $\theta \leftarrow \theta_{\text{init}}$, $v \leftarrow 0$, and $s \leftarrow 0$ (Initialize state vectors)
3 **repeat**
4 $\quad g \leftarrow \nabla C\left(\theta\right)$ (Compute gradient)
5 $\quad v \leftarrow \beta\,v + (1-\beta)\,g$ (Momentum update)
6 $\quad s \leftarrow \gamma\,s + (1-\gamma)\,(g \odot g)$ (Second moment update)
7 $\quad \hat{v} \leftarrow \dfrac{v}{1-\beta^{t+1}}$ (Bias correction)
8 $\quad \hat{s} \leftarrow \dfrac{s}{1-\gamma^{t+1}}$ (Bias correction)
9 $\quad \theta \leftarrow \theta - \alpha\dfrac{1}{\sqrt{\hat{s}}+\varepsilon}\hat{v}$ (Update parameters)
10 $\quad t \leftarrow t + 1$
11 **until** *termination condition* is satisfied
12 **return** θ

With ADAM, α is still called the learning rate and is still the most important parameter which one needs to tune. The other parameters are $\beta \in [0,1)$ for the momentum exponential smoothing as used in (4.16) and $\gamma \in [0,1)$ for the RMSprop exponential smoothing as is used in (4.22). The common defaults for these parameters are $\beta = 0.9$ and $\gamma = 0.999$.

ADAM is presented in Algorithm 4.2 where Step 4 is the gradient computation, Steps 5 and 6 are exponential smoothing (momentum and RMSprop), Steps 7 and 8 are bias corrections, and finally Step 9 is the descent step.

Since the introduction of ADAM in 2014, this algorithm has become the most widely used algorithm (or "optimizer") in deep learning frameworks. In the next section, we focus on what is typically the costliest computation within the algorithm, namely Step 4, where we evaluate the gradient. Further generalizations of ADAM and other variations of first-order methods are presented in Section 4.5.

4.4 Automatic Differentiation

The computation of gradient vectors is a crucial step in the ADAM algorithm or in any other first-order optimization method. Practically, in deep learning, the most common way for computing such gradients is called *automatic differentiation* and is embodied in the *backpropagation algorithm*. In this section we introduce general concepts of automatic differentiation and then in Section 5.4 we specialize it to the backpropagation algorithm for deep neural networks.

All popular methods for computing derivatives, gradients, Jacobians, and Hessians follow one of two approaches. In one approach, the algebraic expressions of the derivatives are computed first and then the derivatives for a particular given point are evaluated. For this approach the expressions of the derivatives are either derived manually, which is generally a tedious process, or using computer-based *symbolic differentiation* methods. For example, we have seen explicit gradient expressions in (2.21) for linear regression and (3.17) for logistic regression. However, for general deep learning models, explicit expressions are not with such compact form and instead suffer from the problem of *expression swell*. Thus an alternative approach is to compute the numerical values of the derivatives at a given point directly. This approach includes standard methods of *numerical differentiation* as well as *automatic differentiation* which is our focus here.

In this section, we first present an overview of numerical and symbolic differentiation. We then outline key ideas of *differentiable programming* which is a programming paradigm that encompasses automatic differentiation. We then present forward mode automatic differentiation which is a stepping stone toward understanding backward mode automatic differentiation. Finally, we present backward mode automatic differentiation.

Numerical and Symbolic Differentiation

Suppose we would like to compute the gradient of the objective function $C(\theta)$ at a given parameter vector θ. Numerical differentiation methods use the definition of partial derivative, see (A.5) in Section A.2, to compute the gradient $\nabla C(\theta)$ approximately. In particular, the most basic form of numerical differentiation approximates the partial derivative via

$$\frac{\partial C(\theta)}{\partial \theta_i} \approx \frac{C(\theta + h e_i) - C(\theta)}{h}, \tag{4.26}$$

for a small constant $h > 0$, where e_i is the i-th unit vector. Thus to obtain a numerical estimate of $\nabla C(\theta)$ we need to evaluate $C(\cdot)$ at θ, and $\theta + h e_i$ for $i = 1, \ldots, d$, and each time use (4.26).

A classic problem with numerical differentiation is selection of the constant h. While mathematically, smaller h provides a better approximation, numerically, small h yields numerical instability due to round-off errors.[19] Further, in the context of deep learning where θ is of very high dimension, the major drawback of numerical differentiation is that it requires us to perform an order of d function evaluations to compute the gradient at a point.

The basic alternative to numerical differentiation is symbolic differentiation. With this paradigm, instead of directly obtaining numerical values of derivatives, we represent expressions using *computer algebra systems* and obtain mathematical expressions for the derivatives using automated algorithms based on the rules of calculus. At its core, symbolic differentiation is a very useful tool. However, with deep learning, trying to rely on symbolic differentiation solely is not practical. A key problem is expression swell where the exact mathematical representation of partial derivatives of loss functions associated with deep neural networks may often require excessive resources due to the complexity of the resulting expressions.

Since in deep learning we are mainly concerned with numerical values of the derivatives but not with their analytical expressions, the application of symbolic differentiation can be unwieldy. However, instead of trying to find the expression of the objective function directly, if it is decomposed into several elementary operations, then we can numerically compute the gradients of the objective function by obtaining symbolic derivative expressions of these elementary operations while keeping track of the intermediate numerical values in a sequential manner. This idea of interleaving between symbolic expressions and numerical evaluations at elementary levels is the basis of automatic differentiation.

Overview of Differentiable Programming

Let us now introduce basic terminology of automatic differentiation within the world of *differentiable programming*. While for deep learning, the central application is associated with the multi-input scalar-output loss $C : \mathbb{R}^d \to \mathbb{R}$, to get a general feel for differentiable programming and automatic differentiation, let us first consider general functions $g : \mathbb{R}^d \to \mathbb{R}^m$. That is, when $m > 1$, the outputs are vector valued.

Differentiable programming refers to a programming paradigm where numerical computer programs, or code for computer functions, are transformed into other functions which execute the derivatives, gradients, Jacobians, Hessians, or higher-order derivatives of the original computer functions. For example, consider a case with $d = 3$ and $m = 2$ and some hypothetical computer programming language with some programmed function g() which appears in code as follows:

```
g(t1, t2, t3) = (t1 - 7*t2 + 5*t3, t2*t3)
```

We interpret this code as a function which operates on three input numbers θ_1, θ_2, and θ_3 and then returns two numbers of which the first is a linear combination of the three inputs, $\theta_1 - 7\theta_2 + 5\theta_3$, and the second is a product of two of the inputs, $\theta_2\theta_3$. From a programming perspective it is a *pure function* in the sense that it does not depend on any other variables and does not change any other variables in the system. Thus, the computer function g()

---

[19]There are other similar schemes that generally exhibit less numerical error than (4.26), yet any numerical differentiation scheme requires tuning of h.

implements a mathematical function $g : \mathbb{R}^3 \to \mathbb{R}^2$ with

$$g(\theta_1, \theta_2, \theta_3) = \big(g_1(\theta_1, \theta_2, \theta_3), \; g_2(\theta_1, \theta_2, \theta_3)\big),$$

where $g_1 : \mathbb{R}^3 \to \mathbb{R}$ and $g_2 : \mathbb{R}^3 \to \mathbb{R}$.

With differentiable programming and automatic differentiation, we are interested in evaluation of partial derivatives such as $\partial g_i / \partial \theta_j$ for $i = 1, 2$ and $j = 1, 2, 3$, at specific values of θ_1, θ_2, and θ_3. For this simple example these derivatives are obviously known analytically and constitute the Jacobian matrix of g; see (A.9) in Appendix A. In our case, the Jacobian is

$$J_g(\theta_1, \theta_2, \theta_3) = \begin{bmatrix} 1 & -7 & 5 \\ 0 & \theta_3 & \theta_2 \end{bmatrix},$$

where the i, j element of the Jacobian J_g is $\partial g_i / \partial \theta_j$.

In its most basic form, a differentiable programming system provides constructs such as `partial_derivative()` which gives us the ability to write code such as for example:

```
partial_derivative(target=g, t=[2.5, 2.1, 1.4], i=2, j=3)
```

The meaning of such code is a request of the computer to evaluate the numerical value of the Jacobian of the target function g at $\theta = (2.5, 2.1, 1.4)$ for $i = 2$ and $j = 3$. That is, we expect such code to compute $\partial g_2 / \partial \theta_3$ and return 2.1 as a result in this case. In contrast to symbolic or numerical differentiation, this computation is to be carried out through mechanisms of automatic differentiation.

In addition to such general partial derivatives via `partial_derivative()`, we may also expect differentiable programming frameworks to expose software constructs such as `gradient()` for computing gradient vectors of functions $g : \mathbb{R}^d \to \mathbb{R}$, `jacobian()` for computing the Jacobian matrix of functions $g : \mathbb{R}^d \to \mathbb{R}^m$, and `hessian()` for Hessian matrices of functions $g : \mathbb{R}^d \to \mathbb{R}$. In any case, the mechanism of computation of these operations typically involves a transformation of the source code of the original functions, or some data structures that represent those functions, into new code or data structures that encode how derivative evaluation is carried out. See the notes and references at the end of the chapter for references to contemporary concrete software frameworks that execute such operations.

Beyond `partial_derivative()`, `gradient()`, `jacobian()`, and `hessian()`, there are also two other basic constructs in a differentiable programming framework. These are *jacobian vector products*, `jvp()`, and *vector Jacobian products*, `vjp()`. A Jacobian vector product is based on $\theta \in \mathbb{R}^d$ and $v \in \mathbb{R}^d$ and is the evaluation of $J_g(\theta)v$ which results in a vector in $\mathbb{R}^m$. Similarly a vector Jacobian product is based on $\theta \in \mathbb{R}^d$ and $u \in \mathbb{R}^m$ and is the evaluation of $u^\top J_g(\theta)$ which results in a row vector whose transpose is in $\mathbb{R}^d$. Due to the way in which automatic differentiation operates, `jvp()` and `vjp()` are in fact the basic operations that one may expect from a differentiable programming framework, whereas the other operations, `partial_derivative()`, etc. are all built on top of either `jvp()` or `vjp()`.

Automatic differentiation generally works in two types of modes, namely *forward mode* which is the algorithm behind `jvp()` and *backward mode*, also known as reverse mode, which is the algorithm behind `vjp()`. In both cases the multivariate chain rule plays a key role; see Section A.3 in the appendix for a review of the chain rule. The general principle of automatic differentiation is based on computing intermediate partial derivatives and incorporating these computed values in an iterative computation. Forward mode automatic differentiation is a straightforward application of the multivariable chain rule while backward mode is slightly more involved. Note that the backpropagation algorithm for deep learning is a special case of backward mode. The subsections below focus on the inner workings of forward mode and backward mode automatic differentiation, but first let us see how `jvp()` and `vjp()` are generally used in differentiable programming.

As stated above, the basic service provided by forward mode automatic differentiation is the computation of Jacobian vector products. For instance, returning to the simple $d = 3$, $m = 2$ example above, we may invoke forward mode automatic differentiation via code such as

```
jvp(target=g, t=[2.5, 2.1, 1.4], v=[0,1,0]).
```

Here we choose $\theta = (2.5, 2.1, 1.4)$ as before and v to be the unit vector e_2. The result is in this case is `[-7, 1.4]`. Setting v as a unit vector[20] can be useful because in general when using $v = e_j$, the output $J_g(\theta)e_j$ is the j-th column of the Jacobian matrix. Hence in this example of `jvp()`, we obtain as output the effect of θ_2 on the two outputs $g_1(\theta)$ and $g_2(\theta)$ which is the vector $(\partial g_1/\partial \theta_2, \ \partial g_2/\partial \theta_2) = (-7, 1.4)$.

This shows that to implement `partial_derivative` for some $\partial g_i/\partial \theta_j$ at θ, we can compute $J_g(\theta)e_j$ (the j-th column of the Jacobian matrix) using `jvp()` and take the i-th index of the output. Specifically a single application of `jvp()` on $v = e_j$ provides us with the partial derivatives $\partial g_i/\partial \theta_j$ for all $i = 1, \ldots, m$. Hence vector Jacobian products provided by forward mode automatic differentiation are useful when the output m is large and we wish to only see the effect of a single, or a few input variables θ_j on the output.

However, note that in the deep learning loss function application where $m = 1$ and d is large, vector Jacobian products are not an efficient means for evaluating the gradient. Specifically, when we wish to evaluate `gradient()` using `jvp()` we need to invoke `jvp()` d separate times, each time with $v = e_j$ for $j = 1, \ldots, d$. Thus using forward mode automatic differentiation (Jacobian vector products) is not efficient for deep learning loss landscape gradient evaluation.

The alternative to `jvp()` is to use vector Jacobian products as exposed via backward mode automatic differentiation. With this approach we may invoke code such as

```
vjp(target=g, t=[2.5, 2.1, 1.4], u=[1,0])
```

In this example we use the same θ as above and set $u = e_1 \in \mathbb{R}^2$. Since the vector Jacobian product evaluates $u^\top J_g(\theta)$, the output is a row vector of dimension $d = 3$ which is the first row of $J_g(\theta)$, namely $\nabla g_1(\theta)^\top$. More generally applying Jacobian vector products on $u = e_i$ yields the transposed gradient $\nabla g_i(\theta)^\top$.

In the deep learning loss minimization scenario d is very large and $m = 1$. Thus the Jacobian is simply the transpose of the gradient and is a long row vector. In such a case, computing the Jacobian vector product with $u = 1$ (scalar) yields the gradient of the loss function directly. Thus since backward mode automatic differentiation implements `vjp()`, using it on the loss function is directly suited for deep learning.

---

[20]We can also use more general (non-unit vectors) and obtain directional derivatives, yet in the remainder of our discussion inputs to Jacobian vector products (v) and vector Jacobian products (u) are unit vectors.

We now explore the inner workings of forward mode automatic differentiation followed by backward mode automatic differentiation. In both cases we restrict to $m = 1$. Understanding forward mode is a useful stepping stone to understanding backward mode. Note that forward mode and backward mode involve internal implicit computations of Jacobian vector products or vector Jacobian products, respectively.[21]

The Computational Graph and Forward Mode Automatic Differentiation

Computational graphs play a key role in the implementation of automatic differentiation. A *computational graph* for a multivariate function $C : \mathbb{R}^d \to \mathbb{R}$ is a directed graph where the nodes correspond to elementary operations involved in the mathematical expression of $C(\theta)$. Inputs to the graph are the variables $\theta_1, \ldots, \theta_d$ and constants if required, and the output is the function value $C(\theta)$.

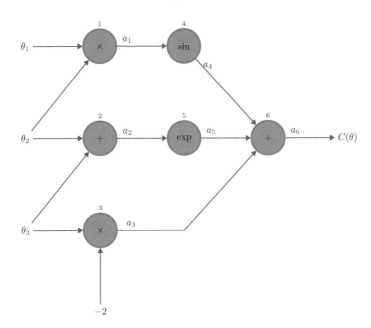

Figure 4.7: The computational graph of (4.27) with nodes numbered 1 to 6. The primal a_i of each node is the result of an elementary operation applied on its inputs.

As a simple example with $d = 3$, consider,

$$C(\theta) = \sin(\theta_1 \theta_2) + \exp(\theta_2 + \theta_3) - 2\theta_3. \tag{4.27}$$

The computational graph for this function, presented in Figure 4.7, is composed of nodes with inputs and outputs. The output of each node is an elementary operation on the inputs to the node. These elementary operations include summation, multiplication, $\sin(\cdot), \cos(\cdot), \exp(\cdot)$, etc. Nodes are numbered $1, 2, \ldots$, and we denote the output of node i with $a_i = f_i(u_1, \ldots, u_{\ell_i})$ when there are ℓ_i inputs to node i, denoted as $u_1, \ldots, u_{\ell_i}$. In this example there are 6 nodes, and taking node 6, as an illustration, we have $a_6 = f_6(a_3, a_4, a_5)$ where $\ell_6 = 3$ and $f_6(\cdot)$ is

---

[21]It may be also useful to see the mathematical representations of Jacobian vector products and vector Jacobian products through a composition of functions. For this we may see (A.16), (A.17), and (A.18), in Appendix A.

the summation of its input arguments. Note that a_i is sometimes called the *primal* of that node in a given computation.

The evaluation at each node in the computational graph is executed at the instance in which inputs to the node become available. This then implies the notion of iterations, where a node is considered to be evaluated at a given iteration if the inputs to that node are all available at that iteration but not previously. As an example, with $\theta = (\pi/6, 2, 5)$, the iterations for (4.27) are summarized in Table 4.1.

Table 4.1: Evaluation of the value of $C(\theta)$ in (4.27) at an example θ via the computational graph in Figure 4.7 computing the primals a_i for $i = 1, \ldots, 6$.

	General expressions of primals	Values of primals at specific θ
Input	$\theta = (\theta_1, \theta_2, \theta_3)$	$\theta = (\pi/6, 2, 5)$
Iteration 1	$a_1 = \theta_1 \theta_2$ $a_2 = \theta_2 + \theta_3$ $a_3 = -2\theta_3$	$a_1 = \pi/3$ $a_2 = 7$ $a_3 = -10$
Iteration 2	$a_4 = \sin(a_1)$ $a_5 = \exp(a_2)$	$a_4 = \sqrt{3}/2$ $a_5 = e^7$
Iteration 3 (output)	$a_6 = a_4 + a_5 + a_3$ $= C(\theta)$	$a_6 = C(\theta)$ $= \sqrt{3}/2 + e^7 - 10$

As mentioned above, forward mode automatic differentiation implements Jacobian vector products. In our exposition here, since the target function $C(\cdot)$ is scalar valued ($m = 1$), the Jacobian is the transpose of the gradient and with $v \in \mathbb{R}^d$, the Jacobian vector product, $J_C(\theta)v$, is a scalar.[22] Further, our exposition focuses on $v = e_j$, implying that the output of forward mode automatic differentiation is $\partial C(\theta)/\partial \theta_j$. Hence for our exposition here, j is fixed.

The key idea of forward mode automatic differentiation is to maintain a record of intermediate derivatives as computation progresses through the computational graph. For this we denote

$$\dot{a}_i = \frac{\partial a_i}{\partial \theta_j}, \tag{4.28}$$

and call $\dot{a}_i$ the *tangent* at node i, for fixed j. Now since $a_i = f_i(u_1, \ldots, u_{\ell_i})$, using the multivariable chain rule (see Appendix A.3), we have

$$\dot{a}_i = \sum_{k=1}^{\ell_i} \frac{\partial f_i(u_1, \ldots, u_{\ell_i})}{\partial u_k} \dot{u}_k = \nabla f_i(u_1, \ldots, u_{\ell_i})^\top \dot{u}, \tag{4.29}$$

with $\dot{u}_k$ denoting the tangent of u_k and with $\dot{u}$ denoting the vector of these tangents. Returning to Figure 4.7 as an example, we have that $\nabla f_6(u_1, u_2, u_3) = (1, 1, 1)$ and thus, $\dot{a}_6 = \dot{a}_3 + \dot{a}_4 + \dot{a}_5$. Similarly, in that figure, $\nabla f_4(u_1) = \cos(u_1)$ and thus $\dot{a}_4 = \cos(a_1)\dot{a}_1$.

---

[22]It is in fact the directional derivative of $C(\cdot)$ in the direction v.

Since (4.29) defines a recursive computation of the tangents, we also require the inputs to the function $\theta_1, \ldots, \theta_d$ to have tangents. We set these as the elements of the vector $v \in \mathbb{R}^d$ used in the Jacobian vector product. Specifically, here we focus on Jacobian vector products with $v = e_j$ since we are seeking the derivative with respect to a specific input θ_j. Hence we set the tangent values for the inputs of the function as $\dot{\theta}_j = 1$ and $\dot{\theta}_i = 0$ for $i \neq j$.

For forward mode automatic differentiation, we progress on the computational graph in the forward direction from the input to the output, evaluating both the primal a_i and the tangent $\dot{a}_i$ at each node. As we progress, in each iteration, we consider the set of all nodes i whose predecessors already have primal values and tangent values available, and for each of these nodes we compute the primal values as was already shown in Table 4.1, and importantly also compute the tangent values via (4.29). This computation may be broken into iterations, in a similar way to the iterations of the function evaluation computed in Table 4.1.

In particular, since $f_i(\cdot)$ denotes an elementary operation, such as the function $\exp(\cdot)$ at node 5 in Figure 4.7, we assume the availability of a symbolic expression for each $\frac{\partial f_i(u_1, \ldots, u_{\ell_i})}{\partial u_k}$, or alternatively an ability to numerically compute it exactly. Therefore, since we know the numerical primal values $u_1, \ldots, u_{\ell_i}$ and the numerical tangent values $\dot{u}_1, \ldots, \dot{u}_{\ell_i}$, we can easily compute $\dot{a}_i$ using (4.29). Consequently, the numerical value of $\frac{\partial C(\theta)}{\partial \theta_j}$ can be computed at the end of forward mode as it is the tangent value of the output node of the computational graph.

Table 4.2 illustrates forward mode automatic differentiation using the example function $C(\theta)$ in (4.27). The first column focuses on the general expression and the other column computes the partial derivative of $C(\theta)$ with respect to θ_2 at a specific point θ.

Table 4.2: Forward mode automatic differentiation for the example function (4.27) and $j = 2$, i.e., we seek the derivative with respect to θ_2. The right column illustrates computation[23] of the derivative $\frac{\partial C(\theta)}{\partial \theta_2}$ at $\theta = (\pi/6, 2, 5)$ which is obtained via the final tangent value $\dot{a}_6 = \pi/12 + \mathrm{e}^7$.

	General expressions of primals and tangents	Values of primals and tangents at specific θ for $j = 2$
Start	$\theta = (\theta_1, \theta_2, \theta_3)$ $\dot{\theta} = (\dot{\theta}_1, \dot{\theta}_2, \dot{\theta}_3)$	$\theta = (\pi/6, 2, 5)$ $\dot{\theta} = (0, 1, 0)$
Iteration 1	$(a_1, \dot{a}_1) = (\theta_1 \theta_2, \theta_1 \dot{\theta}_2 + \dot{\theta}_1 \theta_2)$ $(a_2, \dot{a}_2) = (\theta_2 + \theta_3, \dot{\theta}_2 + \dot{\theta}_3)$ $(a_3, \dot{a}_3) = (-2\theta_3, -2\dot{\theta}_3)$	$(a_1, \dot{a}_1) = (\pi/3, \pi/6)$ $(a_2, \dot{a}_2) = (7, 1)$ $(a_3, \dot{a}_3) = (-10, 0)$
Iteration 2	$(a_4, \dot{a}_4) = (\sin(a_1), \cos(a_1)\dot{a}_1)$ $(a_5, \dot{a}_5) = (\exp(a_2), \exp(a_2)\dot{a}_2)$	$(a_4, \dot{a}_4) = (\sqrt{3}/2, \pi/12)$ $(a_5, \dot{a}_5) = (\mathrm{e}^7, \mathrm{e}^7)$
Iteration 3	$(a_6, \dot{a}_6) = \begin{bmatrix} a_3 + a_4 + a_5 \\ \dot{a}_3 + \dot{a}_4 + \dot{a}_5 \end{bmatrix}$	$(a_6, \dot{a}_6) = \begin{bmatrix} \sqrt{3}/2 + \mathrm{e}^7 - 10 \\ \pi/12 + \mathrm{e}^7 \end{bmatrix}$

---

[23]To verify the computation recall the basic linearity and product rules of scalar differentiation. Also, remember that the derivative of $\sin(\cdot)$ is $\cos(\cdot)$, that the derivative of $\exp(\cdot)$ is $\exp(\cdot)$, and that $\cos(\pi/3) = 1/2$.

Backward Mode Automatic Differentiation

As mentioned above, backward mode automatic differentiation implements vector Jacobian products. This allows us to use a single run of backward mode automatic differentiation to compute the gradient vector for $C : \mathbb{R}^d \to \mathbb{R}$. For deep learning, this is a significant improvement over forward mode automatic differentiation which would require d executions to obtain such a gradient.

For a given point θ, backward mode automatic differentiation is executed in two phases often called the *forward pass* and *backward pass*. The forward pass phase is executed before the backward pass phase. In the *forward pass*, the algorithm simply evaluates the computational graph to compute the values of the intermediate primal variables a_i while recording the dependencies of these variables in a bookkeeping manner, similar to the evaluation in Table 4.1.

In the backward pass, all the derivatives of the objective function are computed by using intermediate derivatives called *adjoints*. Specifically, for each intermediate variable a_i, the adjoint is

$$\zeta_i = \frac{\partial C}{\partial a_i}. \tag{4.30}$$

Adjoint ζ_i captures the rate of change in the final output C with respect to variable a_i. Contrast this with the tangent $\dot{a}_i$ from forward mode automatic differentiation, as in (4.28), which captures the rate of change of a_i with respect to an input variable.

In the backward pass we populate the values of the adjoint variables ζ_i. Specifically, we progress on the computational graph in the backward direction starting from the output node and propagating toward the input. For this, say that the output of node i is input to $\tilde{\ell}_i$ nodes with indices $i_1, \ldots, i_{\tilde{\ell}_i}$. Then the only way that a_i can influence the output C is by influencing $a_{i_1}, \ldots, a_{i_{\tilde{\ell}_i}}$. Now using the multivariable chain rule,

$$\frac{\partial C}{\partial a_i} = \sum_{k=1}^{\tilde{\ell}_i} \frac{\partial a_{i_k}}{\partial a_i} \frac{\partial C}{\partial u_{i_k}},$$

and thus using the notation of (4.30),

$$\zeta_i = \sum_{k=1}^{\tilde{\ell}_i} \frac{\partial a_{i_k}}{\partial a_i} \zeta_{i_k}. \tag{4.31}$$

The relationship (4.31) is used in iterations of the backward pass. In every iteration, the adjoints ζ_{i_k} in the right hand side of (4.31) are available from previous iterations and are used to compute ζ_i. In particular at the start of the backward pass we initialize the adjoint of the output node at 1. Further, recalling the notation used for forward mode automatic differenation, $a_{i_k} = f_{i_k}(u_1, \ldots, u_{\ell_{i_k}})$ where $u_1, \ldots, u_{\ell_{i_k}}$ are the inputs to node i_k and $f_{i_k}(\cdot)$ has readily computable derivatives with respect to its inputs. Hence, since the primals of $u_1, \ldots, u_{\ell_{i_k}}$ were already populated during the forward pass, we compute $\frac{\partial a_{i_k}}{\partial a_i}$ locally at node i and use it in (4.31).

Table 4.3: The backward pass of backward mode automatic differentiation for the example function (4.27) to compute its gradient at $\theta = (\pi/6, 2, 5)$. The result is $\nabla C(\theta) = (1, \pi/12 + e^7, e^7 - 2)$. This is computed after the forward pass is executed; see Table 4.1.

	General expressions of adjoints	Values of adjoints at specific θ
Start	$\zeta_6 = \dfrac{\partial C(\theta)}{\partial a_6}$	$\zeta_6 = 1$
Iteration 1	$\zeta_4 = \dfrac{\partial a_6}{\partial a_4}\zeta_6 = 1 \times \zeta_6$ $\zeta_5 = \dfrac{\partial a_6}{\partial a_5}\zeta_6 = 1 \times \zeta_6$ $\zeta_3 = \dfrac{\partial a_6}{\partial a_3}\zeta_6 = 1 \times \zeta_6$	$\zeta_4 = 1$ $\zeta_5 = 1$ $\zeta_3 = 1$
Iteration 2	$\zeta_1 = \dfrac{\partial a_4}{\partial a_1}\zeta_4 = \cos(a_1)\zeta_4$ $\zeta_2 = \dfrac{\partial a_5}{\partial a_2}\zeta_5 = \exp(a_2)\zeta_5$	$\zeta_1 = \cos(\pi/3) = 1/2$ $\zeta_2 = e^7$
Iteration 3	$\begin{aligned}\dfrac{\partial C(\theta)}{\partial \theta_1} &= \dfrac{\partial a_1}{\partial \theta_1}\zeta_1 \\ &= \theta_2\zeta_1\end{aligned}$ $\begin{aligned}\dfrac{\partial C(\theta)}{\partial \theta_2} &= \dfrac{\partial a_1}{\partial \theta_2}\zeta_1 + \dfrac{\partial a_2}{\partial \theta_2}\zeta_2 \\ &= \theta_1\zeta_1 + 1 \times \zeta_2\end{aligned}$ $\begin{aligned}\dfrac{\partial C(\theta)}{\partial \theta_3} &= \dfrac{\partial a_2}{\partial \theta_3}\zeta_2 + \dfrac{\partial a_3}{\partial \theta_3}\zeta_3 \\ &= 1 \times \zeta_2 - 2 \times \zeta_3\end{aligned}$	$\begin{aligned}\dfrac{\partial C(\theta)}{\partial \theta_1} &= 2 \times 1/2 = 1\end{aligned}$ $\begin{aligned}\dfrac{\partial C(\theta)}{\partial \theta_2} &= \pi/6 \times 1/2 + e^7 \\ &= \pi/12 + e^7\end{aligned}$ $\begin{aligned}\dfrac{\partial C(\theta)}{\partial \theta_3} &= e^7 - 2\end{aligned}$

The elements of the gradient $\nabla C(\theta)$ depend on the adjoints in a similar manner to (4.31). Specifically,

$$\frac{\partial C}{\partial \theta_j} = \sum_k \frac{\partial a_{j_k}}{\partial \theta_j}\zeta_{j_k}, \tag{4.32}$$

where the summation is over all nodes indexed via j_k that take θ_j as input in the computational graph. Thus, in the final iteration of the backward pass, we use (4.32) to obtain the gradient $\nabla C(\theta)$.

Let us return to the example in (4.27) with computational graph (4.7) evaluated at $\theta = (\pi/6, 2, 5)$. Evaluation of $\nabla C(\theta)$ via backward mode automatic differentiation first executes a forward pass as previously illustrated in Table 4.1. This populates the primals $a_1, \ldots, a_6$. Then the backward pass progresses via repeated application of (4.31) and with application of (4.32) in the final iteration. This backward pass computation is summarized in Table 4.3.

4.5 Additional Techniques for First-Order Methods

In Section 4.3 we have seen one of the most popular deep learning optimization algorithms, ADAM, as well as key milestone techniques that have led to its development. Then in

Section 4.4 we explored ways to compute the gradient $\nabla C(\theta)$. In this section, we explore other first-order optimization ideas that are less popular in practical deep learning, but still embody useful principles.

We begin the discussion by considering a modification of the momentum technique, to a variant called *Nesterov momentum*.[24] The idea of this method has also found its way into an algorithm called *Nadam* (Nesterov ADAM). We continue by considering variants of Adagrad and RMSProp for adaptive learning rates per coordinate. The variants we describe are called *Adadelta* and *Adamax*. While as of today, these additional techniques are not common in deep learning practice, they have made a significant impact earlier on and may be fruitful in the future as well. In a sense, momentum (covered in Section 4.3), ADAM, Nesterov momentum, Nadam, Adagrad (covered in Section 4.3), Adadelta, and Adamax are all variants and improvements of basic gradient descent. Thus, our exposition in this section aims to complete the picture on how basic gradient descent may be improved.

We close this section with an overview of *line search* techniques. With such techniques, once a descent direction is determined, we seek to move in that direction with a step size which minimizes loss over the next step in that direction. There are several variants for this approach, and we present an overview of these variants.

Nesterov Momentum and the Nadam Algorithm

We have seen that the momentum updates as in (4.16) and (4.17) accelerate like a ball rolling downhill. An issue with this method is that the steps do not slow down after reaching the bottom of a valley, and hence, there is a tendency to overshoot the valley floor. Therefore, if we know approximately where the ball will be after each step, we can slow down the ball before the hill slopes up again.

The idea of *Nesterov momentum* tries to improve on standard momentum by using the gradient at the predicted future position instead of the gradient of the current position. The update equations are,

$$v^{(t+1)} = \beta v^{(t)} + (1 - \beta)\nabla C(\underbrace{\theta^{(t)} - \alpha\beta v^{(t)}}_{\substack{\text{predicted} \\ \text{next point}}}), \tag{4.33}$$

$$\theta^{(t+1)} = \underbrace{\theta^{(t)} - \alpha v^{(t+1)}}_{\substack{\text{actual} \\ \text{next point}}}, \tag{4.34}$$

for constants $\beta \in [0, 1)$ and $\alpha > 0$, with $v^{(1)} = 0$. Compare (4.33) and (4.34) with (4.16) and (4.17). The difference here is that the gradient is computed at a predicted next point $\theta^{(t)} - \alpha\beta v^{(t)}$, which is a proxy for the (unseen) actual next point $\theta^{(t+1)} = \theta^{(t)} - \alpha v^{(t+1)}$ when $\beta \approx 1$.

Implementing Nesterov momentum via (4.33) and (4.34) requires evaluation of the gradient at the predicted points instead of actual points. This may incur overheads, especially when incorporated as part of other algorithms. For example, if we also require $\nabla C(\theta^{(t)})$ at each iteration for purposes such as RMSprop, then the gradient needs to be computed twice instead of once per iteration; once for $\nabla C(\theta^{(t)})$ and once for $\nabla C(\theta^{(t)} - \alpha\beta v^{(t)})$. For this

---

[24]Nesterov momentum is also sometimes called *Nesterov acceleration* or *Nesterov accelerated gradient*.

reason, and also for simplicity of implementing gradients only at $\theta^{(t)}$, a *look-ahead momentum* mechanism is sometimes used in place of (4.33) and (4.34).

To see how this mechanism works, first revisit the basic momentum update equations (4.16) and (4.17) and observe that the parameter update can be represented as

$$\theta^{(t+1)} = \theta^{(t)} - \alpha\big(\beta v^{(t)} + (1-\beta)\nabla C(\theta^{(t)})\big). \tag{4.35}$$

Observe that $v^{(t)}$ is based on gradients at points $\theta^{(0)}, \ldots, \theta^{(t-1)}$, but not on the gradient at $\theta^{(t)}$. Hence a way to incorporate this last gradient is with look-ahead momentum where we replace $v^{(t)}$ of (4.35) by $v^{(t+1)}$. This achieves behavior similar to Nesterov momentum.

With such a replacement, we arrive at update equations of the form,

$$v^{(t+1)} = \beta v^{(t)} + (1-\beta)\nabla C\big(\theta^{(t)}\big), \tag{4.36}$$

$$\theta^{(t+1)} = \theta^{(t)} - \alpha\big(\underbrace{\beta v^{(t+1)} + (1-\beta)\nabla C\big(\theta^{(t)}\big)}_{\substack{\text{look-ahead}\\\text{momentum}}}\big). \tag{4.37}$$

Using (4.36) and (4.37) aims to provide similar behavior to the Nesterov momentum equations (4.33) and (4.34), yet only requires evaluation of gradients at $\theta^{(t)}$ as opposed to a shifted point as in (4.33). While look-ahead momentum is not equivalent to Nesterov momentum, the gist of both methods is similar.

Now with update equations like (4.36) and (4.37), it is straightforward to adapt ADAM to include behavior similar to Nesterov momentum. This is done using similar steps to those used to construct ADAM using momentum, RMSprop, and bias correction in Section 4.3. The difference here is that we use the look-ahead momentum equations (4.36) and (4.37). This algorithm is called Nadam.

In short, Nadam can be viewed exactly as the ADAM procedure in Algorithm 4.2, yet with line 7 replaced by,

$$\hat{v} \leftarrow \frac{\beta}{1-\beta^{t+2}} v + \frac{1-\beta}{1-\beta^{t+1}} g.$$

In this case, $\hat{v}$ is not just a bias corrected version of v as in ADAM since it also incorporates look-ahead momentum, which is the key extra feature that Nadam adds to ADAM.

Adadelta

Recall the computation of $s^{(t+1)}$ for the Adagrad method as in (4.22). When Adagrad's $s^{(t+1)}$ is used in (4.21), it has the problem of monotonically decreasing effective learning rates. This is one of the reasons that eventually RMSprop became more popular than Adagrad. However, there are other very popular alternatives. One such alternative is the *Adadelta* method which uses exponential smoothing of the squared gradients as in RMSprop, but also uses another exponentially smoothed sequence of the descent step's squares.

A key motivation for Adadelta is the observation that the update equation (4.21) for RMSprop or Adagrad uses a unit-less quantity as the descent step. Specifically, in (4.21) the only unit-full quantity in the coefficient multiplying $\nabla C(\theta^{(t)})$ is the reciprocal of $\sqrt{s^{(t+1)}}$.

Hence that coefficient has units which are the inverse of the gradient and these cancel out the units of the gradient implying that the descent step is unit-less.

With Adadelta, the update equation (4.21) is modified so that the descent step maintains the same units of the gradient, via

$$\theta^{(t+1)} = \theta^{(t)} - \underbrace{\frac{\sqrt{\Delta\theta^{(t)}} + \varepsilon}{\sqrt{s^{(t+1)}} + \varepsilon} \odot \nabla C(\theta^{(t)})}_{\text{Descent step } \widetilde{\nabla} C(\theta^{(t)})}, \tag{4.38}$$

where $\Delta\theta^{(t)}$ is adaptively adjusted yet has the same units as $s^{(t+1)}$, making the coefficient of the gradient unit-free. Compare (4.38) with (4.21) to observe that $\sqrt{\Delta\theta^{(t)}} + \varepsilon$ replaces the learning rate α. This also means that Adagrad is "learning rate free". Instead, a potentially more robust parameter $\rho \in [0, 1)$ is used similarly to the γ parameter for RMSProp. This parameter specifies how to exponentially smooth squares of the descent steps.

Using the descent step of (4.38), the update equation for $\Delta\theta^{(t)}$ is

$$\Delta\theta^{(t)} = \rho\Delta\theta^{(t-1)} + (1-\rho)\left(\widetilde{\nabla}C(\theta^{(t-1)}) \odot \widetilde{\nabla}C(\theta^{(t-1)})\right),$$

starting with $\Delta\theta^{(0)} = 0$. Then at iteration t, updating $\theta^{(t+1)}$ via (4.38), we use both $\Delta\theta^{(t)}$ and $s^{(t+1)}$, where the latter is updated via the RMSprop update equation as in (4.22).

Other Norms and Adamax

We have already seen that the quantities $s_i^{(t)}$ used in RMSprop or Adadelta as in (4.22) are useful for adaptive learning rates per coordinate. With these methods, we may loosely interpret $\sqrt{s_i^{(t)}}$ as an estimate of the standard (L_2) Euclidean norm of the sequence of gradients for coordinate i, smoothed up to time t.

In principle, one may wish to use other L_p norms (see Appendix A.1) in place of the Euclidean norm. For this we may modify the RMSprop recursion as in (4.22) to the form

$$s_i^{(t+1)} = \gamma^p\, s_i^{(t)} + (1-\gamma^p)|g_i^{(t)}|^p, \qquad \text{with} \qquad g_i^{(t)} = \frac{\partial C(\theta^{(t)})}{\partial\theta_i}, \tag{4.39}$$

and then interpret $\left(s_i^{(t)}\right)^{\frac{1}{p}}$ as an estimate of the smoothed L_p norm of the sequence of gradients for coordinate i up to time t. Note that for convenience of the calculations below, we reparameterize the exponential smoothing parameter to be γ^p in place of γ.

With a recursion such as (4.39) we can obtain an "ADAM like" algorithm that updates parameters via an update rule such as

$$\theta^{(t+1)} = \theta^{(t)} - \alpha\frac{1}{\left(\hat{s}^{(t+1)}\right)^{1/p} + \varepsilon}\,\hat{v}^{(t+1)}, \tag{4.40}$$

where $\hat{v}^{(t+1)}$ is some bias-corrected momentum value and $\hat{s}^{(t+1)}$ is a bias-corrected value obtained from (4.39); compare (4.40) to (4.25). This additional hyper-parameter p, determining which L_p norm to use, can then potentially be tuned. However, for non-small p

computations may often exhibit excessive numerical error due to overflow since we are raising the gradient coordinate value to high powers and then taking a low power of it.

Adamax is a variant of this approach which instead of using the L_p norm uses the L_∞ norm. Specifically in place of $(s_i^{(t)})^{\frac{1}{p}}$ we use $r_i^{(t)}$ which is recursively defined as

$$r_i^{(t+1)} = \max \left[\gamma\, r_i^{(t)},\ g_i^{(t)} \right], \qquad \text{with} \qquad r_i^{(0)} = 0. \tag{4.41}$$

Then in place of (4.40) we use

$$\theta^{(t+1)} = \theta^{(t)} - \alpha \frac{1}{r^{(t+1)} + \varepsilon}\, \hat{v}^{(t+1)}. \tag{4.42}$$

Note that the updates (4.41) and (4.42) are well justified as approximations of (4.39) and (4.40). To see this recall that a sequence of L_p norms converges to the L_∞ norm and thus,

$$
\begin{aligned}
r_i^{(t+1)} &= \lim_{p\to\infty} \left(s_i^{(t+1)} \right)^{\frac{1}{p}} \\
&= \lim_{p\to\infty} \left[(1-\gamma^p)^{1/p} \left(\sum_{\tau=0}^{t} \gamma^{p(t-\tau)} |g_i^{(\tau)}|^p \right)^{\frac{1}{p}} \right] \\
&= \lim_{p\to\infty} \left(\sum_{\tau=0}^{t} \left(\gamma^{(t-\tau)} |g_i^{(\tau)}| \right)^p \right)^{\frac{1}{p}} \\
&= \max_{\tau=0,\dots,t} \gamma^{(t-\tau)} |g_i^{(\tau)}| \\
&= \max \left[\gamma \max_{\tau=0,\dots,t-1} \gamma^{(t-1-\tau)} |g_i^{(\tau)}|,\ |g_i^{(t)}| \right].
\end{aligned}
$$

Here, moving from the first line to the second line, we made use of the general representation of exponential smoothing (4.15), and moving from the third line to the fourth line we use the fact that L_p vector norms converge to the L_∞ vector norm as $p \to \infty$. This then justifies (4.41) as a limiting case.

Line Search

So far all the methods we considered were variants of gradient descent. When these are viewed as a special case of the general descent direction method of Algorithm 4.1, updates are of the form $\theta^{(t+1)} = \theta^{(t)} + \theta_s^{(t)}$. In the case of basic gradient descent we have descent steps $\theta_s^{(t)} = -\alpha \nabla C(\theta^{(t)})$, yet in general we can represent the descent step as $\theta_s^{(t)} = \alpha \theta_d^{(t)}$ where we call $\theta_d^{(t)}$ a *descent direction*. With basic gradient descent $\theta_d^{(t)} = -\nabla C(\theta^{(t)})$. In any case, for some prescribed descent direction, the step size $\|\theta_s^{(t)}\|$ is determined by the choice of α.

Line search is an approach where we seek the best step size for a given descent direction in each iteration. Specifically, in iteration t, given a descent direction $\theta_d^{(t)}$, we determine the α for that iteration, denoted as $\alpha^{(t)}$, via

$$\alpha^{(t)} = \arg\min_{\alpha} C\left(\theta^{(t)} + \alpha \theta_d^{(t)} \right). \tag{4.43}$$

That is, each iteration involves a one-dimensional minimization problem determining $\alpha^{(t)}$ which is used for the update

$$\theta^{(t+1)} = \theta^{(t)} + \alpha^{(t)}\theta_{\mathrm{d}}^{(t)}.$$

Hence with line search, the next point $\theta^{(t+1)}$ is a minimizer of $C(\cdot)$ along the ray,

$$\{\theta^{(t)} + \alpha\theta_{\mathrm{d}}^{(t)} \ : \ \alpha > 0\}. \tag{4.44}$$

When we pick $\alpha^{(t)}$ to exactly optimize (4.43), the method is called *exact line search*. In very specific cases, exact line search can be achieved explicitly with closed-form formulas. However, in most cases, if we wish to carry out exact line search, we need to numerically optimize the one-dimensional optimization problem (4.43) for the best α. An alternative to exact line search is *inexact line search* where we only probe the loss on the ray (4.44) for a few values of α. Details are in the sequel.

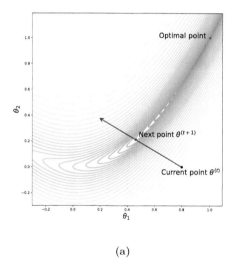

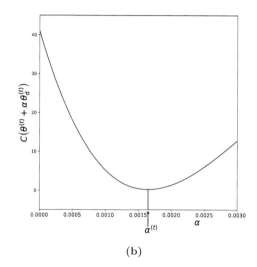

(a)   (b)

Figure 4.8: One iteration of line search applied to a Rosenbrook function of (4.9). (a) With the current point $\theta^{(t)} = (0.8, 0)$ and the descent direction $\theta_{\mathrm{d}}^{(t)} = -\nabla C(\theta^{(t)})$, we optimize over a given ray. (b) The value of the loss function along the ray with optimal $\alpha^{(t)} \approx 0.00165$.

Note that in practice, for deep learning, neither exact nor inexact line search is used directly. One reason is that loss function evaluations $C(\cdot)$ are computationally demanding and are of a similar order to the computational cost of gradient evaluation. Hence practice has shown that one might as well update descent directions instead of optimizing loss over a fixed ray (4.44). A second reason is that with deep learning we almost always have noisy gradients using stochastic gradient descent or mini-batches, and hence searching for an optimal α on a noisy direction can be practically less useful. Empirical evidence has shown that in such cases, line search techniques are seldom useful in their own right. Nevertheless, understanding line search is important as part of a general background for optimization and is also used in quasi-Newton methods described in Section 4.6 which are sometimes used in deep learning.

As a basic illustration of line search, return to Rosenbrock function of (4.9) with contours plotted in Figure 4.8 (a). In this plot we are at a point $\theta^{(t)}$ with the descent direction $\theta_{\mathrm{d}}^{(t)} = -\nabla C(\theta^{(t)})$. The plot in Figure 4.8 (b) presents the one-dimensional cross section along the ray (4.44) with that descent direction where we see that the optimal (exact numerical line search) α is at $\alpha^{(t)}$.

It is interesting to analyze the case $\theta_{\mathsf{d}}^{(t)} = -\nabla C(\theta^{(t)})$ which we call *basic gradient descent with exact line search*. In this case, trajectories of such optimization result in "zig-zagging". Specifically we have that every two consecutive descent steps are orthogonal, namely,

$$\theta_{\mathsf{s}}^{(t)\top} \theta_{\mathsf{s}}^{(t+1)} = 0 \qquad \text{or} \qquad g^{(t)\top} g^{(t+1)} = 0,$$

with $g^{(t)}$ denoting $\nabla C(\theta^{(t)})$. This property is illustrated in Figure 4.9 where we consider gradient descent with exact line search applied to the Rosenbrock function (4.9). We start with initial condition $\theta^{(0)} = (0.4, -0.2)$ from which the first two steps take very long jumps. However, afterward once the search is in a narrow valley, the zig-zagging phenomenon is apparent and incurs a significant effective slow down on the rate of advance toward the optimal point.

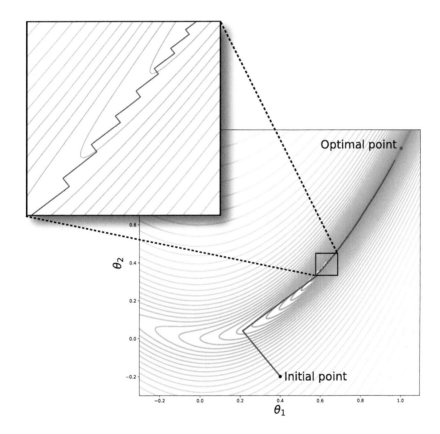

Figure 4.9: Evolution of basic gradient descent with exact line search. We see the zig-zagging property which slows down progress in narrow valleys.

To see the derivation of the zig-zagging property, observe that at α which optimizes (4.43), we have

$$\frac{\partial C\left(\theta^{(t)} - \alpha g^{(t)}\right)}{\partial \alpha} = 0,$$

or written in terms of the basic definition of the derivative we have

$$\lim_{h \to 0} \frac{C\left(\theta^{(t)} - (\alpha + h)g^{(t)}\right) - C\left(\theta^{(t)} - \alpha g^{(t)}\right)}{h} = 0. \qquad (4.45)$$

We now use the first-order Taylor series expansion as in (A.22) of $C(\cdot)$ around the point $\theta^{(t)} - \alpha g^{(t)}$ for the point $\theta^{(t)} - \alpha g^{(t)} - hg^{(t)}$. This is

$$C(\theta^{(t)} - \alpha g^{(t)} - hg^{(t)}) = C(\theta^{(t)} - \alpha g^{(t)}) - h\, g^{(t)^\top} \nabla C\left(\theta^{(t)} - \alpha g^{(t)}\right) + O(h^2),$$

where $O(h^2)$ is a function such that $O(h^2)/h^2$ goes to a constant as $h \to 0$. Substituting in (4.45) we obtain

$$\begin{aligned}
0 &= \lim_{h \to 0} \frac{-h\, g^{(t)^\top} \nabla C\left(\theta^{(t)} - \alpha g^{(t)}\right) + O(h^2)}{h} \\
&= -g^{(t)^\top} \nabla C\left(\theta^{(t)} - \alpha g^{(t)}\right) \\
&= -g^{(t)^\top} g^{(t+1)},
\end{aligned}$$

where the last equality holds because $g^{(t+1)} = \nabla C\left(\theta^{(t)} - \alpha g^{(t)}\right)$.

We now visit a powerful numerical technique called the *conjugate gradient method*. This is a topic which may receive special treatment in a general optimization text, yet here we only present it in brief as an application of exact line search. The conjugate gradient is not typically used directly for deep learning, yet since it is useful for approximately solving certain large systems of linear equations, it sometimes finds its way as an aid to other deep learning algorithms.

Suppose the objective function is of the following quadratic form,

$$C(\theta) = \frac{1}{2}\theta^\top A\, \theta - b^\top \theta, \qquad (4.46)$$

where A is a $d \times d$ dimensional symmetric positive definite matrix, and $b \in \mathbb{R}^d$. We have the Hessian $\nabla^2 C(\theta) = A$ for every θ and since A is positive definite, $C(\cdot)$ is strictly convex and there is a unique global minimum. We further have $\nabla C(\theta) = A\theta - b$ and this means that the equation $\nabla C(\theta) = 0$ with the unique solution $\theta^* = A^{-1}b$ minimizes $C(\theta)$. Hence the system of equations,

$$A\theta = b, \qquad \text{with } A \text{ positive definite}, \qquad (4.47)$$

can be solved by solving the optimization problem of minimizing (4.46). Thus one may view the conjugate gradient method as an algorithm for (approximately) solving a linear system of equations such as (4.47). In particular when the dimension d is excessively large, the conjugate gradient method can be efficient while standard techniques such as Gaussian elimination may fail. We have already seen one application of such equations in Chapter 2, with the normal equations (2.15) where $A = X^\top X$ and $b = X^\top y$.

When carrying out line search on an objective function as (4.46), the cross section above the ray (4.44) for any direction is a parabola and hence exact line search has an explicit solution for (4.43). Conjugate gradient carries out iterations of such line search, where in each iteration the search direction $\theta_d^{(t)}$ is wisely chosen.

Specifically, at iteration t with current point $\theta^{(t)}$ and search direction $\theta_d^{(t)}$, the next point determined via (4.43) with (4.46) is

$$\theta^{(t+1)} = \theta^{(t)} + \alpha^{(t)}\theta_d^{(t)}, \qquad \text{with} \qquad \alpha^{(t)} = -\frac{\theta_d^{(t)}{}^\top \left(A\,\theta^{(t)} - b\right)}{\theta_d^{(t)}{}^\top A\,\theta_d^{(t)}}.$$

Then the next search direction is determined as a linear combination of the next gradient and the current search direction. Specifically,

$$\theta_d^{(t+1)} = -\nabla C(\theta^{(t+1)}) + \beta^{(t)}\theta_d^{(t)}, \qquad \text{with} \qquad \beta^{(t)} = \frac{\nabla C(\theta^{(t+1)})^\top A\,\theta_d^{(t)}}{\theta_d^{(t)}{}^\top A\,\theta_d^{(t)}}.$$

This coefficient $\beta^{(t)}$ is designed such that the current search direction $\theta_d^{(t)}$ and the next search direction $\theta_d^{(t+1)}$ are *conjugate* with respect to A in the sense that, $\theta_d^{(t+1)}{}^\top A\,\theta_d^{(t)} = 0$. This conjugacy ensures desirable properties for the method and in particular implies that a minimum is reached in d iterations. Analysis and further motivation of conjugate gradient are beyond our scope. Our presentation here is merely to show that conjugate gradient is an application of exact line search.[25]

Inexact Line Search

As discussed above, one may use exact line search to solve the univariate optimization (4.43) optimally in each iteration. An alternative is only probing a few specific values of $\alpha^{(t)}$ according to some predefined set of rules and stop at the best probed value. This is called *inexact line search* and especially attractive when there is no analytical solution of (4.43). Inexact line search may significantly reduce the computational cost per iteration in comparison to the repetitive application of a univariate numerical optimization technique.

A typical approach for inexact line search is *backtracking* (also known as *backtracking line search*). The idea is to start with some predetermined maximal $\alpha_0 > 0$ and then decrease it multiplicatively until a stopping condition is met. Specifically, we evaluate $C(\cdot)$ on the ray (4.44) with candidate α values such as

$$\alpha_0, \quad \frac{2}{3}\alpha_0, \quad \left(\frac{2}{3}\right)^2\alpha_0, \quad \left(\frac{2}{3}\right)^3\alpha_0, \dots,$$

where the decrease factor of $\frac{2}{3}$ can be tuned to be any factor in $(0, 1)$. We then stop at the instant at which the stopping condition is met. Note that this search takes place at every iteration t.

---

[25] Note that in each iteration, the conjugate gradient traverses the steepest direction under the norm $\|u\|_A := \sqrt{u^\top A u}$, $u \in \mathbb{R}^d$.

One common stopping condition is known as the *Armijo condition* or the *first Wolfe condition*. Under this condition, at each iteration t, we decrease α values as above until it satisfies

$$C\left(\theta^{(t)} + \alpha\theta_{\mathrm{d}}^{(t)}\right) \leq C(\theta^{(t)}) + \beta\,\alpha\,\nabla_{\mathrm{d}}C(\theta^{(t)}), \tag{4.48}$$

where $\beta \in [0, 1]$ is a parameter and ∇_{d} is shorthand notation for the directional derivative in the direction $\theta_{\mathrm{d}}^{(t)}$; for directional derivatives recall (A.8) from Appendix A. That is,

$$\nabla_{\mathrm{d}}C(\theta^{(t)}) = \nabla_{\theta_{\mathrm{d}}^{(t)}}C(\theta^{(t)}) = \theta_{\mathrm{d}}^{(t)\top}\nabla C(\theta^{(t)}).$$

To get an intuition for (4.48), suppose that $\beta = 1$, then the right hand side of (4.48) is equal to the first-order Taylor approximation of $C\left(\theta^{(t)} + \alpha\,\theta_{\mathrm{d}}^{(t)}\right)$. Therefore, the decrease in the objective suggested by the first Wolfe condition is at least as good as the prediction by the first-order approximation. On the other extreme, if $\beta = 0$, we select α such that $C\left(\theta^{(t)} + \alpha\,\theta_{\mathrm{d}}^{(t)}\right) \leq C(\theta^{(t)})$, which is also acceptable as it guarantees that the next step at least does not increase the objective.

Beyond the condition (4.48), there are additional common stopping conditions that one may employ together with backtracking line search. One such condition known as the *second Wolfe condition* is

$$\nabla_{\mathrm{d}}C\left(\theta^{(t)} + \alpha\,\theta_{\mathrm{d}}^{(t)}\right) \geq \gamma\nabla_{\mathrm{d}}C(\theta^{(t)}), \tag{4.49}$$

where $\gamma \in (0, 1)$. Another common condition known as the *strong Wolfe condition* is

$$\left|\nabla_{\mathrm{d}}C\left(\theta^{(t)} + \alpha\,\theta_{\mathrm{d}}^{(t)}\right)\right| \leq -\gamma\nabla_{\mathrm{d}}C(\theta^{(t)}). \tag{4.50}$$

This condition forces $\alpha^{(t)}$ to lie close to the solution of exact line search. The first Wolfe condition and the above strong Wolfe condition together are called the *strong Wolfe conditions*. These conditions are sometimes used with quasi-Newton methods described below.

4.6 Concepts of Second-Order Methods

All the optimization methods we studied so far were based on gradient evaluation and implicitly use the first-order Taylor approximation of the objective function $C(\theta)$. In particular, the gradient of the objective function helps us determine the direction of the next step. However, by using curvature information captured via the second derivatives, or Hessian $\nabla^2C(\theta)$, algorithms can take steps that move farther and are more precise.

In this section, we present some well-known second-order Taylor approximation-based optimization methods, simply referred to as *second-order methods*. These methods also fall within the general descent direction framework of Algorithm 4.1, yet their evaluation of the descent step $\theta_{\mathrm{s}}^{(t)}$ either uses the Hessian explicitly or employs some approximated estimate for Hessian vector products.

While today's deep learning practice rarely involves second-order methods, these methods have huge potential. Indeed, if employed efficiently, their application in general deep learning training can yield significant performance improvement. Hence, understanding basic principles of second-order techniques is useful for the mathematical engineering of deep learning.

This section starts by exploring simple concepts of second-order univariate optimization where we introduce *Newton's method* which is based on the first and second derivative of the objective function. We also present the *secant method* which is only based on the first derivative. We then move on to Newton's method in the general multivariate case where the Hessian matrix plays a key role. Indeed, for simple neural networks such as logistic regression of Chapter 3, a method such as Newton is highly applicable.

For general deep learning with $\theta \in \mathbb{R}^d$, where d is typically huge and the loss function is complicated, it is hopeless to use explicit expressions or automatic differentiation for the Hessian $\nabla^2 C(\theta)$. In such a case, one may use *quasi-Newton* methods that only approximate the Hessian via gradients. Such application of second-order methods to deep learning is a contemporary active area of research which we are not able to cover fully. Instead, our presentation builds up toward a popular method called the *limited-memory BFGS (Broyden–Fletcher–Goldfarb–Shanno)* algorithm, or *L-BFGS* for short. With this we aim to open up horizons for the reader to the multitude of optimization technique variations that one may consider.

The Univariate Case

Newton's method is an iterative method that uses the quadratic approximation, also called the second-order Taylor approximation (A.22), of the objective function around the current point and finds the next step by optimizing the quadratic function. For ease of understanding, first consider the univariate case where $\theta \in \mathbb{R}$. Assume that the objective function $C(\theta)$ is twice differentiable. Suppose we are at $\theta^{(t)}$ in the t-th iteration and let $Q^{(t)}(\theta)$ be the quadratic approximation of the objective function $C(\theta)$ around $\theta^{(t)}$, given by

$$C(\theta) \approx Q^{(t)}(\theta) = C(\theta^{(t)}) + (\theta - \theta^{(t)})C'(\theta^{(t)}) + \frac{(\theta - \theta^{(t)})^2}{2}C''(\theta^{(t)}),$$

where $C'(\cdot)$ and $C''(\cdot)$ are the first and second derivatives, respectively.

Note that $Q^{(t)}(\theta)$ depends on $\theta^{(t)}$. Further, observe that $C(\theta)$ and $Q^{(t)}(\theta)$ take the same value when $\theta = \theta^{(t)}$. To find the next step, $\theta^{(t+1)}$, we minimize $Q^{(t)}(\theta)$ by setting its derivative to zero and obtain

$$C'(\theta^{(t)}) + (\theta - \theta^{(t)})C''(\theta^{(t)}) = 0,$$

with the corresponding solution (for the variable θ)

$$\theta^{(t+1)} = \theta^{(t)} - \frac{C'(\theta^{(t)})}{C''(\theta^{(t)})}, \qquad (4.51)$$

whenever $C''(\theta^{(t)}) \neq 0$. Newton's method (for univariate function optimization) starts at some initial point $\theta^{(0)}$ and then iterates (4.51) until some specified stopping criterion is met. If converged to some point θ^*, then $C'(\theta^*) = 0$ implying that the point is a local minimum, local maximum, or saddle point.[26]

Any quadratic function is either strictly convex or strictly concave depending on the leading coefficient. Therefore, if $C(\theta)$ itself a strictly convex quadratic function, it has a unique global minimum, and since the third-order derivative of $C(\theta)$ is zero, the quadratic approximation

---

[26]Note that when considered in the context of general root finding of the form $F(\theta) = 0$ beyond the case of $F(\cdot) = C'(\cdot)$, the method is sometimes called *Newton-Raphson*.

$Q^{(t)}(\theta)$ is identically equal to $C(\theta)$. In such a case, Newton's method finds the minimum in a single iteration of (4.51) for any initial point $\theta^{(0)}$. This property hints at the speed at which Newton's method can move toward the minimum since any smooth function is well approximated by a quadratic function in the neighborhood of a local minimum.

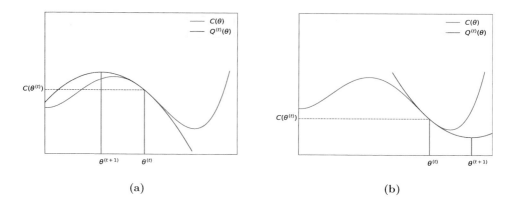

(a)   (b)

Figure 4.10: Illustration of updates in Newton's method. (a) The approximating quadratic function $Q^{(t)}(\theta)$ is concave. (b) The approximating quadratic function $Q^{(t)}(\theta)$ is convex.

While potentially very efficient, in its basic form, Newton's method exhibits several instability issues. First observe that the method can end up in a local maximum even though the goal is minimization. This is made apparent in Figure 4.10 which illustrates single step updates with Newton's method on a function that has both a local maximum and a local minimum. In (a) a step is taken and it approaches (yet overshoots) the local maximum. In (b) a step is taken (and also overshoots) the local minimum. As is evident, the initial point $\theta^{(t)}$ for both (a) and (b) are near each other, yet the subtle difference in the sign of $C''(\theta^{(t)})$ implies movement in completely different directions. In both cases, with another iteration the extremum point would have been nearly reached. That is (a) would end up very near the local maximum and similarly with another iteration (b) ends up very near the local minimum.

In general, the Newton update (4.51) is very sensitive to the second derivative and if $C''(\theta^{(t)})$ is very close to zero, then $\theta^{(t+1)}$ can be too far from $\theta^{(t)}$, and thus, the quadratic approximation is not appropriate. That is, *inflection points*[27] on the $\theta \in \mathbb{R}$ axis cause havoc with Newton's method. Similarly, as we see in the sequel for the multivariate case, points $\theta \in \mathbb{R}^d$ where the Hessian $\nabla^2(\theta)$ is nearly singular (eigenvalues near zero) are also problematic in the same way.

To gain insights into some of the issues that may occur with Newton's method, let us consider some examples. The sensitivity of the method can cause *overshoot* of the minimum. Consider for example

$$C(\theta) = (\theta - 3)^2(\theta + 3). \tag{4.52}$$

We have $C'(\theta) = 3(\theta - 3)(\theta + 1)$ and $C''(\theta) = 6(\theta - 1)$, and the function has a local minimum at $\theta^* = 3$ with $C(\theta^*) = 0$. The descent step resulting from (4.51) is thus

$$\theta_s^{(t)} = -\frac{C'(\theta^{(t)})}{C''(\theta^{(t)})} = \frac{1}{2}\frac{(\theta^{(t)} - 3)(\theta^{(t)} + 1)}{1 - \theta^{(t)}}.$$

---

[27]An *inflection point* for a twice differentiable function $C : \mathbb{R} \to \mathbb{R}$ is a point θ where $C''(\theta) = 0$.

Say we start at $\theta^{(0)} = 1.5$ where $C(\theta^{(0)}) = 10.125$. At this point the descent step is $\theta_s^{(0)} = 3.75$ implying that $\theta^{(1)} = \theta^{(0)} + \theta_s^{(0)} = 5.25$. This overshoot of the local minimum yields $C(\theta^{(1)}) \approx 41.77$ which is about a four-fold increase in the loss value. Note that in this case (as the reader may readily verify via simple calculations), the iterations that follow (presented to 4 decimal points precision) are

$$\theta^{(2)} = 3.5956, \quad \theta^{(3)} = 3.0683, \quad \theta^{(4)} = 3.0011,$$

and in fact the error for $\theta^{(6)}$ is already less than 10^{-13}. Thus we see in this example that there is an initial significant overshoot which is later corrected and is followed by very quick convergence.

Worse than an overshoot is a situation which results in moving away from the local minimum of interest. We have already seen an example of this in Figure 4.10 (a) where starting at the left of the inflection point yields convergence to a local maximum. Further, for the example function (4.52) this will occur if $\theta^{(0)} < 1$.

An additional phenomenon that may occur is *oscillation*. As an extreme example, consider,

$$C(\theta) = -\frac{1}{4}\theta^4 + \frac{5}{2}\theta^2, \quad \text{which implies} \quad \theta_s^{(t)} = \frac{\theta^{(t)}(\theta^{(t)^2} - 5)}{5 - 3\theta^{(t)^2}}.$$

If $\theta^{(t)} = 1$, then we get $\theta_s^{(t)} = -2$ and thus $\theta^{(t+1)} = -1$. Further, in the next step we get $\theta_s^{(t+1)} = 2$ which brings us back to $\theta^{(t+2)} = 1$. Such undesirable perfect oscillation is a singular phenomenon and is not likely to occur in practice. However approximate oscillation may sometimes occur and persist for multiple iterations until dampening.

In practice, and especially when considering multi-dimensional generalizations, *trust region methods* can mitigate problems such as overshoot, movement in the wrong direction, or oscillations. These methods incorporate conditions that disallow steps of large step size beyond some predefined or adaptive thresholds. Trust region methods also incorporate first-order methods as a backup. We do not cover the details further.

A final concern with Newton's method is that it requires both the first derivative $C'(\cdot)$ and the second derivative $C''(\cdot)$. In some scenarios, these derivatives are easy to obtain, yet in some other cases, the univariate objective may be a complicated function and even the application of automatic differentiation techniques may be inadequate for evaluating $C''(\cdot)$. This motivates the introduction of the *secant method* which is a modification of Newton's method where the second derivative in each iteration is approximated by the last two iterations. Namely, we use the approximation,

$$C''(\theta^{(t)}) \approx \frac{C'(\theta^{(t)}) - C'(\theta^{(t-1)})}{\theta^{(t)} - \theta^{(t-1)}}, \tag{4.53}$$

which becomes precise when $\theta^{(t)}$ and $\theta^{(t-1)}$ are not too far from each other. With this approximation, updates of the secant method are given by

$$\theta^{(t+1)} = \theta^{(t)} - \frac{\theta^{(t)} - \theta^{(t-1)}}{C'(\theta^{(t)}) - C'(\theta^{(t-1)})} C'(\theta^{(t)}),$$

where we initialize $\theta^{(0)}$ and $\theta^{(1)}$ to be close but not equal. In general, the secant method may require more iterations for convergence than Newton's method. The study of general numerical analysis and optimization theory contains an in-depth analysis of such convergence rates. This is a topic that we do not cover further.

The secant method also suffers from the problems associated with Newton's method mentioned above, including overshoot, moving in the wrong direction, and oscillation. Nevertheless, like Newton's method, the secant method is generally very powerful and arrives to a local minimum as long as the initial points are within the vicinity of the minimum.

Importantly, note that the secant method can be viewed as a *quasi-Newton method* since it approximates the Hessian (second derivative of a scalar function) with an expression that only depends on the first derivative. As we now generalize to multivariate optimization we should keep in mind that in the context of deep learning, almost any second-order method that one may wish to use should be a quasi-Newton method based on first-order derivatives (gradients), as opposed to explicit or automatically differentiated Hessians (second derivatives). Nevertheless, let us first study Newton's method for multivariate functions.

The Multivariate Case and Hessians

For a twice differentiable multivariate objective function $C : \mathbb{R}^d \to \mathbb{R}$ the update of Newton's method is very similar to the univariate case of (4.51). Specifically, the gradient and the Hessian play the roles of the first and the second-order derivatives, respectively. We can represent this update as,

$$\theta^{(t+1)} = \theta^{(t)} - H_t^{-1} \nabla C(\theta^{(t)}), \tag{4.54}$$

where $H_t = \nabla^2 C(\theta^{(t)})$ is the Hessian of the objective function at $\theta^{(t)}$, and when H_t is non-singular; compare (4.54) with (4.51).

To see the development of (4.54), as in the univariate case consider the quadratic approximation $Q^{(t)}(\theta)$ of $C(\theta)$ around $\theta^{(t)}$ given by

$$C(\theta) \approx Q^{(t)}(\theta) - C(\theta^{(t)}) + (\theta - \theta^{(t)})^\top \nabla C(\theta^{(t)}) + \frac{1}{2}(\theta - \theta^{(t)})^\top H_t (\theta - \theta^{(t)}). \tag{4.55}$$

Then, similar to the univariate case, the next point $\theta^{(t+1)}$ can be determined as θ such that the gradient of $Q^{(t)}(\theta)$ is zero. Observe that this gradient is

$$\nabla Q^{(t)}(\theta) = \nabla C(\theta^{(t)}) + H_t (\theta - \theta^{(t)}),$$

and by equating $\nabla Q^{(t)}(\theta)$ to 0 and representing $\theta - \theta^{(t)}$ as the descent step $\theta_{\mathrm{s}}^{(t)}$,

$$H_t \, \theta_{\mathrm{s}}^{(t)} = -\nabla C(\theta^{(t)}). \tag{4.56}$$

This linear systems of equations for $\theta_{\mathrm{s}}^{(t)}$ can be represented in terms of the elegant update equation (4.54) with descent step $\theta_{\mathrm{s}}^{(t)} = -H_t^{-1} \nabla C(\theta^{(t)})$, in case H_t is non-singular. However, in practice, we typically try to solve (4.56) using a suitable linear solver.[28] Sometimes

---

[28]Suitability of an algorithm depends on the size of d and any special structure in H_t that we can expect. For example for small d, one may use LU decomposition (Gaussian elimination) or QR factorization-based methods, yet for large d conjugate gradient or other *Krylov subspace methods* may be suitable.

approximate solutions for (4.56) are also an option. For example one may use the conjugate gradient method when d is large and run for a small number of iterations to only approximately solve (4.56).

In summary, Newton's algorithm for multivariate $C(\cdot)$ starts with some $\theta^{(0)} \in \mathbb{R}^d$. Then at each iteration t in the algorithm with current point $\theta^{(t)}$, we (in principle) evaluate the gradient and Hessian at that point. We then solve or approximately solve (4.56) to obtain $\theta_{\mathsf{s}}^{(t)}$ for the update $\theta^{(t+1)} = \theta^{(t)} + \theta_{\mathsf{s}}^{(t)}$.

In principle, in each iteration, solving (4.56) requires the Hessian matrix $H_t = \nabla^2 C(\theta^{(t)})$ either via a closed-form formula as for example in the case of logistic regression (4.7), or more generally via automatic differentiation. However with large d, computing the Hessian or Hessian vector products via automatic differentiation is generally not feasible. Hence in general, Newton's method on its own is not suitable for deep learning. Nevertheless, quasi-Newton methods such as the L-BFGS method, presented below, or various adaptations can be employed.

We also mention that all of the potential problems discussed above for the univariate case can appear in the multivariate case. Overshoot, moving in the wrong direction, or oscillation can occur due to similar reasons as in the univariate case. Specifically at points $\theta^{(t)}$ that are far from the local minimum, the quadratic approximation $Q^{(t)}(\theta)$ may be far from $C(\theta)$. Further, there are problems that may occur due to a singular, near-singular, or ill conditioned Hessian matrix H_t. In this case the descent step solution to (4.56) may be very noisy.[29] Nevertheless, Newton's method is very powerful and often when improved via trust region methods and similar techniques, it outperforms first-order methods very well.

To build intuition of why Newton's method often outperforms first-order methods, let us first explore a relationship between the two approaches. If we use the constant diagonal matrix $\frac{1}{\alpha}I$ in place of the Hessian H_t, then Newton's method reduces to the basic gradient descent method with update (2.20) where the descent step is $\theta_{\mathsf{s}}^{(t)} = -\alpha \nabla C(\theta)$. To see this, return to the quadratic approximation (4.55) with H_t replaced by $\frac{1}{\alpha}I$, to obtain

$$\widetilde{Q}^{(t)}(\theta) = C(\theta^{(t)}) + (\theta - \theta^{(t)})^\top \nabla C(\theta^{(t)}) + \frac{1}{2\alpha}(\theta - \theta^{(t)})^\top (\theta - \theta^{(t)}), \qquad (4.57)$$

and thus,

$$\nabla \widetilde{Q}^{(t)}(\theta) = \nabla C(\theta^{(t)}) + \frac{1}{\alpha}(\theta - \theta^{(t)}).$$

Equating this gradient expression to 0 yields the update (2.20). Hence, we notice that the inverse of H_t in Newton's method plays the role of the learning rate α of basic gradient descent. As a consequence, with this basic form of Newton's method, there is no need to calibrate the learning rate, as we do in most gradient descent approaches.[30]

Finally, let us get an intuition of why Newton's method generally converges to a (local) minimum θ^* faster than gradient descent when starting at a point close to θ^*. For this, return to Figure 2.8 which compares two loss landscapes, one of which is circular and one of which is elliptical. While that figure is created via a quadratic form where Netwon's method

---

[29]In practice one often monitors the level of ill-conditioning via the condition number of H_t, or similar measures.

[30]Note that Newton's method can also be cast as a steepest descent method like basic gradient descent, but with descent direction chosen with respect to the Hessian norm, $\|u\|_{H_t} = \sqrt{u^\top H_t\, u}$, $u \in \mathbb{R}^d$.

would reach the minimum in a single iteration, in more realistic cases when functions are not quadratic forms, they can still be locally well approximated by quadratic forms in the vicinity of θ^* and hence considering Figure 2.8 as a general plot is valid.

It is known that with gradient descent, highly elliptical loss landscapes are difficult for optimization since descent steps may be trapped in valleys, and thus as argued in Section 2.4 in view of Figure 2.8, one often tries to carry out variable transformations to alleviate such problems. In a sense the gradient descent quadratic approximation $\widetilde{Q}^{(t)}(\cdot)$ in (4.57) assumes that the loss landscape is already perfectly circular since it uses a constant diagonal matrix in place of the Hessian matrix. This is the best that basic gradient descent can offer. However, when using Newton's method, the correct local quadratic approximation $Q^{(t)}(\cdot)$ in (4.55) already adapts to the curvature and potential true elliptic nature of the loss landscape. With such an approximation $Q^{(t)}(\cdot)$, minimization over quadratic loss landscapes occurs in a single iteration, and since locally all smooth functions are well approximated by quadratic loss landscapes, when $\theta^{(t)}$ is in the vicinity of θ^*, converge is very quick.

Quasi-Newton Methods

In deep learning, the dimension of θ is very large and it is thus difficult to solve the linear equations (4.56) as required in each iteration of Newton's method. Even in dimensions where these equations are solvable, there is still the basic problem of computing the Hessian matrix, a feat which is typically intractable for mid-size problems and beyond. A popular alternative is offered by the use of quasi-Newton methods where either the Hessian or the inverse of the Hessian matrix are approximated with each iteration. While the study of quasi-Newton methods is mature, and these are used in many areas of applied mathematics and engineering, the development of best practices for application in deep learning is still an active area of research. Here we simply outline the key ideas of such methods.

The central idea in the type of quasi-Newton methods that we cover is based on an evolving sequence of $d \times d$ matrices, $B_0, B_1, B_2, \ldots$. At each iteration, some *update rule* specifies how B_t is updated based on B_{t-1} and other quantities from the current and previous iteration. By the design of the update rule, each such B_t is kept as positive semidefinite (and hence also non-singular), and in our presentation, B_t^{-1} is ideally a good approximation[31] for the Hessian matrix $H_t = \nabla^2 C(\theta^{(t)})$ associated with the current point $\theta^{(t)}$ of iteration t.

With such a B_t matrix available, we now use the quadratic approximation

$$Q^{(t)}(\theta) = C(\theta^{(t)}) + (\theta - \theta^{(t)})^\top \nabla C(\theta^{(t)}) + \frac{1}{2}(\theta - \theta^{(t)})^\top B_t^{-1}(\theta - \theta^{(t)}), \qquad (4.58)$$

which is similar to Newton's method quadratic approximation (4.55) with the difference that now B_t^{-1} plays the role of the Hessian. Similarly to the case in Newton's method, $Q^{(t)}(\theta)$ of (4.58) can be minimized where now the unique minimizer is $-B_t \nabla C(\theta^{(t)})$. Hence using such a quadratic approximation we have a descent direction,

$$\theta_\mathsf{d}^{(t)} = -B_t \nabla C(\theta^{(t)}). \qquad (4.59)$$

With quasi-Newton methods, we take a step in the direction specified by $\theta_\mathsf{d}^{(t)}$, but we also scale the step with a scalar $\alpha^{(t)} > 0$ which controls the step size. Hence, the key update rule

---

[31]We note that alternative presentations may use B_t as an approximation for the Hessian itself.

for quasi-Newton methods can be stated as

$$\theta^{(t+1)} = \theta^{(t)} - \alpha^{(t)} B_t \nabla C(\theta^{(t)}), \tag{4.60}$$

which is similar to (4.54). The determination of $\alpha^{(t)}$ is typically obtained via a line search technique such as inexact line search using the second Wolfe condition (4.49).

Algorithm 4.3: Quasi-Newton

> **Input:** Dataset $\mathcal{D} = \{(x^{(1)}, y^{(1)}), \ldots, (x^{(n)}, y^{(n)})\}$,
> objective function $C(\cdot) = C(\cdot\,; \mathcal{D})$, and
> initial parameter vector θ_{init}
> **Output:** Approximately *optimal* θ

1   $\theta \leftarrow \theta_{\mathsf{init}}$
2   $B \leftarrow B_{\mathsf{init}}$
3   **repeat**
4     $g \leftarrow \nabla C(\theta)$ (Compute gradient)
5     $\theta_{\mathsf{d}} \leftarrow -Bg$ (Determine descent direction)
6   Determine α using θ and θ_{d} (Line search)
7     $\theta \leftarrow \theta + \alpha\theta_{\mathsf{d}}$ (Update parameters)
8   Update B (Update approximation of $\nabla^2 C(\theta)^{-1}$)
9   **until** *termination condition* is satisfied
10   **return** θ

The typical quasi-Newton method is presented in Algorithm 4.3. As in any gradient-based algorithm, computation of the gradient is needed, and this is specified in Step 4. The computation of the descent direction (4.59) is in Step 5. In Step 6 we use line search, and Step 7 updates the current parameter based on (4.60). Finally, Step 8 is the update rule, specific to the type of quasi-Newton method used. In the sequel we present BFGS and L-BFGS, each with their own specification of this step.

As this method falls within the realm of a descent direction approach as in Algorithm 4.1, ideally we wish to set θ_{init} in Step 1, at a point close to a minimum, just like all the other descent direction approaches in this chapter. As for the initialization matrix B_{init} in Step 2, ideally we would have a matrix whose inverse is the Hessian at θ_{init}. However, since evaluation of the Hessian matrix or its inverse is typically computationally costly or intractable, we often initialize with $B_{\mathsf{init}} = I$, a $d \times d$ identity matrix.

Now with the general outline of the quasi-Newton approach presented, one may seek good update rules for Step 8. Before we see concrete formulas for these updates within the BFGS and L-BFGS algorithms, let us touch on theoretical aspects that helped develop these methods, and can be used for further analysis of their performance and properties.

Since the Hessian H_t captures the curvature information of the objective function, we may consider the latest two points $\theta^{(t-1)}$ and $\theta^{(t)}$ together with the gradients at these points, $\nabla C(\theta^{(t-1)})$ and $\nabla C(\theta^{(t)})$, for the approximate equality,

$$H_t \left(\theta^{(t)} - \theta^{(t-1)}\right) \approx \nabla C(\theta^{(t)}) - \nabla C(\theta^{(t-1)}). \tag{4.61}$$

This approximation follows from (A.24) of Appendix A, and is a good approximation when $\theta^{(t)}$ and $\theta^{(t-1)}$ are close. It generalizes (4.53) used in the univariate secant method. Note that (4.61) holds with equality if $C(\theta)$ is a quadratic function. Further, with the notation,

$$\theta_{\mathsf{s}}^{(t-1)} = (\theta^{(t)} - \theta^{(t-1)}) \qquad \text{and} \qquad g_{\mathsf{s}}^{(t-1)} = \nabla C(\theta^{(t)}) - \nabla C(\theta^{(t-1)}), \qquad (4.62)$$

we have that (4.61) is represented as $H_t \theta_{\mathsf{s}}^{(t-1)} \approx g_{\mathsf{s}}^{(t-1)}$. Now inspired by this approximation, if we are able to select the $d \times d$ positive definite matrix B_t, such that B_t^{-1} is an approximation of H_t, then by considering the approximation as an equality

$$\theta_{\mathsf{s}}^{(t-1)} = B_t\, g_{\mathsf{s}}^{(t-1)}. \qquad (4.63)$$

The equation (4.63) is known as the *secant equation* and it implies that B_t maps the change of gradients $g_{\mathsf{s}}^{(t-1)}$ into the descent direction $\theta_{\mathsf{s}}^{(t-1)}$. A condition related to the secant equation is

$$g_{\mathsf{s}}^{(t-1)\top} \theta_{\mathsf{s}}^{(t-1)} > 0, \qquad (4.64)$$

which is known as the *curvature condition*. Importantly, it can be shown that when the curvature condition is satisfied, there is always a positive definite B_t such that the secant equation is satisfied.[32] In fact, there can be infinitely many solutions. In general, quasi-Newton methods are designed to maintain the curvature condition (4.64) and to provide a unique update rule for B_t.

The curvature condition (4.64) allows us to also gain insight into why line search as in Step 6 of Algorithm 4.3 is needed. In simple or synthetic cases where the loss function $C(\cdot)$ is strictly convex, it can be shown that the curvature condition (4.64) is always satisfied. However, loss functions in deep learning are typically highly non-convex, and thus there is no guarantee of existence of B_t such that the secant equation is satisfied, unless we impose an additional condition on $\alpha^{(t)}$ of (4.60).

If in each iteration t we determine $\alpha^{(t)}$ using inexact line search with the second Wolfe condition (4.49), then the curvature condition (4.64) is guaranteed to hold.[33] To see this, suppose that the second Wolfe condition is applied, then from $\theta^{(t)} = \theta^{(t-1)} + \alpha^{(t-1)}\theta_{\mathsf{d}}^{(t-1)}$, we obtain

$$\nabla C(\theta^{(t)})^\top \theta_{\mathsf{d}}^{(t-1)} \geq \gamma\, \nabla C(\theta^{(t-1)})^\top \theta_{\mathsf{d}}^{(t-1)}, \qquad (4.65)$$

where we used the relationship between the gradient and the directional derivative shown in (A.8) of Appendix A. Now the inner product of the curvature condition (4.64) satisfies,

$$
\begin{aligned}
g_{\mathsf{s}}^{(t-1)\top} \theta_{\mathsf{s}}^{(t-1)} &= \nabla C(\theta^{(t)})^\top \theta_{\mathsf{s}}^{(t-1)} - \nabla C(\theta^{(t-1)})^\top \theta_{\mathsf{s}}^{(t-1)} \\
&\geq \gamma \alpha^{(t-1)} \nabla C(\theta^{(t-1)})^\top \theta_{\mathsf{d}}^{(t-1)} - \nabla C(\theta^{(t-1)})^\top \theta_{\mathsf{s}}^{(t-1)} \\
&= (\gamma - 1)\alpha^{(t-1)} \nabla C(\theta^{(t-1)})^\top \theta_{\mathsf{d}}^{(t-1)} \\
&= -(\gamma - 1)\alpha^{(t-1)} \nabla C(\theta^{(t-1)})^\top B_{t-1} \nabla C(\theta^{(t-1)}) \\
&> 0.
\end{aligned}
$$

---

[32] It can be shown that if for $u, v \in \mathbb{R}^d$, $u^\top v > 0$, then there exists a symmetric positive definite matrix $B \in \mathbb{R}^{d \times d}$ such that $Bu = v$.

[33] This also holds for the strong Wolfe condition (4.50).

The first equality uses the definition of $g_{\mathsf{s}}^{(t-1)}$ and the step size resulting from (4.60), $\theta_{\mathsf{s}}^{(t-1)} = \alpha^{(t-1)}\theta_{\mathsf{d}}^{(t-1)}$. The inequality that follows is due to (4.65). Finally, the last inequality is because $\gamma < 1$ and B_{t-1} is assumed to be positive definite based on the update rule. We thus see that the second Wolfe condition is a means to ensure that the curvature condition is satisfied.

The BFGS and L-BFGS Update Rules

Let us now focus on the *Broyden-Fletcher-Goldfarb-Shanno* (BFGS) algorithm, and its variant, the *limited-memory BFGS* (L-BFGS) algorithm. In practice, some small deep learning applications already make efficient use of L-BFGS. We mention that one common approach in training, is to first carry out training using standard gradient descent or a variant, and then apply L-BFGS to "complete" the training process. Nevertheless, to date, the application of L-BFGS and other variants of advanced quasi-Newton methods is still not widespread in large scale deep learning.

We present the update rules for these two algorithms which specify Step 8 of Algorithm 4.3, keeping in mind that $B_0 = B_{\mathsf{init}}$. Each quasi-Newton iteration of Algorithm 4.3 already bears the computational cost of computing the gradient, the cost of several function evaluations within the line search, and other costs associated with determining the descent direction and updating of parameters. The computational cost of the update rule in Step 8 is generally heavy. We contrast the computational cost of the BFGS update rule and that of the L-BFGS update rule, below.

The BFGS update rule is

$$B_t = V_{t-1}^{\top} B_{t-1} V_{t-1} + \rho^{(t-1)}\theta_{\mathsf{s}}^{(t-1)}\theta_{\mathsf{s}}^{(t-1)^{\top}}, \tag{4.66}$$

where the positive scalar $\rho^{(t-1)}$ and the matrix V_{t-1} are respectively given by,

$$\rho^{(t-1)} = \frac{1}{g_{\mathsf{s}}^{(t-1)^{\top}}\theta_{\mathsf{s}}^{(t-1)}} \quad \text{and} \quad V_{t-1} = I - \rho^{(t-1)}g_{\mathsf{s}}^{(t-1)}\theta_{\mathsf{s}}^{(t-1)^{\top}}. \tag{4.67}$$

Observe the use of the rank one matrices $\theta_{\mathsf{s}}^{(t-1)}\theta_{\mathsf{s}}^{(t-1)^{\top}}$ and $g_{\mathsf{s}}^{(t-1)}\theta_{\mathsf{s}}^{(t-1)^{\top}}$ using (4.62). The positivity of $\rho^{(t-1)}$ is maintained by the curvature condition (4.64). Importantly, it can be shown that each matrix in the sequence $B_0, B_1, \ldots$, resulting from this update rule is positive definite. We omit the details.

Note that due to the matrix multiplication $V_{t-1}^{\top} B_{t-1} V_{t-1}$, the computational cost of updating B_t using (4.66) is $O(d^2)$. This is a reasonable cost for small d, yet for larger d as in some deep learning applications, it renders BFGS too costly.

We mention that the development of the update rule in (4.66) and (4.67) can be cast as a minimization problem where we seek the closest possible positive definite B_t to the previous B_{t-1} under a specific matrix norm. The minimization is subject to the constraint that the secant equation (4.63) holds. We omit the details of this derivation yet mention that the matrix norm used for BFGS is a *weighted Frobenius norm*. See the notes and references at the end of the chapter for more information.

The L-BFGS algorithm overcomes the computational cost of BFGS. Its update rule is not for maintaining the matrix B_t in memory directly, but is rather based on keeping a limited history (or limited memory) of the constituent vectors $\theta_s^{(t)}$ and $g_s^{(t)}$. Observe that in principle, one can unroll the BFGS update rules (4.66) and (4.67) to obtain,

$$
\begin{aligned}
B_t = {}& V_{t-1}^\top V_{t-2}^\top \cdots V_0^\top \, B_{\text{init}} \, V_0 \cdots V_{t-2} V_{t-1} \\
& + \rho^{(t-1)} \theta_s^{(t-1)} \theta_s^{(t-1)\top} \\
& + \rho^{(t-2)} V_{t-1}^\top \theta_s^{(t-2)} \theta_s^{(t-2)\top} V_{t-1} \\
& \;\; \vdots \\
& + \rho^{(0)} V_{t-1}^\top V_{t-2}^\top \cdots V_0^\top \, \theta_s^{(0)} \theta_s^{(0)\top} \, V_0 \cdots V_{t-2} V_{t-1}.
\end{aligned}
\tag{4.68}
$$

Further, observe that this representation of B_t is a function of the sequence $\{\theta_s^{(k)}, g_s^{(k)}\}$ for $k = 0, 1, \ldots, t-1$ and B_{init}. Hence in principle, assuming that B_{init} is diagonal, if we wish to compute the matrix-vector product $B_t \nabla C(\theta^{(t)})$, we can do so with all multiplications being vector-vector multiplications. This is due to the fact that each V_t matrix is composed of the outer products $g_s^{(t-1)} \theta_s^{(t-1)\top}$. Yet obviously, for non-small t such multiplications become inefficient as the number of terms in the expression (4.68) grows.

The L-BFGS idea is to use an approximate version $\widetilde{B}_t$ of B_t that can be stored indirectly by storing only the latest m pairs of vectors $\{g_s^{(k)}, \theta_s^{(k)}\}$, for $k = t-m, \ldots, t-1$. The hyper-parameter m is the length of the history (or memory) and is typically chosen to be much smaller than d. We modify (4.68) with an approximation which limits the history used to the last m iterations. For this we drop the last $t-m$ terms of (4.68), and in the first term $V_{t-1}^\top V_{t-2}^\top \cdots V_0^\top \, B_{\text{init}} \, V_0 \cdots V_{t-2} V_{t-1}$, we replace the middle multiplicands, $V_{t-m-1}^\top \cdots V_0^\top \, B_{\text{init}} \, V_0 \cdots V_{t-m-1}$, with a symmetric positive definite matrix $B_{\text{init}}^{(t)}$. This modification yields $\widetilde{B}_t \approx B_t$, with the form,

$$
\begin{aligned}
\widetilde{B}_t = {}& V_{t-1}^\top V_{t-2}^\top \cdots V_{t-m}^\top \, B_{\text{init}}^{(t)} \, V_{t-m} \cdots V_{t-2} V_{t-1} \\
& + \rho^{(t-1)} \theta_s^{(t-1)} \theta_s^{(t-1)\top} \\
& + \rho^{(t-2)} V_{t-1}^\top \theta_s^{(t-2)} \theta_s^{(t-2)\top} V_{t-1} \\
& \;\; \vdots \\
& + \rho^{(t-m)} V_{t-1}^\top V_{t-2}^\top \cdots V_{t-m}^\top \theta_s^{(t-m-1)} \theta_s^{(t-m-1)\top} V_{t-m} \cdots V_{t-2} V_{t-1}.
\end{aligned}
\tag{4.69}
$$

Note that $B_{\text{init}}^{(t)}$ is selected for each iteration t and is usually set as a diagonal matrix to reduce the storage and computational costs. A practically effective choice is

$$
B_{\text{init}}^{(t)} = \frac{g_s^{(t-1)\top} \theta_s^{(t-1)}}{g_s^{(t-1)\top} g_s^{(t-1)}} I.
$$

With this choice for $B_{\text{init}}^{(t)}$, the expression (4.69) allows us to compute $\widetilde{B}_t \nabla C(\theta^{(t)})$ efficiently. In particular, $\widetilde{B}_t \nabla C(\theta^{(t)})$ can be computed using $O(m)$ inner products between d-dimensional vectors, and thus the per iteration computational cost for the L-BFGS algorithm is $O(md)$, while requiring to store only the latest m pairs of vectors $\{g_s^{(k)}, \theta_s^{(k)}\}$, $k = t-m, \ldots, t-1$, which takes $O(md)$ storage space. With m much smaller than d, the L-BFGS algorithm requires far less storage than the BFGS algorithm. Finally, we note that in terms of Algorithm 4.3, the update rule in Step 8 simply needs to push the vectors $\{g_{\mathsf{s}}^{(t)}, \theta_{\mathsf{s}}^{(t)}\}$ onto a queue while popping the vectors $\{g_{\mathsf{s}}^{(t-m)}, \theta_{\mathsf{s}}^{(t-m)}\}$ out of that queue. The bulk of the algorithm is in Step 5 with the efficient matrix-vector multiplication using the expressions (4.69) and (4.67).

Notes and References

Mathematical optimization is a classical subject and some of the general texts on this subject are [37] where linear optimization is covered, [272] where both linear and non-linear optimization methods are covered, and [35] where the focus is on non-linear methods. Further texts focusing on convexity, convex analysis, and convex optimization are [27], [32], and [55]. Many of the algorithms in these texts describe descent direction methods, yet some examples of optimization algorithms that are not descent direction methods include direct methods (also known as zero-order or black box) such as coordinate and pattern search methods; see [238] or [311] for an overview. *The Rosenbrock function* was developed in [354] within the context of multivariate function optimization and since then it became popular for basic optimization test cases and illustrations.

As mentioned in the Chapter 2 notes and references, gradient descent is attributed to Cauchy from 1847 with [72]; see [253] for a historical account. *Stochastic gradient descent* was initially introduced in 1951 with [350] in the context of optimizing the expectation of a random variable over a parametric family of distributions. See [229], [230], [231], [255], and [349] for early references using stochastic gradient descent approaches. After the initial popularity of this approach for general optimization, other numerical optimization methods were developed and stochastic gradient descent's popularity decreased. Years later, it became relevant again due to developments in deep learning; see [48] and [52] for early references in the context of deep learning. Interesting works that attempt to explain the suitability of stochastic gradient descent for deep learning are [49], [50], and [51]. Relationships between the use of mini-batches and stochastic gradient descent are studied in [202], [203], and [261]. Practical recommendations of gradient-based methods can be found in [29].

The ADAM algorithm was introduced in [233]. See [43] and [345] for additional papers that study its performance. Ideas of momentum appeared in optimization early on in [335]. RMSprop appeared in lectures by Geoff Hinton in a 2014 course and has since become popular. Adagrad appeared earlier on in [112]. An interesting empirical study considering ADAM and related methods is [82]. See [238] and [356] for a detailed study of momentum-based methods.

The basic idea of *automatic differentiation* dates back to the 1950s; see [25] and [313]. The idea of computing the partial derivatives using forward mode automatic differentiation is attributed to Wengert [420]. Even though ideas of backward mode automatic differentiation first appeared around 1960, they were used in control theory by the end of that decade. The 1980s paper [386] is generally attributed for presenting backward mode automatic differentiation in the form we know it now. Backward mode automatic differentiation for deep neural networks is referred to as the backpropagation algorithm, a concept that we cover in detail in Chapter 5. In fact, ideas for the backpropagation algorithm were developed independently of some of the automatic differentiation ideas; see [173] for an historical overview as well as the notes and references at the end of Chapter 5 for more details.

In recent years, with the deep learning revolution, automatic differentiation has gained major popularity and is a core component of deep learning software frameworks such as *TensorFlow* [1], *PyTorch* [324], and others. In parallel, the application of automatic differentiation for a variety of scientific applications has also been popularized with technologies such as *Flux.jl* [201] in Julia and *JAX* [57] in Python. See [24] for a survey in the context of machine learning and [147] for a more dated review. Additional differentiable programming frameworks that implement automatic differentiation include CasADi, Autograd, AutoDiff, JAX, Zygote.jl, Enzyme, and Chainer. Much of the complexity of differentiable programming software frameworks is centered around the use of GPUs as well as the placement and management of data in GPU memory.

Nesterov momentum was introduced in [305]. See [362], [363], and [431] for some comparisons of Nesterov momentum with the standard momentum approach. The Nadam algorithm was introduced in [109] to incorporate Nesterov momentum into ADAM and is now part of some deep learning frameworks. The Adadelta method appeared in [440] prior to the appearance and growth in popularity of ADAM. The Adamax method appeared in the same paper as ADAM [233], as an extension. Line search techniques are central in classical first-order optimization theory. Inexact line search techniques using Wolfe conditions first appeared in [422] and [423]. *Backtracking line search*, also known as *Armijo line search*, first appeared in [16]. See for example chapter 3 of [311] and chapter 4 of [238] for overviews of exact and inexact line search algorithms. Concepts of the conjugate gradient method first appeared in [177] for solving linear equations with symmetric positive semidefinite matrices. Since then, the method was incorporated in many numerical algorithms; see for example chapter 11 of [141] and chapter 5 of [311] for more details.

Despite the success of second-order methods in statistics, in general numerical optimization, and in engineering, they are yet to make it into mainstream large scale deep learning. Perhaps one reason for this is that when implemented without care, they are much slower than first-order methods for high-dimensional models such as deep neural networks. Nevertheless, their efficient incorporation in deep learning is still an active area of research. The origins of Newton's method, also known as the *Newton-Raphson method*, can be dated back to 17th century; see chapter 2 of [237] for an account on the history of optimization methods. For a detailed overview see [47].

Even though the secant method can be thought of as an approximation to Newton's method, from an historical perspective, it is several decades older than Newton's method; see for example [321] for origins and an account of the evolution of the secant method. Quasi-Newton methods are more modern with the first technique, the *Davidon-Fletcher-Powell* (DFP) method, appearing in 1959 as a technical report [99] and later in 1991 published as [100]. In the interim, the DFP method was further modified in the early 1960s with [124]. The BFGS method materialized in the 1960s and 1970s, and the L-BFGS extension was introduced in 1980 with [310]. See also [123] for an overview of these methods. See [98] and [266] for a discussion of convergence properties of BFGS and L-BFGS, respectively. Information about the weighted Frobenius norm used for constructing BFGS updates is in chapter 6 of [311]. A related second-order method is the Levenberg–Marquardt algorithm; see chapter 18 of [56] for an introduction. Trust region methods are believed to have originated from [254]; see [439] for a survey. Finally we suggest chapter 5 of [311] as a review of *quasi-Newton methods*.

5 Feedforward Deep Networks

We now enter the heart of deep learning models by presenting key concepts for *feedforward deep neural networks* also known as *general fully connected neural networks*. These models are a powerful extension of shallow neural networks presented previously and are central entities of deep learning. The key mathematical objects used for such networks are highlighted in this chapter. These include layers with neurons, activation function alternatives, and the backpropagation algorithm for gradient evaluation. A universal approximation theorem is stated and further demonstrations highlight the benefit of exploiting deep neural networks for machine learning tasks. Key steps for training a deep neural network are presented. These include weight initialization and batch normalization. We then focus on key strategies for handling overfitting. These are dropout and addition of regularization terms.

In Section 5.1 we present details for the general fully connected architecture. In Section 5.2 we explore the expressive power of neural networks to get a feel for why the general fully connected architecture is a very useful machine learning model. Toward that end, we explore a few theoretical underpinnings for the power of deep learning. In Section 5.3 we introduce activation functions beyond the sigmoid and softmax functions which were already presented in Chapter 3. In Section 5.4 we study the backpropagation algorithm used for gradient evaluation. This algorithm which is a form of backward mode automatic differentiation, is at the core of training deep learning models. In Section 5.5 we study various methods for weight initialization. In Section 5.6 we introduce the idea of batch-normalization. In Section 5.7 we explore methods for mitigating overfitting. These include the addition of regularization terms and the concept of dropout.

5.1 The General Fully Connected Architecture

We refer to the generic deep learning model as *the general fully connected architecture* and also mention that such a model can be called a *fully connected network*, a *feedforward network*, or a *dense network* where each of these terms may also be augmented with the phrases "deep", "neural", and "general". Another name for such a model is a *multi-layer perceptron* (MLP) or a *multi-layer dense network*.

This architecture involves multiple layers, each with non-linear transformations, and is an extension to the shallow neural networks presented in Chapter 3. Schematic illustrations of this architecture are presented in Figure 5.1. In that figure, each circle may be called a *neuron* or *unit* and each vertical set of neurons is a *layer*. A network with a single hidden layer is in (a) and a more general multi-layer network is in (b). Compare these illustrations with Figure 3.3 which has no hidden layers and has a single output (logistic regression), or Figure 3.6 which also has no hidden layers and has multiple outputs (multinomial regression).

Returning to Figure 5.1, the left most layer is called the *input layer* and it has the input features vector x. Each element of this layer is sometimes called an *input neuron* even though

DOI: 10.1201/9781003298687-5

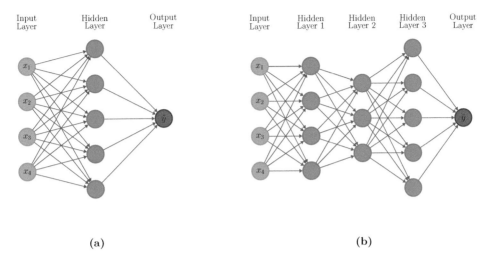

Figure 5.1: Fully connected feedforward neural networks. (a) A network with a single hidden layer. (b) A deep neural network with multiple hidden layers.

it does not involve any computation. The right most or *output layer* contains the output neurons (just one in this example). The layers in the middle are called *hidden layers*, since the neurons in these layers are neither inputs nor outputs. When using the network in production for prediction (regression or classification), we do not directly observe what goes on in the hidden layers, but rather observe network outputs resulting from inputs. However in certain cases, the values of neurons in hidden layers are also called "features" or more precisely, *extracted features* (also known as *computed features* or *derived features*) since they may encode some intermediate summary of the input data.

The design of the input and output layers in a network is often straightforward. There are as many neurons in the input layer as the number of features. As for the output layer, the number of output neurons depends on the application. In certain cases the output can be a scalar, determining either a probability in binary classification, or a response in a regression problem. In other cases, the output may be a vector, such as for example in multi-class classification where there will typically be as many output neurons as the number of possible labels (classes) and the output is a probability vector. In contrast to the input and output layers, when selecting the number and size of the hidden layers, there is much room for variation of model choice.

For example, assume we wish to create a model for classification of a type of plant based on $p = 120$ measured indicators (features). Assume further that our classifier supports 30 different types of plants. Then in this case we may use a model where the input layer has 120 input neurons and the output layer has 30 output neurons which may be interpreted as probabilities, similar to the multinomial regression model of Chapter 3. With this, there still remains a choice of how many hidden layers to use, and how many neurons to use for each of those hidden layers. These choices determine the number of trained parameters, the model expressivity, and the ease/difficulty of training the model.

Finally note that in our terminology, when counting layers in a multi-layer network, we count hidden layers as well as the output layer, but we do not count an input layer. Hence with our terminology, Figure 5.1 (a) has 2 layers, and (b) has 4 layers.

A Model Based on Function Composition

The goal of a feedforward network is to approximate some function $f^* : \mathbb{R}^p \longrightarrow \mathbb{R}^q$. A feedforward network model defines a mapping $f_\theta : \mathbb{R}^p \longrightarrow \mathbb{R}^q$ and learns the value of the parameters θ that ideally result in

$$f_\theta(x) \approx f^*(x).$$

The function f_θ is recursively composed via a chain of functions:

$$f_\theta(x) = f_{\theta^{[L]}}^{[L]}(f_{\theta^{[L-1]}}^{[L-1]}(\ldots(f_{\theta^{[1]}}^{[1]}(x))\ldots)), \qquad (5.1)$$

where $f_{\theta^{[\ell]}}^{[\ell]} : \mathbb{R}^{N_{\ell-1}} \longrightarrow \mathbb{R}^{N_\ell}$ is associated with the ℓ-th layer which depends on parameters $\theta^{[\ell]} \in \Theta^{[\ell]}$, where $\Theta^{[\ell]}$ is the parameter space for the ℓ-th layer. The *depth* of the network is L. We have that $N_0 = p$ (the number of features) and $N_L = q$ (the number of output variables). Note that in case of networks used for classification, we typically have $q = K$, the number of classes. The number of neurons in the network is $\sum_{\ell=1}^{L} N_\ell$.

Affine Transformations Followed by Activations

In deep learning, the function $f_{\theta^{[\ell]}}^{[\ell]}$ is generally defined by an affine transformation followed by an activation function. Activation functions are the means of introducing non-linearity into the model. The output of layer ℓ is represented by the vector $a^{[\ell]}$ and the intermediate result of the affine transformation is represented by the vector $z^{[\ell]}$ (see also Figure 3.3 in Chapter 3). We typically denote the output of the model via $\hat{y}$ and hence,

$$\hat{y} = a^{[L]} = f_\theta(x).$$

The action of $f_{\theta^{[\ell]}}^{[\ell]}$ can be schematically represented as follows:

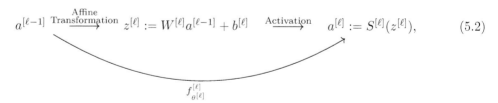

$$a^{[\ell-1]} \xrightarrow{\text{Affine Transformation}} z^{[\ell]} := W^{[\ell]}a^{[\ell-1]} + b^{[\ell]} \xrightarrow{\text{Activation}} a^{[\ell]} := S^{[\ell]}(z^{[\ell]}), \qquad (5.2)$$

where $a^{[0]} = x$. Hence the parameters of the ℓ-th layer, $\theta^{[\ell]}$, are given by the $N_\ell \times N_{\ell-1}$ *weight matrix* $W^{[\ell]} = \left(w_{i,j}^{[\ell]}\right)$ and the N_ℓ dimensional *bias vector* $b^{[\ell]} = (b_i^{[\ell]})$. Thus the parameter space of the layer is $\Theta^{[\ell]} = \mathbb{R}^{N_\ell \times N_{\ell-1}} \times \mathbb{R}^{N_\ell}$.

The activation function $S^{[\ell]} : \mathbb{R}^{N_\ell} \longrightarrow \mathbb{R}^{N_\ell}$ is a non-linear multivalued function. For $\ell = 1, \ldots, L - 1$, it is generally of the form

$$S^{[\ell]}(z) = \left[\sigma^{[\ell]}(z_1) \ \ldots \ \sigma^{[\ell]}(z_{N_\ell}) \right]^\top, \tag{5.3}$$

where $\sigma^{[\ell]} : \mathbb{R} \to \mathbb{R}$ is typically an activation function common to all hidden layers. For the output layer, $\ell = L$, it is often of a different form depending on the task at hand.

In the popular case of multi-class classification, a *softmax* function as (3.25) is used, or more specifically for binary classification, a *sigmoid* function as (3.4) is typically used; see Chapter 3 for background. Thus, in such a classification framework, the output of the network is a vector of probability values determining class membership. In order to get a class label prediction one can convert the predicted probability scores into a class label using a chosen *threshold*, or more simply *maximum a posteriori probability*, as described in (3.34).

The Forward Pass

The *forward pass* equation, (5.1), of a deep neural network can be expanded out as follows:

$$
\begin{aligned}
\text{Layer 1} \quad & \left\{
\begin{aligned}
z^{[1]} &= W^{[1]} \overbrace{x}^{\text{Input}} + b^{[1]} \\
a^{[1]} &= S^{[1]}(z^{[1]})
\end{aligned}
\right. \\[2em]
\text{Layer 2} \quad & \left\{
\begin{aligned}
z^{[2]} &= W^{[2]} a^{[1]} + b^{[2]} \\
a^{[2]} &= S^{[2]}(z^{[2]})
\end{aligned}
\right. \\[1em]
& \qquad \vdots \\[1em]
\text{Layer L} \quad & \left\{
\begin{aligned}
z^{[L]} &= W^{[L]} a^{[L-1]} + b^{[L]} \\
\underbrace{\hat{y}}_{\text{output}} &= a^{[L]} = S^{[L]}(z^{[L]}).
\end{aligned}
\right.
\end{aligned}
\tag{5.4}
$$

Thus, for a given input x, when computing $f_\theta(x)$, we sequentially execute the affine transformations and activation functions from layer 1 to layer L. The computational cost of such a forward pass is at an order of the total number of weights. To see this, observe that at the ℓ-th layer, the cost to compute $z^{[\ell]} = W^{[\ell]} a^{[\ell-1]} + b^{[\ell]}$ and $a^{[\ell]} = S^{[\ell]}(z^{[\ell]})$ are of the order $N_\ell \times N_{\ell-1} + N_\ell$ and N_ℓ, respectively. Hence, the total computational cost is of the order $\sum_{\ell=1}^{L} N_\ell(N_{\ell-1} + 2) \approx \sum_{\ell=1}^{L} N_\ell N_{\ell-1}$, which is the total number of weights.

An Example with Concrete Dimensions

As an illustration, let us return to the networks depicted in Figure 5.1. Consider the network in (a) with one-hidden layer. Here $N_0 = p = 4$, $N_1 = 5$, and $N_2 = q = 1$. Hence the dimension of $W^{[1]}$ is 5×4, the dimension of $b^{[1]}$ is 5, the dimension of $W^{[2]}$ is 1×5, and the dimension of $b^{[2]}$ is 1. Putting these elements together we obtain

$$f_\theta(x) = S^{[2]}(\overbrace{W^{[2]}\underbrace{S^{[1]}(\overbrace{\underbrace{W^{[1]}x + b^{[1]}}_{a^{[1]}}})}^{z^{[1]}} + b^{[2]}}^{z^{[2]}}), \qquad (5.5)$$

where the number of parameters in θ is $5 \times 4 + 5 + 1 \times 5 + 1 = 31$. Similarly, the deeper network on the right of Figure 5.1 is represented via

$$f_\theta(x) = S^{[4]}(W^{[4]}S^{[3]}(W^{[3]}S^{[2]}(W^{[2]}S^{[1]}(W^{[1]}x + b^{[1]}) + b^{[2]}) + b^{[3]}) + b^{[4]}). \qquad (5.6)$$

We may work out that the number of parameters is

$$\underbrace{4 \times 4 + 4}_{\text{Hidden layer 1}} + \underbrace{3 \times 4 + 3}_{\text{Hidden layer 2}} + \underbrace{5 \times 3 + 5}_{\text{Hidden layer 3}} + \underbrace{1 \times 5 + 1}_{\text{Output layer}} = 61.$$

The Scalar-Based View of the Model

It is also instructive to consider the scalar view of the system. The i-th neuron of layer ℓ, with $i = 1, \ldots, N_\ell$, is typically composed of both $z_i^{[\ell]}$ and $a_i^{[\ell]}$. The transition from layer $\ell-1$ to layer ℓ takes the output of layer $\ell-1$, an $N_{\ell-1}$ dimensional vector, and operates on it as follows.

$$\text{Affine}_{\text{Transformation}} : \begin{cases} z_1^{[\ell]} &= {w_{(1)}^{[\ell]}}^\top a^{[\ell-1]} + b_1^{[\ell]} \\ z_2^{[\ell]} &= {w_{(2)}^{[\ell]}}^\top a^{[\ell-1]} + b_2^{[\ell]} \\ &\vdots \\ z_{N_\ell}^{[\ell]} &= {w_{(N_\ell)}^{[\ell]}}^\top a^{[\ell-1]} + b_{N_\ell}^{[\ell]} \end{cases} \Rightarrow \text{Activation}_{\text{Step}} : \begin{cases} a_1^{[\ell]} &= \sigma\left(z_1^{[\ell]}\right) \\ a_2^{[\ell]} &= \sigma\left(z_2^{[\ell]}\right) \\ &\vdots \\ a_{N_\ell}^{[\ell]} &= \sigma\left(z_{N_\ell}^{[\ell]}\right) \end{cases}, \qquad (5.7)$$

where,

$$ {w_{(j)}^{[\ell]}}^\top = \begin{bmatrix} w_{j,1}^{[\ell]} & \cdots & w_{j,N_{\ell-1}}^{[\ell]} \end{bmatrix}, \qquad \text{for} \qquad j = 1, \ldots, N_\ell,$$

is the j-th row of the weight matrix $W^{[\ell]}$ and $b_j^{[\ell]}$ is the j-th element of the bias vector $b^{[\ell]}$. Hence the parameters associated with neuron j in layer ℓ are ${w_{(j)}^{[\ell]}}^\top$ and $b_j^{[\ell]}$.

Vectorizing Across Multiple Samples

So far we have defined our neural network using only one input feature vector x to generate a prediction $\hat{y}$. Namely,

$$x \longrightarrow a^{[L]} = \hat{y}.$$

Let us now consider mini-batches as introduced in Section 4.2. Here we use n_b training samples $x^{(1)}, \ldots, x^{(n_b)}$ where n_b is the size of the mini-batch and $x^{(i)} \in \mathbb{R}^p$. We can use the *feedforward pass* equation to get for each training sample $x^{(i)}$, a prediction

$$x^{(i)} \longrightarrow a^{[L](i)} = \hat{y}^{(i)} \quad i = 1, \dots, n_b.$$

One can clearly iterate using a loop for getting all the predictions. However, a matrix representation is often useful for describing the prediction of the whole mini-batch together. This is particularly useful in GPU implementations when the mini-batch size, n_b, is appropriately chosen to fit GPU memory. Let us first define the matrix $X^\top$ in which every column is a feature vector for one training sample:

$$X^\top = \begin{bmatrix} | & | & & | \\ x^{(1)} & x^{(2)} & \dots & x^{(n_b)} \\ | & | & & | \end{bmatrix}. \tag{5.8}$$

Then, for each ℓ, we define the matrix $Z^{[\ell]}$ with columns $z^{[\ell](1)}, \dots, z^{[\ell](n_b)}$ as

$$Z^{[\ell]} = \begin{bmatrix} | & | & & | \\ z^{[\ell](1)} & z^{[\ell](2)} & \dots & z^{[\ell](n_b)} \\ | & | & & | \end{bmatrix}.$$

The activation matrix $A^{[\ell]}$ for each layer ℓ is defined similarly via

$$A^{[\ell]} = \begin{bmatrix} | & | & & | \\ a^{[\ell](1)} & a^{[\ell](2)} & \dots & a^{[\ell](n_b)} \\ | & | & & | \end{bmatrix},$$

where for example the element in the first row and in the second column of a matrix $A^{[\ell]}$ is an activation of the first hidden unit from the layer ℓ and the second training example. Based on this matrix and the forward pass representation (5.4), we get

$$\begin{cases} Z^{[1]} = & W^{[1]} X^\top + B^{[1]} \\ A^{[1]} = & \sigma^{[1]}(Z^{[1]}) \\ Z^{[2]} = & W^{[2]} A^{[1]} + B^{[2]} \\ A^{[2]} = & \sigma^{[2]}(Z^{[2]}) \\ \quad \vdots \\ Z^{[L]} = & W^{[L]} A^{[L-1]} + B^{[L]} \\ \underbrace{A^{[L]}}_{[\hat{y}^{(1)}, \dots, \hat{y}^{(n_b)}]} = & S^{[L]}(Z^{[L]}) \end{cases} \tag{5.9}$$

where the scalar activation functions, as in (5.3), $\sigma^{[\ell]}(\cdot)$ are applied independently to each element of the matrix $Z^{[\ell]}$ for $\ell = 1, \dots, L-1$, while $S^{[L]}(\cdot)$ is applied independently on each column of $Z^{[L]}$. For this representation, the matrix $B^{[\ell]}$ for each layer ℓ is based on the bias vector $b^{[\ell]}$ and is constructed via

$$B^{[\ell]} = \begin{bmatrix} | & | & & | \\ b^{[\ell]} & b^{[\ell]} & \dots & b^{[\ell]} \\ | & | & & | \end{bmatrix}. \tag{5.10}$$

An Overview of Model Training

Training of feedforward deep neural networks follows the same iterative optimization-based learning paradigm presented in Section 2.4 of Chapter 2. This paradigm was also applied to simple neural networks in Chapter 3. Specifically a standardized training dataset $\mathcal{D} = \{(x^{(1)}, y^{(1)}), \dots, (x^{(n)}, y^{(n)})\}$ is used to find network parameters θ (weights and biases) that minimize a loss function $C(\theta\,;\,\mathcal{D})$. As illustrated in previous chapters, common loss functions include the cross entropy loss (binary or categorical) in the context of classification or the square loss function in the context of regression.

In practice, gradient-based optimization generalizing the gradient descent of Algorithm 2.1 is the typical technique of training. Further, as discussed in detail in Chapter 4, a very common variant is ADAM presented in Algorithm 4.2. In any case, such learning algorithms require the evaluation of gradients, e.g., Step 3 of Algorithm 2.1 or Step 4 of Algorithm 4.2. For deep learning models, such gradient evaluation is carried out via the backpropagation algorithm which we study in detail in Section 5.4.

Another important aspect of iterative optimization involves parameter intialization. This is Step 1 of Algorithm 2.1 or Step 2 of Algorithm 4.2. In the context of deep learning this is called *weight initialization*, a topic we discuss in Section 5.5.

Finally, other aspects of training deep neural networks include batch normalization presented in Section 5.6 and dropout presented in Section 5.7, as well as other methods of regularization also presented in that section.

5.2 The Expressive Power of Neural Networks

Deep neural network models are extremely expressive and versatile. Research about their strength dates all the way back to the early days of artificial intelligence and continues in current times. We now explore the expressivity of such models where we simply touch the tip of the iceberg, attempting to illustrate why the model is sensible and versatile.

Neural networks are known for being able to approximate arbitrarily complex functions. Our exposition in this section aims to illustrate this while also presenting intuition about the benefits of deep models. We begin with a simple constructive example for scalar real-valued functions, and we then progress to present an overview of key results and intuition which highlight why these models are so powerful.

Simple Function Approximation

We now see one possible way to approximate scalar functions via a neural network. Say we have a function $f^* : [\underline{l}, \bar{l}] \to \mathbb{R}$ and we are given the values of the function at $v_i = f^*(u_i)$ for $u_1, \dots, u_r$ with $\underline{l} \le u_1 < u_2 < \dots < u_r \le \bar{l}$.

See for example Figure 5.2 focusing on this arbitrary example,

$$f^*(x) = \cos\left(2\cos\left(x^2\right) + \frac{1}{4}(x-1)^2\right) + \frac{x}{2} + 1, \quad \text{for} \quad x \in [0, 4]. \tag{5.11}$$

Approximation functions obtained for $f^*(\cdot)$ of (5.11) are also illustrated in Figure 5.2 where the case of $r = 10$ is in (b) and the case of $r = 30$ is in (c). Our approximation of $f^*(\cdot)$ uses a feedforward neural network $f_\theta(\cdot)$ as illustrated in (a). In this case we apply a construction of a piecewise continuous affine function $f_\theta : \mathbb{R} \to \mathbb{R}$ with breakpoints at $u_1, \ldots, u_r$ and $f_\theta(u_i) = f^*(u_i)$ for $i = 1, \ldots, r$. For completeness, we now present the details.

Our constructed network has one hidden layer ($L = 2$), and model dimensions $N_0 = 1$, $N_1 = r - 1$, and $N_2 = 1$. In this case, the approximation is obtained by setting the scalar activation functions as in (5.3) to be $\sigma^{[1]}(u) = \max(u, 0)$ and $\sigma^{[2]}(u) = u$, where the former is called the ReLU function, see also (5.17), and the latter is simply the identity function. We set the first $N_1 \times 1$ dimensional weight matrix, $W^{[1]}$, to have all entries 1; the first N_1 dimensional bias vector to have entries $b_i^{[1]} = -u_i$; the second $1 \times N_1$ dimensional weight matrix is set with entries $w_{1,i}^{[2]} = s_i - s_{i-1}$ for $i = 1, \ldots, r-1$ with $s_0 = 0$ and $s_i = \frac{v_{i+1} - v_i}{u_{i+1} - u_i}$; and finally the second scalar bias $b^{[2]}$ is set to be v_1.

In summary, the construction uses a linear combination of shifted ReLU functions (see also Figure 5.6) where each shifted ReLU (shifted by the bias $-u_i$) is constructed to match the interval $[u_i, u_{i+1}]$ with a slope s_i. As the construction moves from left to right, slopes are canceled out via subtraction of the s_{i-1} term in $w_{1,i}^{[2]}$.

With such a construction we see that essentially any continuous real-valued function on a bounded domain can be approximated via a neural network.[1] It is evident that as $r \to \infty$ the approximation becomes exact, and this can be made rigorous for any continuous target function $f^*(\cdot)$ on a bounded interval.

A General Approximation Result

The expressive power of feedforward neural networks generalizes to functions that have p inputs and q outputs. In fact, as the result below shows, similarly to the construction above, a single hidden layer ($L = 2$) and identity activations in the second layer suffices for this.

Theorem 5.1. *Consider a continuous function $f^* : \mathcal{K} \to \mathbb{R}^q$ where $\mathcal{K} \subseteq \mathbb{R}^p$ is a compact set. Then for any non-polynomial activation function $\sigma^{[1]}(\cdot)$ and any $\varepsilon > 0$, there exists an N_1 and parameters $W^{[1]} \in \mathbb{R}^{N_1 \times p}$, $b^{[1]} \in \mathbb{R}^{N_1}$, and $W^{[2]} \in \mathbb{R}^{q \times N_1}$, such that the function*

$$f_\theta(x) = W^{[2]} S^{[1]}(W^{[1]} x + b^{[1]}), \qquad \text{with} \qquad S^{[1]} \quad \text{as in (5.3)},$$

satisfies $\|f_\theta(x) - f^*(x)\| < \varepsilon$ for all $x \in \mathcal{K}$.

Hence this theorem states that essentially all functions can be approximated to arbitrary precision dictated via ε. Practically for complicated functions $f^*(\cdot)$ and small ε one may need large N_1. Yet, the theorem states that it is always possible. A reference to the proof is provided in the notes and references section at the end of this chapter. The constructive example in Figure 5.2 above (for $p = 1$ and $q = 1$) may hint at the validity of the result. Note also that the tanh, sigmoid, and ReLU activation functions described in the next section are some of the valid activation functions for this result.

---

[1]This construction is one of many options one could use.

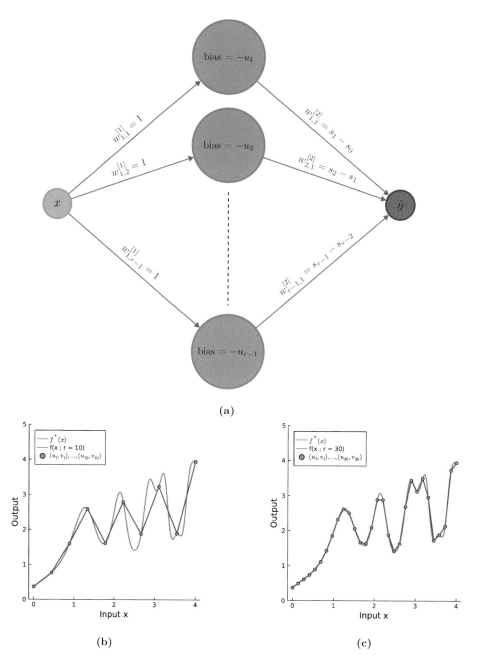

(a)

(b)   **(c)**

Figure 5.2: Approximation of an arbitrary real-valued function on a bounded domain via a piecewise continuous function obtained by a single hidden layer feedforward model with ReLU activation functions. (a) A neural network with one hidden layer that constructs such an approximation. (b) Approximation based on $r = 10$ sampled points of the function. (c) Approximation based on $r = 30$ sampled points of the function.

The Strength of a Hidden Layer

As we saw in Chapter 3, shallow neural networks such as logistic regression or softmax regression can be used to create classifiers with linear decision boundaries. Further, as in Section 3.4, for cases where more general decision boundaries are needed, one may attempt to create additional transformed features while still using these basic models. However, the expressiveness of models with a single hidden layer (or more), as introduced in the current chapter, can yield a versatile alternative to the shallow networks of Chapter 3.

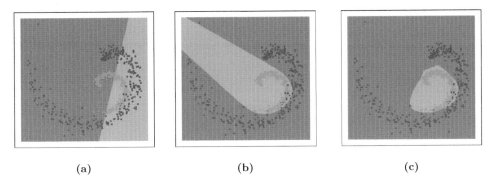

Figure 5.3: Binary classification example with $x \in \mathbb{R}^2$, moving from a shallow neural network to a model with a hidden layer and then increasing the number of neurons. (a) Sigmoid model ($L = 1$). (b) One-hidden layer ($L = 2$) $N_1 = 4$ neurons. (c) One-hidden layer ($L = 2$) $N_1 = 10$ neurons.

Consider Figure 5.3 (a) for a classification task based on two inputs $x \in \mathbb{R}^2$ using logistic regression. By adding a single hidden layer with 4 neurons (*sigmoid* activation function is used for all units), we can move beyond the linear boundaries to obtain Figure 5.3 (b). Then, by increasing the number of neurons in the hidden layer from 4 to 10 units, the model further refines the decision boundary as in Figure 5.3 (c).

We thus see that obtaining non-linear decision boundaries is possible not only via feature engineering as discussed in Section 3.4 and illustrated in Figure 3.9, but can also be obtained via neural networks as hinted by Theorem 5.1 and are illustrated in Figure 5.3.

Stylized Functions via Simple Models

To further appreciate the expressive power of feedforward neural networks we also consider specific stylized functions. For example, historically, in the study of neural networks much effort has gone into characterizing the set of logical functions, sometimes called *gates*, that may be described via certain classes of models. In our exploration we focus on a different type of specific task, *multiplication of two inputs*, and demonstrate a simple constructive network that approximates this task. We thus construct what may be called a *multiplication gate*.

A simple construction of a single hidden layer network with $p = 2$ and $q = 1$, allows us to create a function $f_\theta(\cdot)$, parametrized by $\lambda > 0$, such that for input $x = (x_1, x_2)$, the function approximately implements multiplication of inputs,

$$f_\theta(x_1, x_2) \approx x_1 x_2. \tag{5.12}$$

Importantly, the approximation error vanishes as $\lambda \to 0$ and achieving (5.12) to arbitrary accuracy requires only $N_1 = 4$ neurons in the single hidden layer. Similarly to the model in Theorem 5.1 and to the simple function approximation example of Figure 5.2, the activation function of the output layer is the identity. There are no bias terms, and the weight matrices are

$$W^{[1]} = \begin{bmatrix} \lambda & \lambda \\ -\lambda & -\lambda \\ \lambda & -\lambda \\ -\lambda & \lambda \end{bmatrix}, \qquad \text{and} \qquad W^{[2]} = \begin{bmatrix} \mu & \mu & -\mu & -\mu \end{bmatrix}, \tag{5.13}$$

with $\mu = \left(4\lambda^2 \ddot{\sigma}(0)\right)^{-1}$. Here $\ddot{\sigma}(0)$ represents the second derivative of the scalar activation function of the hidden layer ($\ell = 1$) at 0. Hence the model assumes $\sigma^{[1]}(\cdot)$ is twice differentiable (at 0) with a non-zero second derivative at zero. A schematic representation of $f_\theta(\cdot)$ is in Figure 5.4.

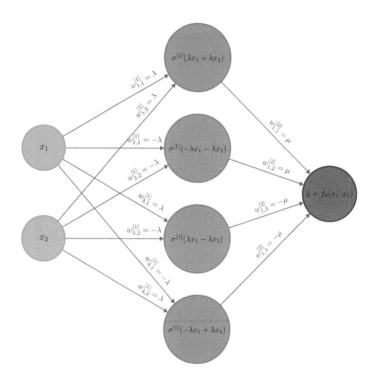

Figure 5.4: A simple neural network model with a single hidden layer composed of 4 neurons. This network approximates multiplication, or a multiplication gate, in the sense that $\hat{y} \approx x_1 x_2$.

We may verify that the model $f_\theta(\cdot)$ evaluates to,

$$f_\theta(x_1, x_2) = \frac{\sigma\left(\lambda(x_1 + x_2)\right) + \sigma\left(\lambda(-x_1 - x_2)\right) - \sigma\left(\lambda(x_1 - x_2)\right) - \sigma\left(\lambda(-x_1 + x_2)\right)}{4\lambda^2 \ddot{\sigma}(0)}, \tag{5.14}$$

where we use $\sigma(\cdot)$ to denote $\sigma^{[1]}(\cdot)$. We may now use a Taylor expansion (see Theorem A.1 in Appendix A) of $\sigma(\cdot)$ around the origin,

$$\sigma(u) = \sigma(0) + \dot{\sigma}(0)u + \ddot{\sigma}(0)\frac{u^2}{2} + O\left(u^3\right), \tag{5.15}$$

with $O(h^k)$ denoting a function such that $O(h^k)/h^k$ goes to a constant as $h \to 0$. We can now use (5.14) and (5.15) to represent the model as

$$f_\theta(x_1, x_2) = x_1 x_2 \left(1 + O\left(\lambda(x_1^2 + x_2^2)\right)\right).$$

Hence as $\lambda \to 0$ the desired goal (5.12) becomes exact. Note that this continuous multiplication gate is mostly a theoretical construct and for many popular activation functions the condition $\ddot{\sigma}(0) = 0$ does not hold. Nevertheless, this problem can be overcome by introducing biases to shift the origin and hence use a different input into the activation. Below we use this construction to further argue about the power of deep learning models.

Feature Focus with Neural Networks

We now contrast the usage of feedforward neural networks with the more classic practice of *feature engineering* where one would consider data and add additional features by transforming existing features. See also Section 3.4 where feature engineering is illustrated for simple neural networks. As an example, we consider a case where there are originally p features $x_1, \ldots, x_p$ and we wish to construct $\tilde{p} = p(p+1)/2$ features based on all possible pairwise interactions[2] (multiplications) $x_i x_j$ for $i, j = 1, \ldots, p$. For instance if $p = 1,000$ then we arrive at $\tilde{p} \approx 500,000$. Clearly for non-small p we quickly arrive at huge number of transformed features $\tilde{p}$.

Let us contrast the usage of two alternatives. On the one hand consider a linear model acting on the transformed features $\tilde{x}$ where for simplicity we ignore the bias (intercept). On the other hand let us consider a neural network with a single hidden layer acting on the original features x. In the linear model, we have $\tilde{f}_\theta(\tilde{x}) = \tilde{w}^\top \tilde{x}$ where the learned weight vector $\tilde{w}$ has $\tilde{p}$ parameters. In the single hidden layer neural network, there are p inputs, $q = 1$ output, and N_1 units in the hidden layer. Thus the number of parameters is $N_1 \times p + N_1 + N_1 + 1$.

It is often the case that not all interactions (product features) are relevant. As an example let us consider that only a fraction α of the interactions are relevant. We now argue that the neural network model with N_1 hidden units sufficiently large, but not necessarily huge, can in principle capture these interactions. To do so, revisit the multiplication example presented above where 4 hidden units were needed to approximate an interaction (multiplication). While the construction using weight parameters as in (5.13) is artificial, the example hints at the fact that with $N_1 \approx 4\alpha p$ hidden units we may be able to capture the key interactions. In such a framework, the basic linear model still requires the full set of parameters. With this, compare the number of parameters,

$$\underbrace{\frac{1}{2}p(p+1)}_{\text{Linear Model}} \qquad \text{vs.} \qquad \underbrace{4\alpha p(p+2) + 1}_{\text{Neural Network}}.$$

---

[2]Typically in statistics interactions refer to terms such as $x_i x_j$ for $i \neq j$. However in this example we also allow for terms such as x_i^2.

Observe that p^2 is the dominant term in both models but for $\alpha < 1/8$ and large p, the neural network model has less parameters. Keep in mind that often α is very small while p may grow. Returning to the numerical case of $p = 1,000$, if $\alpha = 0.02$ (20 significant interactions) then the linear model has an order of 500,000 parameters while the neural network only has order of 80,000 parameters and is thus parisimonious in comparison to the linear model.

Improvement in Expressivity by an Increase of Depth

Despite the fact that Theorem 5.1 states that almost any function can be approximated using a neural network model with a single hidden layer, practice and research has shown that to gain high expressive power, this model might require a very large number of units (N_1 needs to be very large). Hence gaining significant expressive power may require a very large number of parameters. The power of deep learning then arises via repeated composition of non-linear activations functions via an increase of depth (an increase of L).

Note first that if the identity activation function is used in each hidden layer, then the network reduces to a shallow neural network

$$f_\theta(x) = S^{[L]}(\tilde{W}x + \tilde{b}),$$

where,[3]

$$\tilde{W} = W^{[L]}W^{[L-1]} \cdot \ldots \cdot W^{[1]}, \qquad \text{and} \qquad \tilde{b} = \sum_{\ell=1}^{L} \left(\prod_{\tilde{\ell}=\ell+1}^{L} W^{[\tilde{\ell}]} \right) b^{[\ell]}.$$

In the case where the identity function is also used for the output layer, the model reduces to be a linear (affine) model. Thus, we have no gain by going deeper and adding multiple layers with identity activations. The expressivity of the neural network comes from the composition of non-linear activation functions. The repeated compositions of such functions has significant expressive power and can reduce the number of units needed in each layer in comparison to a network with a single hidden layer. A consequence is that the parameter space is reduced as well.

Let us consider an artificial example to demonstrate why exploiting the depth of the network is crucial for modeling complex relationships between input and output. We revisit the previous example involving models using interactions of order 2 (i.e., $x_i x_j$). Let us consider a higher complexity model by exploiting potential high-order interactions, namely products of r inputs, similarly to the discussion in Section 3.4.

Consider a fully connected network with p inputs, $q = 1$ output, and $L \geq 2$ layers with same size ($N^{[\ell]} = N$ in each layer). We re-use the idea appearing in (5.13) that with $N \approx 4 \alpha p$ hidden units we may be able to capture the αp relevant interactions of order $r = 2$. Then by moving forward in the network, the subsequent addition of a layer with $N \approx 4 \alpha p$ hidden units will capture interactions of order $r = 2^2$ and so on, until we capture interaction of order $r = 2^L$ at the output layer. Hence, to achieve interactions of order r we may require $L \approx \log_2 r$, e.g., $L = \lceil \log_2 r \rceil$. Such network is depicted in Figure 5.5 where the number of

---

[3]Note that in the expression $\prod_{\tilde{\ell}=\ell+1}^{L} W^{[\tilde{\ell}]}$ we assume that the product multiplies the matrices in the left to right order $W^{[L]}W^{[L-1]}W^{[L-2]} \ldots$. Further, for $\ell = L$ the product is taken as the identity matrix.

parameters is

$$\underbrace{N \times p + N}_{\text{First hidden layer}} + \underbrace{(L-2) \times (N^2 + N)}_{\text{Inner hidden layers}} + \underbrace{N+1}_{\text{Output layer}} \approx LN^2.$$

For example, assume we wish to have a model for $p = 1{,}000$ features that supports about 20 meaningful interactions of order $r = 500$. Hence we can consider $\alpha = 0.02$. With a model not involving hidden layers (e.g., a linear model or a logistic model), as shown in Section 3.4, we cannot specialize for an order of 20 interactions and thus we require a full model of order $p^r/r! \approx 10^{365}$ parameters. In contrast, with a deep model we require about $L = 10 \approx \log_2 500$ layers and $N = 4\alpha p = 80$ neurons per layer, and thus the total number of parameters is at the order of $LN^2 = 57{,}600$. This deep construction which can capture a desired set of meaningful interactions is clearly more feasible and efficient than the shallow construction of astronomical size.

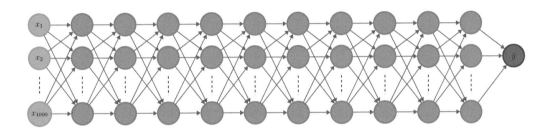

Figure 5.5: A deep learning model with $L = 10$ hidden layers may express many meaningful interactions for the inputs.

5.3 Activation Function Alternatives

As presented in (5.2), each layer ℓ of the network incorporates an activation function which is generally a non-linear transformation of $z^{[\ell]}$ to arrive at $a^{[\ell]}$. For most layers of the network, the activation function $S^{[\ell]}(\cdot)$ is composed of a sequence of identical scalar valued activation functions as in (5.3). That is, the ℓ-th layer incorporates a scalar activation function $\sigma^{[\ell]} : \mathbb{R} \to \mathbb{R}$ and it is applied to each of the N_ℓ coordinates of $z^{[\ell]}$ separately. In some models, all the scalar activation functions across all layers will be of the same form, while in other models different layers will sometimes incorporate different forms of scalar activation functions.

Scalar Activations and their Derivatives

When it comes to the choice of the scalar activation functions, there are different heuristic considerations that depend on model expressivity and learning ability. These considerations are often not based on theory, yet practice and experience over the years has shown that some scalar activation functions perform much better than others. In Section 5.5 we outline how such choices interface with optimization and the associate vanishing gradient problem is discussed in Section 5.5. Here we simply outline the key activation functions.

At the onset of the development of deep learning, in the late 1950s, the *step* scalar activation function was used. That is,

$$\sigma_{\text{Step}}(u) = \begin{cases} -1 & u < 0, \\ +1 & u \geq 0. \end{cases} \tag{5.16}$$

However, σ_{Step} is not used in modern neural network models, primarily because its derivative $\dot{\sigma}_{\text{Step}}(u) = 0$ for all $u \neq 0$. Indeed the derivative of activation functions is important since it is used in the backpropagation algorithm (see the next section) to compute the gradient of the loss function with respect to the model parameters.

The *sigmoid* activation function (also known as the *logistic* function), used heavily in Chapter 3, see (3.4), is a much more popular choice. Note also that its derivative $\dot{\sigma}_{\text{Sig}}$ can be expressed in terms of the function σ_{Sig} itself,

$$\sigma_{\text{Sig}}(u) = \frac{e^u}{1+e^u} = \frac{1}{1+e^{-u}}, \qquad \text{with} \qquad \dot{\sigma}_{\text{Sig}}(u) = \sigma_{\text{Sig}}(u)\big(1-\sigma_{\text{Sig}}(u)\big).$$

A similar popular function is *hyperbolic tangent*, denoted *tanh*,

$$\sigma_{\text{Tanh}}(u) = \frac{e^u - e^{-u}}{e^u + e^{-u}}, \qquad \text{with} \qquad \dot{\sigma}_{\text{Tanh}}(u) = 1 - \sigma_{\text{Tanh}}(u)^2.$$

Both σ_{Sig} and σ_{Tanh} share similar qualitative properties as σ_{Step}. They are non-decreasing and bounded. At $u \to \infty$ both functions converge to unity just like σ_{Step} and at $u \to -\infty$ the sigmoid function converges to 0 while the tanh function converges to -1 like σ_{Step}. This minor quantitative difference is not significant as it may generally be compensated by learned weights and biases or a shifting and scaling of the functions, yet in practice when the output $\hat{y}$ is a probability in $[0,1]$ using σ_{Sig} is much more common. See Figure 5.6 for plots of several activation functions.

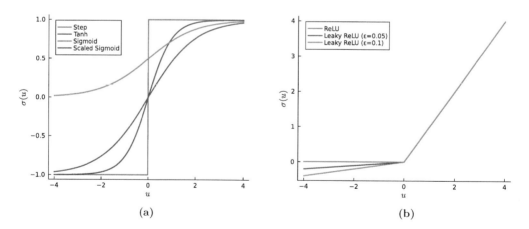

Figure 5.6: Several common scalar activation functions. (a) The step and tanh function have a range of $(-1,1)$ while the sigmoid function has a range of $(0,1)$. We also plot a scaled sigmoid to the range $(-1,1)$ so it can be compared to tanh. (b) The ReLU activation function and the leaky ReLU variant with different leaky ReLU parameters.

In earlier applications of deep learning, it was a matter of empirical research, practice, and heuristics to choose between models or layers that use σ_{Sig}, σ_{Tanh}, or similar forms. However in more recent years, a completely different type of scalar activation function became popular, namely the *rectified linear unit* [4] or *ReLU*,

$$\sigma_{\text{ReLU}}(u) = \max(0, u) = \begin{cases} 0 & u < 0, \\ u & u \geq 0, \end{cases} \qquad \text{with,} \qquad (5.17)$$

$$\dot{\sigma}_{\text{ReLU}}(u) = \mathbf{1}\{u \geq 0\} = \begin{cases} 0 & u < 0, \\ 1 & u \geq 0. \end{cases}$$

Note that while $\sigma_{\text{ReLU}}(u)$ is not differentiable at $u = 0$ we still arbitrarily define the derivative at $u = 0$ to be the same as the derivative for $u > 0$.

As we describe in Section 5.5, when parameters are initialized properly, the unboundedness of σ_{ReLU} often presents an advantage in training over bounded activation functions such as σ_{Sig} since it overcomes a training problem known as the vanishing gradient problem. However, in certain cases, it also introduces a problem called *dying ReLU* since the derivative is 0 for negative inputs. While in practice, this is often not considered a major problem, to handle dying ReLU, one may use a related activation function, *leaky ReLU*, parameterized by a fixed small $\varepsilon \geq 0$ (e.g., $\varepsilon = 0.01$) and defined via,

$$\sigma_{\text{LeakyReLU}}(u) = \max(0, u) + \min(0, \varepsilon u) = \begin{cases} \varepsilon u & u < 0, \\ u & u \geq 0, \end{cases} \qquad \text{with,}$$

$$\dot{\sigma}_{\text{LeakyReLU}}(u) = \mathbf{1}\{u \geq 0\} + \varepsilon\mathbf{1}\{u < 0\} = \begin{cases} \varepsilon & u < 0, \\ 1 & u \geq 0. \end{cases}$$

Observe that when $\varepsilon = 0$, this is just the ReLU activation function. Another variant called *PReLU* (parametric ReLU) considers the leaky ReLU parameter ε as a learned parameter. That is, the gradient-based optimization for the parameters of the network also includes improvement steps of ε, incorporating it as part of the parameters for ℓ-th layer, $\theta^{[\ell]}$.

Note that feedforward models that only use piecewise affine scalar activation functions such as the step, ReLU, leaky ReLU, or PReLU have some mathematical appeal. When only such activations are used in a deep learning model, these functions imply that the complete model can be described as a *piecewise affine function* of the input. That is, the trained model implicitly partitions the input space $\mathbb{R}^p$ into polytopes (regions defined by intersections of half spaces), and for each polytope, when the input x is in that polytope, the response of the model is a fixed affine transformation of x. This property is simply a consequence of the fact that compositions of piecewise affine functions remain a piecewise affine function. There are no immediate practical applications for this property but it further hints at the expressive power of neural network models further to the description in Section 5.2.

Also note that many other forms of scalar activation functions have been introduced and experimented with. These include the *arctan, softsign, softplus, elu, selu,* and *swish* among others. Figure 5.6 presents the key activation functions σ_{Step}, σ_{Sig}, σ_{Tanh}, σ_{ReLU}, and $\sigma_{\text{LeakyReLU}}$ (with $\epsilon = 0.01$), along with a few of these alternatives.

---

[4] Note that source of the name is from electrical engineering where a rectifier is a device that converts alternating current to direct current.

Non-Scalar Activations and their Derivatives

Some layers also use non-scalar activation functions. That is, $S^{[\ell]} : \mathbb{R}^{N_\ell} \to \mathbb{R}^{N_\ell}$ is a vector to vector function that cannot be decomposed as in (5.3). The most common example of this is the softmax activation function, typically used for classification in the last layer $\ell = L$. We now denote $N_L = K$ since we often deal with classification of K classes as in Section 3.3. In such a case,

$$a^{[L]} = S_{\text{Softmax}}(z^{[L]}). \tag{5.18}$$

The $\mathbb{R}^K \to \mathbb{R}^K$ softmax activation function is defined as

$$S_{\text{Softmax}}(z) = \frac{1}{\sum_{i=1}^{K} e^{z_i}} \begin{bmatrix} e^{z_1} & \cdots & e^{z_K} \end{bmatrix}^\top,$$

which was also defined in (3.25) of Section 3.3. Note that other examples of non-scalar activations include max pooling layers, introduced in Section 6.4 of the next chapter.

As will become evident in the next section, denoting the training loss by C, the gradient $\partial C / \partial z^{[L]}$ needs to be evaluated as part of the backpropagation algorithm. One approach for this evaluation is

$$\frac{\partial C}{\partial z^{[L]}} = \underbrace{\frac{\partial a^{[L]}}{\partial z^{[L]}}}_{N_L \times N_L} \underbrace{\frac{\partial C}{\partial a^{[L]}}}_{N_L \times 1},$$

where we use the multivariate chain rule (see also Appendix A) and keep in mind that $a^{[L]}$ is a vector, $z^{[L]}$ is a vector, and C is a scalar. With this approach, when the last layer is a softmax as (5.18), one may essentially need to evaluate $\partial a^{[L]} / \partial z^{[L]}$ which is the transpose of the Jacobian of $S_{\text{Softmax}}(\cdot)$. The elements of this Jacobian can be represented using the (scalar) quotient rule for differentiation, and are,

$$[J_{S_{\text{Softmax}}}(z)]_{ij} = \frac{\partial}{\partial z_j} \frac{e^{z_i}}{\sum_{k=1}^{K} e^{z_k}} = \begin{cases} [S_{\text{Softmax}}(z)]_i (1 - [S_{\text{Softmax}}(z)]_i), & i = j, \\ -[S_{\text{Softmax}}(z)]_i [S_{\text{Softmax}}(z)]_j, & i \neq j. \end{cases} \tag{5.19}$$

However, one rarely uses (5.19) since in the typical case where the loss function is the cross entropy loss (see also Section 3.3), we have a direct expression for $\partial C / \partial z^{[L]}$. To see, this suppose the label y equals k, an element from $\{1, \ldots, K\}$. In this case, as was shown in (3.29) from Section 3.3, the loss for a specific observation is

$$C = -\log [S_{\text{Softmax}}(z^{[L]})]_k = \log \sum_{i=1}^{K} e^{z_i^{[L]}} - z_k^{[L]}.$$

Now to obtain $\partial C / \partial z^{[L]}$ we compute the derivative with respect to j for every $j = 1, \ldots, K$. This yields

$$\frac{\partial C}{\partial z_j^{[L]}} = \frac{e^{z_j^{[L]}}}{\sum_{i=1}^{K} e^{z_i^{[L]}}} - \mathbf{1}\{j = k\} = [S_{\text{Softmax}}(z^{[L]})]_j - \mathbf{1}\{j = k\}.$$

Hence in this case the direct expression is

$$\frac{\partial C}{\partial z^{[L]}} = S_{\text{Softmax}}(z) - e_k, \tag{5.20}$$

where e_k is the K-dimensional vector with 1 in the k-th position and 0 in other coordinates. Note that in practice when using deep learning frameworks, it is often recommended to use (5.20) directly in such a case.

5.4 The Backpropagation Algorithm

Now that we understand the model, we focus on gradient computation so as to facilitate learning using variants of gradient descent as covered in Chapter 4. A key algorithm is the *backpropagation algorithm* which implements backward mode automatic differentiation which is overviewed in Section 4.4 of Chapter 4. Now we build the related backpropagation algorithm of deep learning. We start with a general recursive model and then to specialize in feedforward neural networks where the parameters are weights and biases.

Backpropagation for the General Recursive Model

It is instructive to first consider a general recursive feedforward model as appearing in (5.1). For such a model, the recursive step is of the form $a^{[\ell]} = f_{\theta^{[\ell]}}^{[\ell]}(a^{[\ell-1]})$. However, in this section it is convenient to use notation that treats the function $f^{[\ell]}$ separately as a function of $a^{[\ell-1]}$ and of the parameter $\theta^{[\ell]}$:

$$f^{[\ell]}(\cdot\,;\,\theta^{[\ell]}) : \mathbb{R}^{N_{\ell-1}} \longrightarrow \mathbb{R}^{N_\ell}, \qquad \text{and} \qquad f^{[\ell]}(a^{[\ell-1]}\,;\,\cdot) : \Theta^{[\ell]} \longrightarrow \mathbb{R}^{N_\ell}.$$

Using this notation, the recursive step is

$$a^{[\ell]} = f^{[\ell]}(a^{[\ell-1]}\,;\,\theta^{[\ell]}), \qquad \text{for} \qquad \ell = 1, \ldots, L, \tag{5.21}$$

where $a^{[0]} = x$ and $\hat{y} = a^{[L]}$. Given a single data sample (x, y) we assume there is a loss function which depends on the given parameters θ, on the label value y, and on the output of the model, $a^{[L]}$. We denote this loss via $C(a^{[L]}, y\,;\,\theta)$.

Our purpose is to optimize the loss with respect to $\theta = (\theta^{[1]}, \ldots, \theta^{[L]})$. For this, we require the gradient with respect to θ and we denote its components via

$$g_\theta^{[\ell]} := \frac{\partial C(a^{[L]}, y\,;\,\theta)}{\partial \theta^{[\ell]}}. \tag{5.22}$$

A key aspect of the automatic differentiation setup is that we are presented with computable expressions or code, for evaluation of certain derivatives. In our context these are the derivative of the loss C with respect to $a^{[L]}$ and the derivatives of $f^{[\ell]}(\cdot\,;\,\cdot)$ with respect to the input arguments (both layer input and parameters). These known derivatives are denoted via

$$\dot{C}(u) := \frac{\partial C(u, y\,;\,\theta)}{\partial u}, \quad \dot{f}_a^{[\ell]}(u) := \frac{\partial f^{[\ell]}(u\,;\,\theta^{[\ell]})}{\partial u}, \quad \dot{f}_\theta^{[\ell]}(u) := \frac{\partial f^{[\ell]}(a^{[\ell-1]}\,;\,u)}{\partial u}. \tag{5.23}$$

Note that the shape of these expressions varies based on the domain and co-domain of $C(\cdot)$ and $f(\cdot\,;\,\cdot)$. The derivative $\dot{C}(u)$ is typically a gradient and thus vector is valued with length N_L. The derivative $\dot{f}_a^{[\ell]}(u)$ is typically an $N_{\ell-1} \times N_\ell$ matrix obtained via a transpose of a Jacobian and thus matrix valued. This is because the input to the layer is $N_{\ell-1}$ dimensional and the output argument is N_ℓ dimensional. Finally, the derivative $\dot{f}_\theta^{[\ell]}(u)$ may take on various shapes depending on the form of θ. See Appendix A for a review of matrix derivatives.

On route to compute the desired gradients (5.22) we require intermediate gradients of the loss with respect to the activation values $a^{[1]}, \ldots, a^{[L]}$. Keeping in mind that $a^{[L]} = \hat{y}$ is a function of these activation values, the intermediate gradients[5] are denoted by

$$\zeta^{[\ell]} := \frac{\partial C(a^{[L]}, y\,;\,\theta)}{\partial a^{[\ell]}}, \qquad \ell = 1, \ldots, L. \tag{5.24}$$

Now based on the multivariate chain rule, the recursive step (5.21), and the definitions above, we observe

$$\zeta^{[\ell]} = \frac{\partial a^{[\ell+1]}}{\partial a^{[\ell]}} \frac{\partial C}{\partial a^{[\ell+1]}} = \dot{f}_a^{[\ell+1]}(a^{[\ell]})\,\zeta^{[\ell+1]} \text{ and } \qquad g_\theta^{[\ell]} = \frac{\partial a^{[\ell]}}{\partial \theta^{[\ell]}} \frac{\partial C}{\partial a^{[\ell]}} = \dot{f}_\theta^{[\ell]}(\theta^{[\ell]})\,\zeta^{[\ell]}. \tag{5.25}$$

Note that in the application of the chain rule above, we use the notation specified in Appendix A. Hence once the activation values $a^{[1]}, \ldots, a^{[L]}$ are populated via forward propagation of (5.21), backward computation can be carried out via

$$\zeta^{[\ell]} = \begin{cases} \dot{C}(a^{[L]}), & \ell = L, \\ \dot{f}_a^{[\ell+1]}(a^{[\ell]})\,\zeta^{[\ell+1]}, & \ell = L-1, \ldots, 1, \end{cases} \tag{5.26}$$

and at each step the gradient can be obtained via $g_\theta^{[\ell]} = \dot{f}_\theta^{[\ell]}(\theta^{[\ell]})\,\zeta^{[\ell]}$.

This process is summarized in Algorithm 5.1. See also Figure 5.7 for an illustration of the flow of information.

Algorithm 5.1: Backpropagation for the general recursive model

Input: Dataset $\mathcal{D} = \{(x^{(1)}, y^{(1)}), \ldots, (x^{(n)}, y^{(n)})\}$,
objective function $C(\cdot) = C(\cdot\,;\mathcal{D})$, and
parameter values $\theta = (\theta^{[1]}, \ldots, \theta^{[L]})$
Output: gradients of the loss $g_\theta^{[1]}, \ldots, g_\theta^{[L]}$
1 Compute $a^{[\ell]}$ for $\ell = 1, \ldots, L$ using (5.21) (Forward pass)
2 Compute $\zeta^{[L]} = \dot{C}(a^{[L]})$
3 Compute $g_\theta^{[L]} = \dot{f}_\theta^{[L]}(\theta^{[L]})\,\zeta^{[L]}$
4 **for** $\ell - L - 1, \ldots, 1$ **do**
5 $\quad$ compute $\zeta^{[\ell]} = \dot{f}_a^{[\ell+1]}(a^{[\ell]})\,\zeta^{[\ell+1]}$
6 $\quad$ compute $g_\theta^{[\ell]} = \dot{f}_\theta^{[\ell]}(\theta^{[\ell]})\,\zeta^{[\ell]}$

An Unrolled Example

To get a better feel for Algorithm 5.1, the operation of backpropagation, and the associated notation, we consider a simple hypothetical example as illustrated in Figure 5.8. This is a feedforward neural network with $L = 4$, arbitrary input dimension p, output dimension $q = 1$, and a single neuron on each hidden layer. That is $N_0 = p$ and $N_1, N_2, N_3, N_4 = 1$. Assume the loss function is

$$C(a^{[L]}, y\,;\,\theta) = \frac{1}{2}(a^{[L]} - y)^2,$$

---

[5]These are also called adjoints. See (4.30) in Chapter 4.

Parameter Values

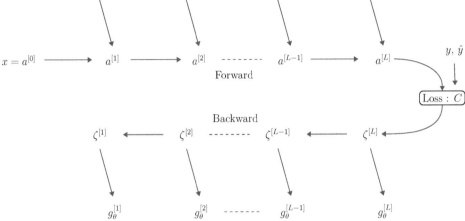

Gradient Values

Figure 5.7: The variables and flow of information in the backpropagation algorithm for the general recursive model.

and assume a model structure

$$a^{[\ell]} = f^{[\ell]}(a^{[\ell-1]}; \theta^{[\ell]}) = \sigma^{[\ell]}(\theta^{[\ell]^\top} a^{[\ell-1]}) \qquad \text{for} \qquad \ell = 1, 2, 3, 4, \qquad (5.27)$$

similar to the neural network structure of Section 5.1 where affine transformations are followed by activation functions, yet without bias terms. Since $N_0 = p$, we have that $\Theta^{[1]} = \mathbb{R}^p$ and since $N_1, \ldots, N_4$ are all unity, we have that $\Theta^{[2]}, \Theta^{[3]}$, and $\Theta^{[4]}$ are each $\mathbb{R}$ and the transpose in (5.27) is not needed for $\ell = 2, 3, 4$.

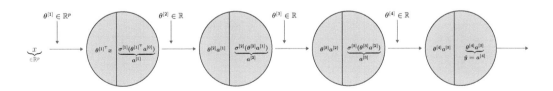

Figure 5.8: A simple hypothetical example with $L = 4$ and scalar hidden units.

In this hypothetical illustrative example, we use various activation functions. Namely,

$$\sigma^{[1]}(\cdot) = \sigma_{\text{ReLU}}(\cdot), \qquad \sigma^{[2]}(\cdot) = \sigma_{\text{Tanh}}(\cdot), \qquad \sigma^{[3]}(\cdot) = \sigma_{\text{Sig}}(\cdot),$$

and for the last step, we use the identity function, i.e., $\sigma^{[4]}(u) = u$. With the model specified we now have computable expressions for known derivatives as in (5.23), see also Section 5.3,

$$
\begin{aligned}
\dot{C}(u) &= u - y, & \dot{f}_\theta^{[4]}(u) &= a^{[3]}, \\[2mm]
\dot{f}_a^{[4]}(u) &= \theta^{[4]}, & \dot{f}_\theta^{[3]}(u) &= a^{[2]} \underbrace{\sigma_{\mathrm{Sig}}(ua^{[2]})(1 - \sigma_{\mathrm{Sig}}(ua^{[2]}))}_{\dot{\sigma}_{\mathrm{Sig}}(ua^{[2]})}, \\[2mm]
\dot{f}_a^{[3]}(u) &= \theta^{[3]}\dot{\sigma}_{\mathrm{Sig}}(u\theta^{[3]}), & \dot{f}_\theta^{[2]}(u) &= a^{[1]} \underbrace{(1 - \sigma_{\mathrm{Tanh}}(ua^{[1]})^2)}_{\dot{\sigma}_{\mathrm{Tanh}}(ua^{[1]})}, \\[2mm]
\dot{f}_a^{[2]}(u) &= \theta^{[2]}\dot{\sigma}_{\mathrm{Tanh}}(u\theta^{[2]}), & \dot{f}_\theta^{[1]}(u) &= a^{[0]} \underbrace{\mathbf{1}\{u^T a^{[0]} \geq 0\}}_{\dot{\sigma}_{\mathrm{ReLU}}(u^T a^{[0]})}.
\end{aligned} \tag{5.28}
$$

Note that in this example we choose to treat $\dot{f}_\theta^{[1]}(u)$ as a p-dimensional vector, while the other functions in this example are scalar-valued both in their domain and co-domain.

Using these computable expressions for the derivatives, Algorithm 5.1 can be used to compute the gradient of the loss function with respect to θ. For illustration we unroll the algorithm where we use the notation $u \leftarrow v$ to indicate assignment of v to the variable u.

Step 1: We compute forward propagation,

$$
\begin{aligned}
a^{[1]} &\leftarrow \sigma_{\mathrm{ReLU}}(x^\top \theta^{[1]}), \\
a^{[2]} &\leftarrow \sigma_{\mathrm{Tanh}}(\theta^{[2]} a^{[1]}), \\
a^{[3]} &\leftarrow \sigma_{\mathrm{Sig}}(\theta^{[3]} a^{[2]}), \\
\hat{y} = a^{[4]} &\leftarrow \theta^{[4]} a^{[3]}.
\end{aligned}
$$

Steps 2 and 3:

$$
\zeta^{[4]} \leftarrow a^{[4]} - y, \qquad g_\theta^{[4]} \leftarrow a^{[3]}\zeta^{[4]}.
$$

Then the loop in steps 4–6 executes for three iterations for $\ell = 3, 2, 1$:

Iteration $\ell = 3$ (steps 5 and 6):

$$
\zeta^{[3]} \leftarrow \theta^{[4]}\zeta^{[4]}, \qquad g_\theta^{[3]} \leftarrow a^{[2]}\dot{\sigma}_{\mathrm{Sig}}(\theta^{[3]} a^{[2]})\zeta^{[3]}.
$$

Iteration $\ell = 2$ (steps 5 and 6):

$$
\zeta^{[2]} \leftarrow \theta^{[3]}\dot{\sigma}_{\mathrm{Sig}}(\theta^{[3]} a^{[2]})\zeta^{[3]}, \qquad g_\theta^{[2]} \leftarrow a^{[1]}\dot{\sigma}_{\mathrm{Tanh}}(\theta^{[2]} a^{[1]})\zeta^{[2]}.
$$

Iteration $\ell = 1$ (steps 5 and 6):

$$
\zeta^{[1]} \leftarrow \theta^{[2]}\dot{\sigma}_{\mathrm{Tanh}}(\theta^{[2]} a^{[1]})\zeta^{[2]}, \qquad g_\theta^{[1]} \leftarrow x\,\dot{\sigma}_{\mathrm{ReLU}}(x^\top \theta^{[1]})\zeta^{[1]}.
$$

By expanding out the resulting expressions and using the chain rule, we may verify that each of the intermediate outputs $g_\theta^{[4]}, g_\theta^{[3]}, g_\theta^{[2]}$, and $g_\theta^{[1]}$ yields the correct gradient expression.

Accounting for $\delta^{[\ell]}$ Instead of $\zeta^{[\ell]}$

Algorithm 5.1 summarizes backpropagation since it deals with gradient evaluation for the general recursive model (5.21). However, it does not make any use of the more specific structure of (5.2) which has an affine transformation followed by a non-linear activation function. It turns out that in the context of deep learning models as in (5.2), one can simplify the algorithm by keeping track of an alternative set of intermediate derivative values, $\delta^{[\ell]} \in \mathbb{R}^{N_\ell}$, instead of $\zeta^{[\ell]}$. These are,

$$\delta^{[\ell]} := \frac{\partial C(a^{[L]}, y; \theta)}{\partial z^{[\ell]}}, \qquad \ell = 1, \ldots, L,$$

where we observe that in contrast to the $\zeta^{[\ell]}$ values from above, these values are derivatives of the loss with respect to $z^{[\ell]}$ values instead of with respect $a^{[\ell]}$ values. Usage of $\delta^{[\ell]}$ is standard in backpropagation for deep learning and serves as the basis for the main backpropagation algorithm that we present below. With this notation, the key recursive relationship for backward computation is

$$\delta^{[\ell]} = \frac{\partial a^{[\ell]}}{\partial z^{[\ell]}} \frac{\partial z^{[\ell+1]}}{\partial a^{[\ell]}} \frac{\partial C}{\partial z^{[\ell+1]}} = \frac{\partial a^{[\ell]}}{\partial z^{[\ell]}} \frac{\partial z^{[\ell+1]}}{\partial a^{[\ell]}} \delta^{[\ell+1]}, \qquad \ell = L-1, \ldots, 1, \qquad (5.29)$$

which in comparison to the left hand side of (5.25) breaks up the step $a^{[\ell]} \Rightarrow a^{[\ell+1]}$ into two steps: $a^{[\ell]} \Rightarrow z^{[\ell+1]}$ followed by $z^{[\ell+1]} \Rightarrow a^{[\ell+1]}$. Further, for the final layer, $\ell = L$, we have,

$$\delta^{[L]} = \frac{\partial a^{[L]}}{\partial z^{[L]}} \frac{\partial C}{\partial a^{[L]}}. \qquad (5.30)$$

Backpropagation for Fully Connected Networks

We now expand and adapt Algorithm 5.1 using the key recursive relationships for $\delta^{[\ell]}$ (5.29) and (5.30). Consider first the component $\partial a^{[\ell]}/\partial z^{[\ell]}$ associated with the (vector) activation function $S^{[\ell]}(\cdot)$. In a typical layer ℓ, the vector activation function is composed of identical scalar activation functions as in (5.3) and hence the transposed Jacobian (see also (A.10) in Appendix A) is

$$\frac{\partial a^{[\ell]}}{\partial z^{[\ell]}} = J_{S^{[\ell]}}(z^{[\ell]})^\top = \mathrm{Diag}\big(\dot{\sigma}^{[\ell]}(z^{[\ell]})\big). \qquad (5.31)$$

Here $\mathrm{Diag}(\cdot)$ transforms a vector into a diagonal matrix and $\dot{\sigma}^{[\ell]}(\cdot)$ is the derivative of the scalar activation which is interpreted as operating element-wise on the input vector. Examples of scalar activation functions and their derivatives are in Section 5.3.

In other cases, $S^{[\ell]}(\cdot)$ is not separable as in (5.3) and the transposed Jacobian of $S^{[\ell]}$ does not have a simple diagonal form as in (5.31). One such potential Jacobian is computed for the softmax in (5.19). However, the most common application of softmax is in the last layer $\ell = L$ together with cross entropy loss. In such a case, using (5.20) directly in place of (5.30) is more efficient and practical.

Continuing now with the recursive relationship (5.29), we consider the component $\partial z^{[\ell+1]}/\partial a^{[\ell]}$. Here, since $z^{[\ell+1]} = W^{[\ell+1]}a^{[\ell]} + b^{[\ell+1]}$, we have

$$\frac{\partial z^{[\ell+1]}}{\partial a^{[\ell]}} = W^{[\ell+1]^\top}.$$

Hence, similarly to (5.26), putting the pieces together we have the recursive relationship,

$$\delta^{[\ell]} = \begin{cases} \frac{\partial C}{\partial z^{[L]}}, & \ell = L, \\ \mathrm{Diag}\big(\dot{\sigma}^{[\ell]}(z^{[\ell]})\big) W^{[\ell+1]\top} \delta^{[\ell+1]}, & \ell = L-1, \dots, 1. \end{cases} \tag{5.32}$$

As discussed above, the case $\ell = L$ can be computed using the two components of (5.30) separately or can be computed according to (5.20) in case of softmax combined with cross entropy loss.

Now with a recursive relationship supporting backpropagation of the $\delta^{[\ell]}$ values, we are ready to deal with the desired derivative values of the parameters $\theta^{[\ell]} = (W^{[\ell]}, b^{[\ell]})$. Define

$$g_W^{[\ell]} = \frac{\partial C}{\partial W^{[\ell]}} \quad \text{and} \quad g_b^{[\ell]} = \frac{\partial C}{\partial b^{[\ell]}}, \qquad \ell = 1, \dots, L.$$

Here $g_W^{[\ell]}$ is an $N_\ell \times N_{\ell-1}$ matrix and $g_b^{[\ell]}$ is an N_ℓ-vector. Paralleling the right hand equation in (5.25) (which involves $\zeta^{[\ell]}$ for the more general model), we seek equations to retrieve these target values (gradient components) in terms of $\delta^{[\ell]}$. This is done via

$$g_W^{[\ell]} = \frac{\partial C}{\partial z^{[\ell]}} \frac{\partial z^{[\ell]}}{\partial W^{[\ell]}} = \delta^{[\ell]} a^{[\ell-1]\top} \quad \text{and} \quad g_b^{[\ell]} = \frac{\partial z^{[\ell]}}{\partial b^{[\ell]}} \frac{\partial C}{\partial z^{[\ell]}} = \delta^{[\ell]}. \tag{5.33}$$

where the first expression results from (A.15) and the second expression from (A.14), both in Appendix A.

All of these relationships are packaged in the backpropagation algorithm for fully connected networks. See also Figure 5.9 for an illustration of the flow of information.

Algorithm 5.2: Backpropagation for fully connected networks

Input: Dataset $\mathcal{D} = \{(x^{(1)}, y^{(1)}), \dots, (x^{(n)}, y^{(n)})\}$, objective function $C(\cdot) = C(\cdot\,; \mathcal{D})$, and parameter values $\theta = (\theta^{[1]}, \dots, \theta^{[L]})$

Output: derivatives of the loss $(g_W^{[1]}, g_b^{[1]}), \dots, (g_W^{[L]}, g_b^{[L]})$

1   Compute $a^{[\ell]}$ and $z^{[\ell]}$ for $\ell = 1, \dots, L$ (Forward pass)

2   Compute $\delta^{[L]} = \frac{\partial C}{\partial z^{[L]}}$

3   Set $g_W^{[L]} = \delta^{[L]} a^{[L-1]\top}$ and set $g_b^{[L]} = \delta^{[L]}$

4   **for** $\ell = L-1 \dots, 1$ **do**

5   Compute $\delta^{[\ell]} = \mathrm{Diag}\big(\dot{\sigma}^{[\ell]}(z^{[\ell]})\big) W^{[\ell+1]\top} \delta^{[\ell+1]}$

6   Set $g_W^{[\ell]} = \delta^{[\ell]} a^{[\ell-1]\top}$ and set $g_b^{[\ell]} = \delta^{[\ell]}$

Backpropagation on a Whole Mini Batch

In Section 4.2 we discussed the concept of iterative optimization using mini-batches. With this approach, instead of considering all of the data points, we only use n_b samples and use the gradient $\sum_i \nabla C_i / n_b$, summed over the samples in the mini-batch i. See also equation (4.13) in Chapter 4.

Parameter Values

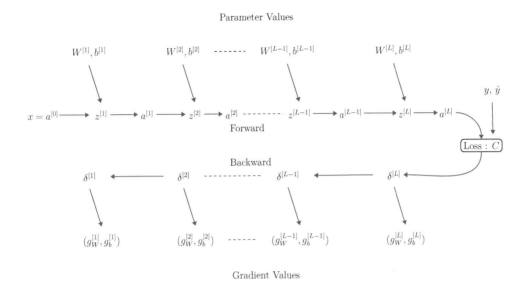

Figure 5.9: The variables and flow of information in the backpropagation algorithm for fully connected networks.

In fact, earlier in this chapter, in (5.9), we introduced notation for executing the forward pass simultaneously on a whole mini-batch. The key here is to properly organize the activation values and the data values in matrices. This notation can be further extended to adapt the backpropagation algorithm for computation of the mini-batch gradient. Hence one may also specify a form of Algorithm 5.2, suitable for mini-batches.

While specific implementation details are not our focus, we mention that in most practical deep learning frameworks, the common interface for gradient evaluation using backpropagation is based on input tensors, where each tensor is associated with a mini-batch. Then one of the indices of the tensor indexes specific data samples within the mini-batch. For example common formats for image data are based on four-dimensional tensors, with a format referred to as *NHWC* or *NCHW*. Here "N" represents the index within the mini-batch, while "H", "W", and "C" are for height, width, and channels, respectively, as appropriate for color image data. Note that the concept of channels is taken from the context of convolutional neural networks which are the topics of study in Chapter 6.

Vanishing and Exploding Gradients

The key backpropagation recursions are the forward step $a^{[\ell+1]} = S^{[\ell]}(W^{[\ell]}a^{[\ell-1]} + b^{[\ell]})$ as in (5.4) and the backward step $\delta^{[\ell]} = \mathrm{Diag}(\dot{\sigma}^{[\ell]}(z^{[\ell]}))W^{[\ell+1]^\top}\delta^{[\ell+1]}$, as in (5.32). From a practical perspective, these steps are sometimes subject to instability when the number of layers L is large.

To see this, let us first simplify the situation by ignoring the activation functions and assuming that $W^{[\ell]}$ is with a fixed square dimension and the same weight matrix W, for $\ell = 1, \ldots, L - 1$, and the last weight matrix is $W^{[L]}$. Further, ignore the bias terms $b^{[\ell]}$. In this simplified case,

$$\hat{y} = a^{[L]} = W^{[L]}W^{[L-1]}W^{[L-2]} \cdots W^{[3]}W^{[2]}W^{[1]}x = W^{[L]}W^{L-1}x, \qquad (5.34)$$

where W^{L-1} is the $L-1$ power of W.

As is well known in linear algebra and systems theory, unless the maximal eigenvalues of W are exactly with a magnitude of unity, as L grows we have that $\hat{y}$ either vanishes (toward 0) or explodes (with values of increasing magnitude). As a further simplified illustration of this, if $W = wI$ (a constant multiple of the identity matrix), then $\hat{y} = W^{[L]}w^{L-1}x$, and for any $w \neq 1$, a vanishing $\hat{y}$ or exploding $\hat{y}$ phenomena persists. This illustration shows that for non-small network depths (large L), instability issues may arise in the forward pass.

The same type of instability problem can then also persist in the backward pass since the backward recursion $\delta^{[\ell]} = \text{Diag}\big(\dot{\sigma}^{[\ell]}(z^{[\ell]})\big)W^{[\ell+1]\top}\delta^{[\ell+1]}$ also includes repeated matrix multiplications, and if for simplicity we ignore the activation functions and again take a constant matrix W, then,

$$\delta^{[\ell]} = \left(W^\top\right)^{L-\ell}\delta^{[L]}. \qquad (5.35)$$

Hence there is often a vanishing or exploding nature of $\delta^{[\ell]}$ for large L and low values of ℓ (the first layers of the network). Now with (5.33) in mind, we see that in general, the gradient values $g_W^{[\ell]}$ and $g_b^{[\ell]}$ may get smaller and smaller (vanishing) or larger and larger (exploding) as we go backward with every layer during backpropagation. These are respectively called the *vanishing gradient* or *exploding gradient* problems.

In the worst case, vanishing gradients, may completely stop the neural network from training, or exploding gradients may throw parameter values toward arbitrary directions. This may result in oscillations around the minima or even overshooting the optimum again and again. Another impact of exploding gradients is that huge values of the gradients may cause number overflow resulting in incorrect computations or introductions of NaN floating point values ("not a number").

Gradient descent improvements such as RMSProp, integrated in ADAM (see Section 4.3), can help normalize such variation in the gradients. Nevertheless, numerical instability can still persist. Further, with activation functions such as sigmoid or tanh, in cases of inputs far from 0 the gradient components of $\text{Diag}\big(\dot{\sigma}^{[\ell]}(z^{[\ell]})\big)$ may also vanish. Activation functions such as ReLU or leaky ReLU handle such problems, yet the overarching phenomenon associated with repeated matrix multiplication as exemplified in (5.34) and (5.35) still persists. A key strategy for mitigating such a problem is based on weight initialization as we discuss in the following section.

To handle exploding gradients, sometimes *gradient clipping* is employed. This approach adjusts the gradient so that it, or its individual components, are not too big in magnitude. One approach is to clip each coordinate of the gradient so that it does not exceed a pre-specified value in absolute value. This approach can obviously change the direction of the gradient since some coordinates may be clipped while others not. Another approach is to scale the whole gradient by its norm and to multiply by a fixed factor. This maintains the direction of the gradient as originally computed and ensures its magnitude is at a fixed threshold.

5.5 Weight Initialization

While gradient-based optimization generally works in advancing toward local minima, as highlighted above, vanishing gradients or exploding gradients may significantly hinder progress. In fact, starting with initial values that are either constant or 0 for the weights and bias parameters may throw the learning process off. Such constant initial parameters may impose symmetry on the activation values of the hidden units and in turn prohibit the model from exploiting its expressive power.

Random initialization enables us to break any potential symmetries and is almost always preferable. Below we outline specific principles of random parameter initialization, yet even if these are not applied, the most basic random initialization approach is to set all parameters of the weight matrices $W^{[1]}, \ldots, W^{[L]}$ as independent and identically distributed standard normal random variables and to set all the entries of the bias vectors $b^{[1]}, \ldots, b^{[L]}$ at 0.

In view of the potential vanishing gradient and exploding gradient problems highlighted above, there is room for smarter weight intialization techniques. Specifically, a general principle that is followed focuses on the activation values $a^{[1]}, \ldots, a^{[L]}$ associated with the initial weight and bias parameters. If we momentarily consider these activation values as random entries, an overarching goal is that the distribution of each entry of $a^{[\ell+1]}$ is approximately similar to that of each entry of $a^{[\ell]}$. In such a case, recursing the forward pass (5.4), a form of distributional stability can persist when the number of layers L is large. With such an approach, if we are to choose the distribution of the initial weight matrix parameters judiciously, vanishing and exploding gradients may be mitigated at the onset of learning.

More concretely, the goal of equating the distribution of $a^{[\ell+1]}$ entries with $a^{[\ell]}$ entries is viewed via the first two moments of the distribution (mean and variance). Specifically, with activation function $\sigma(\cdot)$ we have in the ℓ-th layer,

$$a_i^{[\ell]} = \sigma\left(z_i^{[\ell]}\right), \qquad \text{where} \qquad z_i^{[\ell]} = \sum_{j=1}^{N_{\ell-1}} w_{i,j}^{[\ell]} a_j^{[\ell-1]} + b_i^{[\ell]}. \tag{5.36}$$

To develop an initialization approach, we assume that all $a_j^{[\ell-1]}$ values for $j = 1, \ldots, N_{\ell-1}$, are identically distributed with mean 0, some variance $\breve{v}$, and are statistically independent. Further, we wish to initialize parameters randomly such that $a_i^{[\ell]}$ is also with a mean of 0 and the same variance $\breve{v}$. Our strategy for this is to initialize the bias $b^{[\ell]}$ with entries 0 and the weight parameters $w_{i,j}^{[\ell]}$ as normally distributed random variables with a mean of 0 and some specified variance.

It turns out, that when the activation function, $\sigma(\cdot)$, is tanh, a sensible variance to use for $w_{i,j}^{[\ell]}$ is $1/N_{\ell-1}$. This form of initialization is called *Xavier initialization*. Further, when the activation function is ReLU, a sensible variance to use is $2/N_{\ell-1}$ and this form of initialization is called *He initialization*. Another commonly used heuristic is to use variance $2/(N_{\ell-1} + N_\ell)$ which is the harmonic mean of $\frac{1}{N_{\ell-1}}$ and $\frac{1}{N_\ell}$. In any case, all of these are heuristics, often implemented in practical deep learning frameworks.

Derivation of Xavier initialization

To get a feel for the considerations of the $1/N_{\ell-1}$ variance rule of Xavier initialization we now present the derivation of this rule. The approach is to equate activation variances at some $\breve{v}$, based on (5.36). To do so, we use an approximation of $\sigma_{\text{Tanh}}(\cdot)$ as $\sigma(u) = u$. This is a first-order Taylor approximation of the tanh activation function. With this approximation and with 0 bias entries, (5.36) yields

$$\text{Var}\left(a_i^{[\ell]}\right) = \text{Var}\left(z_i^{[\ell]}\right) = \text{Var}\left(\sum_{j=1}^{N_{\ell-1}} w_{i,j}^{[\ell]} a_j^{[\ell-1]}\right) = \sum_{j=1}^{N_{\ell-1}} \text{Var}\left(w_{i,j}^{[\ell]} a_j^{[\ell-1]}\right). \qquad (5.37)$$

This is based on the assumption that all random variables, both activation values and weight parameters, are statistically independent. Now to evaluate the variance summands on the right hand side, we use the following property of the variance of a product of two independent random variables X and Y,

$$\text{Var}(XY) = \mathbb{E}[X]^2 \text{Var}(Y) + \text{Var}(X)\mathbb{E}[Y]^2 + \text{Var}(X)\text{Var}(Y).$$

With this we obtain,

$$\text{Var}\left(w_{i,j}^{[\ell]} a_j^{[\ell-1]}\right) = \mathbb{E}\left[w_{i,j}^{[\ell]}\right]^2 \text{Var}\left(a_j^{[\ell-1]}\right) + \text{Var}\left(w_{i,j}^{[\ell]}\right)\mathbb{E}\left[a_j^{[\ell-1]}\right]^2 + \text{Var}\left(w_{i,j}^{[\ell]}\right)\text{Var}\left(a_j^{[\ell-1]}\right).$$

Now since we seek activation values with a mean of 0 and a variance of $\breve{v}$, it turns out that $\text{Var}\left(w_{i,j}^{[\ell]} a_j^{[\ell-1]}\right) = \text{Var}(w_{i,j}^{[\ell]})\breve{v}$. Now also assuming this variance $\breve{v}$ for activations of layer ℓ, and combining in (5.37) we have,

$$\breve{v} = \sum_{j=1}^{N_{\ell-1}} \text{Var}(w_{i,j}^{[\ell]})\,\breve{v}.$$

By further requiring that all weight initialization variance entries for layer ℓ are constant, say at $\text{Var}(w_{i,j}^{[\ell]}) = \breve{w}_\ell$, we have $\breve{v} = N_{\ell-1}\breve{w}_\ell\,\breve{v}$. Then, this shows that setting $\breve{w}_\ell = 1/N_{\ell-1}$ achieves the desired result.

Further Insight Regarding Vanishing or Exploding Values

The derivation of the Xavier initialization offers further insight about the nature of vanishing or exploding values. Assume that the input features vector x is also random with each entry having the same variance $\breve{v}_x$. Then by recursing the forward pass (5.4), continuing to approximate the activation function as identity, and using the variance calculations above we have

$$\text{Var}\left(a_j^{[L]}\right) = \left[\prod_{\ell=1}^{L} N_{\ell-1}\breve{w}_\ell\right]\breve{v}_x.$$

This further highlights that setting $\breve{w}_\ell = 1/N_{\ell-1}$ yields stability of the variance of the outputs which is especially important when the number of layers L is large. Other choices where $N_{\ell-1}\breve{w}_\ell < 1$ across all layers ℓ may yield vanishing activations, and similarly if $N_{\ell-1}\breve{w}_\ell > 1$ we may observe exploding activations.

5.6 Batch Normalization

In Section 2.1 we discussed standardization of data. As seen in equations (2.1) and (2.2), this is simply a transformation for each feature of the input data based on subtraction of the mean and division by the standard deviation. There are multiple benefits in carrying out standardization, one of which is a reshaping of the loss landscape. As an illustration, in Section 2.4 we explored the effect that such standardization has on simple problems. It turns out that with much more complicated deep neural networks, standardization also called *normalization*, may also be very beneficial. In this section we present an extended idea called *batch normalization* where the outputs of intermediate hidden layers are also normalized in an adaptive manner.

The overarching idea of batch normalization is to normalize (or standardize) not just the input data but also individual neuron values within the intermediate hidden layers or final layer of the network. In a nutshell returning to the display (5.2) and taking j as an index of a neuron in layer ℓ, we may wish to have either $z_j^{[\ell]}$ or $a_j^{[\ell]}$ exhibit near-normalized values over the input dataset. Such normalization of the neuron values then yields more consistent training and mitigates vanishing or exploding gradient problems. It also has a slight regularization effect which may prevent overfitting.

Our exposition here outlines normalization of the $z_j^{[\ell]}$ values, yet the reader should keep in mind that one may choose to do so for the $a_j^{[\ell]}$ values instead. When applying the batch normalization technique that we describe here, the output of the training process involves further parameters, some of which are trained via optimization (gradient descent), and others are based on running averages in the training process. We now present the details.

The Idea of Per Unit Normalization

The main idea of batch normalization is to consider neuron j in layer ℓ and instead of using $z_j^{[\ell]}$ as in (5.2), to use a transformed version $\tilde{z}_j^{[\ell]}$. Such a transformation takes place both in training time and when using the model in production, with subtle differences between the two cases as we describe below. The transformation aims to position the $\tilde{z}_j^{[\ell]}$ values so that they have approximately zero mean and unit standard deviation over the data. Further, the transformation involves a correction using trainable parameters.

The transformation requires estimates of the unit's mean and standard deviation so that the unit value can be normalized. During training time, at a given training epoch and for a given mini-batch of size n_b, such estimates are obtained via

$$\hat{\mu}_j^{[\ell]} = \frac{1}{n_b} \sum_{i=1}^{n_b} z_j^{[\ell](i)} \quad \text{and} \quad \hat{\sigma}_j^{[\ell]} = \sqrt{\frac{1}{n_b} \sum_{i=1}^{n_b} (z_j^{[\ell](i)} - \hat{\mu}_j^{[\ell]})^2}, \tag{5.38}$$

where $z_j^{[\ell](i)}$ is the value at unit j, at layer ℓ, and sample i within the mini-batch, prior to carrying out normalization.

With $\hat{\mu}_j^{[\ell]}$ and $\hat{\sigma}_j^{[\ell]}$ available, we compute

$$\bar{z}_j^{[\ell](i)} = \frac{z_j^{[\ell](i)} - \hat{\mu}_j^{[\ell]}}{\sqrt{(\hat{\sigma}_j^{[\ell]})^2 + \varepsilon}}, \tag{5.39}$$

where $\varepsilon > 0$ is a small fixed quantity that ensures we do not divide by zero. At this point $\bar{z}_j^{[\ell](i)}$ has nearly zero mean and nearly unit standard deviation for all data samples i in the mini-batch (it is nearly and not exactly only due to ε).

Now finally, an additional transformation takes place in the form,

$$\hat{z}_j^{[\ell](i)} = \gamma_j^{[\ell]} \bar{z}_j^{[\ell](i)} + \beta_j^{[\ell]}, \tag{5.40}$$

where $\gamma_j^{[\ell]}$ and $\beta_j^{[\ell]}$ are trainable parameters. A consequence of (5.39) and (5.40) is that $\hat{z}_j^{[\ell](i)}$ has a standard deviation of approximately $\gamma_j^{[\ell]}$ and a mean of approximately $\beta_j^{[\ell]}$ over the data samples i in the mini-batch. These parameters are respectively initialized at 1 and 0, and then as training progresses, $\gamma_j^{[\ell]}$ and $\beta_j^{[\ell]}$ are updated using the same learning mechanisms applied to the weights and biases of the network. Namely they are updated using backpropagation, and gradient-based learning. Specific backpropagation details are presented at the end of this section.

As presented above, each unit or neuron has their own set of trainable parameters, $\gamma_j^{[\ell]}$ and $\beta_j^{[\ell]}$. However, in practice, multiple neurons in the same layer often share the same batch normalization parameters. This implies adjusting the mean and standard deviation estimates (5.38) to have summations over multiple neurons j. For example in convolutional neural networks as presented in the next chapter, all neurons of the same channel typically have the same batch normalization applied to them.

Batch Normalization in Production

When using the model in production we need to be able to apply the model to a single input data sample in which case evaluation of $\hat{\mu}_j^{[\ell]}$ and $\hat{\sigma}_j^{[\ell]}$ as in (5.38) is impossible. Instead, averages from the training set are collected during train time and these are supplied and deployed with the model. Practically, since parameters are updated during a training run and this updating in turn affects the $z_j^{[\ell]}$ values, averages are often collected in parallel to training. A common technique is to use exponential smoothing as in (4.14) in Chapter 4, and apply it to the sequence of computed values $\hat{\mu}_j^{[\ell]}$ and $\hat{\sigma}_j^{[\ell]}$ between mini-batches during training. The result of this exponential smoothing, denoted here as $\hat{\bar{\mu}}_j^{[\ell]}$ and $\hat{\bar{\sigma}}_j^{[\ell]}$, is then deployed together with the model.

As a summary, in a deployed model, each unit (j in layer ℓ), or set of units, to which batch normalization is applied is deployed with the trained $\gamma_j^{[\ell]}$ and $\beta_j^{[\ell]}$ values as well as with the exponentially smoothed estimates $\hat{\bar{\mu}}_j^{[\ell]}$ and $\hat{\bar{\sigma}}_j^{[\ell]}$. Then in production,

$$\hat{z}_j^{[\ell]} = \gamma_j^{[\ell]} \frac{z_j^{[\ell]} - \hat{\bar{\mu}}_j^{[\ell]}}{\sqrt{\left(\hat{\bar{\sigma}}_j^{[\ell]}\right)^2 + \varepsilon}} + \beta_j^{[\ell]} \tag{5.41}$$

is used in place of $z_j^{[\ell]}$.

Interestingly, returning to (5.2) and observing that the vector $z^{[\ell]} = W^{[\ell]}a^{[\ell-1]} + b^{[\ell]}$ is an affine function of the previous activation vector, we see that when combined with (5.41) we are left with an affine transformation $\tilde{z}^{[\ell]} = \tilde{W}^{[\ell]}a^{[\ell-1]} + \tilde{b}^{[\ell]}$ with a modified weight matrix $\tilde{W}^{[\ell]}$ and bias vector $\tilde{b}^{[\ell]}$. Hence at least in principle, deployment of batch normalization of this sort can be done without the production model needing to be aware of batch normalization at all since we can encode it in the weight matrices and bias vectors.[6]

Backpropagation of Batch Normalization Parameters

We close this section with an exposition of how the batch normalization parameters $\gamma_j^{[\ell]}$ and $\beta_j^{[\ell]}$ can be updated as part of the backpropagation algorithm. To simplify the notation we omit the superscript $[\ell]$ and the subscript j as the batch learning parameters are either neuron-specific or shared by a group of neurons. Our focus is thus on one pair γ and β for which we require $\partial C/\partial \gamma$ and $\partial C/\partial \beta$, respectively. We also require computing the intermediate derivative value $\partial C/\partial z^{(i)}$ for backpropagation to operate.

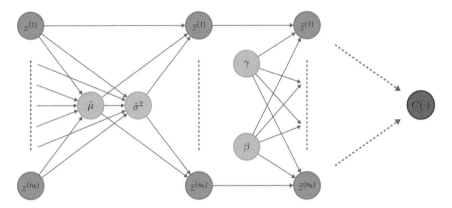

Figure 5.10: A schematic of the computational graph for batch normalization at an arbitrary layer. The goal is to compute the gradients of the loss with respect to γ, β, and each $z^{(i)}$.

At each iteration, the network estimates the mean $\hat{\mu}$ and the standard deviation $\hat{\sigma}$ corresponding to the current batch. In the forward pass, given inputs (output of the previous layer) $z^{(i)}$ for $i = 1, \ldots, n_b$, we calculate the outputs of the batch normalization procedure $\tilde{z}^{(i)}$. Then, the backpropagation pass will propagate the gradient of the loss function $C(\cdot)$ through this transformation and compute the gradients with respect to the parameters γ and β. We then also compute the intermediate derivatives $\partial C/\partial z^{(i)}$.

A schematic of the computational graph associated to the batch normalization steps is presented in Figure 5.10. Backpropagation of the higher layers yields the gradient $\frac{\partial C}{\partial \tilde{z}^{(i)}}$. With this we want to compute $\frac{\partial C}{\partial \gamma}$, $\frac{\partial C}{\partial \beta}$, and $\frac{\partial C}{\partial z^{(i)}}$.

---

[6]This consolidation of batch normalization into the weight matrix and bias vector is often not done in practice, especially due to the fact that W is often a convolution matrix as described in the next chapter.

The gradients $\partial C / \partial \gamma$ and $\partial C / \partial \beta$ are simple. By applying the chain rule with (5.40), we get

$$\frac{\partial C}{\partial \gamma} = \sum_{i=1}^{n_b} \frac{\partial C}{\partial \bar{z}^{(i)}} \bar{z}^{(i)} \qquad \text{and} \qquad \frac{\partial C}{\partial \beta} = \sum_{i=1}^{n_b} \frac{\partial C}{\partial \bar{z}^{(i)}}. \tag{5.42}$$

In order to compute $\frac{\partial C}{\partial z^{(i)}}$, we need also to evaluate $\frac{\partial C}{\partial \hat{\mu}}$, $\frac{\partial C}{\partial \hat{\sigma}^2}$, and $\frac{\partial C}{\partial \bar{z}^{(i)}}$ since

$$\frac{\partial C}{\partial z^{(i)}} = \frac{\partial C}{\partial \bar{z}^{(i)}} \frac{\partial \bar{z}_{(i)}}{\partial z^{(i)}} + \frac{\partial C}{\partial \hat{\sigma}^2} \frac{\partial \hat{\sigma}^2}{\partial z^{(i)}} + \frac{\partial C}{\partial \hat{\mu}} \frac{\partial \hat{\mu}}{\partial z^{(i)}}. \tag{5.43}$$

With the multivariate chain rule and (5.39) and (5.38), we have

$$\frac{\partial C}{\partial \bar{z}^{(i)}} = \frac{\partial C}{\partial \bar{z}^{(i)}} \gamma,$$

$$\frac{\partial C}{\partial \hat{\sigma}^2} = \sum_{i=1}^{n_b} \frac{\partial C}{\partial \bar{z}^{(i)}} \frac{\partial \bar{z}^{(i)}}{\partial \hat{\sigma}^2} = \sum_{i=1}^{n_b} \frac{\partial C}{\partial \bar{z}^{(i)}} \left(z^{(i)} - \hat{\mu} \right) \left(-\frac{1}{2} \right) \left(\hat{\sigma}^2 + \varepsilon \right)^{-3/2},$$

$$\frac{\partial C}{\partial \hat{\mu}} = \sum_{i=1}^{n_b} \frac{\partial C}{\partial \bar{z}^{(i)}} \frac{\partial \bar{z}^{(i)}}{\partial \hat{\mu}} + \frac{\partial C}{\partial \hat{\sigma}^2} \frac{\partial \hat{\sigma}^2}{\partial \hat{\mu}} = \sum_{i=1}^{n_b} \frac{\partial C}{\partial \bar{z}^{(i)}} \frac{-1}{\sqrt{\hat{\sigma}^2 + \varepsilon}} + \frac{\partial C}{\partial \hat{\sigma}^2} \frac{\sum_{i=1}^{n_b} -2 \left(z^{(i)} - \hat{\mu} \right)}{n_b}.$$

These formulas present us with an explicit expression by transforming (5.43) to

$$\frac{\partial C}{\partial z^{(i)}} = \frac{\partial C}{\partial \bar{z}^{(i)}} \frac{1}{\sqrt{\hat{\sigma}^2 + \varepsilon}} + \frac{\partial C}{\partial \hat{\sigma}^2} \frac{2(z^{(i)} - \hat{\mu})}{n_b} + \frac{\partial C}{\partial \hat{\mu}} \frac{1}{n_b}. \tag{5.44}$$

This representation of (5.44) can then be integrated with backpropagation through the neural network.

5.7 Mitigating Overfitting with Dropout and Regularization

In Section 2.5 we discussed the challenges and tradeoffs associated with overfitting and the need for generalization. On the one hand we seek a model that will make use of the available data and properly capture the dependence on the input features, while on the other hand we wish to avoid overfitting. As presented in Section 2.5, one general approach for mitigating overfitting is called *regularization*, whereas in (2.33) we may augment the loss function $C(\cdot)$ with a regularization term $R_\lambda(\theta)$, and optimize $\min_\theta C(\theta; \mathcal{D}) + R_\lambda(\theta)$, in place of just minimization of the loss function. Such practice using additive regularization is possible with deep neural networks as well. However, there is also an alternative popular approach called *dropout* which is specific to deep neural networks. We first describe the dropout approach and then highlight relationships between dropout, ensemble methods, as well as addition of regularization terms.

Dropout

The idea of *dropout* is to randomly zero out certain neurons during the training process. This allows training to focus on multiple random subsets of the parameters and yields a form of regularization.

With dropout, at any backpropagation iteration (forward pass and backward pass) on a mini-batch, only some random subset of the neurons is active. Practically neurons in layer ℓ,

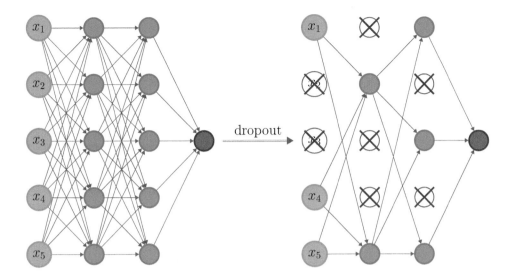

Figure 5.11: An illustration of dropout during training for a network with $L = 3$ layers, $p = 5$ input features, and $q = 1$ output. In each iteration during training, the network is transformed to the network on the right where the dropped out units are randomly selected.

for $\ell = 0, \ldots, L - 1$, have a specified probability $p_{\text{keep}}^{[\ell]} \in (0, 1]$ where if $p_{\text{keep}}^{[\ell]} = 1$ dropout does not affect the layer, and otherwise each neuron i of the layer is "dropped out" with probability $1 - p_{\text{keep}}^{[\ell]}$. This is simply a zeroing out of the neuron activation $a_i^{[\ell]}$ as we illustrate in Figure 5.11.

In the forward pass, when we get to the neurons of layer $\ell + 1$, all the neurons in layer ℓ that were zeroed out do not participate in the computation. Specifically, the update for a neuron j in the next layer, assuming a scalar activation function $\sigma(\cdot)$, becomes

$$a_j^{[\ell+1]} = \sigma\big(b_j^{[\ell+1]} + \sum_{i \text{ kept}} w_{i,j}^{[\ell+1]} a_i^{[\ell]}\big). \tag{5.45}$$

If $p_{\text{keep}}^{[\ell]} = 1$ all neurons are kept and the summation is over $i = 1, \ldots, N_\ell$ as in (5.7), but otherwise the summation is over the random subset of kept neurons. The Bernoulli coin flips determining which neurons are kept and which are not are all carried out independently within the layer, across layers, and throughout the training iterations.

In the backward pass, the effect of dropping out a neuron is evident via (5.33). When neuron i is dropped out in layer ℓ, the weights $w_{i,j}^{[\ell+1]}$ for all neurons $j = 1, \ldots, N_{\ell+1}$ are updated based on the gradient $[g_W^{[\ell]}]_{i,j}$ which is set at 0. With a pure gradient descent optimizer this means that weights $w_{i,j}^{[\ell+1]}$ are not updated at all during the given iteration, whereas with a momentum-based optimizer such as ADAM it means that the descent step for those weights has a smaller magnitude; see Section 4.3 for a review of such optimizers.

At the end of training, even with dropout implemented, the trained model still has a complete set of weight matrices without any zeroed out elements, similar to the case in which we do not have dropout. Hence to account for the fact that neuron i in layer ℓ only took part in a proportion of iterations $p_{\text{keep}}^{[\ell]}$ during the training, when using the model in production (test time), we would like to use a weight matrix $\tilde{W}^{[\ell+1]} = p_{\text{keep}}^{[\ell]} W^{[\ell+1]}$ in place of $W^{[\ell+1]}$. The rationale here is to have the production forward pass similar to the training pass. Namely such a production forward pass has

$$a_j^{[\ell+1]} = \sigma\Big(b_j^{[\ell+1]} + \sum_{i=1}^{N_\ell} p_{\text{keep}}^{[\ell]} w_{i,j}^{[\ell+1]} a_i^{[\ell]}\Big), \tag{5.46}$$

and this serves as an approximation to (5.45). To see the basis of this approximation, treat the summands in (5.45), $w_{i,j}^{[\ell+1]} a_i^{[\ell]}$, as identically distributed random variables, say with expected value μ_j. Now observe that the expected value of both the summation (over a random number of elements) in (5.45) and the expected value of the summation in (5.46) are both $N_\ell \, p_{\text{keep}}^{[\ell]} \, \mu_j$.

In practice, the training–production pair (5.45)–(5.46) is not typically used per-se. The more practical alternative is instead of remembering $p_{\text{keep}}^{[\ell]}$ and deploying it with the production model, the training forward pass is modified to have the reciprocal of $p_{\text{keep}}^{[\ell]}$ as a scaling factor of the weights. Namely, the training forward pass is

$$a_j^{[\ell+1]} = \sigma\Big(b_j^{[\ell+1]} + \sum_{i \text{ kept}} \frac{1}{p_{\text{keep}}^{[\ell]}} w_{i,j}^{[\ell+1]} a_i^{[\ell]}\Big). \tag{5.47}$$

This form allows to use the resulting model normally in production without having to take dropout into consideration at all. Namely, the production forward pass is

$$a_j^{[\ell+1]} = \sigma\Big(b_j^{[\ell+1]} + \sum_{i=1}^{N_\ell} w_{i,j}^{[\ell+1]} a_i^{[\ell]}\Big), \tag{5.48}$$

With this setup, the expected value of the summations in (5.47) and (5.48) agree and further the forward step (5.48) agrees with the standard model as in (5.2), (5.4), and (5.7).

In practice, this simple and easy to implement idea of dropout has improved the performance of deep neural networks in many empirically tested cases. It is now an integral part of deep learning training. We now explore the idea a bit further through the viewpoint of ensemble methods.

Viewing Dropout as an Ensemble Approximation

Dropout can be viewed as an approximation of an *ensemble method*, a general concept from machine learning. Let us first present an overview of ensemble methods or *ensemble learning* and then argue why dropout serves as an approximation.

In general when we seek a model $\hat{y} = f_\theta(x)$, we may use the same dataset to train multiple models that all try to achieve the same task. We may then combine the models into an *ensemble* (model). The latter is usually more accurate than each of the individual models.

Let us illustrate this in the case of a scalar output model. We can choose to use M models and denote each model via $\hat{y}^{\{i\}} = f_{\theta_{\{i\}}}^{\{i\}}(x)$ for $i = 1, \ldots, M$, where $\theta_{\{i\}}$ is taken here as the set of parameters of the i-th model. The ensemble model on an input x is then the average,

$$f_\theta(x) = \frac{1}{M} \sum_{i=1}^{M} f_{\theta_{\{i\}}}^{\{i\}}(x), \qquad \text{where} \qquad \theta = (\theta_{\{1\}}, \ldots, \theta_{\{M\}}). \qquad (5.49)$$

Clearly $f_\theta(\cdot)$ is more computationally costly since it requires M models instead of a single model. This implies storing M parameter sets, training M times instead of once, and evaluating M models in (5.49) instead of once during production. Nevertheless, there are benefits.

To illustrate the strength of this technique assume for simplicity that the models are homogenous in nature and only differ due to randomness in the training process and not the model choice or hyper-parameters. In this case, for some fixed unseen input $\tilde{x}$ we may treat the output of model i, denoted $\hat{y}_{\theta_{\{i\}}}^{\{i\}}(\tilde{x})$, as a random variable that is identical in distribution to every other model output $\hat{y}_{\theta_{\{j\}}}^{\{j\}}(\tilde{x})$, yet generally not independent. We further assume that any pair of model outputs is identically distributed to any other pair. In this case we may denote

$$\mathbb{E}\left[\hat{y}_{\theta_{\{i\}}}^{\{i\}}(\tilde{x})\right] = \mu, \qquad \text{Var}\left(\hat{y}_{\theta_{\{i\}}}^{\{i\}}(\tilde{x})\right) = \sigma^2, \qquad \text{and} \qquad \text{cor}\left(\hat{y}_{\theta_{\{i\}}}^{\{i\}}(\tilde{x}), \hat{y}_{\theta_{\{j\}}}^{\{j\}}(\tilde{x})\right) = \rho,$$

respectively, where $\text{cor}(\cdot, \cdot)$ is the correlation between two models $i \neq j$ and is assumed to be the same for all i, j pairs. Such an assumption on the correlation also imposes[7] a lower bound on the correlation,

$$-\frac{1}{M-1} \leq \rho, \qquad \text{or} \qquad 0 \leq \rho + \frac{1-\rho}{M}. \qquad (5.50)$$

We can now evaluate the mean and variance of the ensemble model (5.49). Namely,

$$\mathbb{E}[f_\theta(\tilde{x})] = \frac{1}{M} \mathbb{E}\left[\sum_{i=1}^{M} f_{\theta_{\{i\}}}^{\{i\}}(\tilde{x})\right] = \mu, \qquad (5.51)$$

and further noting that $\rho\sigma^2$ is the covariance between any two models we obtain[8]

$$\text{Var}\left(f_\theta(\tilde{x})\right) = \frac{1}{M^2} \text{Var}\left(\sum_{i=1}^{M} f_{\theta_{\{i\}}}^{\{i\}}(\tilde{x})\right) = \frac{1}{M^2}\left(M\sigma^2 + M(M-1)\rho\sigma^2\right) = \left(\rho + \frac{1-\rho}{M}\right)\sigma^2. \qquad (5.52)$$

With (5.50) it is confirmed that as required this variance expression in (5.52) is non-negative. Further, we see that as the number of models in the ensemble, M, grows, the variance of the ensemble model converges to $\rho\sigma^2$. Since in addition to (5.50), $\rho \leq 1$ and practically $\rho < 1$, this limiting variance is less than σ^2. For example if $\rho = 0.5$ as M grows the variance of the estimator drops by 50%.

Putting the computational costs aside, these properties of ensemble models make them very attractive because the bias does not change as shown in (5.51), but the variance decreases;

---

[7] This may be shown based on the constraint that the covariance matrix is a positive semi-definite matrix.
[8] The variance of a sum of random variables is the sum of the elements of the covariance matrix.

recall also the general discussion of the bias variance tradeoff in Section 2.5. Nevertheless, deep learning models are not easily amenable for ensemble models in the form of (5.49) because the number of parameters and computational cost (both for training and production) is too high. Training a single model may sometimes take days and the computational costs of a single evaluation $f^{\{i\}}_{\theta_{\{i\}}}(\tilde{x})$ are also non-negligible. This is where dropout comes in.

We may loosely view dropout as an ensemble of M models where M is the number of training iterations. In a practical training scenario, M can be on the order of hundreds, thousands, or tens of thousands, depending on the number of epochs during training and the size of the mini-batch in comparison to the number of training samples. For example, if $n = 10^5$ training samples and the mini-batch size is $n_b = 100$, then each epoch has 1,000 iterations and if we execute, say, 1,000 epochs of training then $M = 10^6$. Then, since the production model involves weights accumulated throughout all 10^6 iterations, we may loosely view the production model as an ensemble model and we may expect its variance to be reduced from σ^2 to approximately $\rho\sigma^2$. Now clearly each of the M iterations did not execute training of the model fully but rather only trained for a single iteration. Hence this ensemble view of dropout is merely a heuristic description.

Addition of Regularization Terms and Weight Decay

In addition to dropout, as already introduced in Section 2.5, addition of a regularization term is another key approach to prevent overfitting and improve generalization performance. Augmenting the loss with a regularization term $R_\lambda(\theta)$ restricts the flexibility of the model, and this restriction is sometimes needed to prevent overfitting. In the context of deep learning, and especially when ridge regression style regularization is applied, this practice is sometimes called *weight decay* when considering gradient-based optimization. We now elaborate.

Take the original loss function $C(\theta)$ and augment it to be $\tilde{C}(\theta) = C(\theta) + R_\lambda(\theta)$. In our discussion here, let us focus on the ridge regression type regularization with parameter $\lambda > 0$, and,

$$R_\lambda(\theta) = \frac{\lambda}{2}R(\theta), \quad \text{with} \quad R(\theta) = \|\theta\|^2 = \theta_1^2 + \ldots + \theta_d^2.$$

Here for notational simplicity, we consider all the d parameters of the model as scalars, θ_i for $i = 1, \ldots, d$ (not to be confused with the notation used above for the parameters of a model as part of an ensemble). Nevertheless, note that typically regularization is only applied to the weight matrix parameters and not to the bias vectors. Further, we may even restrict regularization to certain layers and not others.

Now assume we execute basic gradient descent steps as in (2.20). With a learning rate $\alpha > 0$, the update at iteration t is

$$\theta^{(t+1)} = \theta^{(t)} - \alpha\nabla\tilde{C}(\theta^{(t)}).$$

In our ridge regression style penalty case we have $\nabla\tilde{C}(\theta) = \nabla C(\theta) + \lambda\theta$, and hence the gradient descent update can be represented as

$$\theta^{(t+1)} = (1 - \alpha\lambda)\theta^{(t)} - \alpha\nabla C(\theta^{(t)}). \tag{5.53}$$

Now the interesting aspect of (5.53), assuming that $\alpha\lambda < 2$, is that it involves shrinkage or weight decay directly on the parameters in addition to gradient-based learning. That is, independently of the value of the gradient $\nabla C(\theta^{(t)})$, in every iteration, (5.53) continues to decay the parameters, each time multiplying the previous parameter by a factor $1 - \alpha\lambda$.

This weight decay phenomena can then be extended algorithmically to enforce regularization not directly via addition of a regularization term, but rather simply by augmenting the gradient descent updates to include weight decay. For example we may consider popular gradient-based algorithms such as ADAM in Section 4.3 and the other algorithms in Section 4.5, and in each case add an additional step which incorporates multiplying the weights by a constant less than but close to unity.

Notes and References

The origins of deep learning date back to the same era during which the digital computer materialised. In fact, early ideas of artificial neural networks were introduced first in 1943 with [283]. Then, in the post WWII era, Frank Rosenblatt's *perceptron* was the first working implementation of a neural network model [353]. The perceptron and follow-up work drew excitement in the 1960s yet with the 1969 paper [291], there was formal analysis that shone negative light on limited capabilities of single-layer neural networks and this eventually resulted in a decline of interest, which is sometimes termed the "AI winter" of 1974–1980. Before this period there were even implementations of deep learning architectures with [206], and with [205] where 8 layers were implemented. In 1967 such *multi-layer perceptrons* were even trained using *stochastic gradient descent* [12].

Some attribute the end of the 1974–1980 AI winter to a few developments that drew attention and resulted in impressive results. Some include *Hopfield networks* which are recurrent in nature (see Chapter 7), and also formalism and implementation of the *backpropagation algorithm* in 1981, [421]. In fact, early ideas of backpropagation can be attributed to a PhD thesis in 1970, [265] by S. Linnainmaa. See also our notes and references on early developments of reverse mode automatic differentiation at the end of Chapter 4. In parallel to this revival of interest in artificial intelligence in the early 1980s there were many developments that led to today's contemporary convolutional neural networks. See the notes and references at the end of Chapter 6. Historical accounts of deep learning can be found in [367] and [370] as well as a website by A. Kurenkov.[9] The book [169] was a key reference of neural networks, summarizing developments up to the turn of the 21st century. The 2015 Nature paper by Yann LeCun, Yoshua Bengio, and Geoffrey Hinton, [249] captures more contemporary developments.

Positive results about the universal approximation ability of neural networks, such as Theorem 5.1 presented in this chapter, appear in [97] for a class of sigmoid activations and in [186] for a larger class of non-polynomial functions. With such results, it became evident that with a single hidden layer, neural networks are very expressive. Still, the practical insight to add more hidden layers to increase expressivity arose with the work of Geoffrey Hinton et al. in 2006 [180], Yoshua Bengio et al. 2006 [30], and other influential including works such as [28], [86], and [248]. The big explosion was in 2012 with *AlexNet* in [239].

Our example of a multiplication gate as in Figure 5.4 comes from [259]. Our motivation to use this example and the analysis we present around Figure 5.5 dealing with the expressivity of deeper networks is motivated by a 2017 talk of Niklas Wahlström.[10] Beyond our elementary presentation, many researchers have tried to provide theoretical justifications to why deep neural networks are so powerful. Some justifications of the power of deep learning are in [376] using *information theory*, and [90], [116], [334], and [399], using other mathematical reasoning approaches. See also [285] and [333] for surveys of theoretical analysis of neural networks.

In terms of activation functions, the sigmoid function was the most popular function in early neural architectures with the *tanh* function serving as an alternative. The popularity of *ReLU* grew with [301], especially after its useful application in the AlexNet convolutional neural network [239]. Note that ReLU was first used in 1969, see [126]. Other activation functions such as *leaky ReLU* were introduced in [276], as well as parameterized activation functions such as *PReLU* studied in [171] in order to mitigate vanishing and exploding gradient issues. A general survey of activation functions is in [111].

The backpropagation algorithm can be attributed to [357], yet has earlier origins with general *backward mode automatic differentiation* surveyed in [24] (see also notes of Chapter 4). Our presentation in Algorithm 5.2 is specific to the precise form of feedforward networks that we considered, yet variants can be implemented. Importantly, with the advent of automatic differentiation frameworks such as *TensorFlow* [1] followed by *PyTorch* [324], *Flux.jl* [201], *JAX* [57], and others, the use of backpropagation as part of deep learning has become standard. Such software frameworks automatically implement backpropagation as a special case of backward mode automatic differentiation where the computational graph is constructed, often automatically based on code. Early deep learning work up to about 2014 did not have such software frameworks available; hence "hand coding" of backpropagation was more delicate and time consuming. It is fair to say that with the proliferation of deep learning software frameworks, innovations in research and industry accelerated greatly. We

---

[9]https://www.skynettoday.com/overviews/neural-net-history.
[10]See https://www.it.uu.se/katalog/nikwa778/talks/DL_EM2017_online.pdf.

also note that properly considering matrix analysis aspects of backpropagation often requires *matrix calculus* for which useful references are [141] and [330].

Extensive discussion of *vanishing* and *exploding gradients* is in [323]. Related aspects of training including *weight initialization* are discussed in chapter 7 of [336]. The *Xavier initialization* technique was introduced in [137]. Related initialization techniques and their analysis are studied in [171]. The idea of *batch normalization* was initially introduced in [203] and analyzed in [17] and [202]. Since then, batch normalization has been extended in several ways including instance normalization in [406]. A survey is in [194]. Our analysis of backpropagation of batch-normalization parameters is from [203]. *Dropout* was initially introduced in [181] and [388]. Analysis of dropout as an ensemble approximation is in [159]. See also [263] for an overview of ensemble methods. Further study of dropout and its implications is in [128]. *Regularization* in feedforward networks was reviewed in [243] and analysis of weight decay is in [241] and [270].

6 Convolutional Neural Networks

While offering generality and versatility, the fully connected feedforward neural networks described in the previous chapter are often too general to be effective on their own right. For many applications, such dense architectures can have too many parameters and are not able to generalize well. This is especially the case when considering vision, sound, or similar data for which the spatial orientation of pixels or features is a key defining attribute. For such data, learned rules associated with certain features often need to be repeated systematically. Convolutional neural networks offer an ability to do so by training convolutional filters that can be applied in a spatially homogenous manner. Such networks yield a significant reduction in the number of trained parameters. Understanding convolutional neural networks requires a grasp of the convolution operation and how it is incorporated in a deep learning architecture together with the concept of channels and additional operations such as max-pooling. This chapter covers the main details of such convolutional neural networks as well as specific convolutional architectures that have by now become standard. As the main application of convolutional neural networks is images, we also outline key tasks of deep learning in image processing applications.

We start the chapter with an overview in Section 6.1 where we introduce general concepts of convolutional neural networks. We first touch filtering in signal processing and then consider a high-level view of the VGG19 network as a concrete example. In Section 6.2 we study basics of the convolution operation in one dimension as well as more generally. Toward that end, we relate convolutions to systems theory, to probability distributions, and to multiplication of polynomials. In Section 6.3 we focus on a single convolutional layer. First, we motivate such a layer and then focus on details such as padding, stride, and dilation. We then introduce the concept of channels and the way that volume convolutions are carried out. In Section 6.4 we put the pieces together and discuss how multiple convolutional layers, and other layers such as max-pooling and fully connected layers, are combined into a network model. Here again, the VGG19 serves as a complete example. In Section 6.5 we describe common landmark architectures and key ideas of convolutional neural networks. The ideas and architectures surveyed include inception networks, ResNets, as well as ways for interpreting the meaning of internal features of the networks. We close with Section 6.6 where we discuss the various tasks that one may consider for vision problems beyond classification. These tasks include object localization, face identification, segmentation models, and others.

6.1 Overview of Convolutional Neural Networks

Convolutional neural networks (CNNs) are designed to handle *grid-structured data*, such as image data, where there is a strong local dependency between the neighboring items of the grid. For instance, in an image, there is a high chance that adjacent pixels carry similar properties. Such a grid-based structure is also present in many sequential data formats such as text and sound, where a strong correlation exists among adjacent items. Even though this chapter as well as most of the literature on convolutional neural networks focuses on image

DOI: 10.1201/9781003298687-6

data, these networks are suitable for working with any temporal, spatial, or spatiotemporal data.

Convolutional neural networks are computationally more efficient than the fully connected neural networks studied in Chapter 5 and are more suitable for grid-structured data. This is primarily because convolutional neural networks require fewer parameters than fully connected networks, with a parameter structure focused on feature (pixel) locality. For instance, consider a classification task for detecting cats within a dataset consisting of images of different animals. Such data exhibits two key properties:

Translation invariance: The classification decision for each image is independent of the position of the animal on the image. A cat is a cat irrespective of whether it appears at the top or at the bottom of an image.

Locality: The classification decision does not really depend on a pixel that is far away from the animal on the image. A cat is still a cat irrespective of whether far away pixels correspond to a building or a tree.

Ideally, we want our neural network to take advantage of these two properties. When using the fully connected neural networks of Chapter 5 for images, the first step is to convert each image to an input features vector. By doing this we may lose both the above mentioned structural properties of images. On the other hand, convolutional neural networks retain and exploit these properties while generally using fewer parameters.

Filtering

To understand convolutional neural networks it is helpful to have a basic understanding of *filtering*, a well-established field in signal processing, and particularly in the subfields of image processing and computer vision. Filtering is a method or process that removes certain unwanted information from a signal or an image, or alternatively enhances it by accentuating certain information. Taking image processing as an example domain, filtering applies mathematical operations on input images, with the most common operation being the *convolution*. A convolution can be viewed as an operation between two mathematical objects, such as two matrices or two functions, where one object represents an image and the other a filter. Section 6.2 is completely devoted to a basic introduction of convolutions.

In the field of signal processing, each filter is often custom designed depending on the specific task at hand. For example, a popular filter, called the *Sobel filter*, is useful for the task of edge detection. Figure 6.1 illustrates filtering for extracting edges in an image using two Sobel filters applied on an input image appearing in (a). In (b) we detect the vertical edges and in (c) we detect the horizontal edges. By adding the outputs of these two filtering operations, we get the final image shown in (d) which captures most of the vertical and horizontal edge information. More details on edge detection using Sobel filters are in Section 6.2. Beyond edge detection, there are several other tasks that traditional image processing filters can offer, including blurring, smoothing, sharpening, and accentuating images, and each of these is achieved using a custom-designed filter.

Convolutional neural networks build upon the classic ideas of filtering using *convolutional layers*. Each convolutional layer is made up of one or more *filters*, also known as *kernels*, each of which aims to extract a particular feature of the input to the layer. Early layers of the network usually aim to detect lower-level features such as edge detection while the latter layers focus on higher-level features such as identifying cats, dogs, cars, etc. The final hidden layer, ultimately, provides a summary of the input image. Then for example, when considering classification tasks, the final hidden layer is used to classify the input image into different classes.

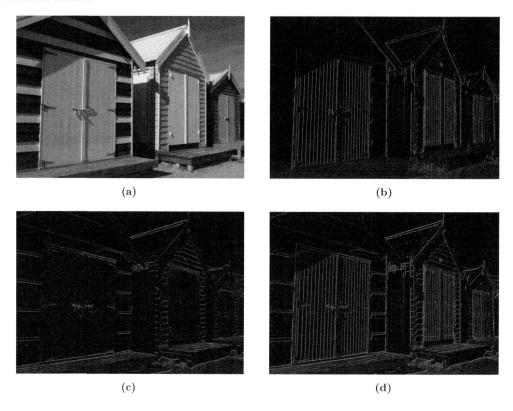

(a)

(b)

(c)

(d)

Figure 6.1: Edge detection using the Sobel filter.

The filtering operation at each layer in a convolutional neural network is similar to classical filtering illustrated with edge detection above. However, unlike classical filtering, filters in the convolutional neural network are learned rather than designed. A training dataset is used for learning the filters before using the network for image processing. Here, learning a filter means learning the entries of the matrix that represents the filter, and these entries are called *weights* as the filters play a similar role to the weight matrices of a fully connected neural networks, covered in the previous chapter.

An Example: VGG19

To get a feel for convolutional neural networks, let us consider the task of classifying color images. As an example, assume input images are of dimension $3 \times 224 \times 224$, where 3 is the number of channels (red, green, and blue), and 224×224 specifies the pixel dimensions. Hence the number of input features is $p = 3 \times 224 \times 224 \approx 150{,}000$. Assume we wish to use such networks for classification of $K = 1{,}000$ possible classes. If we are to use a fully connected network with no hidden layer, as in the multinomial regression model of Section 3.3, we already use $p \times K + K$ learned parameters. This is an order of 150 million parameters. Further, deeper networks that extend the multinomial regression model by adding more layers as in Chapter 5, generally require even more parameters. Yet limiting the number of parameters in any machine learning model is important since it bounds computational time, limits usage of computational resources, and reduces overfitting while respecting training data limitations. We now explore what can be done with a convolutional neural network for such a task using approximately the same number of parameters.

The $3 \times 224 \times 224$ input dimensions agree with inputs of the VGG19[1] model, first touched upon in Section 1.1. The VGG19 model has about 144 million parameters (similar to the dense $p \times K + K$ multinomial regression case). However, in contrast to the dense single-layer model, VGG19 spans 19 trainable layers! This depth makes the model much more expressive; see Section 5.2 for a discussion on the benefits of depth in networks. Indeed, convolutional neural networks such as VGG19 are specifically suited for image tasks and have a relatively low number of parameters which allow them to be deep.

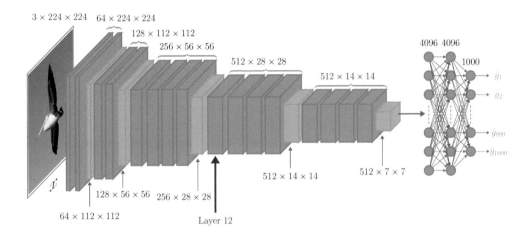

Figure 6.2: The VGG19 network architecture. An input x is a $3 \times 224 \times 224$ color image. It is processed through a series of convolutional layers followed by fully connected layers. The resulting output, $\hat{y}_1, \ldots, \hat{y}_{1000}$ is a vector of probabilities indicating the class of the image.

While Sections 6.2, 6.3, and 6.4, introduce the components of convolutional neural networks in detail, at this point let us informally explore the VGG19 model illustrated in Figure 6.2. Like the feedforward networks of Chapter 5, it is composed of layers where data flows between layers down stream from the input x to the output $\hat{y}$. However, unlike feedforward fully connected layers, most layers are not composed of a dense matrix multiplication as in equation (5.2) of Chapter 5, but are rather made of filtering operations implemented via a combination of convolutions and non-linear activation functions. Only the final layers are dense layers.

The rectangular boxes in Figure 6.2 represent neurons, also known as *internal features*, that are computed via the successive application of convolutional layers. Each such box, is in a sense an image or a tensor, yet unlike the input with 3 channels, these internal representations generally have a different number of channels (also known as *feature maps*), not directly corresponding to color values but rather to internal features. As an example

---

[1]VGG stands for *Visual Geometry Group*, the group at Oxford University that created the network.

consider the layer $\ell = 12$, pointed at with a red arrow in the figure. That layer has 512 channels each containing a 28×28 pixel "image".

The network also incorporates operations called max-pooling without learned parameters, discussed in detail in Section 6.3. These operations are generally used to reduce dimensions as data flows downstream in the network. Importantly, and quite characteristically of convolutional networks, the VGG19 network starts with a succession of convolutional layers with interleaved max-pooling operations, and toward the end, has dense layers that are similar to the layers of feedforward networks of Chapter 5.

6.2 The Convolution Operation

The convolution operation is a key component of convolutional neural networks. In this section we study convolutions via various mathematical and engineering viewpoints. We consider linear time invariant systems, probability distributions, multiplication of polynomials, and the general representation of a convolution as a linear operation. We then consider multi-dimensional convolutions and focus on an engineering filtering example, the Sobol filter, used above in Figure 6.1.

A convolution can be viewed as an operation on two functions which creates a third function. In finite domains, these functions may be represented as vectors, matrices, or tensors. We begin the presentation by focusing on one-dimensional convolutions. Suppose $f, g : \mathbb{Z} \rightarrow \mathbb{R}$ are two functions (or sequences) with discrete domains. Then the *convolution* between f and g is defined as

$$(f \star g)(t) = \sum_{\tau \in \mathbb{Z}} f(t - \tau)g(\tau), \quad t \in \mathbb{Z}. \tag{6.1}$$

In other words, the convolution $f \star g$ between f and g at a point t is obtained by taking summation of the product of the two functions after one of them is *flipped* at the origin and then *shifted* by t. The convolution is *commutative*, namely, $(f \star g)(t) = (g \star f)(t)$. This property can be easily observed via a change of variables in the summation of (6.1).

In case f and g have continuous domains, say $f, g : \mathbb{R} \rightarrow \mathbb{R}$, the convolution between f and g is defined as

$$(f \star g)(t) = \int_{\mathbb{R}} f(t - \tau)g(\tau)d\tau, \quad t \in \mathbb{R}. \tag{6.2}$$

In both the discrete convolution (6.1) and the continuous convolution (6.2) we assume that the summation or integral, respectively, converges. In the context of deep learning we focus on convolutions of vectors, matrices, and tensors, in which case (6.2) is used on a finite domain and hence always converges. We now present a few viewpoints of one-dimensional convolutions before stepping up to multi-dimensional cases.

Convolutions in LTI Systems

To get a feel for the importance of convolutions we consider *Linear Time Invariant* (LTI) systems. These objects are key in classic control theory and signal processing, and they have influenced machine learning, eventually leading to the development of convolutional neural networks. An LTI system, denoted here by $\mathcal{L}(\cdot)$, maps an input signal $x = \{x(t) : t \in \mathbb{R} \text{ or } \mathbb{Z}\}$

to an output signal y via $y = \mathcal{L}(x)$. LTI systems satisfy the linearity and time invariance properties:

Linearity: For any two input signals $x_1(t)$ and $x_2(t)$ and scalars α_1 and α_2,

$$\mathcal{L}(\alpha_1 x_1 + \alpha_2 x_2) = \alpha_1 \mathcal{L}(x_1) + \alpha_2 \mathcal{L}(x_2).$$

Time invariance: When the shifted (delayed by τ) signal $\tilde{x}(t) = x(t - \tau)$ is given as an input, then the corresponding output signal $\tilde{y} = \mathcal{L}(\tilde{x})$ is $\tilde{y}(t) = y(t - \tau)$, where $y = \mathcal{L}(x)$. Namely, the output of the shifted input is the shifted output of the original input.

An important input signal to consider for any LTI system is the *impulse signal*. In the discrete time case, the impulse signal, denoted by $\delta(t)$, takes 1 at $t = 0$ and 0 for any other t, that is,

$$\delta(t) = \begin{cases} 1, & \text{if } t = 0, \\ 0, & \text{otherwise.} \end{cases}$$

When the input is the impulse signal, the output of the system is called the *impulse response* and is denoted here as $w = \mathcal{L}(\delta)$. It turns out that the operation of any LTI system on any input signal x is equivalent to a convolution of x with the impulse response w. That is, $\mathcal{L}(x) = w \star x$.

To see this, using this impulse function, any signal $x = \{x(t) : t \in \mathbb{Z}\}$ can be represented as

$$x(t) = \sum_{\tau = -\infty}^{\infty} x(\tau) \delta(t - \tau),$$

where observe that $\delta(t - \tau)$ takes 1 at $t = \tau$ and 0 otherwise. Consequently, $y(t)$ is equal to

$$\mathcal{L}(x)(t) = \sum_{\tau = -\infty}^{\infty} x(\tau) \mathcal{L}(\delta(t - \tau)) = \sum_{\tau = -\infty}^{\infty} x(\tau) \mathcal{L}(\delta)(t - \tau) = \sum_{\tau = -\infty}^{\infty} x(\tau) w(t - \tau) = (w \star x)(t),$$

where the first and second equalities respectively follow from the linearity and the time invariance properties of LTI systems.

A similar result exists for continuous time LTI systems, where the impulse response is the output of the system when the input is a generalized impulse function, $\delta(t)$, called the *Dirac delta function*.[2] Generally, convolutional neural networks do not rely on such continuous time representations. Nevertheless, we mention it here for completeness because most treatments of LTI systems use the delta function.

---

[2]The Dirac delta function is not an $\mathbb{R} \to \mathbb{R}$ function in the standard sense but is rather a generalized function. It is a mathematical abstraction which allows one to describe an object, $\delta(t)$, that satisfies $\delta(t) = 0$ for $t \neq 0$ as well as $\int_{-\infty}^{\infty} \delta(t)\, dt = 1$. No such standard $\mathbb{R} \to \mathbb{R}$ function exists, but the formalism of generalized functions allows us to treat $\delta(t)$ as though it was standard function. Conceptually, we may also consider $\delta(t)$ as the limit of a Gaussian density centered at zero, with the standard deviation approaching zero.

Convolutions in Probability

Convolutions also appear naturally in the context of probability. This is when considering the distribution of a random variable which is a sum of two independent random variables. For example consider $\xi = \xi_1 + \xi_2$ for two discrete valued independent random variables ξ_1 and ξ_2 with probability mass functions $f_1(t)$ and $f_2(t)$, respectively. Then manipulating the probabilities and noting that $\mathbb{P}(A \mid B)$ is the conditional probability of event A given event B, we have

$$
\begin{aligned}
\mathbb{P}(\xi = t) &= \mathbb{P}(\xi_1 + \xi_2 = t) \\
&= \sum_{\tau=-\infty}^{\infty} \mathbb{P}(\xi_1 = \tau)\mathbb{P}(\xi_1 + \xi_2 = t \mid \xi_1 = \tau) \\
&= \sum_{\tau=-\infty}^{\infty} \mathbb{P}(\xi_1 = \tau)\mathbb{P}(\xi_2 = t - \tau) \\
&= \sum_{\tau=-\infty}^{\infty} f_1(\tau)f_2(t - \tau) \\
&= (f_1 \star f_2)(t).
\end{aligned}
$$

In other words, the probability mass function[3] of ξ is equal to the convolution of the probability mass functions of ξ_1 and ξ_2.

Multiplication of Polynomials and the Convolution Matrix

The convolution also arises when multiplying polynomials. Consider two example polynomials $f(r) = f_0 + f_1 r + f_2 r^2$ and $g(r) = g_0 + g_1 r + g_2 r^2 + \ldots + g_5 r^5$, and their product polynomial $z(r) = f(r)g(r)$. The degree of $z(r)$ is $2 + 5 = 7$ with coefficients $z_0, \ldots, z_7$, as follows:

$$
\begin{aligned}
z_0 &= f_0 g_0, & z_4 &= f_0 g_4 + f_1 g_3 + f_2 g_2, \\
z_1 &= f_0 g_1 + f_1 g_0, & z_5 &= f_0 g_5 + f_1 g_4 + f_2 g_3, \\
z_2 &= f_0 g_2 + f_1 g_1 + f_2 g_0, & z_6 &= f_1 g_5 + f_2 g_4, \\
z_3 &= f_0 g_3 + f_1 g_2 + f_2 g_1, & z_7 &= f_2 g_5.
\end{aligned}
\tag{6.3}
$$

With these expressions it is evident that, if we were to set $f_t = 0$ for $t \notin \{0, 1, 2\}$ and similarly $g_t = 0$ for $t \notin \{0, 1, 2, 3, 4, 5\}$, then we could denote

$$
z_t = \sum_{\tau=-\infty}^{\infty} f_\tau g_{t-\tau} = (f \star g)_t.
$$

Further, it is also useful to consider the following alternative finite sum representation of z_t given by

$$
z_t = \sum_{i+j=t} f_i g_j,
$$

where the sum is over (i, j) pairs with $i + j = t$ and further requiring $i \in \{0, 1, 2\}$ and $j \in \{0, 1, 2, 3, 4, 5\}$. In this context we can also view z as a vector of length 8, f as a vector

---

[3] Analogous results exist for continuous random variables where probability density functions are used in place of probability mass functions and a continuous convolution such as (6.2) is applied.

of length 3 and g as a vector of length 6. We can then create an 8×6 *Toeplitz matrix*[4] $T(f)$ that encodes the values of f such that $z = T(f)g$. More specifically,

$$
\begin{bmatrix} z_0 \\ z_1 \\ z_2 \\ z_3 \\ z_4 \\ z_5 \\ z_6 \\ z_7 \end{bmatrix} = \underbrace{\begin{bmatrix} f_0 & 0 & 0 & 0 & 0 & 0 \\ f_1 & f_0 & 0 & 0 & 0 & 0 \\ f_2 & f_1 & f_0 & 0 & 0 & 0 \\ 0 & f_2 & f_1 & f_0 & 0 & 0 \\ 0 & 0 & f_2 & f_1 & f_0 & 0 \\ 0 & 0 & 0 & f_2 & f_1 & f_0 \\ 0 & 0 & 0 & 0 & f_2 & f_1 \\ 0 & 0 & 0 & 0 & 0 & f_2 \end{bmatrix}}_{T(f)} \begin{bmatrix} g_0 \\ g_1 \\ g_2 \\ g_3 \\ g_4 \\ g_5 \end{bmatrix}.
$$

This shows that the convolution of f with g is a linear transformation given by the matrix-vector product $T(f)g$. Since convolutions are commutative operations, we can also represent the output z as $z = T(g)f$ where $T(g)$ is an 8×3 Toeplitz matrix. In this case,

$$
\begin{bmatrix} z_0 \\ z_1 \\ z_2 \\ z_3 \\ z_4 \\ z_5 \\ z_6 \\ z_7 \end{bmatrix} = \underbrace{\begin{bmatrix} g_0 & 0 & 0 \\ g_1 & g_0 & 0 \\ g_2 & g_1 & g_0 \\ g_3 & g_2 & g_1 \\ g_4 & g_3 & g_2 \\ g_5 & g_4 & g_3 \\ 0 & g_5 & g_4 \\ 0 & 0 & g_5 \end{bmatrix}}_{T(g)} \begin{bmatrix} f_0 \\ f_1 \\ f_2 \end{bmatrix}.
$$

At this point, having seen that convolutions may be encoded via Toeplitz matrices such as $T(f)$ or $T(g)$, we see that the convolution operation is a linear operation. The same also holds for multi-dimensional generalizations which we discuss now.

Multi-Dimensional Generalizations

The convolution operation (6.1) can be generalized to multivariate functions. In fact, for deep learning, convolutions are almost always multivariate. Suppose $f, g : \mathbb{Z}^d \to \mathbb{R}$ are two multivariate functions with discrete domains. Then the *convolution* between f and g is a commutative operation given by

$$
(f \star g)(u) = \sum_{v \in \mathbb{Z}^d} f(u - v)\, g(v) = \sum_{v \in \mathbb{Z}^d} f(v)\, g(u - v), \quad u \in \mathbb{Z}^d. \tag{6.4}
$$

This is a direct extension of (6.1) with shifting and flipping of the functions carried out across all dimensions. Also, similarly to the univariate case over a continuous domain, shown in (6.2), multivariate convolutions have continuous domain representations and are not presented here because convolutional neural networks use discrete domain convolutions.

The applications presented above for univariate convolutions, namely LTI systems, addition of independent random variables, and multiplication of polynomials, also extend to multivariate

---

[4]This is a matrix with constant values on the diagonals. Observe that an $n \times m$ Toeplitz matrix requires at most $n + m - 1$ parameters.

cases. Specifically, the probability law of the sum of two independent random vectors can be obtained via a convolution, the coefficients of the product of multivariate polynomials can be obtained via a convolution, and the action of an LTI system operating on a multivariate input signal can be represented via a convolution. This last case is particularly important for this chapter since one often considers a multivariate convolution.

Any vector, matrix, or tensor can be seen as a function from $\mathbb{Z}^d$ to $\mathbb{R}$ with $d = 1$ for vectors, $d = 2$ for matrices, and $d \geq 3$ for general tensors. As a result, the *convolution* between two vectors, two matrices, or two tensors respectively returns a third vector, matrix, or tensor. In particular for the $d = 2$ case, suppose $W = [w_{i,j}]$, for $i = 1, \ldots, K_h$ and $j = 1, \ldots, K_v$, is a $K_h \times K_v$ matrix and $x = [x_{i,j}]$, for $i = 1, \ldots, M_h$ and $j = 1, \ldots, M_v$, is an $M_h \times M_v$ matrix.[5] In this scenario, f and g in (6.4) can be seen as functions from $\mathbb{Z}^2$ to $\mathbb{R}$ by using the matrices W and x respectively by assigning zeros outside the range of their indices. Then we denote the convolution $f \star g$ as $W \star x$. By ignoring obvious zeros in $W \star x$, we can represent this convolution as a matrix of dimension

$$(M_h + K_h - 1) \times (M_v + K_v - 1). \tag{6.5}$$

To see how such output dimensions arise, refer to the analogy in the one-dimensional polynomial multiplication (6.3), where we consider an example with input dimensions 3 and 6 (for second and fifth-degree polynomials), and thus the output dimension is $3 + 6 - 1 = 8$, matching a seventh degree polynomial.

While (6.5) describes the dimensions of such classical convolutions, the convolution operation used in deep learning differs. In convolutional neural networks, the dimensions of one matrix are smaller than the corresponding dimensions of the other matrix; namely, $K_h \leq M_h$ and $K_v \leq M_v$. In this special case, taking the dimension of W as $K_h \times K_v$ and the dimension of x as $M_h \times M_v$, the convolution $W \star x$ is usually defined to be a matrix of dimension

$$(M_h - K_h + 1) \times (M_v - K_v + 1), \tag{6.6}$$

where now for output at $i' = 1, \ldots, M_h - K_h + 1$ and $j' = 1, \ldots, M_v - K_v + 1$, the convolution action is

$$z_{i',j'} = (W \star x)_{i',j'} = \sum_{i=0}^{K_h-1} \sum_{j=0}^{K_v-1} w_{K_h-i,\, K_v-j} \; x_{i'+i,\, j'+j}. \tag{6.7}$$

Observe that the convolution in (6.7) is a submatrix of the result one would get if applying (6.4). Further note that in this case, $\star$ is not a commutative operation. Figure 6.3 (a) presents a schematic of the convolution operation (6.7) where W is of dimension $K_h \times K_v = 3 \times 3$ and x is of dimension $M_h \times M_v = 6 \times 7$. Here the output $z = W \star x$ is a 4×5 dimensional matrix according to (6.6).

The green entry $z_{1,1}$ in Figure 6.3 (a) is based on the green values in x and all of W. In more detail, Figure 6.3 (b) presents the computation of the first element $z_{1,1}$. For this we consider the flipped W to obtain another 3×3 matrix and then take the element-wise product with the sub-matrix of dimension 3×3 at the top left corner on the matrix x shown in (a) or also shown in green in (b). Similarly, in (a), the red entry $z_{1,2}$ is obtained by sliding the window to the right by one pixel on the matrix x to consider the next 3×3 sub-matrix (denoted in red).

---

[5]We use the subscripts h and v in (M_h, M_v) or (K_h, K_v) throughout this chapter. These subscripts stand for "horizontal" (rows) and "vertical" (columns), respectively.

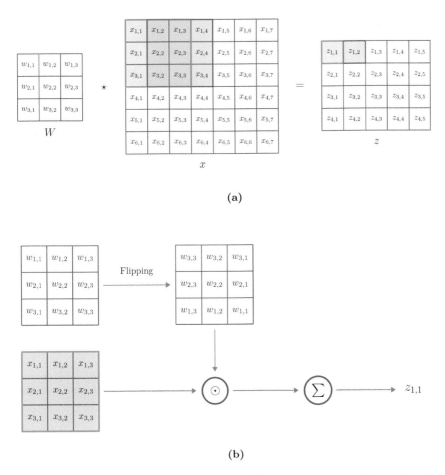

Figure 6.3: (a) Convolution between two matrices W and x to get $z = W \star x$. The dimensions of W and x are $K_h \times K_v = 3 \times 3$ and $M_h \times M_v = 6 \times 7$, respectively. The dimension of the output z is $(M_h - K_h + 1) \times (M_v - K_v + 1) = 4 \times 5$. (b) Computation of the first element $z_{1,1}$ of the convolution $W \star x$. Here, $\odot$ denotes the element-wise product between two matrices of the same dimension and $\sum$ denotes the summation of all the elements of a matrix.

The convolution operation continues with this process until we reach the top right corner on x to obtain the last element $z_{1,5}$ of the first row of z. To obtain the second row, $z_{2,1}, \ldots, z_{2,5}$, we repeat the same process by moving the window one row down on x. Ultimately, after the window is at 4 horizontal positions and 5 vertical positions the 4×5 dimensional output z is obtained. Note that from an implementation perspective, especially when using graphical processing units (GPUs), the computation of z can be parallelized by carrying out the operations as illustrated in Figure 6.3 (b) simultaneously for each output $z_{i,j}$.

Now suppose W and x are three-dimensional tensors with respective dimensions $K_c \times K_h \times K_v$ and $M_c \times M_h \times M_v$, where similarly to before W is smaller than x in the sense that $K_c \leq M_c$, $K_h \leq M_h$, and $K_v \leq M_v$. Here the new dimension sizes K_c and M_c are referred to as the *depth* of the corresponding tensor. For instance, if x denotes a color image, then the depth $M_c = 3$ is attributed to the red, blue, and green components of the image. In this three-dimensional setup, (6.7) can be generalized to provide a *volume convolution*, $W \star x$, with output dimension,

$$(M_c - K_c + 1) \times (M_h - K_h + 1) \times (M_v - K_v + 1). \tag{6.8}$$

Here, for $k' = 1, \ldots, M_c - K_c + 1$, $i' = 1, \ldots, M_h - K_h + 1$, and $j' = 1, \ldots, M_v - K_v + 1$,

$$(W \star x)_{k',i',j'} = \sum_{k=0}^{K_c-1} \sum_{i=0}^{K_h-1} \sum_{j=0}^{K_v-1} W_{K_c-k,\,K_h-i,\,K_v-j}\; x_{k'+k,\,i'+i,\,j'+j}. \tag{6.9}$$

For deep learning, an important special scenario is the case in which the depths of both W and x are the same, namely $K_c = M_c$. In this case the depth is also called the number of *input channels*. In such a scenario, the dimensions in (6.8) have a depth of 1 and thus the output of the volume convolution $W \star x$ defined by (6.9) can be viewed as a matrix of dimension (6.6). This convolution can also be represented as

$$W \star x = \sum_{i=1}^{K_c} W_{(i)} \star x_{(i)}, \tag{6.10}$$

where the $\star$ on the right hand side denotes the matrix convolution as in (6.7) and the summation is element-wise. Here the matrices that are convolved are $W_{(i)}$ which is the ith matrix along the depth of W and $x_{(i)}$ which is the ith matrix along the depth of x (also called the i'th input channel).

Edge Detection Revisited

From an engineering viewpoint, convolutions implement filters, and in the context of image processing (of monochrome images) these are often two-dimensional convolutions. We now explore the operation of one such engineered filter, the Sobel filter for edge detection, first mentioned in Section 6.1 and applied in Figure 6.1.

Suppose an input image $x = [x_{i,j}]$ is of dimension $M_h \times M_v$. As we have seen earlier, edge detection involves two separate operations, namely, vertical edge detection and horizontal edge detection, exemplified in Figure 6.1 (b) and (c), respectively. With Sobel filtering, each of these operations is a convolution of x with a 3×3 kernel matrix given by either

$$W^{(\leftrightarrow)} = \begin{bmatrix} -1 & -2 & -1 \\ 0 & 0 & 0 \\ 1 & 2 & 1 \end{bmatrix} \quad \text{or} \quad W^{(\updownarrow)} = \begin{bmatrix} -1 & 0 & 1 \\ -2 & 0 & 2 \\ -1 & 0 & 1 \end{bmatrix},$$

for horizontal or vertical edge detections, respectively. The actual entries of $W^{(\leftrightarrow)}$ and $W^{(\updownarrow)}$ are part of the Sobel filter design and were engineered[6] to achieve edge detection. Such filters were developed via engineering intuition, trial and error, and experimentation. From our perspective, the actual entries of $W^{(\leftrightarrow)}$ and $W^{(\updownarrow)}$ are merely an example since in convolutional neural networks, the values of filters (weights in convolutional layers) are automatically learned during training.

Suppose $y^{(\leftrightarrow)}$ and $y^{(\updownarrow)}$ are the outputs corresponding to horizontal edge detection and vertical edge detection, respectively. These outputs are each computed using (6.7) with W

---

[6]Sobel filters work by approximating the gradient of the image intensity via a discrete differentiation operation.

replaced by $W^{(\leftrightarrow)}$ and $W^{(\updownarrow)}$, respectively, both using the same input image x. The overall edge detection can be obtained by superimposing the two outputs as the pixel-wise sum $y^{(\leftrightarrow)} + y^{(\updownarrow)}$, or average $(y^{(\leftrightarrow)} + y^{(\updownarrow)})/2$. In case of color images, one may apply Sobel filters separately on each color component, or seek other generalizations and use the convolution formulas (6.9) or (6.10).

6.3 Building a Convolutional Layer

In Chapter 5, we have seen the construction of general fully connected neural networks, each of which consists of a series of layers where every neuron in a given layer is connected to every neuron in the next layer. These networks are general in the sense that they are *structure agnostic*, that is, there are no specific assumptions made about the structure of the input. This property makes fully connected neural networks versatile. However, they are inadequate when dealing with specific applications, such as image classification, where the input has rich structural properties.

Convolutional neural networks make use of the aforementioned two key properties of grid-structured data, namely *translation invariance* and *locality*. As a result, the number of parameters to learn in convolutional neural networks is significantly smaller than that of corresponding fully connected neural networks. Convolutional layers are based on the convolution operation, and in this section we focus on building a single convolutional layer.

Motivating a Convolutional Layer

Convolutional neural networks are designed so that the spatial properties of the image data are inherited from one layer to the next. Therefore, for image processing, it is better to represent both the input and output of a convolutional layer as images. As we are familiar from Chapter 5 with fully connected neural networks, to build a convolutional layer, we begin with a fully connected layer and then we show how the number of learned parameters is reduced using translation invariance and locality.

Consider an input dataset consisting of two-dimensional grey scaled images x of dimension $M_h^{[0]} \times M_v^{[0]}$. For the time being, we focus on the first hidden layer of this fully connected network and the superscript $[0]$ denotes that x is an input to this layer. Each input image x is a matrix with the (i,j)-th element denoting the pixel value at the (i,j)-th location on the image. When treating x as an input to a fully connected neural network, it is represented as an $M_h^{[0]} \cdot M_v^{[0]}$ dimensional vector consisting of all the elements of x. Since such a matrix to vector conversion is executed in a consistent manner,[7] without loss of generality we can continue to index the elements of the vector x via tuples $(i,j) \in \{1, \ldots, M_h^{[0]}\} \times \{1, \ldots, M_v^{[0]}\}$.

We wish to represent the output of the first layer also as an image,[8] in this instance having dimension $M_h^{[1]} \times M_v^{[1]}$. Thus, as with the input, the output vector $a^{[1]}$ can also be represented as a matrix, indexed by tuples $(i',j') \in \{1, \ldots, M_h^{[1]}\} \times \{1, \ldots, M_v^{[1]}\}$. As described in Chapter 5, the output $a^{[1]}$ is composed of an affine transformation of x parameterized by $W^{[1]}$ and $b^{[1]}$ composed with a non-linear activation function $S^{[1]}(\cdot)$; see (5.2). Here, with our image based indexing we represent each element of $W^{[1]}$ as $w_{(i',j'),(i,j)}^{[1]}$ and each element

---

[7]This can be in column major or row major form, and the specific choice between the two is insignificant as long as consistency is maintained.

[8]This requires a non-prime number of neurons in the first layer.

of $b^{[1]}$ as $b^{[1]}_{i',j'}$. With this notation, the output of the layer is

$$a^{[1]} = S^{[1]}(z^{[1]}), \qquad \text{where} \qquad z^{[1]}_{i',j'} = \sum_{(i,j)} w^{[1]}_{(i',j'),(i,j)} \, x_{i,j} + b^{[1]}_{i',j'}. \qquad (6.11)$$

It is useful to represent each element of $z^{[1]}$ slightly differently. For this fix (i',j') and reindex the terms in the summation by setting (i'',j'') for each (i,j) such that

$$i = i' + i'', \quad \text{and} \quad j = j' + j''.$$

Now $z^{[1]}_{i',j'}$ can be represented as

$$z^{[1]}_{i',j'} = \sum_{(i'',j'')} w^{[1]}_{(i',j'),(i'+i'',j'+j'')} \, x_{i'+i'',j'+j''} + b^{[1]}_{i',j'}, \qquad (6.12)$$

where in the summation, $(i'',j'') \in \{1 - i', \ldots, M^{[0]}_h - i'\} \times \{1 - j', \ldots, M^{[0]}_v - j'\}$. Observe that generally these indices, i'' and j'', take on both positive and negative values as they reflect the offset relative to i' and j', respectively.

We now return to the first structural property of image data, namely, translation invariance. With this property, we expect that any shift in x results only as a shift in the output. As an illustration, let us revisit edge detection and consider a pelican in flight as shown in Figure 6.4. In (a) we see an input to an edge detection filter and in (b) we have the output. Similarly, (c) and (d) are input–output pairs of a similar image. Observe that the pairs (a)-(b) and (c)-(d) are essentially the same, except for the fact that the position of the pelican in the output depends only on its position in the input. In other words, the filtering operation's action on the object is generally independent of the location of the object in the image.

In mathematical terms, such translation invariance implies that the weights $w^{[1]}_{(i',j'),(i'+i'',j'+j'')}$ must be independent of the output indices (i',j') because (i',j') is the pixel location in the output image. With the change of variables, we can use i'' and j'' as relative offsets to that pixel coordinate instead of absolute coordinates. We can then define a smaller set of parameters made of weights $w_{i'',j''}$ and a scalar bias b such that for all output coordinates (i',j'), the original parameters are

$$w^{[1]}_{(i',j'),(i'+i'',j'+j'')} = w_{i'',j''} \qquad \text{and} \qquad b^{[1]}_{i',j'} = b.$$

This simplifies the expression for $z^{[1]}_{i',j'}$ in (6.12) to be

$$z^{[1]}_{i',j'} = \sum_{(i'',j'')} w_{i'',j''} \, x_{i'+i'',j'+j''} + b. \qquad (6.13)$$

The expression (6.13) already indicates a significant reduction in the number parameters to learn in comparison to (6.12). To see this observe that in (6.12) our weights potentially vary based on i' and j' whereas in (6.13) they do not.

We now see further reduction of the parameters by invoking the second structural property, namely locality. Viewed in terms of pixels, this property states that a pixel $x_{i,j}$ is not significantly influenced by far away pixels. A motivational illustration is in Figure 6.5

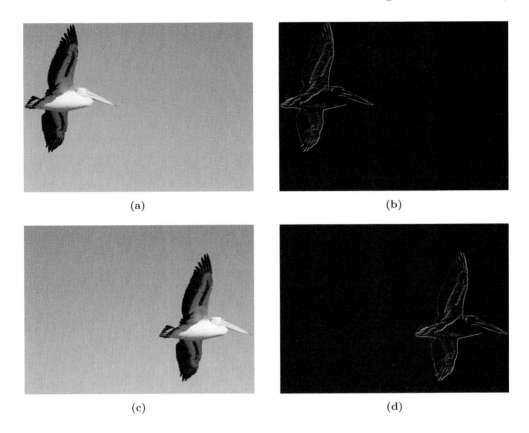

(a)　　　　　　　　　　　　　　(b)

(c)　　　　　　　　　　　　　　(d)

Figure 6.4: Edge detection of images with a pelican to illustrate the property of translation invariance.

consisting of pelicans and seagulls, with each individual bird enclosed in a red box. Generally, the structural property of locality implies that if we are seeking information about one of these specific birds, then it is sufficient to know the pixel information only within the box that is enclosing the bird. Similarly, at a much finer level when we seek information about edges or similar features, it is often enough to consider 1, 2, or 3, neighboring pixels in each direction – yielding convolution kernels of size 3×3, 5×5, or 7×7, respectively.

(a)　　　　　　　　　　　　　　(b)

Figure 6.5: Images of birds to illustrate the property of locality. The pixel information within each red box is typically sufficient for understanding the characteristics of the bird inside the box.

To mathematically enforce locality for the evaluation of $z_{i',j'}^{[1]}$, we ignore the pixel values $x_{i'+i'',j'+j''}$ for $i'' < 0$, $j'' < 0$, $i'' \geq K_h$, and $j'' \geq K_v$ for some chosen $K_h, K_v > 0$; e.g., K_h, K_v at 3, 5, or 7. Equivalently, we set $w_{i'',j''} = 0$ for all (i'', j'') with $i'' \notin \{0, \ldots, K_h - 1\}$ and $j'' \notin \{0, \ldots, K_v - 1\}$. This further reduces the layer's affine transformation to be

$$z_{i',j'}^{[1]} = \underbrace{\sum_{i''=0}^{K_h-1} \sum_{j''=0}^{K_v-1} w_{i'',j''}\, x_{i'+i'',j'+j''}}_{\star} \; + b, \qquad (6.14)$$

where the first term marked by $\star$ is essentially a convolution $W \star x$ with W denoting a kernel matrix[9] of dimension $K_h \times K_v$. Hence, the operation of the layer can be represented as

$$a^{[1]} = S^{[1]}(z^{[1]}), \qquad \text{where} \qquad z^{[1]} = (W \star x) + b, \qquad (6.15)$$

where the addition of the scalar bias b is element-wise to each element of the matrix $W \star x$. Note that the $\star$ convolution operation in (6.14) and (6.15) is slightly different from (6.7) studied in the previous section. To see this difference, recall that the (i', j')-th element of the convolution operation (6.7) is given by

$$\sum_{i''=0}^{K_h-1} \sum_{j''=0}^{K_v-1} w_{K_h-i'',K_v-j''}\, x_{i'+i'',j'+j''},$$

and compare this with the summation marked by $\star$ in (6.14). Hence in our case, $W \star x$ is the conventional convolution if we replace each $w_{i'',j''}$ with $w_{K_h-i'',K_v-j''}$; i.e., flipping at the origin. In the context of neural networks, such a replacement only implies reindexing of the learned parameters and has no effect on the network structure or its performance. For instance, if we observe the edge detection operation illustrated in Figure 6.1, the filter w is flipped only once, and after that for any input x we obtain an element-wise product between the flipped w and sub-matrices of x. Therefore, learning a filter and learning its flipped version are equivalent. As a result, in deep learning, the flipping operation is avoided for simplicity. In any case, the kernel matrix W is still called a *convolutional kernel*.

In summary we have seen that at its core, a single convolutional layer involves the following actions on the input x. First it is convolved with a convolution kernel W. Then the result is shifted by a scalar bias b. Finally an activation function $S^{[1]}(\cdot)$ is applied. These actions are summarized in (6.15). Note that when the input dimension is $M_h^{[0]} \times M_v^{[0]}$, using (6.6), the dimension of the output is

$$M_h^{[1]} \times M_v^{[1]} = (M_h^{[0]} - K_h + 1) \times (M_v^{[0]} - K_v + 1). \qquad (6.16)$$

For an illustration of the reduction in the number of parameters that a convolutional layer has in comparison to a fully connected layer, consider an example with input dimension $M_h^{[0]} \times M_v^{[0]} = 224 \times 224$ and a case with kernel dimension $K_h \times K_v = 3 \times 3$. Here with (6.16), the output dimension is $M_h^{[1]} \times M_v^{[1]} = 222 \times 222$. If we were to seek the same size of output dimension with a fully connected layer, we have $222 \times 222 = 49,284$ neurons. Since the input size is $224 \times 224 = 50,176$, the dimension of the weight matrix is the product of the input size and output size (number of neurons), and together with the bias vector (one entry

---

[9]In Chapter 5 the notation W is used for weight matrices whereas here it is a (generally) smaller kernel matrix. Note that it implicitly defines a weight matrix, not directly used in computation.

for each neuron) we have $2,472,923,268$ parameters. In contrast, in the convolutional layer there are only $3 \times 3 + 1 = 10$ parameters. While on its own, such a single convolutional layer is certainly not as expressive as the fully connected layer with 2.5 billion learned parameters, as we see below, combining convolutional layers in tandem yields very powerful networks with much fewer parameters than their fully connected counterparts.

Alterations to the Convolution: Padding, Stride, and Dilation

The convolution appearing in (6.14) is often tweaked and modified in the context of image data. Specifically, alterations to the convolution operation, known as *padding*, *stride*, and *dilation*, are sometimes employed. For a fixed kernel of dimension $K_h \times K_v$, the combination of these modifications allows us to control the output size as well as the effective input size. Before diving into the details, we mention that these alterations are parameterized by non-negative integer pairs, (p_h, p_v) for padding, (s_h, s_v) for stride, and (d_h, d_v) for dilation, where the subscript h is for height and the subscript v is for width.

In the basic convolution operation above, the absence of padding, stride, and dilation is via a selection of $(0, 0)$ for padding and stride, as well as a selection of $(1, 1)$ for dilation. Such a choice yields output dimension as in (6.16). However, when increasing these integers (typically by small single-digit numbers), the output dimension formula (6.16) is generalized to

$$M_h^{[1]} \times M_v^{[1]} = \left(1 + \left\lfloor \frac{M_h^{[0]} - d_h(K_h - 1) - 1 + p_h}{s_h} \right\rfloor\right) \times \left(1 + \left\lfloor \frac{M_v^{[0]} - d_v(K_v - 1) - 1 + p_v}{s_v} \right\rfloor\right),$$
$$(6.17)$$

where $\lfloor u \rfloor$ represents the largest integer not greater than u. We now introduce and motivate each of these alterations separately and develop (6.17). The reader may verify that with the aforementioned default settings (0 for padding and stride, and 1 for dilation), (6.17) reduces to (6.16).

To motivate padding, recall the edge detection example above. Due to the convolution operation, the output image dimension is smaller than the input image dimension. In particular, since the filter dimension $K_h \times K_v$ is 3×3 (Sobel filter), when the input dimension is $M_h^{[0]} \times M_v^{[0]}$, the output dimension is equal to $(M_h^{[0]} - 2) \times (M_v^{[0]} - 2)$ as in (6.16). Hence we see a slight reduction of the image size at the output. Since convolutional neural networks typically consist of several convolutional layers, the dimension reductions in each of these layers can accumulate, making the overall downstream dimension undesirably small. *Padding* is a simple solution to overcome this problem by adding extra zero-valued pixels around the input so that the effective input dimension is higher, and the desired output dimension is obtained.

To illustrate padding consider the example in Figure 6.3 (a). Here a convolutional layer with a kernel of dimension 3×3 is applied to inputs of dimension 6×7. Without padding, for each input we get an output of dimension 4×5. Now suppose we increase the dimension of the input to 8×9 by adding zeros around the input image. Then when we apply the convolution on the modified input, the output dimension is 6×7, which is equal to the unpadded input image dimension. Figure 6.6 illustrates this operation.

More generally, again suppose that the input dimension is $M_h^{[0]} \times M_v^{[0]}$ and the kernel dimension is $K_h \times K_v$. Further suppose that each input image is modified by adding p_h rows

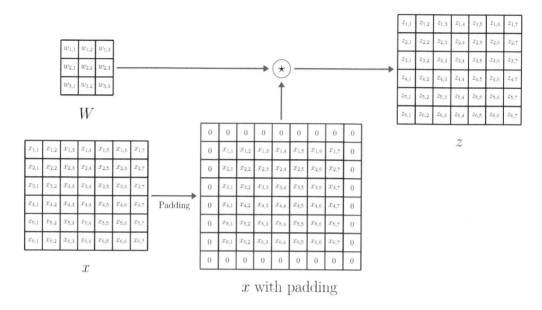

Figure 6.6: Illustration of convolution with padding. In this example a 3×3 convolution with a padding setting of $(p_h, p_v) = (2, 2)$ maintains the same dimensions for the output z as the input x.

roughly half on the top and half on the bottom, and p_v columns roughly half on the left and half on the right. Then it is easy to check that (6.16) is modified so that the output dimension is

$$M_h^{[1]} \times M_v^{[1]} = (M_h^{[0]} - K_h + p_h + 1) \times (M_v^{[0]} - K_v + p_v + 1). \tag{6.18}$$

Note that setting $(p_h, p_v) = (K_h - 1, K_v - 1)$ is a mechanism for ensuring that the input and the output are of the same dimension. Also note that typically convolutional neural networks are designed to have kernels of odd height and odd width. Hence it is common to pad with exactly $p_h/2$ rows of zeros on the left and $p_h/2$ rows of zeros on the right, and similarly for the vertical dimension as shown in (6.6). This helps maintain spatial symmetry while conducting convolutions.

The convolutions we presented up to now involved shifts of the convolution kernel by one pixel at a time. This is called a convolution with a *stride of one*, or $(s_h, s_v) = (1, 1)$. However, in many applications, we may wish to slide the convolution kernel with bigger steps in order to either reduce the computational cost, or to reduce the dimension of the output of the convolutional layer. This is achieved by adjusting the stride size (s_h, s_v) to be greater than one.

As a toy example consider Figure 6.7 where the dimension of the input is 10×10 (potentially after padding), and the kernel is of dimension 3×3. For this example, let us use a hypothetical stride setting of $(s_h, s_v) = (5, 4)$. This setting implies that the convolution kernel is shifted in each step with 5 pixels down, or 4 pixels to the right. As usual we start from the top-left corner, placing the 3×3 convolution kernel on the input image to compute the first element of the output. After computing each element, we move the convolution kernel by 4 pixels to

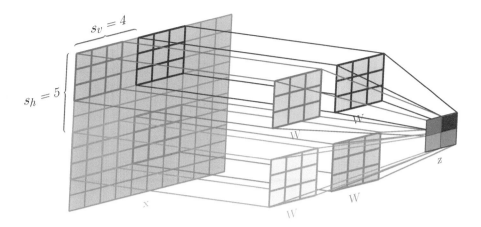

Figure 6.7: Illustration of a convolution with stride settings $(s_h, s_v) = (5,4)$. In this hypothetical example there is no overlap, yet in practice one often uses smaller stride settings.

the right and compute the next element of the row. Once a row of the output is finished, we move the convolution kernel downward by 5 pixels and repeat the horizontal shifting for the next row of the output. Each time we compute an element of the output, we make sure there are enough selected input pixels for the convolution kernel.

Note that in this example, for ease of presentation in the figure, we chose stride settings greater than the size of the convolution kernel and this implies no overlap of the sliding windows. However, in practice one typically uses stride settings of size 2, 3, or similar small steps, smaller than K_h and K_v, and this yields overlap in the convolution multiplications. In general the effect of a stride is in data reduction allowing us to create outputs that are smaller in dimensions than the input, yet capture the essential information. A second mechanism for such reductions is pooling, a concept described in Section 6.4.

The alternation of convolutions, with stride settings (s_h, s_v), modifies the output dimension equation from (6.18) to

$$M_h^{[1]} \times M_v^{[1]} = \left(1 + \left\lfloor \frac{M_h^{[0]} - K_h + p_h}{s_h} \right\rfloor\right) \times \left(1 + \left\lfloor \frac{M_v^{[0]} - K_v + p_v}{s_v} \right\rfloor\right). \tag{6.19}$$

The expression results from the fact that the number of elements computed in each row of the output after computing the first element of the row is equal to the number of rightward moves allowed. With an effective input row size of $M_v^{[0]} + p_v$, this number is $\lfloor (M_v^{[0]} - K_v + p_v)/s_v \rfloor$. Due to the first element, adding 1 yields the width of the output; similarly for the height.

We now focus on *dilation*, a technique for increasing the *receptive field*. The receptive field of an individual filter is marked by the dimensions of the window in the input x that affect a single pixel in the output. For example with a standard 3×3 convolution, the receptive field is 3×3 since each pixel in $W \star x$ is influenced by a 3×3 window in x. When layers are composed, the receptive field has a more general meaning since as data propagates down the network, the receptive field grows.

Dilation increases the respective field by spreading out the elements of the kernel matrix W via the insertion of zeros between elements. This alteration allows the kernel to cover a larger area of the input image without increasing the number of learned parameters. The level of dilation is determined by the settings (d_h, d_v) where dilation converts a kernel of size $K_h \times K_v$ to a kernel of size $K'_h \times K'_v = (d_h(K_h - 1) + 1) \times (d_v(K_v - 1) + 1)$. Specifically, dilation adds $d_h - 1$ all-zero rows between each pair of adjacent rows from the original kernel, and similarly adds $d_v - 1$ columns. Thus the number of all zero rows added is $(K_h - 1)(d_h - 1)$, and similarly $(K_v - 1)(d_v - 1)$ for columns. See Figure 6.8 for an example with $(d_h, d_v) = (2, 2)$.

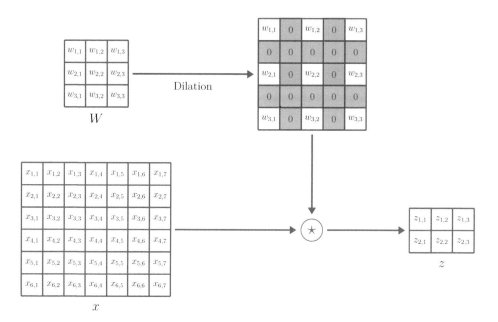

Figure 6.8: Illustration of dilation operation with $(d_h, d_v) = (2, 2)$ extending a 3×3 convolution filter to create a receptive field of 5×5.

Overall, together with a padding of size (p_h, p_v) and a stride of size (s_h, s_v), a dilation factor of (d_h, d_v) implies that the output dimension is determined by (6.17). To see this, replace K_h and K_v in (6.19) with the effective kernel sizes K'_h and K'_v, respectively.

Inputs with Multiple Channels

So far in this section we have looked at the case where each input is a matrix, usually representing a gray scale image. However, convolutional networks often deal with inputs comprised of multiple *channels*. For instance, a color image has 3 channels representing the red, green, and blue components. When we have such data with multiple channels, input to a convolutional layer is no longer a matrix but is rather represented as a three-dimensional tensor. We denote this tensor's dimensions via $M_c^{[0]} \times M_h^{[0]} \times M_v^{[0]}$, where the *depth* $M_c^{[0]}$ denotes the number of channels, and the other two numbers are for the height and width dimensions as used previously. Hence, for color images we have $M_c^{[0]} = 3$ and further, as we describe in the sequel, for hidden layers we often have more than 3 input channels to the layer.

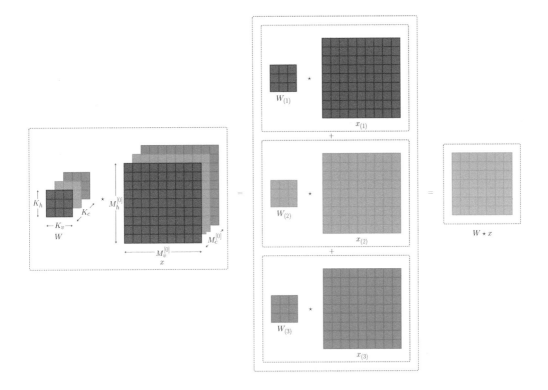

Figure 6.9: Graphical representation of the typical convolution operation with $K_c = M_c^{[0]}$. In this example $M_c^{[0]} = 3$ input channels and we use an $M_c^{[0]} \times 3 \times 3$ convolution kernel.

To deal with inputs with multiple channels we often conduct a volume convolution as in (6.9). For this we use a kernel W with depth greater than one which is a three-dimensional tensor with dimensions denoted via $K_c \times K_h \times K_v$, such that $K_c \leq M_c^{[0]}$, $K_h \leq M_h^{[0]}$, and $K_v \leq M_v^{[0]}$. In fact, the typical case is to set $K_c = M_c^{[0]}$ where the output is a matrix and the convolution is as in (6.10).

Namely for input tensor x, the convolution $W \star x$ is a matrix which is computed via an element-wise sum of the two-dimensional convolutions $W_{(i)} \star x_{(i)}$ for $i = 1, \ldots, M_c^{[0]}$. Each $W_{(i)} \star x_{(i)}$ is a matrix of dimension $M_h^{[1]} \times M_v^{[1]}$ as in (6.17). This two-dimensional convolution is based on the ith channel in the input tensor, denoted $x_{(i)}$, and on $W_{(i)}$ which is the corresponding $K_h \times K_v$ dimensional matrix matching channel i in the convolution kernel tensor W. Note that the same settings of padding, stride, and dilation are applied across all the channels. Figure 6.9 illustrates such a volume convolution for the case of $M_c^{[0]} = K_c = 3$.

After the volume convolution is carried out, a single scalar bias term, b, is added to each element of the matrix $W \star x$. Then a (generally) non-linear activation function $S^{[1]}(\cdot)$ is applied. Hence the action of the convolution on multiple input channels parallels (6.15) and is

$$a^{[1]} = S^{[1]}(z^{[1]}), \qquad \text{where} \qquad z^{[1]} = \left(\sum_{i=1}^{M_c^{[0]}} W_{(i)} \star x_{(i)} \right) + b. \tag{6.20}$$

Outputs with Multiple Channels

Until now, regardless of the number of input channels, the output is a matrix, denoted via $a^{[1]}$ in (6.20). This is because, so far there is only one kernel, possibly a tensor, operating on the input to the convolutional layer. However, the most popular convolutional neural networks have convolutional layers with multiple kernels operating on the input simultaneously. In this case, the output of the layer is a collection of matrices denoted by $a_{(j)}^{[1]}$ for $j = 1, \ldots, M_c^{[1]}$, where $M_c^{[1]}$ is the *number of output channels* (also known as *feature maps*). Consequently, the output can be viewed as a three-dimensional tensor of dimension $M_c^{[1]} \times M_h^{[1]} \times M_v^{[1]}$.

In this case, the convolutional layer is parameterized by multiple kernels $W_{(j)}$ for $j = 1, \ldots, M_c^{[1]}$, each with a scalar bias term $b_{(j)}$. In particular, the kernel $W_{(j)}$ and bias term $b_{(j)}$ correspond to the output channel j. With this notation, the operation of the layer can be represented as

$$a_{(j)}^{[1]} = S^{[1]}(z_{(j)}^{[1]}), \qquad \text{where} \qquad z_{(j)}^{[1]} = \left(\sum_{i=1}^{M_c^{[0]}} W_{(j),(i)} \star x_{(i)} \right) + b_{(j)}, \tag{6.21}$$

for $j = 1, \ldots, M_c^{[1]}$, where similarly to (6.20), $W_{(j),(i)}$ is the matrix corresponding to the ith input channel for the jth kernel. See Figure 6.10 for an illustration in the case of $M_c^{[0]} = 3$ and $M_c^{[1]} = 2$ (3 input channels and 2 output channels).

It is a common practice to use the same dimension $K_c \times K_h \times K_v$ for all kernels $W_{(j)}$ of the layer with the same settings of padding (p_h, p_v), stride (s_h, s_v), and dilation (d_h, d_v) for all the channels. In that case, the dimension $M_h^{[1]} \times M_v^{[1]}$ of each output channel is given by (6.17).

As an illustrative hypothetical example of multiple output channels, assume that the input to the first layer is a color image with 3 channels. One kernel can be used to extract horizontal edges in each input channel of the image while another kernel of the same size extracts vertical edges. In that case, the output has 2 channels where one consists of horizontal edges and the other consists of vertical edges. More generally, in trained networks, we can think of different channels of the output as different feature extractions from the input. These channels jointly help in overall feature extraction for the whole network.

6.4 Building a Convolutional Neural Network

We have now acquired all the crucial elements necessary for constructing convolutional neural networks, such as the VGG19 model depicted in Figure 6.2. We now put the pieces together for constructing a convolutional neural network that, in addition to convolutional layers, includes fully connected layers, as studied in Chapter 5, and *pooling* layers described

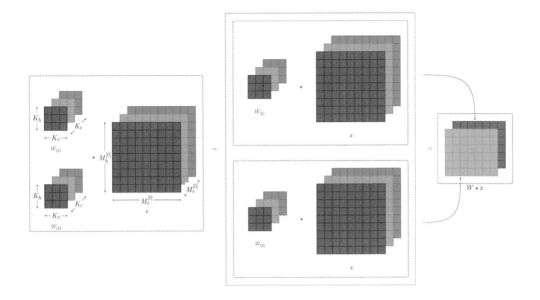

Figure 6.10: Illustration of a convolutional layer with 3 input channels and 2 output channels.

in this section. This section also offers complete details of the previously introduced VGG19 network, serving as an illustrative example. It also introduces *fully convolutional networks*, an architecture that uses convolutional layers in place of fully connected layers.

A convolutional neural network is generally deep with multiple layers, similar to feedforward networks studied in Chapter 5. Unlike feedforward networks which consist of only fully connected layers, convolutional neural networks have different types of layers, of which some are trainable and the others are not, and the trainable layers are further broken up into *convolutional layers* and *dense layers*. Using the notation of Chapter 5, we use L for the number of layers, and decompose L to

$$L = L_{\text{train}} + L_{\text{pool}}, \qquad \text{where} \qquad L_{\text{train}} = L_{\text{conv}} + L_{\text{dense}}.$$

Here L_{train}, counts the number of trainable layers, whereas L_{pool} counts the number of layers that do not have trainable parameters. Further, the trainable layers are either convolutional layers, counted by L_{conv}, or fully connected layers, counted by L_{dense}. It is important to note that in terms of naming conventions, in some instances the depth of the network is taken as L, whereas in others it is taken as L_{train}. For example, in the VGG19 network,

$$L = 24, \qquad L_{\text{train}} = 19, \qquad L_{\text{pool}} = 5, \qquad L_{\text{conv}} = 16, \qquad L_{\text{dense}} = 3, \qquad (6.22)$$

yet the network is called VGG19 and not "VGG24".

Similar to a feedforward network, the goal of a convolutional neural network is to approximate some unknown function $f^*(\cdot)$. For instance, for classification of image data with animal faces, the function value $f^*(x)$ for any given image x may yield a probability vector with the highest weight on the index associated with the label of the image x. A convolutional neural network defines a mapping $f_\theta(\cdot)$ and learns the values of the unknown parameters θ that ideally result in $f^*(x) \approx f_\theta(x)$ for as many input images x as possible. In general, similar to equation (5.1) for feedforward networks, the approximating function $f_\theta(\cdot)$ is recursively composed as

$$f_\theta(x) = f^{[L]}_{\theta^{[L]}}(f^{[L-1]}_{\theta^{[L-1]}}(\ldots(f^{[1]}_{\theta^{[1]}}(x))\ldots)),$$

where for each ℓ, the function $f^{(\ell)}_{\theta^{[\ell]}}(\cdot)$ is associated with the ℓth layer which depends on the layer's parameters $\theta^{[\ell]} \in \Theta^{[\ell]}$. Note that for layers that are not trainable (as counted via L_{pool}), the parameter space $\Theta^{[\ell]}$ is empty.

In general, similarly to feedforward networks, it is useful to denote the neuron activations of the network via $a^{[1]}, a^{[2]} \ldots, a^{[L]}$ where $a^{[L]} = \hat{y}$ is the output, and for $\ell = 1, \ldots, L-1$,

$$a^{[\ell]} = f^{[\ell]}_{\theta^{[\ell]}}(a^{[\ell-1]}),$$

with $a^{[0]} = x$. We mention that the shape of the neurons per layer $a^{[\ell]}$ varies as it is sometimes a tensor (of order 3) and sometimes a vector, depending on the type of layer.

Convolutional Layers

When the ℓ-th layer of the network is a convolutional layer, then $f^{(\ell)}_{\theta^{[\ell]}}(\cdot)$ uses (6.21), treating $a^{[\ell-1]}$ as the input. In this case the input and output are generally 3-tensors as we have seen in the previous sections. In particular,

$$f^{(\ell)}_{\theta^{[\ell]}} : \mathbb{R}^{M^{[\ell-1]}_c \times M^{[\ell-1]}_h \times M^{[\ell-1]}_v} \longrightarrow \mathbb{R}^{M^{[\ell]}_c \times M^{[\ell]}_h \times M^{[\ell]}_v},$$

maps $a^{[\ell-1]}$ of dimension $M^{[\ell-1]}_c \times M^{[\ell-1]}_h \times M^{[\ell-1]}_v$ to $a^{[\ell]}$ of dimension $M^{[\ell]}_c \times M^{[\ell]}_h \times M^{[\ell]}_v$. Now (6.21) is implemented for the $M^{[\ell]}_c$ output channels and this operation can be represented as

$$f^{(\ell)}_{\theta^{[\ell]}}(a^{[\ell-1]}) = S^{[\ell]}\Big(\underbrace{\big[b^{[\ell]}_{(j)} + \sum_{i=1}^{M^{[\ell-1]}_c} W^{[\ell]}_{(j),(i)} \star a^{[\ell-1]}_{(i)} \big]_{j=1,\ldots,M^{[\ell]}_c}}_{z^{[\ell]}_{(j)}} \Big),$$

where the input tensor has $M^{[\ell-1]}_c$ channels and the output tensor has $M^{[\ell]}_c$ channels. Using similar notation to (6.21), the kernel $W^{[\ell]}_{(j)}$ is of dimension $K^{[\ell]}_c \times K^{[\ell]}_h \times K^{[\ell]}_v$ (the same for all output channels j) where the kernel matrix for the i-th input channel and j-th output channel is denoted $W^{[\ell]}_{(j),(i)}$. The tensor after the volume convolutions and bias term is denoted using the notation $[z^{[\ell]}_{(j)}]_{j=1,\ldots,M^{[\ell]}_c}$ where each $z^{[\ell]}_{(j)}$ is a matrix of dimension $M^{[\ell]}_h \times M^{[\ell]}_v$.

Note that the activation function $S^{[\ell]}(\cdot)$ is now considered as a function applied on a tensor of dimension $M^{[\ell]}_c \times M^{[\ell]}_h \times M^{[\ell]}_v$. It is typically an element-wise application of scalar activation functions $\sigma^{[\ell]}(\cdot)$ similarly to previous feedforward examples. In fact, the common activation function is $\sigma_{\text{ReLU}}(\cdot)$; see Section 5.3.

Observe that the number of learned parameters for the layer is

$$M^{[\ell]}_c \cdot \left(M^{[\ell-1]}_c \cdot K^{[\ell]}_h \cdot K^{[\ell]}_v + 1 \right), \tag{6.23}$$

since there are $M_c^{[\ell]}$ kernels (one per output channel) each of dimension $K_c^{[\ell]} \times K_h^{[\ell]} \times K_v^{[\ell]}$ where $K_c^{[\ell]} = M_c^{[\ell-1]}$ (the number of input channels to the layer) and since each output channel adds a scalar bias term.

Pooling Layers

As mentioned above, there are also non-trainable layers counted by L_{pool} and these are typically pooling layers. The main idea of a pooling layer is to reduce the height and width of the input tensor $a^{[\ell-1]}$ to achieve a lower-dimensional output tensor $a^{[\ell]}$ while retaining the same number of channels. The operation of the layer can be summarized with a function,

$$f_{\text{pool}}^{(\ell)} : \mathbb{R}^{M_c^{[\ell-1]} \times M_h^{[\ell-1]} \times M_v^{[\ell-1]}} \longrightarrow \mathbb{R}^{M_c^{[\ell]} \times M_h^{[\ell]} \times M_v^{[\ell]}},$$

$$\text{with} \quad M_c^{[\ell-1]} = M_c^{[\ell]}, \ M_h^{[\ell-1]} > M_h^{[\ell]}, \ \text{and} \ M_v^{[\ell-1]} > M_v^{[\ell]}.$$

Generally for some fixed channel j, and pixel coordinates of the output (i, k), a pooling operation operates on pixels from a window in the input denoted via $\mathcal{I}_{(i,k)}$. Here $\mathcal{I}_{(i,k)}$ is a set of pixel coordinates in the input that are mapped to the specific output pixel (i, k). Two popular pooling techniques are used in practice, namely, *max-pooling* and *average-pooling*. For each channel j, the pooling operation can be summarized as

$$\left[a_{(j)}^{[\ell]} \right]_{i,k} = \begin{cases} \max_{(i',k') \in \mathcal{I}_{(i,k)}} \left[a_{(j)}^{[\ell-1]} \right]_{i',k'}, & \text{(max-pooling)} \\ \frac{1}{|\mathcal{I}_{(i,k)}|} \sum_{(i',k') \in \mathcal{I}_{(i,k)}} \left[a_{(j)}^{[\ell-1]} \right]_{i',k'}. & \text{(average-pooling)}. \end{cases}$$

As is evident, max-pooling takes the maximal pixel value within the window as the output, while average pooling averages pixel values within the window for the output.

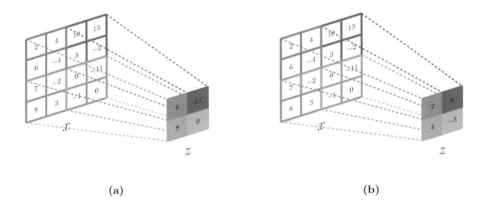

(a)   (b)

Figure 6.11: An example of pooling with a 2×2 window. (a) Max-pooling. (b) Average-pooling.

The specifics of the pooling operation define exactly how $\mathcal{I}_{(i,k)}$ is determined. Generally, similar to the convolution operation and its alternations with stride and padding, we may view pooling as moving a small window over the input to compute an output. The way in which this window moves implicitly defines $\mathcal{I}_{(i,k)}$. As a concrete example see Figure 6.11 which illustrates a case of pooling with a window of dimensions 2×2. With this, $|\mathcal{I}_{(i,k)}| = 4$, and then each output pixel (i, k) is computed based on all 4 pixels $(i', k') \in \mathcal{I}_{(i,k)}$ from the input image which form a 2×2 window in $a_{(j)}^{[\ell-1]}$. A typical *pooling stride* of the window which covers all input pixels while forming non-overlapping windows is to shift each time with the size of the window as in Figure 6.11. In general other pooling stride settings are also possible.

The idea of pooling interplays with the notion that the initial layers of a convolutional network focus on pixel level features similar to edge detection, and as we progress toward the final layers of the network, the information is aggregated to address general questions about the whole image. Thus deeper layers are less sensitive to translation changes on the input image compared to the initial layers. For instance, the answer to a question "is there a bird in the photo?" is the same for both images in Figure 6.5, even though the corresponding outputs from the initial layers look different. Pooling layers are applied after convolutional layers to help achieve such aggregation by reducing the spatial dimension of the outputs. In addition, the dimension reduction which pooling layers offer is important from a computational perspective.

We now return to the notion of a receptive field, previously discussed in the context of dilation and with respect to a single convolution. Now we consider it in the context of a whole network. In particular we consider the *receptive field of a derived feature*. Consider a neuron in the network, $\left[a_{(j)}^{[\ell]}\right]_{i,k}$, for layer ℓ, channel j, and pixel coordinates (i, k). This neuron or activation is a derived feature inside the network. Using the dimensions and specifications of the layers up to that neuron, namely $1, \ldots, \ell$, it can be determined which input pixels from the input image x, affect the value of $\left[a_{(j)}^{[\ell]}\right]_{i,k}$. For example if the neuron is at a first layer involving a 5×5 convolution kernel, then the value of the neuron is only determined by 25 pixels in the input image. However, if the layer ℓ is a hidden layer with multiple convolutions and pooling layers prior to it, it may be that $\left[a_{(j)}^{[\ell]}\right]_{i,k}$ is determined by the whole input image x or a significant portion. In general, pooling layers help increase the receptive field of neurons of hidden layers. This allows the derived features toward the end of the network to depend on the whole input image, or significant parts of it.

Fully Connected Layers

When the ℓ-th layer of the network is a fully connected layer, then the operation of $f_{\theta^{[\ell]}}^{(\ell)}(\cdot)$ is as in (5.2) of Chapter 5. Such layers are typically deployed at the end of the network. This is because the typical task of the last layers of convolutional neural network is to address general questions, such as classification of the objects in the image. Note that since fully connected layers operate on vectors as the input, in cases where the previous layer has a tensor as output, the tensor is flattened to a vector.

It is common to adapt the final fully connected layers of convolutional networks for specific tasks. For example, the VGG19 model can have the final layers fine-tuned for tasks such as object localization discussed in Section 6.6. In doing so, we may take the network trained for classification, and then fine tune it for the other task by only training the fully connected layers. This is sometimes called *freezing layers* (the ones not trained) during training.. Similarly, convolutional networks that were trained on generic images from a general domain, such as ImageNet, can be fine-tuned by training the final layers on more specific images from a specific domain (e.g., only on specific animal images of a certain type). This process, also used in other non-convolutional models, is called *transfer learning*.

VGG19 Revisited

We now take a closer look at the architecture of our running example network, VGG19. While this is not the most modern convolutional architecture, it is instructive to consider it here since it falls directly within the paradigms discussed above. Other popular convolutional architectures are surveyed in the next section. We have seen in (6.22) the counts of different layer types in VGG19 which has $L = 24$ layers of which 19 are trainable. Table 6.1 provides complete details.

Table 6.1: Specifications of the VGG19 architecture. The number of learned parameters for the convolutional layers is computed using (6.23). The number of learned parameters for a fully connected layer with input size $N^{[\ell-1]}$ and output size $N^{[\ell]}$ is $N^{[\ell-1]} \cdot N^{[\ell]} + N^{[\ell]}$; see Section (5.1) for more details on the learned parameters of fully connected layers.

Layer Number	Type of layer	Output dimension	Number of neurons	Number of learned parameters
0	Input	$3 \times 224 \times 224$	-	-
1	Convolution	$64 \times 224 \times 224$	3,211,264	1,792
2	Convolution	$64 \times 224 \times 224$	3,211,264	36,928
3	Max-pooling	$64 \times 112 \times 112$	802,816	0
4	Convolution	$128 \times 112 \times 112$	1,605,632	73,856
5	Convolution	$128 \times 112 \times 112$	1,605,632	147,584
6	Max-pooling	$128 \times 56 \times 56$	401,408	0
7	Convolution	$256 \times 56 \times 56$	802,816	295,168
8	Convolution	$256 \times 56 \times 56$	802,816	590,080
9	Convolution	$256 \times 56 \times 56$	802,816	590,080
10	Convolution	$256 \times 56 \times 56$	802,816	590,080
11	Max-pooling	$256 \times 28 \times 28$	200,704	0
12	Convolution	$512 \times 28 \times 28$	401,408	1,180,160
13	Convolution	$512 \times 28 \times 28$	401,408	2,359,808
14	Convolution	$512 \times 28 \times 28$	401,408	2,359,808
15	Convolution	$512 \times 28 \times 28$	401,408	2,359,808
16	Max-pooling	$512 \times 14 \times 14$	100,352	0
17	Convolution	$512 \times 14 \times 14$	100,352	2,359,808
18	Convolution	$512 \times 14 \times 14$	100,352	2,359,808
19	Convolution	$512 \times 14 \times 14$	100,352	2,359,808
20	Convolution	$512 \times 14 \times 14$	100,352	2,359,808
21	Max-pooling	$512 \times 7 \times 7$	25,088	0
Flattening to a vector of length 25,088				
22	Fully connected	4,096	4,096	102,764,544
23	Fully connected	4,096	4,096	16,781,312
24	Fully connected	1,000	1,000	4,097,000
			Total: 16,391,656	Total: 143,667,240

Each input to the network is a color image x of dimension $M_c^{[0]} \times M_h^{[0]} \times M_v^{[0]} = 3 \times 224 \times 224$. In the basic form we present here, the network is configured for a classification task with

$K = 1,000$ classes. Thus, the output $\hat{y}$ of the network is a probability vector of length $1,000$, where the ith element, $\hat{y}_i$, denotes the probability of x is of class $i \in \{1, \ldots, K\}$.

In this architecture all the convolutional kernels in the network are of the same dimension, $K_h \times K_v = 3 \times 3$. The padding and stride settings are the same for all the convolutional layers with $(p_h, p_v) = (2, 2)$ for padding and $(s_h, s_v) = (1, 1)$ for stride. There is no dilation, i.e., $(d_h, d_v) = (1, 1)$. With these settings, it is evident from (6.17) that for each convolutional layer, the input height and width dimensions are identical to the output height and width dimensions. Thus with this network, height and width dimensions are reduced only via pooling. All the pooling layers are max-pooling using 2×2 windows that are moved with a stride of $(2, 2)$ without padding. Thus each such pooling layer halves the height and width dimensions. The dimensions start at 224×224 and are halved using the sequence, $224, 112, 56, 28, 14$, and 7. Yet as layers progress, more channels are added where we start with 3 channels in the input and increase to 64 channels in the first layer. Then after some of the pooling layers, we double the number of channels so that eventually by layer $\ell = 12$ there are 512 channels.

We see from Table 6.1 that the tensor output of the 21st layer, which is a max-pooling layer, is flattened to a vector that is given as an input to the first fully connected layer, layer $\ell = 22$; namely $512 \times 7 \times 7 = 25,088$. In terms of activation function, the architecture uses the Rectified Linear Unit (ReLU) activation function for all the hidden trainable layers and soft-max for the output layer so that each output assigns a probability to each of the possible 1,000 classes.

In the original VGG19 paper,[10] the network was trained on the ImageNet dataset and nowadays when one uses this network, one often uses a pretrained version. In the original paper the input images were preprocessed by subtracting the mean red, green, and blue value, computed over the entire ImageNet training set, from each pixel. This type of *preprocessing* is needed in production (test time) as well. Note that in the original paper, to obtain the input size 224×224, input images were randomly cropped from rescaled training images, one crop per image per each iteration of the stochastic gradient descent optimization algorithm. This type of data augmentation is further discussed in Chapter 8.

One by One Convolutions and Fully Convolutional Networks

A *one by one convolutional layer* is a special case of a convolutional layer where we apply $K_h^{[\ell]} \times K_v^{[\ell]} = 1 \times 1$ dimensional kernel matrices on all the input channels. At first glance, if one returns to the basics of two-dimensional convolutions as in (6.7), it may seem like a one by one convolution is nothing but a scalar multiplication. However, since now there are $K_c^{[\ell]}$ (or $M_c^{[\ell-1]}$) channels at play, the one by one convolution allows us to create a linear combination of the input channels. For example, in image processing when one converts a red, green, and blue color image into a monochrome (black and white) image, one way to do so is to define each monochrome pixel as a linear combination of the three color pixel values, and this is a one by one convolution.

One obvious application of one by one convolutions is for the reduction of depth (number of channels) inside convolutional neural networks without changing the spatial dimension.

---

[10]The VGG19 architecture achieved state-of-the-art performance on the ImageNet classification task in 2014, with a top-5 error rate of 7.3%. This network is often used as a pre-trained model for transfer learning tasks, where the lower layers are fixed and the higher layers are fine-tuned for a specific task.

Return to the VGG19 architecture in Table 6.1 and observe that from layer 0 to layer 21 depth either stays the same or grows (starting at 3 and reaching 512). However, in contrast to VGG19 that flattens layer 21, say we wanted to have a layer, which we call a *depth reduction layer*, straight after layer 21, which reduces the depth from 512 channels to a lower number. This can be viewed as a non-linear projection of the 512 channels in layer 21 onto a tensor of lower dimension with less channels. Clearly, one by one convolutions offer a natural way for such depth reduction where we set the number of one by one convolution kernels as the desired number of output channels of the reduction layer. So for example in VGG19 if we would have wanted layer 22 to be a tensor of dimension $8 \times 7 \times 7$ instead of the fully connected layer as in Table 6.1, then we would introduce 8 one-dimensional convolutions for that layer. The total parameter count for that layer would be $8 \times 512 + 8 = 4{,}112$, where the additional $+8$ is for the bias term of each of the 8 one by one convolutions.

Importantly, one by one convolutions also allow us to represent fully connected layers as convolutional layers. To see this, recall that a fully connected layer relies on an affine transformation on some input vector, say x of length N, to obtain $z = Wx + b$, where W is the weight matrix with N columns, and b is the bias vector. In that case, the j-th element of z is,

$$z_j = \left(\sum_{i=1}^{N} W_{j,i}\, x_i \right) + b_j. \tag{6.24}$$

Now return to (6.21) and consider a one by one convolution on a volume x of dimension $N \times 1 \times 1$. In this case $x_{(i)}$ can simply be represented as x_i and the $\star$ operation can be replaced by multiplication. Omitting the superscript "[1]" in (6.21), we have

$$z_{(j)} = \left(\sum_{i=1}^{N} W_{(j),(i)} \cdot x_i \right) + b_{(j)}. \tag{6.25}$$

Hence, we see that the fully connected operation (6.24) and the one by one convolution operation (6.25) are essentially identical.

In general a convolutional network that does not have fully connected layers and has all trained weights and biases associated with convolutional layers is called a *fully convolutional network*. In essence a non-fully convolutional network such as VGG19 may be transformed into a fully convolutional network by replacing the fully connected layers using one by one convolutions. This process sometimes termed *convolutionalization*. For example for VGG19 this means transforming layers 22, 23, and 24, as in Table 6.1, into convolutional layers. There are multiple reasons for convolutionalization and multiple advantages to fully convolutional architectures. Primarily, the representation of fully connected layers as convolutional layers allows us to stack multiple parallel outputs or intermediate channels in a single tensor.

Dropout, Batch Normalization, and Group Normalization

Some of the techniques introduced in Chapter 5 for fully connected neural networks are also applicable in convolutional networks. We now discuss two such techniques, namely dropout and batch normalization. We also highlight group normalization which is a variant of batch normalization in the context of convolutional neural networks.

Recall from Section 5.7 that *dropout* is a simple regularization technique where during each forward pass in the training, only a random subset of the neurons (randomly selected for that iteration) is used. In convolutional networks, we can still employ dropout for the fully connected layers but not for the convolutional layers. This is because in convolutional layers, the neurons have spatial orientation, and dropping out individual neurons could disrupt the spatial structure.

Batch normalization, introduced in Section 5.6, often accelerates learning. The key idea is a shifting and scaling transformation using additional learned parameters as in (5.40) of Chapter 5 which generally maintains the activation values in a dynamic range near 0. For convolutional neural networks, batch normalization at a convolutional layer ℓ is usually applied on each channel $z_{(j)}^{[\ell]}$ of the convolution output $z^{[\ell]}$ before the corresponding activation is applied. That is, for two learned scalar parameters $\gamma_j^{[\ell]}$ and $\beta_j^{[\ell]}$, the jth channel matrix of dimension $M_h^{[\ell]} \times M_v^{[\ell]}$ after the batch normalization is given by

$$\tilde{z}_{(j)}^{[\ell]} = \gamma_j^{[\ell]} \bar{z}_{(j)}^{[\ell]} + \beta_j^{[\ell]}, \tag{6.26}$$

for each $j = 1, \ldots, M_c^{[\ell]}$, where as before we use "+" for addition of the scalar to every element of the matrix. Here, the matrix being transformed has (i', j')-element $[\bar{z}_{(j)}^{[\ell]}]_{i',j'}$ that is computed similar to (5.39) by subtracting the mean and then dividing it by the square-root of the variance plus a small constant ε, where the mean and variance are computed for the same element (i', j') of jth channel of the convolution output $z_{(j)}^{[\ell]}$ over the entire mini-batch, similar to (5.38).

A variant that has gained popularity is called *group normalization*. Here, instead of normalizing each channel (applying (6.26) on a standardized $\bar{z}_{(j)}^{[\ell]}$), the channels of the convolution output are divided into a set of groups, and then the mean and variance values are computed for each group over a mini-batch and similarly a form of (6.26) is applied per-group. Hence the learned parameters (γs and βs) are per group in a layer. Note that the group normalization is identical to the batch normalization when the number of groups is equal to the number of channels, but otherwise it reduces the number of learned parameters.

Understanding Inner Layers and Derived Features

Recall an elementary example from Section 2.2 where we estimated a simple linear regression coefficient $\hat{\beta}_1$ to have a value of 8.27. In that simple example, the *interpretation* of the estimated parameter was clear: A unit increase of the feature implies an increase in the output by 8.27. Thus with linear models, beyond using the model for prediction, the actual learned parameters have meaning. Ideally, for deep learning models in general, and particularly for convolutional neural networks, we would also like to have such an interpretation of the learned parameters. Namely, what information do we know based on the learned convolution kernels, weight matrices, and bias vectors. However, convolutional (and deep) models with millions of parameters are much more involved, and simple direct interpretability is typically not attainable.[11]

While direct interpretability is not possible, there are multiple techniques for visualizing convolutional neural networks. We now briefly summarize some overarching approaches which we dichotomize as either *weight-based* or *feature-based*. The weight-based approach

---

[11]We mention that there is a whole field dealing with *interpretable machine learning*. In this subsection our goal is to only present a glimpse of the area.

focuses on visualizing the learned convolution kernels of the network. The feature-based approach focuses on the activation values in specific channels and has several variants.

Starting with the *weight-based approach*, visualizing the weights of convolution kernels with K_c at most 3 is possible just by treating the kernel as a red, green, and blue image and displaying it. For many architectures, this is possible at the first layer since the input has 3 channels (hence $K_c = 3$). In fact, for many trained models, the color image visualization of first layer convolution kernels shows that these filters are similar in nature to simple engineered filters such as edge detectors. On the other hand, for layers down the network, there are often more than 3 channels, and while we may try to use data reduction techniques to visualize the associated filters (each with $K_c > 3$), such a visualization is typically not fruitful.

Continuing to the *feature-based approach*, we focus on the values of activations in specific channels in the network. A simple mechanism is to apply different categories of images and examine which neurons or activations are most excited by which category. A slightly more sophisticated feature-based approach is via the application of *occlusions* (covering part of the view). The basic idea is to first consider a non-occluded image, and then occlude the image by covering up some interesting part such as a face of a person. We then compare the difference in neuron activation values for the non-occluded and occluded inputs. Neurons for which the difference in activation values is significant may then be interpreted as being sensitive to the occluded part of the image (e.g., to a face).

All the aforementioned approaches are simple in the sense that they do not rely on an additional model, but rather just use the trained model under study. However, there are multiple approaches that execute additional optimization for better interpretability insights. As an illustration, let us see one such approach stemming from a landmark paper.[12] In addition to the methodological contribution, the work of this paper also highlighted important structural aspects of trained convolutional neural networks. Specifically, it was shown that initial layers of the network generally seek simple visual features such as corners, colors, and edges, while later layers of the network find much more refined artifacts such as faces, or specific objects.

Consider Figure 6.12 which illustrates a visual interpretation of some channels within a trained convolutional network. The network has many channels across multiple layers, and here we present only a few of those channels, focusing on a pair of arbitrary channels within each of the layers 2, 3, 4, and 5. Before we outline how the visualization in this figure was created, let us interpret it. Each channel that we visualize has a 3×3 grid of synthesized images (channel visualization) as well as a matching 3×3 grid of parts of images from a dataset (original receptive field). These channel visualizations and original receptive fields can serve as a visual interpretation of what the specific channel detects.

For example, we see that the 3 channels visualized in layer 2 detect simple features with one channel focusing on edges and another channel focusing on circles. As we advance deeper in the network we see that the type of visual patterns detected are much more complex. For example, the 2 channels presented for layer 4 detect parts of animals, and the channels of layer 5 detect such representations as well. Note however, that one of the channels in layer 5 that we present appears to detect either faces or car wheels even though these are very different objects. Hence any attempt to categorize channels based on their "meaning"

---

[12]See "Visualizing and Understanding Convolutional Networks" by M. Zeiler and R. Fergus, [441].

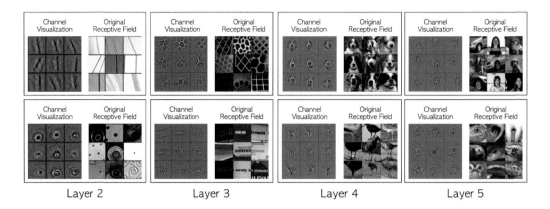

Layer 2 Layer 3 Layer 4 Layer 5

Figure 6.12: Visualization of the meaning of channels of a trained model.[13] We present two arbitrary channels for each of layers, 2, 3, 4, and 5 and for each channel we see the 9 images that yield 9 highest activation values. The gray background images (channel visualization) are processed via a deconvolution network from feature space back to pixel space. The original receptive field color images are the associated receptive fields in the images that excite those activations. It can be seen that initial layers search for more elementary features and layers deeper in the network search for more refined features. Importantly, it appears the type of features searched for in each channel are generally homogenous (although this not always the case, as is evident with the top channel presented for layer 5).

alone is far from absolute. Nevertheless, a visual representation such as that in Figure 6.12 can help to understand the function of individual channels within the network.

Let us now indicate how a visualization such as that in Figure 6.12 can be created. We may focus on any arbitrary specific channel j in layer ℓ. A validation set of images is run though the network and for each image we consider the activation matrix $a_{(j)}^{[\ell]}$ of the channel. We compute $\eta_{(j)}^{[\ell]} = \max_{i,k} \left[a_{(j)}^{[\ell]} \right]_{i,k}$, where the maximum is over the pixel coordinates $(i,k) \in \{1, \ldots, M_h^{[\ell]}\} \times \{1, \ldots, M_v^{[\ell]}\}$. We also keep the coordinates that attain this maximum, denoted here via i_* and k_*. The idea is to find the neuron or activation, that is maximally activated by each image in the validation set. Doing so for all images in the validation set, we then take the 9 images that achieve the maximal $\eta_{(j)}^{[\ell]}$ values and these 9 images are used for visualization of that channel. Now for each image out of those 9 images we take the coordinate (i_*, k_*) of the maximally activated neuron, and determine the receptive field of that neuron within the input image. We then present the receptive field part of the input image for each of the 9 images. This visualization then illustrates the 9 most significant image patches for the channel at question.

As for the channel visualization part (gray background images) of Figure 6.12, a more sophisticated process is carried out on each of the 9 selected images per channel. A type of network, called a *deconvolution architecture* is constructed in parallel to the original convolutional network. This combined architecture enables transforming "feature space" back to "pixel space" for individual input images and specific neuron locations. That is, an input image to the original network is first processed. Then with a specific neuron (i_*, k_*) in channel j of layer ℓ, specified, the deconvolution architecture returns an image associated

---

[13]Image is adapted from figure 2 of "Visualizing and understanding convolutional networks", [441] with thanks to M. D. Zeiler and R. Fergus.

with the receptive field of that particular neuron in "pixel space". While we do not specify the details of this particular deconvolution architecture, let us mention how it is used for channel visualization. Each gray background channel visualization image in Figure 6.12, is an output in pixel space, resulting from the associated neuron, (i_*, j_*) in channel j of layer ℓ specified to the deconvolution architecture. For this all other neurons in channel j of layer ℓ are set to 0, except for $[a_{(j)}^{[\ell]}]_{i_*,k_*}$ which is activated. The deconvolution architecture, then works backward in the network from layer ℓ to layer $\ell-1$, and back, all the way until presenting the result in pixel space. The benefit of such channel visualization images is that they allow us to see how a single neuron $[a_{(j)}^{[\ell]}]_{i_*,k_*}$ "appears" in "pixel space". Importantly, we see that the 9 most activated neurons in the same channel are generally of same nature.

6.5 Inception, ResNets, and Other Landmark Architectures

So far we covered the key components of convolutional neural networks and considered the VGG19 model as one concrete network example. In this section we highlight other landmark architectures within the world of convolutional neural networks. Our main goal is to highlight ideas stemming from these architectures.

In general, the book avoids historical accounts as much as possible, yet in the context of network architectures, some knowledge of the historical progression might be practically useful. We thus begin with a brief historical account naming key architectures. We then focus on three architectural ideas, namely the *network within a network* (*inception network* also known as *GoogLeNet*), *residual connections* (ResNets), and efficient model scaling as in *EfficientNet*. See also the notes and references at the end of the chapter for further information.

A Brief Historical Account

As with many ideas in deep learning, with convolutional networks one can find quite early roots. In this case, early convolutional networks include the *Neocognitron* from the late 1970s and early 1980s and *LeNet-5* worked on during the mid-1980s until the late 1990s. Both of these networks already encompass many of the ideas presented in this chapter, yet in those days computation power was lacking and ease of implementation with software was much less advanced.

The architecture that really advanced deep learning as a whole, and particularly convolutional neural networks is *AlexNet* from 2012. At the time, it was a breakthrough in image classification, achieving state-of-the-art performance on the ImageNet dataset. The architecture consists of five convolutional layers followed by three fully connected layers. It also uses two parallel computation streams allowing the network to execute parallel forward propagation and backpropagation, using two state-of-the-art GPUs of the time. The work on AlexNet also introduced several innovations that are now commonplace, such as the use of ReLU activation functions and dropout regularization. While today, AlexNet is probably not the first off-the shelf model that one would use, it can still be cast as the first "modern convolutional neural network". From a research and applied perspective, it was the success of AlexNet that sparked the start of the deep learning era. After the introduction and success of AlexNet, hundreds (and now many thousands) of researchers, both applied and theoretical, shifted focus toward deep learning. This heavy research effort accelerated advances in the field.

Architectures that followed AlexNet include *ZFNet* in 2013, *VGGNet* (including VGG19) in 2014, *GoogLeNet* in 2014, and *ResNet* in 2015. This short sequence of advances marks the main evolution of convolutional architectures to what they are today. In more recent years, vision tasks have also been tackled by non-convolutional networks using *transformers*. For such ideas see Chapter 7, describing transformers in the context of sequence or language models, and see Chapter 8 where we highlight how ideas from different deep learning domains interplay. Nevertheless, convolutional networks remain the bread and butter of modern computer vision. A recent advance that we cover below is *EfficientNet*. This set of models tries to optimally scale models to balance performance and model size.

Inception and Networks within a Network

The *inception network*, also called *GoogLeNet*, works by composing multiple sub-networks into a bigger network. This idea is sometimes called a *network within a network*. Each sub-network is called an *inception module* and such a module uses multiple filter sizes in parallel.

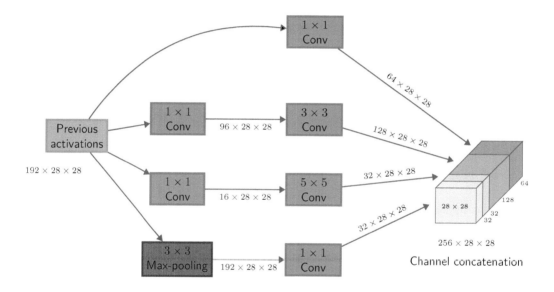

Figure 6.13: One form of an inception module, playing part in an inception network. The key idea is parallel computation of various paths followed by a concatenating of the outputs from all paths.

Figure 6.13 illustrates an example of one such inception module. In this example a volume of previous activations of dimension $192 \times 28 \times 28$ is transformed to an output volume of dimension $256 \times 28 \times 28$ (the number of channels grows from 192 to 256). Inside the inception module, there are four parallel paths, each operating independently and producing its own set of output channels. Then the outputs of these paths are concatenated.

The different paths of the inception module are designed to handle different scales and resolutions. The first path has 1×1 convolutions and yields 64 output channels. The second path has 1×1 convolutions followed by 3×3 convolutions. This path produces 128 output channels (with an intermediate number of 96 channels). The third path is similar, yet uses 5×5 convolutions instead of 3×3. It results in 32 output channels with 16 intermediate channels. Finally the last path starts with a max-pooling operation with a max-pooling stride of 1, such that there is no reduction in spatial dimension, but rather only a non-linear operation. This is then followed by 1×1 convolutions. This path results in 192 output channels. All convolutions have ReLU activations and where needed, there is padding such that the desired spatial dimensions are respected. It should be noted that when the inception network was developed, mass experiments were conducted to seek near-optimal settings for this inception module and similar ones.

The essence of the inception network is to interconnect such inception modules in series. The concatenated channels that result as output from one module are given as input to the next module. One aspect of this interconnection is that the number of channels generally grows down the network. To mitigate such channel explosion, one by one convolutional layers are placed between some of the inception modules. Such layers have been termed *bottleneck layers*. Another aspect introduced with such networks was intermediate loss functions. Here the idea is that in addition to the final loss function at the exit of the network, the loss is also computed at various intermediate "exit points" and the gradient-based optimization uses the sum of all loss functions.

Empirically, in 2014, the introduction of GoogLeNet outperformed other networks at the time and importantly these types of networks appear to strike a balance between accuracy and computational efficiency. The original GoogLeNet has about 6.8 million parameters which is much less than the 143.7 million parameters of VGG19. GoogLeNet can be viewed as having 22 parameterized layers with a total of 9 inception modules, 2 convolutional layers as initial layers, and only a single dense layer at the output. Practically, these days when one wishes to use an off the shelf trained convolutional neural network, some variant of GoogLeNet is often a prime choice.

Residual Connections

Recall early discussions in Section 2.5, and in particular Figure 2.9. There we claimed that in general, as model complexity grows we expect training error to decrease simply because our model is able to capture more complex relationships. With deep learning one would also hope to see this type of phenomena when adding layers. However, this is only partially true. Empirically it has been observed that when deep learning models get extremely deep with dozens or hundreds of layers, training error actually starts to increase. In other words, as we add more layers to a neural network, its training error initially decreases, but after a certain depth, the network's accuracy on the training set starts to saturate and sometimes degrades. One reason for this phenomenon, which is often termed a *degradation problem*, stems from vanishing and exploding gradient issues. When a gradient is backpropagated through multiple layers, it can become extremely small, causing the weights in the earlier layers to receive almost no updates during training. As a result, the network's ability to learn and generalize is reduced. See Section 5.4 for a discussion of vanishing and exploding gradients.

Some further insight into the computational problems for very deep models is as follows. We may hypothesize that good parameters, $\theta^{[\ell]}$, for layer ℓ, are such that the operation of the layer $f_{\theta^{[\ell]}}^{[\ell]}(\cdot)$ is approximately an identity. Namely the input to the layer, $a^{[\ell-1]}$, and the output, $a^{[\ell]}$, are ideally very similar. This can be hypothesized because with deep models we would expect individual layers to only apply minor variations on their inputs. If we accept such a hypothesis, then we immediately get insight into some of the numerical and computational problems that learning entails. Specifically, learning functions close to $f_{\theta^{[\ell]}}^{[\ell]}(u) = u$ is often not trivial — for example, consider a pure convolutional layer and observe that for it to be an identity function, the convolution kernel requires to be all zeros except for a single entry that is 1. For this, iterating over parameters of convolutional layers until they become close to identity requires many gradient descent steps and can run into numerical problems.

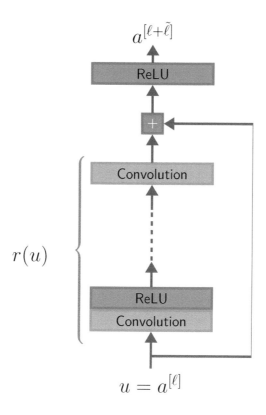

Figure 6.14: A shortcut connection (residual connection) as part of a residual network.

An approach to overcome this problem is to use *shortcut connections* as in Figure 6.14. Here the key idea is to take the input before a given layer (or sequence of layers), and bypass the layer (or the sequence of layers). Then the bypassed information is added to the output down the network, typically before the application of an activation function. Note that as we bypass layers, it may be that channel dimensions are different. In such a case, we use one by one convolutions, and similarly we may use padding, stride, and pooling to adjust the spatial dimensions if needed.

Mathematically, and continuing with the hypothesis that layers should be close to identity functions, we may view the shortcut connection approach as a means to set the bypassed layer (or layers) to a function that approximately outputs zero. To see this return to Figure 6.14 and assume that $r(u) \approx 0$. This then makes the operation of the whole sequence of layers with a bypass close to the identity. Specifically in the figure we bypass $\tilde{\ell}$ layers, and if $r(u) \approx 0$, then $a^{[\ell+\tilde{\ell}]} \approx a^{[\ell]}$. Due to this reason the shortcut connections are also sometimes called *residual connections* and the whole architecture is called a *ResNet*. The usage of the term, "residual" implies that by adding a shortcut connection, we are now learning $r(u)$ to as a deviation from zero, or a residual.

When the ResNet idea was introduced in 2015, networks of depths of dozens and even more than one hundred layers were able to be efficiently trained. This elegant and simple idea allows us to learn residuals instead of actual transformations. Ideas from ResNets propagated to other aspects of deep learning beyond convolutional neural networks, such as some sequence models presented in the next chapter. Several models combine residual connections and inception modules, and these models are near the state of the art of convolutional neural networks.

EfficientNet Models

EfficientNet is a family of convolutional neural network architectures that were developed with the aim of providing better accuracy and efficiency in terms of model size and computation cost. The key idea is to systematically scale up the dimensions of the network's parameters (such as depth, width, and resolution) in a balanced way, while also introducing a new compound scaling method that optimizes these dimensions based on a set of pre-defined constraints. This allows users to choose which form of EfficientNet model they want, in a way that balances the number of parameters and the performance of the model. Figure 6.15 plots the parameter count vs. performance tradeoffs of efficient net models. The models are named B0, B1, ..., B7 where B0 is the most lightweight model in terms of parameter counts, and B7 is the most computationally demanding model. It is seen that EfficientNet dominates other popular models.

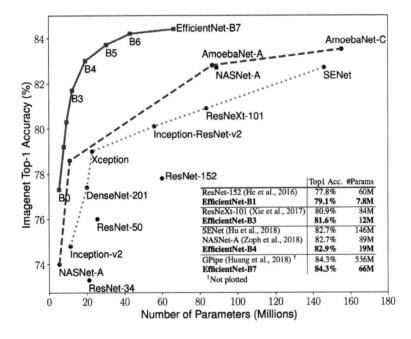

Figure 6.15: Performance of various convolutional models as well as efficient net.[14]

---

[14]Image thanks to M. Tan and Q. V. Le, taken from "EfficientNet: Rethinking Model Scaling for Convolutional Neural Networks", [395]. See also [396].

6.6 Beyond Classification

The sections above focus on the internals of convolutional neural networks. For simplicity in those sections, we discuss the task of image classification, e.g., determining if an image is that of a cat or a dog. However, there are several other important image analysis tasks that are also handled with convolutional neural networks. These tasks deal with analysis and understanding of an image including the location of objects, the count of objects, separating different semantic features of the image, and more. Our purpose in this section is to highlight such tasks. For this we present a brief overview of key *computer vision* developments that use convolutional neural networks for tasks beyond classification.

In terms of the input data, it is important to keep in mind that not all data is of the form of monochrome or color images. Within computer vision, one often deals with *image sequences* (short movies), or images that have more than 3 channels. For example, some images may also have a *distance channel* capturing the distance from the camera per pixel. Further, non-image data can also be handled via convolutional networks. One such example is *fMRI (functional magnetic resonance imaging)* data which is four-dimensional in nature as it records the state of physical locations in three dimensions over time. Nevertheless, most of our attention in this section is restricted to images.

Convolutional Networks and Key Computer Vision Tasks

As mentioned above, classification serves as a simple and useful example. For an input image x, a convolutional neural network $f_\theta(\cdot)$, has output $\hat{y} = f_\theta(x)$ which is a vector of probabilities where the highest probability typically determines the appropriate label for the image. As was evident from our detailed study of the VGG19 model in Section 6.4 and other architectures of Section 6.5, initial layers of the model $f_\theta(\cdot)$ are typically convolutional, and the final layers are typically fully connected layers. These final layers help transform the internal derived features in the network into the output vector of probabilities $\hat{y}$. When one considers tasks other than classification, it is often common to replace the final layers of the network with other layers such that the output $\hat{y}$ suites the desired task. With such a replacement we typically keep in the initial layers as is.

Let us now get a feel for some of these tasks and in each case consider some possible structure for the output $\hat{y}$. Figure 6.16 illustrates key computer vision tasks for images. In (a) we see *object localization* which is the task of identifying the location of an object in an image, as well as possibly the type of object in which case the task is called *localization and classification*. In (b) we see *object detection* which is the task of detecting multiple instances of an object in an image, also separating the objects and classifying their type. In (c) we see *landmark detection* which is the task of identifying the specific pixel locations of landmarks in an image. In (d) we see *semantic segmentation* which is the process of classifying each individual pixel to be of a different class from a finite set of classes (pixel wise classification). In (e) we see *instance segmentation* which finds different instances of objects in the image and separates pixels to be of different instances. Finally, in (f) we see the task of *identification* or more specifically *face recognition* which determines if an image is that of a specific instance (or person).

Let us now consider possible forms of the output $\hat{y}$. For object localization, (a) in Figure 6.16, $\hat{y}$ needs to contain information about a *bounding box* which locates the object. This can be in the form of $(\hat{y}_x, \hat{y}_y, \hat{y}_h, \hat{y}_w)$ where $\hat{y}_x$ and $\hat{y}_y$ are the coordinates of (say) the upper left corner

Figure 6.16: Illustrations[15] of some common computer vision tasks beyond classification: (a) Object localization. (b) Object detection. (c) Landmark detection. (d) Semantic segmentation. (e) Instance segmentation. (f) Identification (face recognition).

of the bounding box and $\hat{y}_h$, $\hat{y}_w$ are the height and width of the bounding box, respectively. This information can also be augmented with probabilities for the respective classes (types of objects) including the possibility of having no object. For object detection, (b) in the figure, a collection of multiple bounding boxes needs to be supplied. For (c), landmark detection, a list of coordinates of the locations of landmarks comprises the output. For (d) semantic

[15]Image (b) is thanks to J. Redmon, S. Divvala, R. Girshick, and A. Farhadi, taken from "You only look once: Unified, real-time object detection", [346]. Image (c) is thanks to H. Lai, S. Xiao, Y. Pan, Z. Cui, J. Feng, C. Xu, J. Yin, and S. Yan, taken from "Deep recurrent regression for facial landmark detection", [246]. Image (d) is attributed to B. Palac under the creative commons license and available via Wikimedia Commons. Image (e) is thanks to K. He, G. Gkioxari, P. Dollár, and R. Girshick, taken from "Mask R-CNN", [170].

segmentation, each pixel location in the input image, x, has an associated probability vector of classes in the output $\hat{y}$. Hence in this case, $\hat{y}$ can be represented as a tensor with width and height dimensions the same as the input image, and a depth dimension which is the number of classes in the segmentation. For (e), instance segmentation, the output is similar to that of semantic segmentation, but instead of recording probabilities of classes, the depth dimension of the output $\hat{y}$ is used for determining the specific instance of any given pixel. Finally, in the case of identification, or face recognition, as in (f) of Figure 6.16, the output is often just a probability as in a binary classifier, since the task is to determine if a face image matches a given pre-stored template or not. Note that in this case, the input x is typically composed of two images, where one image, say x_a, is the template of the person (e.g., a stored image in a security database), and the other image, say x_o, is the other image.

There are many ideas that have gone into developing architectures for handling tasks (a)–(f). Some of these ideas stem from vision analysis research, prior to the era of deep learning, while other ideas evolved in parallel to deep learning in recent years. Object localization and classification as in (a) is a particularly simple example and for this we provide more details below. Similarly, identification (face recognition) is also worth consideration and we provide more details below. Landmark detection (c) is handled easily also in a similar spirit to object localization and classification; we omit the details. The other tasks including object detection (b), semantic segmentation (d), and instance segmentation (e), are each big topics of their own and we leave investigation of these for further reading. See the notes and references at the end of the chapter.

Object Localization

To get a feel for object localization assume that we wish to train a convolutional neural network that operates on an input image x and determines if the image contains a `bird` or a `plane` (classification). The model's second goal is to determine the specific location $(\hat{y}_x, \hat{y}_y, \hat{y}_h, \hat{y}_w)$ of that object (localization). Images with multiple birds or planes are not considered. Images without a bird and without a plane are possible and in this case the output yields `nothing`. One way to encode the output is $\hat{y} = (\hat{p}_{\texttt{nothing}}, \hat{p}_{\texttt{bird}}, \hat{p}_{\texttt{plane}}, \hat{y}_x, \hat{y}_y, \hat{y}_h, \hat{y}_w)$, where as in standard classification examples $(\hat{p}_{\texttt{nothing}}, \hat{p}_{\texttt{bird}}, \hat{p}_{\texttt{plane}})$ is a probability vector, and the other coordinates define a bounding box.

Here an output that has $\hat{p}_{\texttt{nothing}}$ greater than each of $\hat{p}_{\texttt{bird}}$ and $\hat{p}_{\texttt{plane}}$ implies a prediction of no bird and no plane. On the contrary if $\hat{p}_{\texttt{bird}}$ is the highest probability, then the output implies there is a bird, located in the bounding box $(\hat{y}_x, \hat{y}_y, \hat{y}_h, \hat{y}_w)$. Similarly for the other class, `plane`.

In terms of training data, for each input image we denote the output as y where images without a bird or a plane are labeled as, $y = (1, 0, 0, \emptyset, \emptyset, \emptyset, \emptyset)$, where $\emptyset$ are "do not care" values. Images with a bird are labeled as $y = (0, 1, 0, y_x, y_y, y_h, y_w)$ where the bounding box (y_x, y_y, y_h, y_w) is typically based on a manual determination by a human annotator. Similarly, images with a plane are labeled as $y = (0, 0, 1, y_x, y_y, y_h, y_w)$.

We now construct a loss function that captures closeness of $\hat{y}$ and y. For this we first separate the classification and localization objectives into a loss $C_{\text{classification}}(\theta\,;\hat{y}, y)$ and $C_{\text{localization}}(\theta\,;\hat{y}, y)$. The former depends only on the probability components in $\hat{y}$ and y, and the latter depends only on the bounding box components in $\hat{y}$ and y. For the classification loss, we use categorical cross entropy as in (3.31). For the localization loss, we use a mean squared error as in (2.12) or some variant, applied to the four bounding box components.

The two separate losses are then combined such that the loss for a specific observation is

$$C_{\text{classification}}(\theta\,;\,\hat{y}, y) \;+\; \gamma \cdot (1 - y_1) \cdot C_{\text{localization}}(\theta\,;\,\hat{y}, y),$$

where $\gamma > 0$ is a hyper-parameter used to weigh the two losses and taken as $\gamma = 1$ by default. Observe that $y_1 = 1$ when the label is `nothing` and is otherwise 0 and thus for labels in the training data without a bird or a plane only the classification objective is used.

To perform object localization, say with a model like VGG19, the network can be modified by adding additional layers at the end of the architecture to predict the coordinates of the bounding box. This can be achieved by attaching a regression head to the output of the final convolutional layer of the network. The regression head consists of fully connected layers that predict the coordinates of the bounding box. Such simple modifications of networks that were otherwise designed for classification are always possible.

Face Recognition, Siamese Networks, and Triplet Loss

Let us get a feel for how identification (face recognition) as in Figure 6.16 (f) can be implemented both in production and training. First let us consider the simplified use of such a task. Say a face identification system needs to be able to recognize faces where in production one may have an *anchor* face image x_a stored. With each use, the anchor needs to be compared to another image x_o. For example every "login" is based on a new x_o image and the system needs to determine if x_o is of the same person as x_a or not. In contrast to other tasks discussed in the book, here we do not have the ability to train a network for a particular person (or face), and similarly we do not have many different face images of the same person. Hence this setup requires a slightly different architecture.

One type of architecture useful for this task is a *Siamese network*, illustrated schematically in Figure 6.17. The idea is that two parallel replicas of a convolutional neural network $f_\theta(\cdot)$ are used, one applied on x_a and the other on x_o. The output of each of these networks is an embedding vector. Now since we have two embedding vectors, $f_\theta(x_a)$ and $f_\theta(x_o)$, we can compare them and see if they are likely associated with face images of the same person or not. One approach for this comparison is as in Figure 6.17 using a *comparison network*, $f_{\theta_c}^c(\cdot,\cdot)$ for binary classification (output is a probability) with parameters θ_c. Hence, in production we can determine `same` if the output probability $f_{\theta_c}^c(f_\theta(x_a), f_\theta(x_a))$ is greater than a threshold, or otherwise determine `different`. The comparison network is not too complex, and is often a shallow logistic regression model or similar. With such an architecture, the learned parameters θ and θ_c are not designed for one particular face, but rather for any possible face.

Let us describe a simplified approach for training such a model. We can first treat the parameters θ as known, say from a pertained or fine-tuned model, and focus on learning the parameters of the (smaller) comparison network θ_c. For this, the training data can be of the form $\mathcal{D} = \{(x_a^{(1)}, x_o^{(1)}, y^{(1)}), \ldots, (x_a^{(n)}, x_o^{(n)}, y^{(n)})\}$, where each tuple $(x_a^{(i)}, x_o^{(i)}, y^{(i)})$, has an anchor image $x_a^{(i)}$, another image $x_o^{(i)}$, and a binary label $y^{(i)} \in \{0, 1\}$ with the value based on the images being `different` (0) or `same` (1). With such a dataset all that is required is to train the binary classifier $f_{\theta_c}^c(\cdot)$.

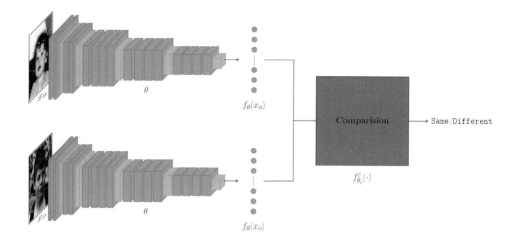

Figure 6.17: A schematic of a Siamese network architecture for identification (face recognition). The two parallel convolutional neural networks both share the same parameters θ, and one operates on x_a while the other operates on x_o. The outputs of these networks are embedding vectors. These are then compared via a comparison module which may be a neural network with parameters θ_c and has output indicating if `same` or `different`.

A related useful concept for training Siamese networks is the *triplet loss*. Say for simplicity that now our goal is to learn θ for $f_\theta(\cdot)$, ignoring the comparison network. For this we can setup a slightly different dataset of the form $\mathcal{D}_{\text{triplet}} = \left\{ (x_a^{(1)}, x_d^{(1)}, x_s^{(1)}), \ldots, (x_a^{(n)}, x_d^{(n)}, x_s^{(n)}) \right\}$ where now each $(x_a^{(i)}, x_d^{(i)}, x_s^{(i)})$ has an anchor face image $x_a^{(i)}$ as before, and also has two additional images with $x_d^{(i)}$ being a face image of a different person, and $x_s^{(i)}$ being a face image of the same person (not the exact same image as $x_a^{(i)}$). Now by applying $f_\theta(\cdot)$ on each element of this dataset, we can construct a loss function for observation i as

$$ C_i(\theta\,;\, x_a^{(i)}, x_s^{(i)}, x_d^{(i)}) = \max \Big\{ \underbrace{\| f_\theta(x_a^{(i)}) - f_\theta(x_s^{(i)}) \|^2}_{d_{\text{same}}} - \underbrace{\| f_\theta(x_a^{(i)}) - f_\theta(x_d^{(i)}) \|^2}_{d_{\text{different}}} + \alpha, 0 \Big\}, \quad (6.27) $$

where $\alpha > 0$ is some hyper-parameter called the *margin* and the Euclidean norm $\| \cdot \|$ can in principle be replaced by a different distance metric as well.

Let us understand the motivation behind the triplet loss (6.27). Our desire is that the embedding associated with the anchor image $x_a^{(i)}$ and embedding associated with the image of the same person $x_s^{(i)}$ be close to each other and hence d_{same} should ideally be small. Similarly we wish to have the embedding of $x_a^{(i)}$ and the embedding $x_d^{(i)}$ to be distant from each other and this motivates the negative sign in front of the $d_{\text{different}}$ term which we ideally want to be large.

Now in general, when we have such an optimization with two competing criteria, d_{same} which we want to be small, and $d_{\text{different}}$ which we want to be large, one approach to capture such a desire via a loss, is by pre-determining a margin α and considering cases where

$$ d_{\text{same}} - d_{\text{different}} \leq -\alpha \qquad (6.28) $$

as being "admissible" and otherwise "inadmissible". We can then assign a loss of 0 to admissible cases, and assign a loss that depends on θ for the inadmissible cases. This is achieved with the $\max\{\cdot, 0\}$ operation since if (6.28) is satisfied, the loss in (6.27) is 0. In contrast, when (6.28) is not satisfied (inadmissible), the loss in (6.27) is $d_{\text{same}} - d_{\text{different}} + \alpha$. Hence when using gradient descent based learning of θ for minimization of $\sum_{i=1}^{n} C_i(\theta; \mathcal{D}_{\text{triplet}})$, at any iteration, we drive loss down for the inadmissible observations.

The triplet loss with a properly curated dataset $\mathcal{D}_{\text{triplet}}$ has been effectively used for state-of-the-art face recognition training. We note that when curating this dataset it is often important to preprocess the images so that $x_a^{(i)}$ and $x_d^{(i)}$ are not acutely different. We also note that with the use of the triplet loss we can add a comparison network $f_{\theta_c}^c(\cdot)$ which is trained as a binary classifier, after training θ with the triplet loss. In other cases, using the cosine distance between the two embedding vectors $f_\theta(x_a)$ and $f_\theta(x_o)$ suffices in production.

Notes and References

Before we outline notes and references associated with explicit details of this chapter, here is a brief description of early *convolutional neural network* developments. Initial ideas originated in the 1950s and 1960s with the study of the visual cortices of animals, primarily by Hubel and Wiesel over a series of publications including [195] and [196]. Early concrete models that have some similarity with modern convolutional neural networks are the 1980 *neocognitron* [127] for pattern recognition, as well as the 1988 *time delay neural network* [247] for speech recognition. In the 1990s convolutional neural networks saw industrial applications for the first time with [154] for handwritten character recognition and [65] for signature verification. Other significant early works include [250] for written digit recognition, [402] for face recognition, and [412] for phoneme recognition. Finally we mention that the *LeNet-5* model developed in the late 1990s by Yann LeCun et al. [252] is recognized as an early form of contemporary convolutional neural networks and it was used for classifying 28×28 size images of grayscale handwritten digits. We also mention that in 1989 with [250] and [251], LeCun et al. developed the first multi-layered convolutional networks for handwritten character recognition trained using *backpropagation*.

The structure of convolutional layers in neural networks as we present in this chapter solidified at around the 2012–2016 period and best fits the *VGG model* [380]. This model followed the pivotal *AlexNet* model [239] from 2012 which was specifically designed for training on two parallel GPUs. Other notable convolutional architectures of this period are the *GoogLeNet* or *inception network* model of [394], the *batch normalization inception* model [203] which uses batch normalization of layer inputs, and *ResNets* which were introduced in [172]. All of these models competed in the *ImageNet challenges* of that era with the results from each model effectively outperforming those that came prior to it. Other developments included the *SqueezeNet* model of [199], which marked a key milestone in reducing parameter size and memory footprint of convolutional network without compromising accuracy; this model achieved the AlexNet-level accuracy with much fewer parameters and a much smaller memory footprint. Also, see the *Network-in-Network* model of [260] which inspired the inception networks and [388] that uses dropout mechanism to reduce overfitting on convolutional layers. See [344] for a comprehensive survey of convolutional neural networks of that time as well as the more recent survey [258]. Closer to the publication of this book, paradigms such as *EfficientNet* appeared in [395], see also the more recent version, EfficientNet v2 in [396].

Ideas of dilation in convolutional networks were introduced in [435] for dense prediction, where the goal is to compute a label for each pixel in the image. Furthermore, dilation for residual networks is introduced in [436]. See also the discussion of *group normalization* in [426].

A general overview of *linear time invariant systems* can be found in standard texts such as [244] which is also useful for understanding basic filtering. The book [14] can provide a more mathematically rigorous foundation and can also be useful for understanding the delta function in continuous time. The probabilistic interpretation of a convolution is standard and can be found in any elementary probability textbook such as [355]. The multiplication of polynomials interpretation, also coupled with the study of the *fast Fourier transform* can be found in [93]. A simple explanation of the representation of discrete convolutions in terms of *Toeplitz matrices* can be found in [56]. For analysis of convolutions of classic image processing applications as well as many other classic image processing techniques see [207]. *The Sobel filter* is one of many convolution-based filtering operations. It was developed by Sobel and Feldman, and presented at a 1968 scientific talk; see [383] for an historical review.

The rise of convolutional neural networks drove the development of many paradigms using these networks for different tasks. In terms of object detection, early works are [135] and [136] and recent work in this direction is [415] where YOLOv7 model enhances the landmark *YOLO* (you only look once) work of [346]. A recent survey on object detection can be found in [454]. The important area of semantic segmentation has received much attention with notable papers being [352] (U-net), as well as [312]. Instance segmentation is studied in [45], [170], and [267]. For additional recent surveys of the subsequent developments in semantic and instance segmentation see [405] and [290]. See also [129] for a survey of video semantic segmentation. Influential work on identification (*face recognition*) is in [368] and early ideas of Siamese networks are from [84]; see also [192] and [425].

Over the years, many effective network visualization methods were developed for understanding inner layers and derived features. Before the era of great popularity of convolutional networks, the work in [119] introduced a technique aimed at optimizing the input to maximize the activity of hidden neurons in a deep neural network. For convolutional neural networks, the *deconvolution architecture* in [441], based on previous work in [442], was a significant one as it was the first work where effective visualization of internal layers was made possible. Other related important papers in this direction are [278] which introduced a technique called *network inversion* and [23] which introduced a framework called *network dissection*. A general useful survey on visual interpretability for deep learning is [447]. Other related ideas that we do not discuss in this book include deep dreaming[16] and directly using convolutional networks for neural style transfer, initially introduced in [131], and further developments reported in [113]. See also the related generative models of Chapter 8.

In terms of real-world applications, these days convolutional neural networks are used in many scenarios. For *image classification* applications of convolutional neural networks, see for example [372] dealing with traffic sign recognition, and [256] for medical image classification, among many others. For a review of advances in image classification, refer to [76]. The most basic application of convolutional neural networks is with three-dimensional tensors as appropriate for color images, yet there are other cases as well. In [450] four-dimensional fMRI data is studied. Also videos are analyzed in [221] by treating the entire video as a bag of short clips. In particular, see [122], [379], and [416] for video-based action recognition; a brief summary of such methods is listed in [434]. In general, techniques for analyzing video data vary depending on the task at hand; see [373] for a brief survey of such tasks and the corresponding methods. We also mention that *transformer models*, as introduced in Chapter 7 have been applied to images and managed to surpass the performance of convolutional networks in certain cases when trained with huge datasets. See [227] for a survey as well as the notes and references at the end of Chapter 7.

---

[16]This blog post is credited for introducing the concept of deep dreaming: https://blog.research. google/2015/06/inceptionism-going-deeper-into-neural.html.

7 Sequence Models

Many forms of data such as text data in the context of natural language processing appear sequentially. In such a case we require deep learning models that can operate on sequences of arbitrary length, and are well adapted to model temporal relationships in the data. The simple first model of this form is the recurrent neural network (RNN) which can be presented as a variation of the feedforward neural network of Chapter 5. In this chapter we explore such models together with many more advanced variants of these models including long short term memory (LSTM) models, gated recurrent unit (GRU) models, models based on the attention mechanism, and in particular transformer models. An archetypical application is end to end natural language translation and we see how encoder-decoder architecture with sequence models can be used for this purpose. The various forms of models including RNN, LSTM, GRU, or transformers can also be used in such an application among others. These models also form the basis for large language models (LLMs) that have been shown to be extremely powerful for general tasks.

In Section 7.1 we consider various forms and application domains of sequence data. As a prime example we consider textual data and ways of encoding textual data as a sequence. In Section 7.2 we introduce and explore basic recurrent neural networks which are naturally suited to deal with sequence data. We present the basic auto-regressive structure of such models and discuss aspects of training. In Section 7.3 we explore generalizations of recurrent neural networks including, stacking and reversing approaches, and importantly long short term memory (LSTM) models, and gated recurrent unit (GRU) models. Prior to the appearance of transformers, LSTMs and GRUs marked the state of the art for sequence modeling. We continue in Section 7.4 where we focus on machine translation applications, and explore how encoder-decoder architectures can be used for end-to-end translation. In the process we introduce the attention mechanism which has become a central pillar of modern sequence models. An encoder-decoder architecture based on attention is also presented. In Section 7.5 we dive into the powerful workhorse of contemporary sequence models, the transformer architecture. Transformers, relying heavily on attention, are presented in detail, culminating with a transformer encoder-decoder architecture.

7.1 Overview of Models and Activities for Sequence Data

Sequence models have been motivated by the analysis of sequential data including text sentences, time-series, and other discrete sequence data such as DNA. These models are especially designed to handle sequential information while convolutional neural networks of Chapter 6 are specialized for processing spatial information. Naturally, most interesting input samples carry some statistical dependence between elements due to the sequential nature of the data. Classical statistical models in time-series such as auto-regressive models are naturally tailored for such data when the sample at each datapoint is a scalar or a low-dimensional vector. In contrast, the deep learning models that we cover here allow one to work with high-dimensional samples as appearing in textual data and similar domains.

DOI: 10.1201/9781003298687-7

Forms of Sequence Data

We denote a data sequence via $x = \left(x^{\langle 1 \rangle}, \ldots, x^{\langle T \rangle}\right)$, where the superscripts $\langle t \rangle$ indicate time or position, and capture the order in the sequence. Each $x^{\langle t \rangle}$ is a p-dimensional numerical data point (or vector). The number of elements in the sequence, T, is sometimes fixed, but is also often not fixed and can be essentially unbounded. A classical example is a numerical univariate data sequences ($p = 1$) arising in time-series of economic, natural, or weather data. Similarly, multivariate time-series data ($p > 1$ but typically not huge) also arise in similar settings.

Most of the motivational examples in this chapter are from the context of textual data. In this case, t is typically not the time of the text but rather the index of the word or token[1] within a text sequence. One way to encode text is that each $x^{\langle t \rangle}$ represents a single word using an *embedding vector* in a manner that we discuss below. If for example x is the text associated with the Bible then T is large,[2] whereas if x is the text associated with a movie review as per the IMDB movie dataset (see Figure 1.6 (d) in Chapter 1), then T is on average 231 words. In data formats similar to the latter case, the data involves a collection of data sequences $\mathcal{D} = \{x^{(1)}, \ldots, x^{(n)}\}$ where each $x^{(i)}$ is an individual movie review and n denotes the total number of movie reviews. While in practice, such data formats often arise, for simplicity, our discussion in this chapter mostly assumes a single (typically long) text sequence x.

To help make the discussion concrete, assume momentarily that we encode input text in the simplest possible manner, where the embedding vector just uses a technique called *one-hot encoding*. With this approach we consider the number of words in the dictionary, vocabulary, or lexicon as $d_{\mathcal{V}}$ (e.g., $d_{\mathcal{V}} \approx 40,000$) and set $p = d_{\mathcal{V}}$. We then associate with each possible word, a unit vector $e_1, \ldots, e_p$ which uniquely identifies the word. At this point, an input data sequence (text) is converted into a sequence of vectors, where $x^{\langle t \rangle} = e_i$ whenever the t-th word in the sequence is the i-th word in lexicographic order in the dictionary. This approach is very simplistic and may appear inefficient. Yet it illustrates that textual data may be easily represented as a numerical input. With more advanced *word embedding* methods discussed below, the dimension of each $x^{\langle t \rangle}$ can be significantly reduced.

Tasks Involving Sequence Data

There are plenty of tasks and applications involving sequence data. In the context of deep learning such tasks are handled by neural network models. The more classical forms of neural networks for sequence data are generally called *recurrent neural networks* (RNN), while more modern forms are called *transformers*. We focus on the more classical RNN forms in the first sections of this chapter and later visit transformers. The basic forms of RNNs are introduced in Section 7.2. At this point assume that each of these models processes an input sequence x to create some output $\hat{y}$, where the creation of the output is sequential in nature.

For our discussion, let us focus on text-based applications and highlight a few of the tasks and applications in this context. The ideas may then be adapted to domains such as time-series, signal processing, and others. Figure 7.1 illustrates schematically how RNNs (or their

---

[1] In a complete treatment of textual data analysis or *natural language processing* (NLP) one requires to define and analyze *tokenizers* which break up text into natural "words" or parts of words known as *tokens*. These details are not our focus and we use "word" and "token" synonymously.

[2] By some counts, there are about half a million sequential words in the old testament of the Bible and more when one considers the new testament and its many variants.

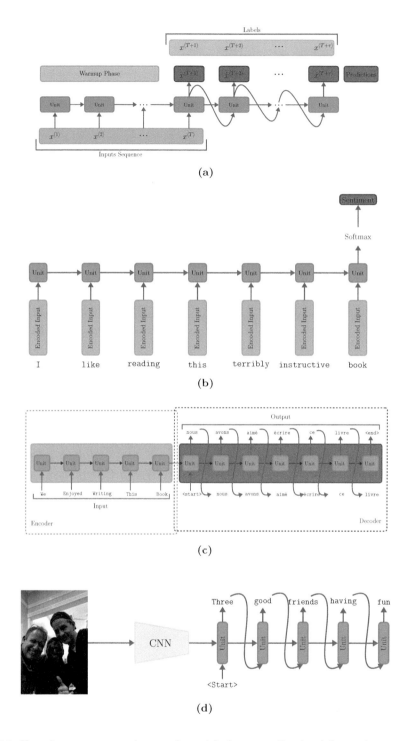

Figure 7.1: Use of recurrent neural network models (or generalizations) for various sequence data and language tasks. The basic building block, called a unit, is recursively used in the computation. (a) Lookahead prediction of the sequence. (b) Classification of a sequence or sentiment analysis. (c) Machine translation. (d) Image captioning.

generalizations) can be used. The building blocks of these types of models are called *units*, and they are recursively used in the computation of input to output. A basic task presented in (a) is *look-ahead prediction* which in the application context of text, implies predicting the next word (or collection of words) in a sequence. Another type of task presented in (b) is sequence regression or classification which can be used for applications such as *sentiment analysis*. An additional major task illustrated in (c) is *machine translation* where we translate the input sequence from one language to another (e.g., Hebrew to Arabic). Another type of task illustrated in (d) involves decoding an input into a sequence. One such example application is *image captioning* where text is generated to describe the input image. Let us now focus on (a)–(d) in more detail.

Consider Figure 7.1 (a) illustrating lookahead prediction. The simple application in the context of text is to predict the next word (or next few words) based on the sequence of input words. Thus, the output is the sequence of inputs shifted by one and the model attempts to predict the next word at any time t. In the context of time-series this is often referred to as an *auto-regressive* analysis. After a warmup phase, the model predicts the next value in the time series which is also used for predicting the subsequent values until a desired horizon. Hence for an input sequence $x = \left(x^{\langle 1 \rangle}, \ldots, x^{\langle t \rangle}\right)$ we have a future prediction $\hat{y} = (\hat{y}^{\langle t+1 \rangle}, \ldots, \hat{y}^{\langle t+\tau \rangle})$ for some *time horizon* τ. Note that the typical use of *large language models* follows this task as well since an input text is given and a response is returned.

Consider now Figure 7.1 (b) illustrating an input sequence processed to produce a single scalar or vector output. An archetypical application in the context of text is *sentiment analysis* where the sentiment or "general vibe" of a sentence is determined. This output may be a vector of probabilities over possible classes, e.g., {happy, sad, indifferent}, and in such a case the output is amenable to classification. Hence $\hat{y}$ is a vector of probabilities and it can also be converted to a categorical output $\widehat{\mathcal{Y}}$ as in (3.34) of Chapter 3.

Moving onto Figure 7.1 (c), consider the application of *machine translation* where the input sentence is from one language and the output sentence is from another language. The architecture of such a model can be composed of two RNNs (or two other types of sequence models) in an *encoder-decoder architecture*. Here the encoder model encodes the input sentence from one language into a *context vector* in a *latent space* and the decoder model decodes from the latent space into a sentence in another language. We describe architectures for such tasks in Section 7.4 and specific transformer models of this form in Section 7.5. Observe that with this task, the input x is a sequence of a certain length while the output $\hat{y}$ is a sequence of a potentially different length. Note that the notion of a latent space was first introduced in a different context of autoencoders in Section 3.5.

Figure 7.1 (d) illustrates the task of *image captioning*. Here for an input image, we wish to output a sentence describing the image. A common way to achieve this is with a convolutional neural network as in Chapter 6 creating a context vector in a latent space. This context vector is then fed into an RNN (or similar sequence model) which acts as a decoder, somewhat similarly to the decoder in the machine translation case. In this application x is an image, and $\hat{y}$ represents an output sentence.

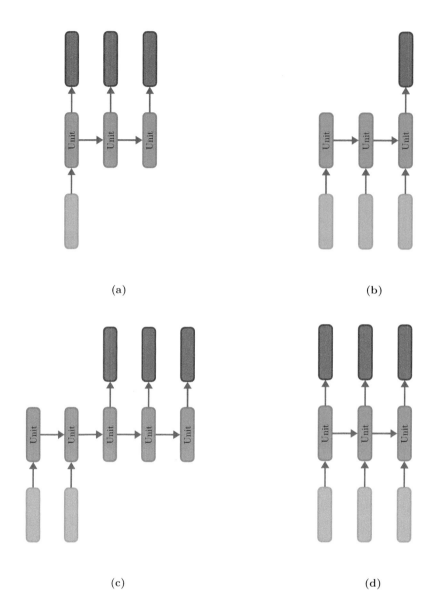

Figure 7.2: Input output paradigms of sequence models. (a) One-to-many. (b) Many-to-one. (c) Many-to-many with partial inputs and outputs. (d) Many-to-many with complete inputs and outputs.

With the tasks and applications highlighted, we see various forms of input x and output $\hat{y}$. Sometimes x and $\hat{y}$ are sequences and at other times they are not. It is often common to describe tasks and models as *one to many, many to one*, or *many to many*; see Figure 7.2. In the *one to many* case, x is simply $x^{\langle 1 \rangle}$ while $\hat{y}$ is a sequence, $\hat{y}^{\langle 1 \rangle}, \hat{y}^{\langle 2 \rangle}, \ldots$ In the *many to one* case, $x = \left(x^{\langle 1 \rangle}, x^{\langle 2 \rangle}, \ldots \right)$ is a sequence while $\hat{y}$ is a single output (scalar or vector). Finally in the *many to many* case, both x and $\hat{y}$ are sequences. Returning to Figure 7.1, observe that the lookahead prediction task (a) falls in either the *many to one* or the *many to many* case, depending on if the time horizon τ is 1 or greater, respectively. The sentiment analysis task (b) is a *many to one* case. The machine translation task (c) is a *many to many* case, while the image captioning task (d) is a *one to many* case.

Word Embedding

One-hot encoding which is the simplest way to encode a word results in a very sparse vector of high-dimensionality, with the dimension being the size of the lexicon, $d_\mathcal{V}$. A popular alternative that has become standard in any application involving text is to use *word embeddings*, where we represent each word (or token) by a vector of real numbers of dimension p, and with p much smaller than $d_\mathcal{V}$.

The essence of word embedding techniques is that words from similar contexts have corresponding vectors which are relatively close. Such closeness is often measured via the cosine of the angle[3] between the two vectors in Euclidean space. As an hypothetical example with $p = 4$, the word king could be represented by the vector $(0.41, 1.2, 3.4, -1.3)$ and the word queen can be represented by a relatively similar vector such as $(0.39, 1.1, 3.5, 1.6)$. Then a completely different word such as mean might be represented by a vector such as $(-0.2, -3.2, 1.3, 0.8)$. One can now verify in this example, that the cosine of the angle between king and queen is about 0.729 while the cosine of the angle between mean and the other two words is lower, and is at about -0.04 for king and 0.156 for queen, respectively.

Hence with such an embedding, beyond the value of reducing the dimension of each $x^{\langle t \rangle}$ from $d_\mathcal{V}$ (in the order of tens of thousands) to p (in the order of hundreds), we also get the benefit of similarity and context groupings. Having such a contextual representation of words plays a positive role in deep learning models since it allows the models to use context more efficiently.

Simple word embedding techniques map individual words into vectors, while more advanced techniques are context-aware and yield a representation of the words based on the context within the rest of the text, enabling models to better deal with homonyms. For example the word mean inside the phrase mean value, is very different than the same word inside the phrase mean person. Hence an advanced word embedding technique will encode each of the occurrences of mean differently.

A popular early word embedding technique is *word2vec*. The creation of this embedding relies on a neural network trained on very large corpora, to build the embedding vectors. The basic idea is to train the neural network for a task, and then use an internal layer of the network as the word embedding. Such an approach of a derived feature vector is common throughout deep learning. There are two common variations of the word2vec training algorithm with one approach called the *bag of words* model, seeking to predict a word from its neighboring words, while the other approach, the *skip-gram model*, seeks to predict the words of the context from a central word. In practice, both with word2vec, and with more advanced algorithms,[4] one may choose if to use a fixed pre-trained version of the word embedding or if to fine tune and adjust the word embedding when used as part of a larger model.

---

[3]See (A.1) in Appendix A.

[4]See references to other word embedding approaches in the notes and references at the end of the chapter.

7.2 Basic Recurrent Neural Networks

Recurrent neural networks are specifically designed for sequences of data and have the ability to: (i) deal with variable-length sequences, (ii) maintain sequence order, (iii) keep track of long-term dependencies, and (iv) share parameters across an input sequence. In order to achieve all of these goals, the *recurrent neural network* (RNN) model introduces an internal loop which allows information to be passed from one step of the network to the next. The RNN maintains a *hidden state*, also termed *cell state*, which allows the model to keep memory as an input sequence is processed. This state evolves over time, as the input sequence is fed into the model. See Figure 7.3 for a schematic illustration of both a *recursive graph* representation and an *unfolded graph* representation of the model. In the figure we schematically see how an RNN transforms an input sequence to an output sequence, with the blue nodes representing units of the model; details follow.

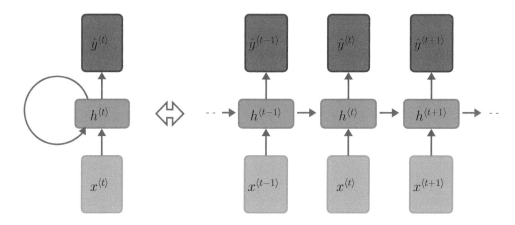

Figure 7.3: A recursive neural network RNN can be represented (left) via a recursive graph and (right) via an unfolded representation of that graph. An input sequence $x^{\langle 1 \rangle}, x^{\langle 2 \rangle} \ldots$ is transformed to an output sequence $\hat{y}^{\langle 1 \rangle}, \hat{y}^{\langle 2 \rangle} \ldots$, where in each step t, the unit, with cell state represented via $h^{\langle t \rangle}$, performs the computation.

RNNs apply a *recurrence relation* where at every time step the next input is combined with the previous state to update the new state and a new output. This internal loop is the key difference between traditional feedforward neural networks of Chapter 5 and RNNs. In the traditional models, the flow of information from input to output via hidden layers is only in the forward direction, whereas in RNNs, the input plays a role as information flows. More specifically, in the traditional models, there is no cyclic connection between neurons; contrast Figure 5.1 of Chapter 5 with Figure 7.3. Moreover, the traditional feedforward neural networks work with fixed length input and fixed length output while the RNN input sequence and output sequence are each allowed to be of variable (essentially unbounded) length.

The neurons inside RNNs implement the cell state and are typically denoted via $h^{\langle t \rangle}$ for the state at time t. Mathematically, the *state evolution* can be represented via the recurrence relation,

$$\underbrace{h^{\langle t \rangle}}_{\text{current state}} = f_{\theta_{hx}, \theta_{hh}}(\underbrace{h^{\langle t-1 \rangle}}_{\text{old state}}, \underbrace{x^{\langle t \rangle}}_{\text{input vector}}),$$

acting on the sequence of input data $x = \left(x^{\langle 1 \rangle}, x^{\langle 2 \rangle}, \ldots\right)$, to create a sequence of cell states $h^{\langle 1 \rangle}, h^{\langle 2 \rangle}, \ldots$, where the initial state $h^{\langle 0 \rangle}$ is typically taken as a zero vector. It is important to note that at every time step t the same function $f_{\theta_{hx}, \theta_{hh}}(\cdot)$ is used with the same fixed (over time) sets of parameters θ_{hx} and θ_{hh}.

The output sequence $\hat{y} = \left(\hat{y}^{\langle 1 \rangle}, \hat{y}^{\langle 2 \rangle}, \ldots\right)$ is defined at each time step by another function $g_{\theta_{yh}}(\cdot)$ with,

$$\hat{y}^{\langle t \rangle} = g_{\theta_{yh}}(h^{\langle t \rangle}),$$

where again, the parameters θ_{yh} are fixed over time.

The recursive loop enables us to express the cell state at time t, $h^{\langle t \rangle}$, in terms of the t first inputs, namely, omitting the parameter subscripts from $f_{\theta_{hx}, \theta_{hh}}(\cdot)$, we have,

$$h^{\langle t \rangle} = f(\ldots f(f(\underbrace{f(\underbrace{h^{\langle 0 \rangle}, x^{\langle 1 \rangle}}_{h^{\langle 1 \rangle}}), x^{\langle 2 \rangle}}_{h^{\langle 2 \rangle}}), x^{\langle 3 \rangle}) \ldots, x^{\langle t \rangle}). \tag{7.1}$$

Thus, since at time t, the output $\hat{y}^{\langle t \rangle}$ is a function of $h^{\langle t \rangle}$, we can also express the output as a function of the inputs up until time step t. Namely,

$$\hat{y}^{\langle t \rangle} = g_\theta^{(t)}(x^{\langle 1 \rangle}, x^{\langle 2 \rangle}, x^{\langle 3 \rangle}, \ldots, x^{\langle t \rangle}),$$

where the function $g_\theta^{(t)}(\cdot)$ is specific to time t and captures the unrolling of the state as in (7.1) or Figure 7.3 (b). This highlights the ability of RNN to deal with variable length input and output sequences. Here $\theta = (\theta_{hx}, \theta_{hh}, \theta_{yh})$ is the collection of all learnable parameters of the RNN.

The functions $f_{\theta_{hx}, \theta_{hh}}(\cdot)$ and $g_{\theta_{yh}}(\cdot)$ are concretely defined via affine transformations and non-linear activations similarly to other common neural network models. Specifically,

$$\begin{cases} h^{\langle t \rangle} = S_h(W_{hh}h^{\langle t-1 \rangle} + W_{hx}x^{\langle t \rangle} + b_h) \\ \hat{y}^{\langle t \rangle} = S_y(W_{yh}h^{\langle t \rangle} + b_y). \end{cases} \tag{7.2}$$

The parameters $\theta = (\theta_{hx}, \theta_{hh}, \theta_{yh})$ are captured via weight matrices and bias vectors.[5] Further, $S_h(\cdot)$ and $S_y(\cdot)$ are vector activation functions typically composed of element-wise scalar activations $\sigma(\cdot)$, similarly to Chapter 5. We denote the dimension of $x^{\langle t \rangle}$ as p, the dimension of $y^{\langle t \rangle}$ as q, and the dimension of the cell state of $h^{\langle t \rangle}$ as m. Hence, $W_{hx} \in \mathbb{R}^{m \times p}$, $W_{hh} \in \mathbb{R}^{m \times m}$, $W_{yh} \in \mathbb{R}^{q \times m}$, $b_h \in \mathbb{R}^m$, and $b_y \in \mathbb{R}^q$.

One variant is to feed the output of the previous time step, $\hat{y}^{\langle t-1 \rangle}$, into the input so that the input at every time is not $x^{\langle t \rangle}$ but rather some merging of $x^{\langle t \rangle}$ and $\hat{y}^{\langle t-1 \rangle}$. We can denote this variant via,

$$\begin{cases} h^{\langle t \rangle} = S_h(W_{hh}h^{\langle t-1 \rangle} + W_{hx}(x^{\langle t \rangle} + \mathcal{T}(\hat{y}^{\langle t-1 \rangle})) + b_h) \\ \hat{y}^{\langle t \rangle} = S_y(W_{yh}h^{\langle t \rangle} + b_y), \end{cases} \tag{7.3}$$

---

[5]The actual mapping of the weight and bias vectors to each of $(\theta_{hx}, \theta_{hh}, \theta_{yh})$ is not important. Specifically, b_h can be viewed as either part of θ_{hx} or θ_{hh}.

where $\mathcal{T}(\cdot)$ is an abstraction[6] of some transformation which results in a vector of dimension p (like $x^{\langle t \rangle}$).

Note that the forms in (7.2) and (7.3) are suitable for many-to-many mappings, as in Figure 7.2 (d), since each input $x^{\langle t \rangle}$ has an associated $h^{\langle t \rangle}$ and $\hat{y}^{\langle t \rangle}$. If we use these recursions for one-to-many tasks, then we only use a first initial $x^{\langle 1 \rangle}$ and then continue the recursion with 0 values for $x^{\langle t \rangle}$ on $t = 2, 3, \ldots$ until some stopping criterion is met. A typical criterion is to have a special `<stop>` token appear within $\hat{y}^{\langle t \rangle}$. A similar adaptation can be done for many-to-one tasks, where we simply ignore all $\hat{y}^{\langle t \rangle}$ for $t < T$.

One often refers to the mechanism of computation described in (7.2) as a *gate*, a *simple gate*, an *RNN cell*, or an *RNN unit*. Figure 7.4 depicts this computation where $\oplus$ and $\otimes$ are the usual vector/matrix addition and multiplication operations respectively. In the sequel we see that the gate structure as in Figure 7.4 can be modified to more complicated forms such as LSTM and GRU gates appearing in Figure 7.8. In terms of applications, the simple structure of RNN gates has already proven useful for many basic tasks such as for example dealing with short sentences for next word prediction as well as for sentiment analysis.

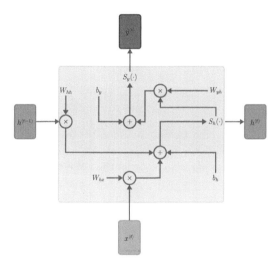

Figure 7.4: An RNN unit, also known as a gate, operating on input $x^{\langle t \rangle}$, and previous cell state $h^{\langle t-1 \rangle}$. The output vector of the unit is $\hat{y}^{\langle t \rangle}$. The unit also determines the cell state $h^{\langle t \rangle}$.

A Simple Concrete Toy Example

To illustrate the application of RNNs let us consider a simple concrete toy example of lookahead text prediction. For simplicity we resort to one-hot encoding (in contrast to more advanced word embedding methods). In our toy example assume a lexicon with $d_\mathcal{V} = 8$ words, appearing here in lowercase alphabetical order as

---

[6]In practice, this variant is often useful in decoders, described in Sections 7.4 and 7.5, where $x^{\langle t \rangle}$ is often set to 0 except for an initial `<start>` token, and the transformation $\mathcal{T}(\cdot)$ typically transforms an output embedding into the desired token (e.g., via argmax) and then transforms the token back into an input word embedding.

$$\left[\text{deep,\quad engineering,\quad learning,\quad machine,\quad mathematical,\quad of,\quad statistics,\quad the}\right],$$

where each word is represented by a unit vector in $\mathbb{R}^8$. For example, the text

the mathematical engineering of deep learning,

is represented via a sequence $x = \left(x^{\langle 1\rangle}, x^{\langle 2\rangle}, x^{\langle 3\rangle}, x^{\langle 4\rangle}, x^{\langle 5\rangle}, x^{\langle 6\rangle}\right)$. Here for example $x^{\langle 1\rangle} = (0, 0, 0, 0, 0, 0, 0, 1)$ because the first word in the sequence, the, is the 8-th word in the lexicon, and similarly $x^{\langle 2\rangle} = (0, 0, 0, 0, 1, 0, 0, 0)$ because the second word in the sequence, mathematical, is the 5-th word in the lexicon, etc. Observe that here with one-hot encoding, $p = d_{\mathcal{V}}$.

For a lookahead prediction application we set the network output to be of dimension $q = 8$ since each output is of the size of the lexicon. Here when the network is fed a partial input $x^{\langle 1\rangle}, \ldots, x^{\langle t\rangle}$, the output at time (step) t, denoted via $\hat{y}^{\langle t\rangle}$ should ideally be equal to or be close to the one-hot encoded target $y^{\langle t\rangle} := x^{\langle t+1\rangle}$ (the next word). Similarly to the classification examples arising in multinomial regression in Section 3.3, our RNN will output $\hat{y}^{\langle t\rangle}$ vectors that are probability vectors over the lexicon. In this case, using maximum a posteriori probability decisions as in (3.34) of Chapter 3, the coordinate of $\hat{y}^{\langle t\rangle}$ with the highest probability can be taken as a prediction $\widehat{\mathcal{Y}}^{\langle t\rangle}$ which is an index into the lexicon, determining the predicted word.

Now following the RNN evolution equations defined in (7.2), we present concrete dimensions for this illustrative example. A design choice is the size of the hidden state m, which in this case we arbitrarily take as $m = 20$. Hence, the weight matrix W_{hh} dealing with state evolution is 20×20, the weight matrix W_{hx} is 20×8 dimensional as it converts the inputs to the state, and the bias vector b_h is a 20-dimensional vector. The vector activation function $S_h(\cdot)$ is composed of scalar activations, which can be of any of the standard types (e.g., sigmoid); see Section 5.3. Further, the matrix translating state to output, W_{yh} is 8×20 since $q = 8$, and finally b_y is an 8-dimensional vector. Importantly, in this case, we take the output activation function $S_y(\cdot)$ as a softmax function since it converts the affine transformation of the state into a probability vector. Hence in summary, such a toy network would have $20 \times 20 + 20 \times 8 + 20 + 8 \times 20 + 8 = 748$ parameters to learn.

Figure 7.5 presents a schematic illustration of the unrolling of this toy network. When the model is used in training, the shifted sequence $y = \left(x^{\langle 2\rangle}, x^{\langle 3\rangle}, \ldots\right)$ serves as the sequence of target labels for comparison; these are one-hot encoded vectors. Then for given weight and bias parameters, we predict $\hat{y} = \left(\hat{y}^{\langle 1\rangle}, \hat{y}^{\langle 2\rangle}, \ldots\right)$ as the predicted labels, where each $\hat{y}^{\langle t\rangle}$ is a vector of probabilities over the lexicon for output word at step t. We then use categorical cross entropy (see equation (3.30) in Chapter 3) to compute the loss. We train using gradient-based learning similarly to all other deep learning models, yet for evaluation of the gradient, we use a variant of backpropagation called *backpropagation through time* which is described below.

In production, the way that the model is used, is by selecting the word at time t, with the highest probability in $\hat{y}^{\langle t\rangle}$. We denote the index of this selection via $\widehat{\mathcal{Y}}^{\langle t\rangle}$. In the illustration of Figure 7.5, most of the words are properly predicted except the fourth word at $t = 4$ which is predicted as deep while the target is engineering.

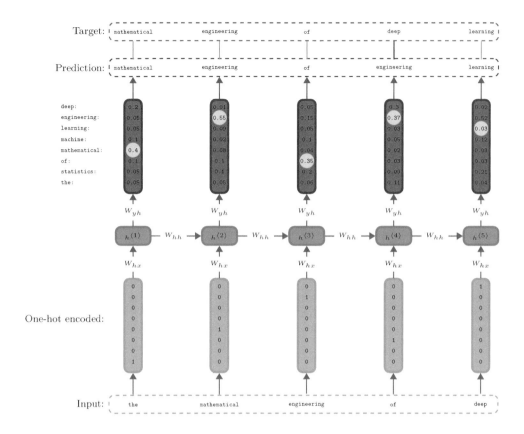

Figure 7.5: A schematic of RNN cells unrolled for language modeling. In this illustration, the input sentence `the mathematical engineering of deep learning`, yields lookahead prediction `mathematical engineering of` *engineering* `learning`, with a single error.

Training an RNN with Backpropagation Through Time

In general when training an RNN, the loss function is accumulated over all time steps t. In particular, during the execution of gradient descent or some generalization of gradient descent, such as ADAM described in Chapter 4, we compute the loss and its derivatives with respect to weight and bias parameters, i.e., θ. A general expression for the loss is

$$C(\theta) = \frac{1}{T} \sum_{t=1}^{T} C^{\langle t \rangle}(\theta). \tag{7.4}$$

Here $C^{\langle t \rangle}(\theta)$ denotes the individual loss associated with time t. For example, continuing with the text language model from above, we may set

$$C^{\langle t \rangle}(\theta) = - \sum_{k=1}^{d_{\mathcal{V}}} y_k^{\langle t \rangle} \log \hat{y}_k^{\langle t \rangle}, \tag{7.5}$$

similarly to the categorical cross entropy in (3.30) of Chapter 3. Here, keep in mind that $y^{\langle t \rangle}$ is one-hot encoded of dimension $q = d_{\mathcal{V}}$, and hence the summation in (7.5) has a single non-zero summand at the index k for which $y_k^{\langle t \rangle} = 1$. Further, note that even if word embeddings are used for the input $x^{\langle t \rangle}$, then the output $\hat{y}^{\langle t \rangle}$ still represents a probability vector over the lexicon of size d_V so that it is comparable to the target output $y^{\langle t \rangle}$.

Gradient computation of $C(\theta)$ with respect to the various weight matrices and bias vectors in θ is somewhat similar to the backpropagation algorithm described in Section 5.4; see also Section 4.4 for automatic differentiation basics. However, a key difference lies in the fact that the same weight and bias parameters are used for all time t; see the unfolded representation of the RNN in Figure 7.3. This difference as well as the fact that inputs to RNN are of arbitrary size T, imposes some hardships on gradient computation. The basic algorithm is called *backpropagation through time*.

One way to view the algorithm is to momentarily return to the feedforward networks of Chapter 5 and assume that the weight matrices and bias vectors of layers are all constrained to be the same with a single set of parameters, W and b, for all layers. Further, momentarily assume that the network depth L, is fixed at the sequence input length T. This form of a feedforward network is essentially an unfolded RNN if we consider every recursive step of the RNN as a layer in the feedforward network, and if we ignore inputs to the RNN beyond the first input $x^{\langle 1 \rangle}$, and impose loss on the RNN only for the last output.

One can also modify feedforward networks to have additional external inputs at each of the hidden layers. In our feedforward analogy, assume now that an external input to the ℓ-th layer is $x^{\langle \ell \rangle}$, where the layer ℓ and the time t play the same role. Further, one may impose loss functions on feedforward networks that not only take the neurons at the last layer as arguments, but rather use all layers, similarly to (7.4). If we also employ such a loss function on the feedforward network analogy then we see that we can treat the unfolded RNN as a feedforward network, where for simplicity we treat the transformation from $h^{\langle t \rangle}$ to $y^{\langle t \rangle}$ in (7.2) as the identity transformation. With this, let us return to the details of the backpropagation algorithm in Section 5.4 and see how it can be adapted for RNNs.

At first a forward pass is carried out to populate the neuron values. In feedforward networks these were denoted $a^{[\ell]}$ whereas in the unfolded RNN they are denoted via $h^{\langle t \rangle}$. Then a backward pass is used to compute the adjoint elements, denoted $\zeta^{\langle t \rangle}$, similarly to the adjoints defined in (5.24) of Chapter 5. In the RNN context, these are

$$\zeta^{\langle t \rangle} = \frac{\partial C(\theta)}{\partial h^{\langle t \rangle}}.$$

The essence of backpropagation is computing $\zeta^{\langle t-1 \rangle}$ based on $\zeta^{\langle t \rangle}$. This computation follows similar lines to (5.26) of Chapter 5, adapted here to be,

$$\zeta^{\langle t \rangle} = \begin{cases} \frac{1}{T}\sum_{\tau=1}^{T} \dot{C}^{\langle \tau \rangle}(h^{\langle \tau \rangle}), & t = T, \\ \frac{\partial h^{\langle t+1 \rangle}}{\partial h^{\langle t \rangle}} \zeta^{\langle t+1 \rangle}, & t = T-1, \ldots, 1, \end{cases} \tag{7.6}$$

where $\dot{C}^{\langle \tau \rangle}(h^{\langle \tau \rangle})$ is the derivative of (7.5) with respect to the prediction $\hat{y}^{\langle \tau \rangle}$ for which we are given computable expressions.

Once the backpropagation through time recursion (7.6) is carried out, we use the computed adjoint sequence $\zeta^{\langle T \rangle}, \ldots, \zeta^{\langle 1 \rangle}$ to evaluate the gradient of the loss with respect to components of θ. For simplicity let us focus only on W_{hh} as appearing in (7.2), denoted here as W for brevity. Specifically, we are interested in evaluating the $m \times m$ derivative matrix,

$$g_W = \frac{\partial C}{\partial W} = \frac{1}{T} \sum_{t=1}^{T} \frac{\partial C^{\langle t \rangle}\left(h^{\langle t \rangle}, y^{\langle t \rangle}\, ;\, W\right)}{\partial W}, \tag{7.7}$$

similarly to the notation in the feedforward case as in (5.22) of Chapter 5. A noticeable difference between (7.7) and (5.22) is that due to the loss function structure in (7.4), g_W is a direct function of the cell state at all times t (all internal layers of the unfolded graph). However, a more important difference is due to the fact that all time steps (unfolded layers) share the same parameter W, and thus the computational graph connecting W and the loss dictates that all adjoints affect g_W. See Figure 7.6 and contrast it with Figure 5.7 of Chapter 5.

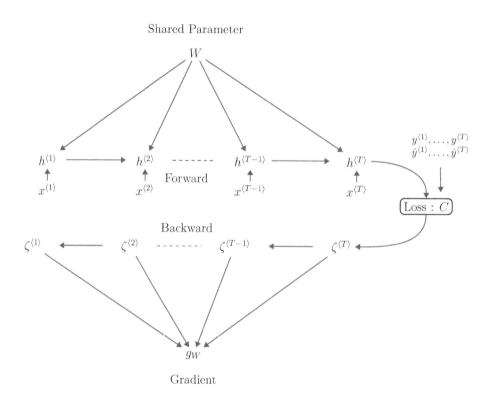

Figure 7.6: The variables and flow of information in the backpropagation through time algorithm. The shared parameter W influences the recursive forward pass computation of all cell states $h^{\langle 1 \rangle}, \ldots, h^{\langle T \rangle}$. Once the backward pass computation is carried out for all adjoints $\zeta^{\langle T \rangle}, \ldots, \zeta^{\langle 1 \rangle}$, they are all used to compute the gradient of the loss, g_W.

While the feedforward case in Chapter 5 with individual parameters per layer has an easy translation of an adjoint into a gradient, as in the right hand side of (5.25), here the translation of adjoints to a gradient is more complicated and more computationally costly.

Specifically, using the multivariate chain rule, we can informally[7] represent the gradient as

$$g_W = \frac{1}{T} \sum_{t=1}^{T} \sum_{\tau=1}^{t} \frac{\partial h^{\langle \tau \rangle}}{\partial W} \underbrace{\frac{\partial C}{\partial h^{\langle \tau \rangle}}}_{\zeta^{\langle \tau \rangle}}. \tag{7.8}$$

To understand the internal summation in (7.8), recall that the output $\hat{y}^{\langle t \rangle}$, used in the individual loss $C^{\langle t \rangle}$, depends on all cell states $h^{\langle 1 \rangle}, \ldots, h^{\langle t \rangle}$, where each cell state is parameterized by a common W. Hence the computational graph for this loss component has to be taken into account when applying the chain rule. This is also illustrated in the top part of Figure 7.6.

Note that formally the expression $\frac{\partial h^{\langle \tau \rangle}}{\partial W}$ in (7.8) is a derivative of a vector with respect to a matrix, and we do not handle such objects in this book. An alternative is to represent each scalar component of $h^{\langle t \rangle}$ separately. Using (7.2) and assuming the vector activation function $S_h(\cdot)$ is composed of scalar activation functions $\sigma(\cdot)$, we have

$$h_j^{\langle \tau \rangle} = \sigma([W_{hh} h^{\langle \tau-1 \rangle} + W_{hx} x^{\langle \tau \rangle} + b_h]_j).$$

Now using (A.15) from Appendix A, we have that the derivative of the scalar $h_j^{\langle \tau \rangle}$ with respect to the weight matrix W_{hh} (abbreviated as W) is given by the matrix,

$$\frac{\partial h_j^{\langle \tau \rangle}}{\partial W} = \dot{\sigma}(W_{hh} h^{\langle \tau-1 \rangle} + W_{hx} x^{\langle \tau \rangle} + b_h) \, e_j \left(h^{\langle \tau-1 \rangle} \right)^{\top}, \tag{7.9}$$

where e_j is the m-dimensional unit vector with 1 at the j-th coordinate, and $\dot{\sigma}(\cdot)$ is the derivative of the scalar activation function; see also Section 5.3.

Continuing with the approach of treating individual neurons $h_j^{\langle \tau \rangle}$, let us now present a more precise version of (7.8). For this consider the fact that in computing each individual loss, $C^{\langle \tau \rangle}$, we rely on the neurons with the cell states $h_j^{\langle 1 \rangle}, \ldots, h_j^{\langle \tau \rangle}$, for all $j = 1, \ldots, m$. In turn, each of these neurons is influenced by W (shorthand for W_{hh}), as in (7.9). Now (also summing up over all individual losses for $t = 1, \ldots, T$), we use the multivariate chain rule to arrive at

$$g_W = \frac{1}{T} \sum_{t=1}^{T} \sum_{\tau=1}^{t} \sum_{j=1}^{m} \frac{\partial h_j^{\langle \tau \rangle}}{\partial W} \zeta_j^{\langle \tau \rangle}, \tag{7.10}$$

which is fully computable using the backpropagated adjoints from (7.6) and (7.9).

To summarize backpropagation through time, we first carry out a forward pass to populate $h^{\langle 1 \rangle}, \ldots, h^{\langle T \rangle}$ using (7.2) or the (7.3) variant. We then carry out a backward pass to populate the adjoints $\zeta^{\langle T \rangle}, \ldots, \zeta^{\langle 1 \rangle}$ using (7.6). We then compute the gradient g_W via (7.10). This summary is for our simplified case focusing only on $W = W_{hh}$ and ignoring the fact that $y^{\langle t \rangle}$ is generally not $h^{\langle t \rangle}$, but rather constructed via the second equation in (7.2). Hence in our simplified presentation we focused on the essence and ignored less complicated details for the complete set of θ parameters.

---

[7]The representation in (7.8) is informal because the vector-matrix derivative $\frac{\partial h^{\langle \tau \rangle}}{\partial W}$ is not a matrix.

Let us also consider the Jacobian $\frac{\partial h^{\langle t+1 \rangle}}{\partial h^{\langle t \rangle}}$ appearing in (7.6). Again, assuming that the vector activation function $S_h(\cdot)$ of (7.2) is composed of element-wise scalar activation functions $\sigma(\cdot)$, this Jacobian can be represented as

$$\frac{\partial h^{\langle t+1 \rangle}}{\partial h^{\langle t \rangle}} = W_{hh}^{\top} \, \text{diag}\Big(\dot{\sigma}(W_{hh}h^{\langle t \rangle} + W_{hx}x^{\langle t \rangle} + b_h) \Big), \tag{7.11}$$

where the derivative of the activation function is denoted via $\dot{\sigma}(\cdot)$ and is applied element-wise to the components of its input.

Computational Challenges

We discussed vanishing and exploding gradient phenomena in Section 5.4, where in equations (5.34) and (5.35) we saw how both the forward pass and the backward pass involve actions of repeated matrix multiplication. More specifically, the backpropagation-based equation (5.35) is based on a simplification of a feedforward neural network that ignores the effect of activation functions, ignores the bias, and assumes that each layer of the network has the same weight matrix. In such a case, it is evident that for deep networks (L large), vanishing or exploding gradient phenomena are likely to occur.

In RNNs, such phenomena are even more problematic than typical deep feedforward networks because the input size T (paralleling the depth of the feedforward network L), can be large. Unrolling (7.6) we get for $t = 1, \ldots, T-1$,

$$\zeta^{\langle t \rangle} = \frac{\partial h^{\langle t+1 \rangle}}{\partial h^{\langle t \rangle}} \frac{\partial h^{\langle t+2 \rangle}}{\partial h^{\langle t+1 \rangle}} \quad \cdots \quad \frac{\partial h^{\langle T-1 \rangle}}{\partial h^{\langle T-2 \rangle}} \frac{\partial h^{\langle T \rangle}}{\partial h^{\langle T-1 \rangle}} \zeta^{\langle T \rangle}.$$

Now using (7.11) and for simplicity ignoring the action of the activation function (treating it as an identity function), ignoring the input $x^{\langle t \rangle}$, and ignoring the bias term, we obtain,

$$\zeta^{\langle t \rangle} = \Big(W_{hh}^{\top} \Big)^{T-t} \zeta^{\langle T \rangle}. \tag{7.12}$$

This representation is similar to (5.35) of Chapter 5, and is even more realistic since in RNNs, the weight matrices of all unrolled layers are the same, whereas in the Chapter 5 analysis of feedforward networks fixing the weight matrix was a simplification. Hence, in RNNs trained on inputs with large sizes T, it is very likely that during backpropagation, the adjoint values $\zeta^{\langle t \rangle}$ vanish or explode. This follows from the matrix power in (7.12), since in most situations, the maximal eigenvalue of W_{hh} is likely to not be at or near unity (see also discussion on the effect of eigenvalues on vanishing and exploding phenomena in Section 5.4).

Considering (7.12) and assuming $W_{hh}^{\top}$ has a maximal eigenvalue less than unity in absolute value, then if T is large, for small t, $\zeta^{\langle t \rangle} \approx 0$. One way to express this is to consider some $T_0 < T$ such for example if $T = 300$, set $T_0 = 250$, and then for $t < T_0$ assume $\zeta^{\langle t \rangle} = 0$. In this case the gradient computation (7.10) can be roughly represented as

$$g_W \approx \frac{1}{T} \sum_{t=T_0}^{T} \sum_{\tau=T_0}^{t} \sum_{j=1}^{m} \frac{\partial h_j^{\langle \tau \rangle}}{\partial W} \zeta_j^{\langle \tau \rangle}. \tag{7.13}$$

Now considering the influence of the input via (7.13) and (7.9), we see that the gradient is only updated based on "near effects", and not based on "long-term effects" since inputs to the sequence $x^{\langle t \rangle}$ for $t < T_0$ do not play a role. For example in language modeling, the contribution of faraway words to predicting the next word at time-step diminishes when the gradient vanishes early on. As an example consider the text

```
Slava grew up in Ukraine before he moved around the world, first to the
United States, and then to Australia. He loves teaching languages and is an
avid teacher of his own mother tongue _.'
```

In this case, completion of the end of the text, marked via _, requires information from the start of the text. Models presented in the sections below were also designed to overcome such difficulties.

Further, with RNNs, computation of the loss and of the gradients across an entire corpus is generally infeasible or too expensive. In practice, a batch of sentences is used to compute the loss to limit the sequence size T. Note also that in cases where $W_{hh}^{\top}$ has eigenvalues greater than unity in absolute value an exploding gradient phenomena is likely to occur. For this, *gradient clipping* may be employed as described at the end of Section 5.4. Another technique is to use *truncated backpropagation through time* (TBPTT) which limits the number of time steps the signal can backpropagate in each forward pass.

Other Aspects of Training

Some practices of training RNNs are very similar to training feedforward or convolutional networks. For example, one uses similar weight initialization techniques to those introduced in Section 5.5 in the context of feedforward networks. However, there are some differences as well. An important aspect to keep in mind is that unlike the supervised setting that prevailed with the models of Chapters 5 and 6, with recurrent models we are often able to train with *self-supervision*. Specifically, as already discussed in the example of Figure 7.5 we may use a shifted sequence $y = \left(x^{\langle 2 \rangle}, x^{\langle 3 \rangle}, \ldots \right)$ as the desired output for the loss, and simply train the model for one step lookahead prediction.

Note however, that not all training is of the self-supervised form. In some cases, often arising in machine translation applications described in the sequel, we are naturally presented with an input sequence $x^{\langle 1 \rangle}, x^{\langle 2 \rangle}, \ldots$ which may result from word embedding of one natural language (e.g., English) and a corresponding output sequence $y^{\langle 1 \rangle}, y^{\langle 2 \rangle}, \ldots$, of one-hot encoded vectors, associated with another natural language (e.g., Arabic). Hence RNNs can be trained in a supervised setting as well.

In both the self-supervised and supervised settings, in cases where we use the formulation (7.3), where the output $\hat{y}^{\langle t-1 \rangle}$ is fed into the input, we sometimes use a training technique called *teacher forcing*. The idea of teacher forcing is to use the actual (correct) one-hot encoded label $y^{\langle t-1 \rangle}$ in place of the model generated (predicted) probability vector $\hat{y}^{\langle t-1 \rangle}$ during training. That is, the recursion (7.3) has now inputs that are based on the actual labels instead of the predictions. Note that in this case, $\mathcal{T}(\cdot)$ in (7.3) can be viewed as also converting the probability vector into a word embedding, if needed. This technique accelerates training by removing the errors in the labels. We revisit the teacher forcing technique both at the end of Section 7.4 in the context of encoder-decoder models, and at the end of Section 7.5 in the context of transformers where it is extremely powerful due to parallelization.

7.3 Generalizations and Modifications to RNNs

The basic recurrent networks of Section 7.2, while powerful, still suffer from some drawbacks in terms of training, vanishing and exploding gradient, and expressivity. In this section we highlight a few generalizations and modifications to RNNs that enable more powerful models for sequence data. An underlying concept is the connection of gates in various creative ways that enable more expressive and robust models. The notion of a gate is already illustrated in Figure 7.4. In this section we see how such gates can be connected in diverse ways, as well as how the internals of the gate can be extended to yield more powerful models.

Stacking and Reversing Gates

Basic extensions to recurrent neural networks are possible by interconnecting gates in more complicated forms than just a forward direction of data flow. In particular, common approaches are to either stack the gates to form deeper networks, reverse the gates, or combine the two approaches. See Figure 7.7 for a schematic representation of such interconnections of gates.

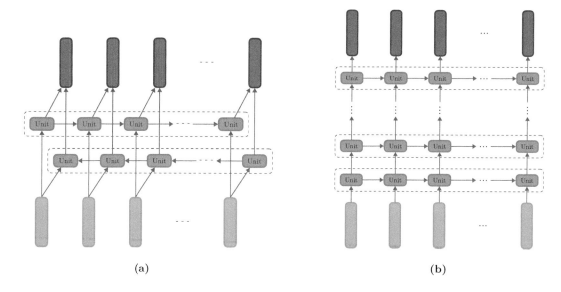

(a) (b)

Figure 7.7: Alternative configurations and extensions of recurrent neural networks. (a) Bidirectional RNN. (b) Stacked RNN.

Let us first consider reversing of gates as in Figure 7.7 (a) to create a *bidirectional recurrent neural network*. For such a modification we extended the RNN evolution equation (7.2) to

$$
\begin{cases}
h_f^{\langle t \rangle} = S_h(W_{hh}^f h_f^{\langle t-1 \rangle} + W_{hx}^f x^{\langle t \rangle} + b_h^f) \\
h_r^{\langle t \rangle} = S_h(W_{hh}^r h_r^{\langle t+1 \rangle} + W_{hx}^r x^{\langle t \rangle} + b_h^r) \\
\hat{y}^{\langle t \rangle} = S_y(W_{yh}^f h_f^{\langle t \rangle} + W_{yh}^r h_r^{\langle t \rangle} + b_y),
\end{cases}
\tag{7.14}
$$

where now for every time t there are two cell states, $h_f^{\langle t \rangle}$ and $h_r^{\langle t \rangle}$, representing the forward direction and reverse direction, respectively. Observe in (7.14) that $h_f^{\langle t \rangle}$ evolves based on the input $x^{\langle t \rangle}$ and $h_f^{\langle t-1 \rangle}$, while $h_r^{\langle t \rangle}$ evolves based on the input $x^{\langle t \rangle}$ and $h_f^{\langle t+1 \rangle}$. Naturally with such an extension there are more trained parameters, superscripted via f and r respectively in (7.14).

As is evident from (7.14), the forward sequence of cell states $h_f^{\langle 1 \rangle}, h_f^{\langle 2 \rangle}, \ldots, h_f^{\langle T \rangle}$ and the reverse sequence of cell states $h_r^{\langle T \rangle}, h_r^{\langle t-1 \rangle}, \ldots, h_r^{\langle 1 \rangle}$ evolve without interaction. Once computed, these sequences are then combined to obtain the output sequence. Such bidirectional data flow enables the model to be more versatile, especially for cases where the entire input sequence is available. This is the setup in applications such as handwritten text recognition, machine translation, speech recognition, and part-of-speech tagging, among others.

Let us now consider stacking of gates as in Figure 7.7 (b) to create a deeper model, also sometimes known as a *stacked recurrent neural network*. With this paradigm we extend the evolution equations (7.2) to,

$$
\begin{cases}
h_{[1]}^{\langle t \rangle} = S_h^{[1]}(W_{hh}^{[1]} h_{[1]}^{\langle t-1 \rangle} + W_{hx}^{[1]} x^{\langle t \rangle} + b_h^{[1]}) \\
h_{[2]}^{\langle t \rangle} = S_h^{[2]}(W_{hh}^{[2]} h_{[2]}^{\langle t-1 \rangle} + W_{hx}^{[2]} h_{[1]}^{\langle t \rangle} + b_h^{[2]}) \\
\quad \vdots \\
h_{[L]}^{\langle t \rangle} = S_h^{[L]}(W_{hh}^{[L]} h_{[L]}^{\langle t-1 \rangle} + W_{hx}^{[L]} h_{[L-1]}^{\langle t \rangle} + b_h^{[L]}) \\
\hat{y}^{\langle t \rangle} = S_y(W_{yh} h_{[L]}^{\langle t \rangle} + b_y),
\end{cases}
\tag{7.15}
$$

where we now use notation such as $[1], \ldots, [L]$ to signify the depth of individual components and L is the number of stacked layers, similarly to the notation of Chapter 5. Observe that the cell state at time t and depth ℓ, denoted via $h_{[\ell]}^{\langle t \rangle}$ is computed based on the cell state at depth $\ell - 1$ and the same time t using the matrix $W_{hx}^{[\ell]}$ (where the notation x here in the subscript implies the previous level). It is also computed using the cell state at the same depth, ℓ, and the previous time, $t - 1$ using the matrix $W_{hh}^{[\ell]}$.

Such stacked RNN models are clearly more expressive and thus they generally outperform single-layer RNNs when trained with enough data. However, they are harder to train as the number of parameters clearly grows proportionally to the number of layers. We also mention that combinations of stacking and reversing are also possible.

Long Short Term Memory Models

Long short term memory (LSTM) models are generalizations of basic RNNs that are designed to preserve information over many time steps. To understand the idea behind LSTM, it is constructive to think in terms of logical operations that are approximated via multiplication of vectors. In particular, as we see below, different components of LSTM interplay in a way that can heuristically be described as computation of a logical circuit. More specifically, some of the neurons inside LSTM can be called *internal gates* and are represented as values in the range $[0, 1]$ and these are then multiplied by other neurons with arbitrary real values. In particular, when a vector of neurons in such an internal gate, say g, has elements in the range $[0, 1]$, and another vector of neurons, say c has general real values, then the element-wise multiplication $g \odot c$ can be viewed as a restriction which approximately zeros out (forgets) entries of c when the corresponding entry of g is near 0. We informally say that the entry of the internal gate is "open" when it is approximately at 1 and similarly "closed" when it is approximately 0. Using internal gates for this type of "approximate logical masking" is common in these models as well as the gated recurrent units described in the sequel.

A key concept in LSTM is to extend the hidden units of RNNs by separating the information flow between units into two groups where one group is called the *cell state* and denoted $c^{\langle t \rangle}$, while the other group is called the *hidden state* and denoted $h^{\langle t \rangle}$. The model is designed so that long-term dependencies are generally retained through $c^{\langle t \rangle}$ while short-term dependencies are carried by $h^{\langle t \rangle}$. The interaction between these groups of neurons is enabled via additional groups of neurons, namely the internal gates, which are generally vectors with entries in the range $[0, 1]$, denoted via $g_f^{\langle t \rangle}$, $g_i^{\langle t \rangle}$, and $g_o^{\langle t \rangle}$. An additional internal group of neurons, denoted via $\tilde{c}^{\langle t \rangle}$, is sometimes called the *internal cell state*.

For the basic RNN models of Section 7.2, we used m for the number of neurons and this is also the dimension of information flow between successive units. However, for LSTM, only some of the neurons are used for information flow between units, namely $c^{\langle t \rangle}$ and $h^{\langle t \rangle}$. In terms of dimension, we retain m as the number of neurons, and assume that $m = 6\,\tilde{m}$ where the dimensions of all vectors $c^{\langle t \rangle}$, $h^{\langle t \rangle}$, $\tilde{c}^{\langle t \rangle}$, $g_f^{\langle t \rangle}$, $g_i^{\langle t \rangle}$, and $g_o^{\langle t \rangle}$ is $\tilde{m}$.

A basic LSTM unit is illustrated in Figure 7.8 (a) which summarizes the evolution associated with this unit. Like the simpler RNN counterpart in Figure 7.4, and equations (7.2), the evolution equations of LSTM describe how the pair $(c^{\langle t \rangle}, h^{\langle t \rangle})$ evolves as a function of the previous pair $(c^{\langle t-1 \rangle}, h^{\langle t-1 \rangle})$ and the input $x^{\langle t \rangle}$. Further, the output $\hat{y}^{\langle t \rangle}$ evolves based on $(c^{\langle t \rangle}, h^{\langle t \rangle})$, directly via $h^{\langle t \rangle}$ and indirectly based on $c^{\langle t \rangle}$. Unlike the RNN (7.2), the LSTM evolution is more complex since it also involves the internal gates and neurons.

The LSTM evolution equations are

$$
\begin{cases}
c^{\langle t \rangle} = g_f \odot c^{\langle t-1 \rangle} \;+\; g_i \odot \underbrace{S_{\text{Tanh}}\left(W_{\tilde{c}h}\, h^{\langle t-1 \rangle} + W_{\tilde{c}x}\, x^{\langle t \rangle} + b_{\tilde{c}}\right)}_{\tilde{c}^{\langle t \rangle}} & \text{(cell state)} \\[3mm]
h^{\langle t \rangle} = g_o \odot S_{\text{Tanh}}(c^{\langle t \rangle}) & \text{(hidden state)} \\[3mm]
\hat{y}^{\langle t \rangle} = S_y(W_{yh} h^{\langle t \rangle} + b_y),
\end{cases} \qquad (7.16)
$$

where for clarity we omit the time superscripts from the internal gates and denote them via g_f, g_i, and g_o. Importantly, at every time t these internal gates are computed as

$$
\begin{cases}
g_f = S_{\text{Sig}}\left(W_{fh} h^{\langle t-1 \rangle} + W_{fx} x^{\langle t \rangle} + b_f\right) & \text{(forget gate)} \\[2mm]
g_i = S_{\text{Sig}}\left(W_{ih} h^{\langle t-1 \rangle} + W_{ix} x^{\langle t \rangle} + b_i\right) & \text{(input gate)} \\[2mm]
g_o = S_{\text{Sig}}\left(W_{oh} h^{\langle t-1 \rangle)} + W_{ox} x^{\langle t \rangle} + b_o\right). & \text{(output gate)}.
\end{cases} \qquad (7.17)
$$

Note that to restrict the value of internal gates to the range $[0, 1]$, sigmoid activation functions are typically used and we denote the associated vector activation function via $S_{\text{Sig}}(\cdot)$. The hidden state, the cell state, and the internal cell state information is not generally restricted to $[0, 1]$ and a typical activation function is tanh where we denote the associated vector activation function via $S_{\text{Tanh}}(\cdot)$.

As evident from (7.16) and (7.17), for an LSTM with input of dimension p and output of dimension q, the trained LSTM parameters include the following. First, there are four $\tilde{m} \times \tilde{m}$

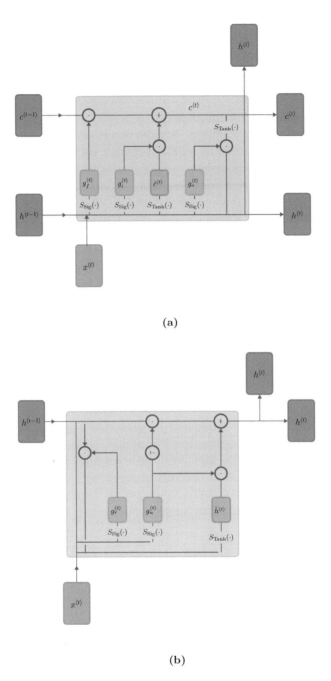

(a)

(b)

Figure 7.8: Representation of the LSTM and the GRU units. Internal gates are represented in yellow and internal states are in gray. The output $\hat{y}^{\langle t \rangle}$ is not presented. (a) In LSTM there are three internal gates and the internal state is called the internal cell state. (b) In GRU there are two internal gates and the internal state is called the internal hidden state.

weight matrices $W_{\tilde{c}h}$, W_{fh}, W_{ih}, and W_{oh}. Further, there are the four $\tilde{m} \times p$ weight matrices $W_{\tilde{c}x}$, W_{fx}, W_{ix}, and W_{ox}. In addition there is the $q \times \tilde{m}$ weight matrix W_{yh}, as well as the five associated bias vectors.

The specific structure of an LSTM unit interconnects the internal gates in a way that enables using both long term and short term memory, captured in $c^{\langle t \rangle}$ and $h^{\langle t \rangle}$, respectively. Specifically, the internal gates help select which information is "forgotten", "used as input", or "used as output". At each time step t the entries in the internal gate vectors $g_f^{\langle t \rangle}$, $g_i^{\langle t \rangle}$, and $g_o^{\langle t \rangle}$ can be "open", "closed", or somewhere in-between where entries that are near 1 are considered open and entries that are near 0 are considered closed. The forget gate $g_f^{\langle t \rangle}$ is multiplied element-wise with the previous cell state $c^{\langle t-1 \rangle}$ to "forget" information from the previous cell state or not, depending on being closed or open respectively. Similarly, the input gate, $g_i^{\langle t \rangle}$ controls what parts of the new cell content are written to the cell and this is applied to the internal cell state $\tilde{c}^{\langle t \rangle}$ which models the "selected information" based on the current input and the previous short term memory. Finally the output gate, $g_o^{\langle t \rangle}$, controls what parts of the cell are written to the hidden state $h^{\langle t \rangle}$ which is then used both for output $\hat{y}^{\langle t \rangle}$ and the short term memory passed onto the next unit.

It is interesting to consider the magnitudes of the LSTM elements, specifically in the first equation of (7.16). At time t, the previous cell state $c^{\langle t-1 \rangle}$ may have entries with general values (not limited to $[-1, 1]$). These values may then be "forgotten" if multiplied by $g_f^{\langle t \rangle}$ in cases where it is approximately at 0. Further, new long term memory is accumulated when $g_i^{\langle t \rangle}$ is approximately at 1. Observe that since the tanh activation function's range is $[-1, 1]$, the accumulation of this new memory is limited at every time step. Specifically, based on the internal cell state, $\tilde{c}^{\langle t \rangle}$, the memory may increase or decrease by at most 1 per time step.

The interconnection of LSTM units follows the same principles as the interconnection of RNN units outlined above. Specifically, one may view an unrolled representation of LSTM in the same manner as an unrolled representation of basic RNNs, presented in Figure 7.3. The difference is that both $c^{\langle t \rangle}$ and $h^{\langle t \rangle}$ are passed between time t and time $t + 1$, and not just $h^{\langle t \rangle}$ as in basic RNN. With this, the same extensions that one may consider for basic RNNs can be applied to LSTM. Specifically, LSTM can be reversed as in Figure 7.7 (a) or stacked into a deeper architecture as in Figure 7.7 (b). In reversing LSTM, the reverse direction LSTM passes $(c^{\langle t+1 \rangle}, h^{\langle t+1 \rangle})$ into the unit computing $(c^{\langle t \rangle}, h^{\langle t \rangle})$. In stacking LSTMs, the hidden state $h_{[\ell]}^{\langle t \rangle}$ of layer ℓ is passed as an input (similar to $x^{\langle t \rangle}$) for the unit above at layer $\ell + 1$ but not the cell state. Note that in stacked LSTM we may view $c_{[1]}^{\langle t \rangle}, \ldots, c_{[L]}^{\langle t \rangle}$ as a representation of long term memory across all layers at step t. This long term memory is passed to the next step, $t + 1$.

Gated Recurrent Unit Models

An alternative to the LSTM architecture is the *gated recurrent unit* (GRU) architecture, with a unit illustrated in Figure 7.8 (b). While LSTMs make an explicit separation of neurons to be long term or short term, with GRUs we return to a somewhat simpler architecture with only one key set of neurons $h^{\langle t \rangle}$, again called the *hidden state*. GRUs store both long-term dependencies and short-term memory in the single hidden state. Like LSTMs, GRUs use internal gates with values in the range $[0, 1]$, this time called the *reset gate* $g_r^{\langle t \rangle}$ and the *update gate* $g_u^{\langle t \rangle}$. Similarly to LSTMs that maintain an internal cell state, GRUs maintain an *internal hidden state* $\tilde{h}^{\langle t \rangle}$. Setting $\tilde{m}$ as the number of neurons in each of these groups, the total number of neurons in a GRU is $m = 4\tilde{m}$. Hence with 4 components instead of 6 components, gated recurrent units provide a simpler architecture in comparison to LSTM as there are only two internal gates (in comparison to three) and a single group of states passed between time units (in comparison to two).

The basic evolution equation for gated recurrent units is

$$
\begin{cases}
h^{\langle t \rangle} = (1 - g_u) \odot h^{\langle t-1 \rangle} \; + \; g_u \odot \underbrace{S_{\text{Tanh}}\left(W_{\tilde{h}h}\,(g_r \odot h^{\langle t-1 \rangle}) + W_{\tilde{h}x}\,x^{\langle t \rangle} + b_{\tilde{h}}\right)}_{\tilde{h}^{\langle t \rangle}} \\[2em]
\hat{y}^{\langle t \rangle} = S_y(W_{yh}h^{\langle t \rangle} + b_y),
\end{cases}
\tag{7.18}
$$

where for clarity we omit the time superscripts from the internal gates and denote them via g_r and g_u. At every time t, these internal gates are computed as

$$
\begin{cases}
g_r = S_{\text{Sig}}\left(W_{rh}h^{\langle t-1 \rangle} + W_{rx}x^{\langle t \rangle} + b_r\right) & \text{(reset gate)} \\[1em]
g_u = S_{\text{Sig}}\left(W_{uh}h^{\langle t-1 \rangle} + W_{ux}x^{\langle t \rangle} + b_u\right). & \text{(update gate).}
\end{cases}
\tag{7.19}
$$

A key attribute of the first equation of (7.18) is that new entries of the cell state $h^{\langle t \rangle}$ are computed as a convex combination of the entries of the previous cell state $h^{\langle t-1 \rangle}$ and the hidden cell state $\tilde{h}^{\langle t \rangle}$. This convex combination is determined by the entries of $g_u^{\langle t \rangle}$ where an entry near 1 implies "update" of the cell state based on the internal cell state, and an entry near 0 implies retaining the previous value (not updating).

As evident from (7.18) and (7.19), for a GRU with input of dimension p and output of dimension q, the trained parameters include the following. First, there are three $\tilde{m} \times \tilde{m}$ weight matrices $W_{\tilde{h}h}$, W_{rh}, and W_{uh}. Further, there are the three $\tilde{m} \times p$ weight matrices $W_{\tilde{h}x}$, W_{rx}, and W_{ux}. In addition there is the $q \times \tilde{m}$ weight matrix W_{yh}, as well as the four associated bias vectors. Again as evident, the number of parameter groups is smaller than that of LSTM.

To gain some intuition about the GRU architecture we may observe that the update gate g_u plays a role similar to both the forget gate, g_f, and input gate, g_i, in LSTM. Specifically compare the first equation in (7.18) with the first equation in (7.16). The simplification offered by GRU is to use a convex combination $(1 - g_u, g_u)$ instead of a general linear combination (g_f, g_i) as in LSTM. In both architectures, this operation controls what parts of long term memory information are updated versus preserved. One may also observe that GRU's internal hidden state $\tilde{h}^{\langle t \rangle}$ is updated via a slightly more complex mechanism than LSTM's internal cell state $\tilde{c}^{\langle t \rangle}$. The innovation in GRUs is to use the reset gate, g_r. Practice has shown that with such an architecture, GRUs are able to maintain both long term and short term memory inside the hidden state sequence, $h^{\langle 1 \rangle}, h^{\langle 2 \rangle}, \ldots$.

Note that the interconnection of GRUs can follow the exact same lines as other RNN architectures. Again, bi-directional connections as well as stacked configurations are possible; see Figure 7.7.

7.4 Encoders Decoders and the Attention Mechanism

One of the great application successes of sequence models is in the domain of *machine translation* tasks, namely the translation of one human language to another. For this, a general paradigm involving an *encoder* neural network and a *decoder* neural network is common. Other applications of encoders and decoders include, *image to text* models and *text to image* models. Yet, the main motivation we consider here is machine translation, since this application was the main driver in the development of *encoder-decoder architectures* within sequence models.

An important machine learning concept attention mechanism has advanced machine translation and other tasks. This idea is incorporated in *transformer models* that currently drive state of the art *large language models*. Transformers are the topic of the next section and in this current section, we first introduce general ideas of encoder-decoder architectures with the motivation of machine translation. We then formally define the attention mechanism. Finally, we see an encoder-decoder architecture that incorporates the attention mechanism at the interface of the encoder and the decoder.

Encoder-Decoder Architectures for Machine Translation

Recall from Section 7.1 that in general, when considering natural language, the input text is converted into a sequence of word embeddings denoted $x^{\langle 1 \rangle}, x^{\langle 2 \rangle}, \dots$. With such a sequence, at some point, an embedding of a word or token such as `<stop>` appears and marks the end of the text. In a machine translation application, our goal is to convert this input sequence to an output sequence $\hat{y}^{\langle 1 \rangle}, \hat{y}^{\langle 2 \rangle}, \dots$, also containing a `<stop>` token representation at its end. Clearly the input is in one natural language, e.g., French, and the output is an another natural language, e.g., Telugu.

Machine translation handled via an encoder-decoder architecture, uses a setup similar to Figure 7.9 (a). First, an encoder model which is a RNN, or a variant such as LSTM, or GRU, is used to convert the input sequence into the latent space by creating a *context vector* also known as the *code*, denoted via $z^\star$. Ideally this code encompasses the meaning and style of the input text. Then, a second sequence model, known as the decoder, takes the code $z^\star$ as input and converts it to the output sentence. Clearly the encoder model in this setup is configured as a many to one model, while the decoder model is configured as a one to many models. In the decoder, the output at each time fed into the input for the next time as in (7.3). Note that the code $z^\star$ is a vector of fixed dimension. Further, the dimension p of each $x^{\langle t \rangle}$, the dimension q of each $\hat{y}^{\langle t \rangle}$ typically differ and each typically has their own encoding. The input and output sequences are of arbitrary length, where the length of the input sequence and the length of the output sequence may differ.

This basic type of encoder-decoder architecture, as in Figure 7.9 (a), has already proven quite useful for early attempts of machine translation using deep sequence models. The choice between basic RNN, LSTM, GRU, or stacked combinations of one of these types of units is a modeling choice that one can make. No matter what type of unit is used, a key weakness is that the impact of the code $z^\star$ on the output $\hat{y}^{\langle 1 \rangle}, \hat{y}^{\langle 2 \rangle}, \dots$, decreases as t grows within the predicted output. Nevertheless, this architecture is a starting point for more advanced architectures.

A natural improvement is to make the context vector accessible for all steps in the decoder. With such a setup, at each time the decoder is fed the concatenation of the previous output and the code vector $z^\star$. A second improvement is to also present the code vector $z^\star$ for the computation of the output,[8] where at this point the output computation is based on

---

[8]Having the code available to the output is in a sense a residual connection similar to the ResNets discussed in Section 6.5. It is particularly useful if a stacked architecture is used in the decoder.

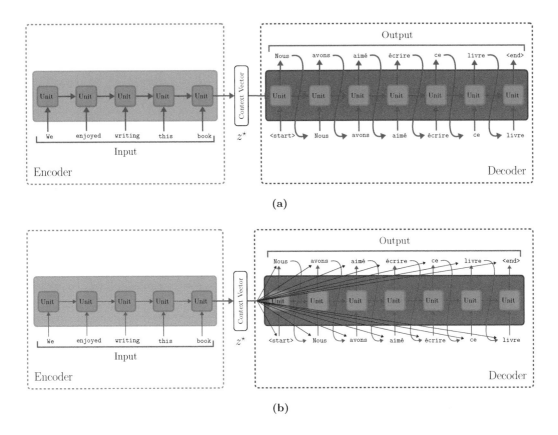

Figure 7.9: Unrolling of basic encoder-decoder architectures for machine translation. (a) A basic architecture where the encoder output context vector $z^\star$ is computed and fed as the initial state to the decoder. (b) A more advanced architecture where $z^\star$ is also presented at the input and output at each time step of the decoder.

a concatenation of the cell state, $h^{\langle t \rangle}$, and the code $z^\star$. This architecture is depicted in Figure 7.9 (b).

In the context of machine translation, it is often useful to modify the encoder-decoder pair such that the encoder accepts the *text in reverse order*. As an example, assume that the input text is

`I am going to read another chapter.`

Then with the text in reverse order paradigm, this input is fed to the encoder as

`chapter another read to going am I.`

The training process then uses reverse order inputs as above, yet outputs are expected in the normal order. Clearly when the model is used operationally, the input text is also reversed.

The benefit of this approach is in keeping inputs and their respective outputs closer on average. For example, assume that we are translating from English to French where the output should be (the non-reversed French text),

<code>je vais lire un autre chapitre.</code>

Note that with this approach the (reversed) input phrase `going am I` is near the output phrase `je vais` (which means "I am going"), and similarly with other pairs. Whereas if the text was not reversed, then generally (assuming inputs and outputs of the same length) the distance between an input and the respective output is in the order of the length of the text.

Even when employing techniques such as reversal of the text, a key drawback of all of the above encoder-decoder architectures is that performance degrades rapidly as the length of the input sentence increases. This is because the encoded vector needs to capture the entire input text, and in doing so, it might skip many important details. The attention mechanism that we describe below, and its application in machine translation architectures, overcome many of these difficulties.

The Attention Mechanism

As we saw above, one of the key considerations in encoder-decoder architectures has to do with the way in which the encoder output enters as input to the decoder. Toward this, we now introduce a general paradigm called the *attention mechanism*. One may view the attention mechanism as a method for "annotating" elements of the input or intermediate sequences, which require more focus, or attention, than others. We first define attention mathematically, and later we see how it can interplay within an encoder-decoder architecture. The attention mechanism defined here is also central to transformer models which encompass Section 7.5, and are used for most contemporary large language models.

In general, an *attention mechanism* can be viewed as a transformation of a sequence of vectors to a sequence of vectors. The input sequence has T_{in} vectors and the output sequence has T_{out} vectors. We assume the vectors are m-dimensional and denote the input sequence via $v^{\langle 1 \rangle}, v^{\langle 2 \rangle}, \ldots, v^{\langle T_{in} \rangle}$ and the output sequence via $u^{\langle 1 \rangle}, u^{\langle 2 \rangle}, \ldots, u^{\langle T_{out} \rangle}$.

As an aid to the attention mechanism, we also have two sequences of vectors, which we call the *proxy vectors*. These are denoted via $z_q^{\langle 1 \rangle}, \ldots, z_q^{\langle T_{out} \rangle}$, and $z_k^{\langle 1 \rangle}, \ldots, z_k^{\langle T_{in} \rangle}$, where the dimension of each vector in the first sequence is m_q, and similarly m_k for the second sequence. The notation using subscripts q and k stems from *query* and *key*, respectively. These terms, query and key, are more common in the application of transformer models in the next section.

One of the components of the attention mechanism is a *score function*, also known as an *alignment function*, $s : \mathbb{R}^{m_q} \times \mathbb{R}^{m_k} \to \mathbb{R}$ which when applied to a pair of proxy vectors, z_q and z_k and denoted via $s(z_q, z_k)$, measures the similarity between the two proxy vectors. A typical simple score function, suitable when $m_q = m_k$, is the inner product. Yet other possibilities, also potentially with learned parameters, can be employed, and in some instances this component is known as an *alignment model*. It is typical to have normalization as part of the score function with a factor such as $\sqrt{\max(m_q, m_k)}$. This normalization maintains score values at a reasonable range for numerical stability.

At the heart of the attention mechanism, for any time $t = 1, \ldots, T_{\text{out}}$, we apply the score function on a fixed $z_q^{\langle t \rangle}$ against all $z_k^{\langle 1 \rangle}, \ldots, z_k^{\langle T_{\text{in}} \rangle}$ and then using softmax we obtain a vector of *attention weights*, also known as *alignment scores*. This vector, denoted $\alpha^{\langle t \rangle}$, is of length T_{in}, and is computed via

$$
\alpha^{\langle t \rangle} = S_{\text{softmax}} \left(\begin{bmatrix} s(z_q^{\langle t \rangle}, z_k^{\langle 1 \rangle}) \\ s(z_q^{\langle t \rangle}, z_k^{\langle 2 \rangle}) \\ \vdots \\ s(z_q^{\langle t \rangle}, z_k^{\langle T_{\text{in}} \rangle}) \end{bmatrix} \right), \qquad \text{for} \qquad t = 1, \ldots, T_{\text{out}}. \tag{7.20}
$$

Here the vector to vector softmax function, $S_{\text{softmax}}(\cdot)$ is as defined in (3.25). The attention weights associated with time t, can also be written as $\alpha^{\langle t \rangle} = \left(\alpha_1^{\langle t \rangle}, \ldots, \alpha_{T_{\text{in}}}^{\langle t \rangle} \right)$. Thus in general, for any $t \in \{1, \ldots, T_{\text{out}}\}$, the attention weight vector $\alpha^{\langle t \rangle}$ captures similarity between the proxy vector $z_q^{\langle t \rangle}$ and all of the proxy vectors $z_k^{\langle 1 \rangle}, \ldots, z_k^{\langle T_{\text{in}} \rangle}$.

With attention weights available, the attention mechanism operates on the input sequence $v^{\langle 1 \rangle}, \ldots, v^{\langle T_{\text{in}} \rangle}$. The mechanism produces an output sequence $u^{\langle 1 \rangle}, \ldots, u^{\langle T_{\text{out}} \rangle}$ where each $u^{\langle t \rangle}$ is computed via the linear combination

$$
\begin{cases} u^{\langle t \rangle} = \sum_{\tau=1}^{T_{\text{in}}} \alpha_\tau^{\langle t \rangle} v^{\langle \tau \rangle} & \text{(Non-causal attention)} \\ \\ \qquad\qquad \text{or} \\ \\ u^{\langle t \rangle} = \sum_{\tau=1}^{t} \alpha_\tau^{\langle t \rangle} v^{\langle \tau \rangle}. & \text{(Causal attention)}. \end{cases} \tag{7.21}
$$

Note that in the causal form the output at time t, $u^{\langle t \rangle}$ only depends on the inputs up to time t, while in the non-causal form $u^{\langle t \rangle}$ depends on inputs at all times $t \in \{1, \ldots, T_{\text{in}}\}$. Also note that the causal form is only possible when $T_{\text{out}} \leq T_{\text{in}}$ (this is the case in the next section where in particular we use $T_{\text{out}} = T_{\text{in}}$).

As we see below, this general mechanism is applied in various sequence model architectures where in each case, the proxy vector sequences $z_q^{\langle 1 \rangle}, \ldots, z_q^{\langle T_{\text{out}} \rangle}$, and $z_k^{\langle 1 \rangle}, \ldots, z_k^{\langle T_{\text{in}} \rangle}$, and the score function can be defined differently. Note that the attention mechanism can also be employed for *graph neural networks* (GNN); see Section 8.5.

Encoder-Decoder with an Attention Mechanism

As a first application of the attention mechanism let us see an encoder-decoder framework. This architecture is described in Figure 7.10 where an attention mechanism is used to tie the encoder output with the decoder. A key attribute of the attention mechanism is to provide more importance to some of the input words, in comparison to others, during machine translation. There are two main ideas in this architecture. The first idea is to use a bi-directional encoder, and importantly the second idea is to incorporate an attention mechanism.

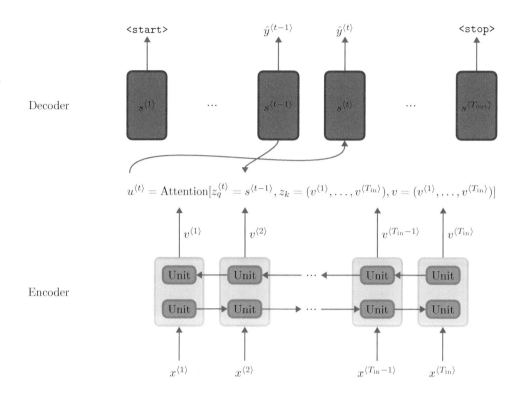

Figure 7.10: An encoder-decoder model with a bi-directional encoder and tying of the encoder and the decoder via an attention mechanism. The output sequence of the encoder is used as input to the attention mechanism. In the attention mechanism the previous decoder state is used as a query and the encoder outputs are used as keys.

The encoder is constructed via a bidirectional RNN, similar to Figure 7.7 (a) and the first two recursions in (7.14). Specifically for an input $x^{\langle 1 \rangle}, \ldots, x^{\langle T_{\text{in}} \rangle}$, we obtain the sequences of encoder hidden states $h_f^{\langle 1 \rangle}, \ldots, h_f^{\langle T_{\text{in}} \rangle}$ and $h_r^{\langle 1 \rangle}, \ldots, h_r^{\langle T_{\text{in}} \rangle}$, resulting from the forward directional and reverse directional recursions, respectively. Note that LSTM or GRU alternatives can be used in place of these forward and reverse recursions as well.

Now the output of the encoder is taken as a concatenation of the forward and reverse direction. Specifically, we denote this encoder output via $v^{\langle 1 \rangle}, \ldots, v^{\langle T_{\text{in}} \rangle}$ where $v^{\langle t \rangle}$ is a concatenation of $h_f^{\langle t \rangle}$ and $h_r^{\langle t \rangle}$. Note that the encoder output is a sequence of vectors of length T_{in}, namely the same length of the input sequence to the whole architecture.

The decoder is a variant of a RNN with hidden state $s^{\langle t \rangle}$, following the recursion,

$$\begin{cases} s^{\langle t \rangle} = f_{\text{decoder}}\left(s^{\langle t-1 \rangle}, \hat{y}^{\langle t-1 \rangle}, u^{\langle t \rangle}\right) \\ \hat{y}^{\langle t \rangle} = f_{\text{decoder-out}}\left(s^{\langle t \rangle}\right), \end{cases} \tag{7.22}$$

where $u^{\langle t \rangle}$ marks the input to the decoder and is computed via an attention mechanism as we describe below. The other two inputs are $s^{\langle t-1 \rangle}$, the hidden decoder state at the previous time step, and $\hat{y}^{\langle t-1 \rangle}$, the output of the decoder at the previous time step.

To see how the attention mechanism is used for computing $u^{\langle t \rangle}$, observe our notation where the encoder output is $v^{\langle t \rangle}$ and the decoder input at time t is $u^{\langle t \rangle}$. This notation agrees with the attention mechanism described above and in this architecture, an attention mechanism tying the encoder and the decoder, converts $v^{\langle t \rangle}$ to $u^{\langle t \rangle}$. Specifically, non-causal attention as in (7.21) is used. For the attention computation, the proxy vectors determining the attention weights via (7.20) are set as $z_q^{\langle t \rangle} = s^{\langle t-1 \rangle}$ and $z_k^{\langle t \rangle} = v^{\langle t \rangle}$.

Observe that when the next decoder output token, $\hat{y}^{\langle t \rangle}$ is created using (7.22), it is based on the decoder state $s^{\langle t \rangle}$ which is based on the previous decoder state, on the previous decoder output, and importantly, on the attention output $u^{\langle t \rangle}$. This attention output is a linear combination of all of the previous decoder outputs, weighted by the attention weights, $\alpha^{\langle t \rangle}$ of time t.

The strength of this attention-based architecture is that any input token can receive attention during the construction of the output sequence, even if the construction is at a location in the sequence far away from the input token. The application of a bi-directional architecture for the encoder enables the model to capture earlier and later information which help to disambiguate the input embedded word. The application of an attention mechanism introduces a form of a *dynamic context vector* between the encoder and decoder, in place of the static context vector $z^\star$ used in the simpler architectures above. Namely, architectures as depicted in Figure 7.9 have a fixed $z^\star$ which does not change during the operation of the decoder. The attention-based approach replaces this fixed $z^\star$ with the sequence $u^{\langle 1 \rangle}, u^{\langle 2 \rangle}, \ldots$, which itself depends on the decoder state (through the proxy $z_q^{\langle t \rangle} = s^{\langle t-1 \rangle}$).

An Illustration of Attention Weights

Let us see parts of an illustrative toy example of English to French translation using the encoder-decoder with attention architecture as in Figure 7.10. Assume the following:

Input: `we love deep learning <stop>`

Output: `nous aimons l' apprentissage en profondeur <stop>`

Observe that in this case, we explicitly use the `<stop>` token, and assume that with our tokenization and word embedding setup, each word or the `<stop>` token is a single vector. Also note that `l'` is considered a word. In this case we have $T_{\text{in}} = 5$ and as resulting from the model, $T_{\text{out}} = 7$.

When the input is processed via the architecture of Figure 7.10, first the encoder creates the sequences $h_f^{\langle 1 \rangle}, \ldots, h_f^{\langle T_{\text{in}} \rangle}$ and $h_r^{\langle 1 \rangle}, \ldots, h_r^{\langle T_{\text{in}} \rangle}$. Then, a sequence where each element is a concatenation of $h_f^{\langle t \rangle}$ and $h_r^{\langle t \rangle}$ is fed into the attention mechanism. The attention and the decoder operation run together, where with each additional time step of the output, (7.22) is applied, and the attention computation of (7.21) takes place. Importantly, with each t, the attention weights $\alpha^{\langle t \rangle}$ are computed via (7.20) where the proxy vector $z_q^{\langle t \rangle}$ is taken as the previous decoder state, and the proxy sequence $z_k^{\langle 1 \rangle}, \ldots, z_k^{\langle T_{\text{in}} \rangle}$ is taken as the concatenated output of the encoder, summarizing all of the input text.

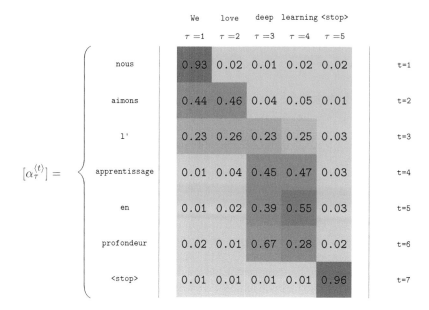

	We	love	deep	learning	<stop>	
	$\tau = 1$	$\tau = 2$	$\tau = 3$	$\tau = 4$	$\tau = 5$	
nous	0.93	0.02	0.01	0.02	0.02	t=1
aimons	0.44	0.46	0.04	0.05	0.01	t=2
l'	0.23	0.26	0.23	0.25	0.03	t=3
apprentissage	0.01	0.04	0.45	0.47	0.03	t=4
en	0.01	0.02	0.39	0.55	0.03	t=5
profondeur	0.02	0.01	0.67	0.28	0.02	t=6
<stop>	0.01	0.01	0.01	0.01	0.96	t=7

$[\alpha_\tau^{\langle t \rangle}] =$ (matrix above)

Figure 7.11: An attention matrix where each row is the attention vector $\alpha^{\langle t \rangle}$. When used in an encoder-decoder, we may consider the attention in a row associated with time t as based on the previous output. So for example for time $t = 4$, the available output from the decoder so far is an encoding of **nous aimons l'** and the attention vector $\alpha^{\langle 4 \rangle}$ dictates which encoded English words to focus on (in this case **deep** at $\alpha_3^{\langle 4 \rangle}$ and **learning** at $\alpha_4^{\langle 4 \rangle}$).

Figure 7.11 illustrates possible values of the attention weights, as they are computed via the machine translation process. The input is in English and the output is in French. Specifically, each row of this matrix is an hypothetical vector of attention weights, $\alpha^{\langle t \rangle}$, computed at time t of the decoder; namely $t = 1, \ldots, T_{\text{out}}$. Observe that each row sum is 1 and entries with higher probabilities are emphasized. This sequence of attention weight vectors shows how attention weights can adjust according to context. For example at time $t = 4$ in the decoder, an encoding of the partial sequence **nous aimons l'** is already available via $z_q = s^{\langle 3 \rangle}$. For creation of the next word, **apprentissage** (which directly means "learning" in English), most of the attention is put on the encoder hidden states associated both with **deep** and with **learning**.

Variants of the Score Function

If we use the inner product score function, then we are constrained that the dimension of the decoder hidden state, $s^{\langle t \rangle}$ be twice the dimension of each of the directional encoder hidden states (this is the dimension of $h_f^{\langle t \rangle}$ and $h_r^{\langle t \rangle}$). However, as already stated, other score functions are also possible. In each alternative, the score function operates on $z_q \in \mathbb{R}^{m_q}$ and $z_k \in \mathbb{R}^{m_k}$. Following are common alternatives:

$$s(z_q, z_k) = \begin{cases} z_q^\top W_s \, z_k, & \text{(General)} \\ w_s^\top \tanh(\widetilde{W}_s[z_q, z_k]), & \text{(Concatenation)} \\ w_s^\top \tanh(W_{sa} z_q + W_{sb} z_k). & \text{(Additive)} \end{cases}$$

Each of these score function alternatives has parameters that are learned during training. In the first case the parameter matrix is $W_s \in \mathbb{R}^{m_q \times m_k}$. In the second case, the parameter vector w_s is $\tilde{m}$-dimensional, and the parameter matrix is $\widetilde{W}_s \in \mathbb{R}^{\tilde{m} \times (m_q + m_k)}$. Note that in this case $[z_q, z_k]$ denotes a concatenation of the two vectors. In third case, $W_{sa} \in \mathbb{R}^{\tilde{m} \times m_q}$ and $W_{sb} \in \mathbb{R}^{\tilde{m} \times m_k}$ and again w_s is a $\tilde{m}$-dimensional vector. In each of these cases tanh is applied element-wise.

Training Encoder-Decoder Models

Continuing with the application of machine translation we now consider various approaches for training encoder-decoder models. Assume we are training models as in the architectures of Figure 7.9 (a) and (b) as well as Figure 7.10. Our discussion here is also relevant for training transformer encoder-decoder models presented in the next section (Figure 7.16).

As input data we have n sequences in the source language (e.g., English), where each sequence, denoted via $x^{(i)}$ is already encoded into vectors $x^{i,\langle 1 \rangle}, \ldots, x^{i,\langle T_x^i \rangle}$, using some word embedding technique (e.g., word2vec or some more advanced variant). Similarly we have n sequences in the target language (e.g., French), where in this case, each sequence $y^{(i)}$ is one-hot encoded into vectors $y^{i,\langle 1 \rangle}, \ldots, y^{i,\langle T_y^i \rangle}$, according to the lexicon of the target language. Clearly we assume that for any i, the pair of sequences $x^{(i)}$ and $y^{(i)}$ have the same semantic meaning.

For a given set of parameters, when feeding an input sequence $x^{(i)}$ into the encoder-decoder architecture, we are presented with an output sequence $\hat{y}^{(i)}$ where each element $\hat{y}^{i,\langle t \rangle}$ is a probability vector in the lexicon of the target language. We then use cross-entropy loss,[9] comparing $\hat{y}^{(i)}$ to $y^{(i)}$, summing over all elements of the sequence, similarly to (7.4) and (7.5); see also the discussion around Figure 7.5 in Section 7.2. One may also use mini-batches over multiple sequences, where for each mini-batch backpropagation (or backpropagation through time) is applied, and then a variant of gradient descent is used, similarly to any other deep learning model.

Teacher forcing, as discussed at the end of Section 7.2, is very commonly used when training encoder-decoder models. For example in the encoder-decoder with an attention architecture of Figure 7.10, during training we replace $\hat{y}^{\langle t-1 \rangle}$ by $y^{\langle t-1 \rangle}$ in (7.22), and similarly for the other encoder-decoder models.

Note also, that once an encoder-decoder model is trained, we may sometimes fine tune either the encoder, or decoder, by *freezing* the layers of one of the components while training the other component. Also, it is common to freeze both the encoder and decoder, and only fine tune an output layer on top of the decoder.

7.5 Transformers

We now introduce a family of models called *transformers*. The transformer architecture was originally introduced to handle machine translation and has since found applications in many other paradigms including large language models, but also non-sequence data

---

[9]A minor technical issue is that often the lengths of $\hat{y}^{(i)}$ and $y^{(i)}$ may differ. In such a case, the shorter sequence is padded with <empty> tokens and no loss is incurred at time t if both the predicted and training sequences have an <empty> token at time t.

domains such as images. The approach we present here continues to focus on the machine translation application, yet the reader should keep in mind that transformers have much wider applicability.

Transformers mark a paradigm shift in dealing with sequence data, as the architecture is no longer of a recurrent nature, but rather works using parallel computation. That is, while RNNs, LSTMs, GRUs, and other variants discussed in earlier sections may seem natural for sequence data, with transformers we return to fixed length input-output schemes. Similarly to the feedforward networks of Chapter 5 or the convolutional networks of Chapter 6, transformers operate on inputs of a fixed length, and yield outputs of a fixed length. Nevertheless, note that transformer decoders are used in an auto-regressive manner with a variable number of iterations.

In the context of sequence data, when using transformers, sequences are converted to have fixed length by padding with representations of `<empty>` tokens at the end of the sequence, when needed. Similarly if the input or output exceeds the dimensions, mechanisms external to the transformer are used to raise an error, break up the computation, truncate the input, or carry out similar workarounds.

As a simple illustration, return to the English to French translation example from the previous section and as a toy example, assume that the transformer input and output dimensions are both 9. In this case, we can expect the padded input and output to be

```
Input:  we    love           deep  learning       <stop> <empty>   <empty> <empty>   <empty>
Output: nous  aimons          l'    apprentissage  en     profondeur<stop>           <empty>   <empty>
```

While the abandonment of variable length inputs and outputs may seem like a step back from RNNs, transformers have shown great benefits in performance. In addition to yielding state of the performance on many language benchmarks, these architectures enable parallel computation which is not possible with RNNs.

The transformer architecture that we introduce here inherits the encoder-decoder pattern used in the previous section, and is well suited for machine translation. Note however that for other tasks, one may sometimes only use the encoder part of the transformer, the decoder part, or slight variants of the architecture that we present here. The key mechanism used in transformers is attention, with various forms of the attention mechanism interconnected in a novel way.

In our overview of transformers, we first describe the notion of *self attention*. We then describe *multi-head self attention*, often called *multi-head attention* in short. We then describe *positional embeddings* and then move onto introducing the *transformer block* which is the basic building block of the transformer architecture both for the encoder and the decoder. We close the section with an outline of the transformer encoder-decoder architecture followed by a discussion of how transformers are used in production and training.

Self Attention

We have already seen the general attention mechanism in equation (7.21) which transforms a sequence $v^{\langle 1 \rangle}, \ldots, v^{\langle T_{\text{in}} \rangle}$ to an output sequence $u^{\langle 1 \rangle}, \ldots, u^{\langle T_{\text{out}} \rangle}$, using linear combinations with attention weights $\alpha_\tau^{\langle t \rangle}$. The attention weights are defined in equation (7.20) and are based on the score function applied to the proxy vectors, denoted $z_q^{\langle 1 \rangle}, \ldots, z_q^{\langle T_{\text{out}} \rangle}$, and $z_k^{\langle 1 \rangle}, \ldots, z_k^{\langle T_{\text{in}} \rangle}$.

In the context of the transformer architecture, we refer to the proxy vectors $z_q^{\langle t \rangle}$ as *queries*, we refer to the proxy vectors $z_k^{\langle t \rangle}$ as *keys*, and we refer to the input vectors $v^{\langle t \rangle}$ as *values*. This terminology is rooted in information retrieval systems and captures the fact that when we compute an attention vector $\alpha^{\langle t \rangle}$, we are "searching" via (7.20) for a query represented via $z_q^{\langle t \rangle}$ against all keys $z_k^{\langle 1 \rangle}, \ldots, z_k^{\langle T_{\text{in}} \rangle}$. Then the attention weights are used to combine the "search results" via (7.21).

The mechanism of *self attention*, illustrated in Figure 7.12, is a form of attention where we convert an input sequence $x^{\langle 1 \rangle}, \ldots, x^{\langle T_{\text{in}} \rangle}$ to an output $u^{\langle 1 \rangle}, \ldots, u^{\langle T_{\text{out}} \rangle}$, with $T = T_{\text{in}} = T_{\text{out}}$. In the simplest form (ignore blue in Figure 7.12) we set the queries, the keys, and the values directly as elements of the input. Namely,

$$z_q^{\langle t \rangle} = x^{\langle t \rangle}, \qquad z_k^{\langle t \rangle} = x^{\langle t \rangle}, \qquad v^{\langle t \rangle} = x^{\langle t \rangle}, \qquad \text{(Simple self attention)}$$

and this implies that in the causal form (ignore red in Figure 7.12), with score function $s(\cdot, \cdot)$, the self attention mechanism yields output for any time $t \in \{1, \ldots, T\}$, via,

$$u^{\langle t \rangle} = \sum_{\tau \leq t} \alpha_\tau^{\langle t \rangle} x^{\langle \tau \rangle}, \qquad \text{with} \qquad \alpha_\tau^{\langle t \rangle} = \frac{e^{s(x^{\langle t \rangle}, x^{\langle \tau \rangle})}}{\sum_{t'=1}^{t} e^{s(x^{\langle t \rangle}, x^{\langle t' \rangle})}}. \qquad (7.23)$$

A more versatile form of self attention, this time involving learned parameters, has queries, keys, and values that are not directly taken as the input, but are rather linear transformations of the input. Namely,

$$z_q^{\langle t \rangle} = W_q x^{\langle t \rangle}, \qquad z_k^{\langle t \rangle} = W_k x^{\langle t \rangle}, \qquad v^{\langle t \rangle} = W_v x^{\langle t \rangle}, \qquad \text{(More versatile self attention)}$$

where the learnable parameter matrices W_q, W_k, and W_v are each $p \times p$ dimensional,[10] with p the dimension of each $x^{\langle t \rangle}$. Hence in this case, the attention mechanism has output for any time $t \in \{1, \ldots, T\}$, via,

$$u^{\langle t \rangle} = \sum_{\tau \leq t} \alpha_\tau^{\langle t \rangle} W_v x^{\langle \tau \rangle}, \qquad \text{with} \qquad \alpha_\tau^{\langle t \rangle} = \frac{e^{s(W_q x^{\langle t \rangle}, W_k x^{\langle \tau \rangle})}}{\sum_{t'=1}^{t} e^{s(W_q x^{\langle t \rangle}, W_k x^{\langle t' \rangle})}}. \qquad (7.24)$$

Observe that that in (7.23) and (7.24) we use the causal form from (7.21). An alternative non-causal form is also applicable (consider also the red part of Figure 7.12). In such a case, the summations are not on $\tau \leq t$ but are rather on $\tau \in \{1, \ldots, T\}$, similarly to the non-causal form appearing in (7.21).

Multi-Head Self Attention

A generalization of self attention is to use multiple self attention mechanisms in parallel and then combine the outputs of these mechanisms. With this parallelism, we can treat each

---

[10]Note that here we use the same dimension for the input, the output, and the proxy vectors. More generally one may set different dimensions for these entities, as we do in the case of multi-head attention below.

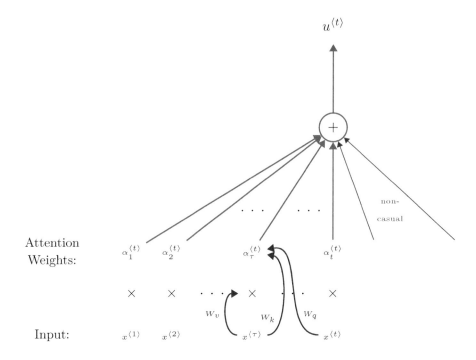

Figure 7.12: The flow of information in self attention. Ignoring the blue and the red, this is causal simple self attention where the output at time t, $u^{\langle t \rangle}$ is a linear combination of $x^{\langle 1 \rangle}, \ldots, x^{\langle t \rangle}$ with attention weights $\alpha_1^{\langle t \rangle}, \ldots, \alpha_t^{\langle t \rangle}$. Considering the blue, this is more versatile self attention where each attention weight $\alpha_\tau^{\langle t \rangle}$ is computed using weighting matrices W_k and W_q of the input, and where the linear combination is of weighted inputs with W_v. Considering also the red, it is non-causal.

individual attention mechanism as searching for a different set of features in the input, and then have information content of the output as a combination of the derived features.

Figure 7.13 illustrates multi-head self attention. Specifically, assume we have H self attention mechanisms, where mechanism $h \in \{1, \ldots, H\}$ has its own set of parameter matrices W_q^h, W_k^h, and W_v^h. Here $W_q^h \in \mathbb{R}^{m \times p}$, $W_k^h \in \mathbb{R}^{m \times p}$, and $W_v^h \in \mathbb{R}^{m_v \times p}$, where m is the dimension of the query and the key ($m = m_q = m_k$), p is the dimension of the input $x^{\langle t \rangle}$ as previously, and m_v is the dimension of the value (and the output of the individual attention head).

At first, each attention head h operates independently similarly to (7.24), yielding an output sequence $u^{h, \langle 1 \rangle}, \ldots, u^{h, \langle T \rangle}$ of m_v dimensional vectors. This operation is via

$$u^{h, \langle t \rangle} = \sum_{\tau=1}^{T} \alpha_\tau^{h, \langle t \rangle} \, W_v^h \, x^{\langle \tau \rangle}, \qquad \text{with} \qquad \alpha_\tau^{h, \langle t \rangle} = \frac{e^{s(W_q^h \, x^{\langle t \rangle}, W_k^h \, x^{\langle \tau \rangle})}}{\sum_{t'=1}^{T} e^{s(W_q^h \, x^{\langle t \rangle}, W_k^h \, x^{\langle t' \rangle})}}. \qquad (7.25)$$

Note that in (7.25) we use a non-causal form of attention, in contrast to the causal form appearing in (7.24). Below we comment on how one may practically convert such a non-causal form to a causal form via a mechanism called *masked self attention*.

Now for any time index t, we have H vectors $u^{1, \langle t \rangle}, \ldots, u^{H, \langle t \rangle}$ which we use to produce the output of the multi-head attention for the specific time index t. For this we apply an

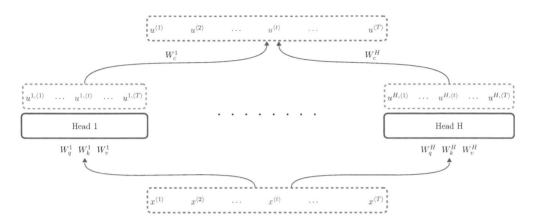

Figure 7.13: Multi-head self attention is the parallel application of H attention heads, where head h has parameter matrices W_q^h, W_k^h, and W_v^h. Each head h operates on the full input, $x^{\langle 1 \rangle}, \ldots, x^{\langle T \rangle}$ and results in output for the head, $u^{h, \langle 1 \rangle}, \ldots, u^{h, \langle T \rangle}$. When determining the output of the whole multi-head self attention mechanism, each output at time t, denoted $u^{\langle t \rangle}$ combines $u^{1, \langle t \rangle}, \ldots, u^{H, \langle t \rangle}$ weighted by $W_c^1, \ldots, W_c^H$.

additional linear transformation to each vector, converting it from dimension m_v back to dimension p. We then sum up over $h = 1, \ldots, H$. Thus, the output of the multi-head self attention mechanism is

$$u^{\langle t \rangle} = \sum_{h=1}^{H} W_c^h u^{h, \langle t \rangle}, \tag{7.26}$$

where for each h, the matrix W_c^h is $p \times m_v$ dimensional. Each W_c^h captures the transformation from the single attention output $u^{h, \langle t \rangle}$ of dimension m_v to dimension p. The combination of these H matrices can also capture weightings between the attention heads.

Multi-head self attention plays a central role in transformer blocks, both in the encoder and the decoder. In certain cases such as the encoder, we use the non-causal form, as in (7.25). However, in other cases, such as the decoder, we use a causal form. Practically we may enforce a causal form via masked self attention, also known as *masking* for short. With this approach, for a computation of attention weights at time t, we set entries of the (key) proxy vectors of times after time t to negative infinity. That is,

$$z_k^{\langle t+1 \rangle} = -\infty, \ \ z_k^{\langle t+2 \rangle} = -\infty, \ \ldots, z_k^{\langle T \rangle} = -\infty. \qquad \text{(Masking)} \tag{7.27}$$

Then with this masking, through the softmax computation, the $-\infty$ values yield zeros for each of $\alpha_{t+1}^{\langle t \rangle}, \alpha_{t+2}^{\langle t \rangle}, \ldots, \alpha_T^{\langle t \rangle}$, and thus, when computing the output at time t, no attention is given to future times.

Implementing causality via masking is especially important when one considers a matrix representation of the multi-head self attention mechanism. Specifically, one may treat the input as a matrix X of dimension $p \times T$ and then represent all of the attention operations as matrix on matrix operations. While we omit the details of such a representation, the reader should keep in mind that unlike RNNs where time t implies a step in the computation, for transformers time t is simply a dimension of the input matrix X and operations can be parallelized based on the equations above.

Positional Embeddings

Sequence models have a natural time index, t, where $x^{\langle t \rangle}$ followed by $x^{\langle t+1 \rangle}$, embodies some relationship between the two vectors in the sequence. As a simple example consider some input text, `deep learning`, and the reverse text, `learning deep`. These two short sequences have different semantic meaning, since the order matters. However, as is evident from the multi-head self attention mechanism (7.25), the order in the input sequence $x^{\langle 1 \rangle}, \ldots, x^{\langle T \rangle}$ is not captured by the mechanism at all. This stands in stark contrast to previous sequence models such as RNNs and their generalizations, where the recurrent nature of the model makes use of sequence order.

Hence, for using non-causal multi-head self attention effectively, we require a mechanism for encoding the order of the sequence in the input data. Such mechanisms are generally called *positional embeddings*. A basic and primitive form of positional embedding is to extend each input vector $x^{\langle t \rangle}$ with an additional one-hot encoded vector that captures its position in the sequence. For example for a sequence of length $T = 4$, we extend $x^{\langle 1 \rangle}$ with $e_1 = (1, 0, 0, 0)$, extend $x^{\langle 2 \rangle}$ with $e_2 = (0, 1, 0, 0)$, and so forth. Then the input sequence is no longer taken as having vectors of length p, but rather as having vectors of length $\tilde{p} = p + T$.

To further illustrate the point using the toy *one-hot encoding positional embedding* example with $p = 2$ and $T = 4$, assume the first vector is $x^{\langle 1 \rangle} = (0.2, -1.3)$ and assume that as a matter of coincidence the last vector $x^{\langle 4 \rangle}$ has the same values. Then after applying such positional embedding, the vectors are transformed to

$$\tilde{x}^{\langle 1 \rangle} = (0.2, -1.3, \underbrace{1, 0, 0, 0}_{e_1}) \qquad \text{and} \qquad \tilde{x}^{\langle 4 \rangle} = (0.2, -1.3, \underbrace{0, 0, 0, 1}_{e_4}),$$

and then when these positionally embedded vectors are processed by non-causal multi-head self attention, the model can distinguish between $\tilde{x}^{\langle 1 \rangle}$ and $\tilde{x}^{\langle 4 \rangle}$. Even in the (more common) case where different vectors will not have repeated values, the order in the sequence is still encoded and this enhances performance.

However, this type of encoding is clearly wasteful and inefficient, yielding an excessively large $\tilde{p}$. A more advanced form is to encode the vectors and the embeddings jointly. For example, one might use a transformation such as

$$\tilde{x}^{\langle t \rangle} = W_1 S \left(W_2 x^{\langle t \rangle} + W_3 e_t + b \right) \in \mathbb{R}^{\tilde{p}}, \tag{7.28}$$

where $W_1 \in \mathbb{R}^{\tilde{p} \times \tilde{p}}$, $W_2 \in \mathbb{R}^{\tilde{p} \times p}$, and $W_3 \in \mathbb{R}^{\tilde{p} \times T}$ are learnable weight matrices, $b \in \mathbb{R}^{\tilde{p}}$ is a learnable bias vector, and $S(\cdot)$ is some vector activation function such as for example ReLU applied element-wise. In this case, we may just use $\tilde{p} = p$, yet larger $\tilde{p}$ are also possible.

With this type of encoding, after training the parameters, positional embeddings are ideally encoded within the word vectors. This is similar to how vector representations encode words from a dictionary into a lower-dimensional space when using word embeddings.

With the introduction of transformers, a different type of positional encoding, motivated by *Fourier analysis*, was popularized. With this approach we set some $\tilde{p} > p$ and denote $p_e = \tilde{p} - p$. That is, p_e is the increase of dimension from the original encoding in $\mathbb{R}^p$ to the new encoding in $\mathbb{R}^{\tilde{p}}$. Assume also that p_e is even. Unlike the trained positional encoding in (7.28), here we just use sines and cosines without any trainable parameters.

Specifically for $i \in \{0, \ldots, \frac{p_e}{2} - 1\}$ and $t \in \{1, \ldots, T\}$, we set

$$r_{2i}^{\langle t \rangle} = \sin \left(\frac{t}{M^{2i/p_e}} \right), \quad r_{2i+1}^{\langle t \rangle} = \cos \left(\frac{t}{M^{2i/p_e}} \right), \tag{7.29}$$

where a common value for M is $10,000$. The vector with positional embedding is then,

$$\tilde{x}^{\langle t \rangle} = (x_1^{\langle t \rangle}, \ldots, x_p^{\langle t \rangle}, \overbrace{r_0^{\langle t \rangle}}^{\sin}, \overbrace{r_1^{\langle t \rangle}}^{\cos}, \overbrace{r_2^{\langle t \rangle}}^{\sin}, \quad \cdots \quad , \overbrace{r_{p_e-1}^{\langle t \rangle}}^{\cos}).$$

$$\underbrace{\phantom{r_0^{\langle t \rangle}, r_1^{\langle t \rangle}, r_2^{\langle t \rangle}, \quad \cdots \quad , r_{p_e-1}^{\langle t \rangle}}}_{\text{positional embedding}}$$

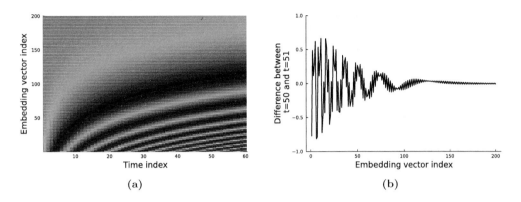

Figure 7.14: The sine and cosine-based positional embedding as in (7.29). (a) A heat map of $r_j^{\langle t \rangle}$ with $p_e = 768$ and $T = 4,000$ plotted for t only over the first 60 time indexes for j only over the first 200 positions in the embedding vector. Positive values are yellow and negative values are blue. (b) A comparison of the positional embeddings at the time index $t = 50$ and $t = 51$ via a plot of the difference. As is evident, there is a significant difference marked by around the first hundred vector positions.

One may then also reduce the dimension back from $\tilde{p}$ to a lower dimension in similar vein to (7.28). Specifically one common simplistic approach that has empirically worked well is

$$\tilde{x}^{\langle t \rangle} = (x_1^{\langle t \rangle} + r_0^{\langle t \rangle}, x^{(2)} + r_1^{\langle t \rangle}, \quad \cdots \quad , x_p^{\langle t \rangle} + r_{p-1}^{\langle t \rangle}) \in \mathbb{R}^p, \tag{7.30}$$

where here $p_e = p$.

The benefit of using positional embedding such as (7.30) first arises from the fact that sines and cosines are bounded within $[-1, 1]$ and is further aided by the idea of a Fourier representation of the position. In particular, consider Figure 7.14 where (a) illustrates embedding values from (7.29). By adding a vertical slice (for fixed t) of the embedding values (7.29) to the original $x^{\langle t \rangle}$ vector, we enable the model to distinguish the time value from typical other values. This is particularly evident by considering (b) where we compare two neighboring time steps by plotting their difference.

The Transformer Block

The basic building block of the transformer architecture is a unit called a *transformer block* which is used multiple times within a transformer, interconnected in series, both in the encoder and the decoder. There are several variations of transformer blocks and here we focus on the basic block used in the encoder. Later when we describe the decoder architecture we highlight differences.

We denote the input of a transformer block as a_{in} and the output as a_{out} where both a_{in} and a_{out} are $p \times T$ matrices. Thus in general, we can view the block as a function $f_\theta : \mathbb{R}^{p \times T} \to \mathbb{R}^{p \times T}$, where θ represents trained parameters of the block and $a_{\text{out}} = f_\theta(a_{\text{in}})$. For example, the encoder of the transformer architecture has a first transformer block that operates on input,

$$a_{\text{in}} = \left[\tilde{x}^{\langle 1 \rangle}, \ldots, \tilde{x}^{\langle T \rangle} \right],$$

where each column is a positional encoding vector as in (7.30). Then the output of this block, a_{out}, is fed into a second transformer block in the encoder, and so forth. Common architectures have an encoder composed of a sequence of half a dozen or more transformer blocks. Thus for example the input of the second transformer block has a_{in} set as the a_{out} of the first block, etc.

A transformer block has several internal layers with the two main layers being a multi-head self attention layer, and downstream to it, a feedforward layer. Each of these also utilizes a residual connection and a normalization layer. A schematic of a typical transformer block is in Figure 7.15 (a). The main idea of the transformer block is to enhance multi-head self attention with further connections using the feedforward layer. The residual connections enhance training performance, similar to ResNets discussed in Section 6.5. Normalization, in the form of *layer normalization*, stabilizes training and production performance by ensuring that values remain in a sensible dynamic range. We now present the details of a block.

Let us denote the columns of the input a_{in} as $a_{\text{in}}^{\langle 1 \rangle}, \ldots, a_{\text{in}}^{\langle T \rangle}$. The first step of the block with multi-head self attention is to employ (7.25) for every head $h = 1, \ldots, H$, and then (7.26) where in these equations $x^{\langle t \rangle}$ is replaced by $a_{\text{in}}^{\langle t \rangle}$. The output of this multi-head self attention layer is then denoted $u_{[1]}^{\langle 1 \rangle}, \ldots, u_{[1]}^{\langle T \rangle}$, where we use the subscript [1] to indicate it as an output of the first layer.

We then apply residual connections and layer normalization to each of $u_{[1]}^{\langle 1 \rangle}, \ldots, u_{[1]}^{\langle T \rangle}$ yielding $u_{[2]}^{\langle 1 \rangle}, \ldots, u_{[2]}^{\langle T \rangle}$. This step can be summarized as

$$u_{[2]}^{\langle t \rangle} = \text{LayerNorm} \big(\underbrace{a_{\text{in}}^{\langle t \rangle} + u_{[1]}^{\langle t \rangle}}_{\text{Residual connection}} ; \gamma, \beta \big). \tag{7.31}$$

Here, the LayerNorm($\cdot$) operator is defined for $z \in \mathbb{R}^p$ with parameters $\gamma, \beta \in \mathbb{R}^p$, via,

$$\text{LayerNorm}(z \, ; \gamma, \beta) = \gamma \odot \frac{(z - \mu_z)}{\sqrt{\sigma_z^2 + \varepsilon}} + \beta,$$

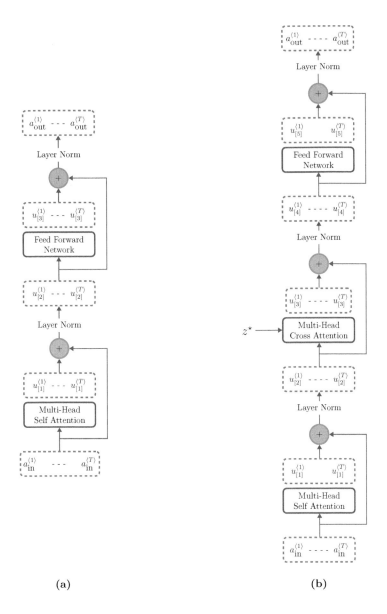

(a)   (b)

Figure 7.15: Architecture of transformer blocks. (a) A single transformer block. The input a_{in} passes through a multi-head self attention layer, followed by a feedforward layer. Layer normalization and residual connections are applied in these steps as well. (b) A transformer decoder block. In addition to the components of a transformer block, it also has a multi-head cross attention layer that is fed the context vector $z^{\star}$.

where

$$\mu_z = \frac{1}{p}\sum_{i=1}^{p} z_i \quad \text{and} \quad \sigma_z = \sqrt{\frac{1}{p}\sum_{i=1}^{p}\left(z_i - \mu_z\right)^2},$$

$\varepsilon > 0$ is a small fixed quantity that ensures that we do not divide by zero, and the addition, division, and square root operations are all element-wise operations.

Layer normalization is somewhat similar to batch normalization, outlined in Section 5.6, and group normalization outlined in Section 6.4. A major difference between layer normalization and batch normalization is that for batch normalization we obtain the mean and standard deviation per feature over a mini batch, whereas with layer normalization we use a single sample, yet compute statistics over all features (of a single feature vector). In both cases, the normalization forces feature values to remain at normalized values, further aided by the learnable parameter vectors γ and β.

The next step in the transformer block is the application of a fully connected neural network on each $u_{[2]}^{\langle t \rangle}$ to yield $u_{[3]}^{\langle t \rangle}$. Note that the same learnable network parameters are used for each $t \in \{1, \ldots, T\}$. Commonly this network has a single hidden layer with non-linear activation, followed by a layer with linear (identity) activation,[11] yet there are other possibilities as well. Sticking with the commonly used architecture we have

$$u_{[3]}^{\langle t \rangle} = W^{[2]} S(W^{[1]} u_{[2]}^{\langle t \rangle} + b^{[1]}) + b^{[2]},$$

where $S(\cdot)$ is commonly an element-wise application of ReLU. Here we denote the dimension of the inner layer as N_1 and the learnable parameters are $b^{[1]} \in \mathbb{R}^{N_1}$, $b^{[2]} \in \mathbb{R}^p$, $W^{[1]} \in \mathbb{R}^{N_1 \times p}$, and $W^{[2]} \in \mathbb{R}^{p \times N_1}$.

Finally, to yield the output of the transformer block, we apply residual connections and layer normalization in the same manner as (7.31). Specifically we use

$$a_{\text{out}}^{\langle t \rangle} = \text{LayerNorm}\left(u_{[2]}^{\langle t \rangle} + u_{[3]}^{\langle t \rangle}; \tilde{\gamma}, \tilde{\beta}\right),$$

where here $\tilde{\gamma}, \tilde{\beta} \in \mathbb{R}^p$ are trainable parameters for this layer normalization.

Note that when considered as a variation of a neural network, one may view a transformer block as "wide and shallow". Even though such a single transformer block is not deep, the residual connections provide direct access to the previous levels of abstraction and enable the levels above to infer more fine-grained features without having to remember or store previous ones.

Let us summarize the learned parameters of a single transformer block with dimensions p for the vector length, T for the sequence length, H for the number of self attention heads, m_v for the dimension inside each self attention block, and m for the dimension of the query and key inside each self attention block. In this case, the total number of parameters is

$$\underbrace{4\,p}_{\gamma,\beta,\tilde{\gamma},\tilde{\beta}} + \underbrace{2\,N_1 p}_{W^{[1]},W^{[2]}} + \underbrace{N_1 + p}_{b^{[1]},b^{[2]}} + H \times (\ \underbrace{2\,m\,p}_{W_k^h \text{ and } W_q^h} + \underbrace{2\,m_v\,p}_{W_v^h \text{ and } W_c^h}\ ). \tag{7.32}$$

As a quantitative example, agreeing with the first transformer architecture introduced in 2017,[12] let us consider a case with $p = 512$, $N_1 = 2048$, $m = m_v = 64$, and $H = 8$. In this case there are just over 3 million learnable parameters for such a transformer block. Specifically (7.32) evaluates to $3,150,336$. As we see now, multiple transformer blocks are typically connected, yielding architectures with many millions of parameters.

---

[11]Incidentally, this two-layer architecture with an activation function only in the single hidden layer, is the same network used in Theorem 5.1 of Chapter 5.

[12]See the "Attention is all you need" paper [410], as well as other notes and references at the end of the chapter.

Putting the Bits Together Into an Encoder-Decoder Framework

Now that we have seen the design of the transformer block, we are ready to interconnect such blocks in an encoder, and also interconnect variations of this block in a decoder. We now describe a basic *transformer encoder-decoder architecture* using such blocks and interconnections. It is useful to recall the more classical encoder-decoder architectures as appearing in (a) and (b) of Figure 7.9. A transformer architecture is somewhat similar to the architecture in (b), since the output of the encoder is fed into each of the decoder steps. However, unlike the architectures in Figure 7.9, transformers do not process one word at a time. A schematic of a transformer encoder-decoder architecture is in Figure 7.16.

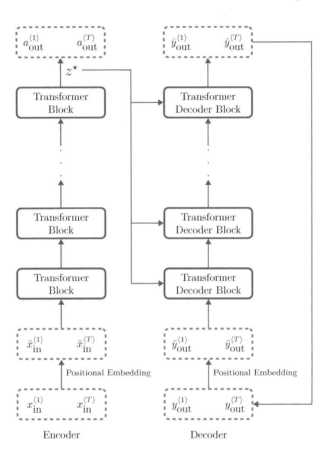

Figure 7.16: An encoder-decoder transformer architecture. The encoder is composed of multiple transformer blocks, and the decoder is composed of multiple transformer decoder blocks. Each block in the decoder is fed the code $z^\star$. The loop from the output of the decoder going back into the input of the decoder illustrates the auto-regressive application of transformer decoders.

Transformer encoders simply stack the transformer blocks in series, where each block has exactly the specifications described above as in Figure 7.15 (a). The first block is fed with the positional encoded input, and the output of that block goes into the second block, and so forth. The output of the encoder is $a_{\text{out}}^{\langle 1 \rangle}, \ldots, a_{\text{out}}^{\langle T \rangle}$ resulting from the last transformer block.

We also denote this output via $z^\star$ as it describes a code, and thus for each position t we denote the encoder output via $z^{\star\langle t\rangle}$.

Transformer decoders use a variation of the transformer block which we call the *transformer decoder block*, illustrated in Figure 7.15 (b). This block architecture differs from the transformer block specified above in two ways. First, the multi-head self attention layer is causal. This is implemented via masking as in (7.27). Such causality prevents attendance of future positions as suitable for auto-regressive prediction. Second, an additional layer, called a *cross attention layer*, is used between the causal multi-head self attention and the feedforward layer. The new cross attention layer is handled with layer normalization and residual connections, similar to the other two layers.

In addition to the flow of information within the transformer decoder block, the cross attention layer is fed with the encoder output $z^\star$. This is similar to the encoder-decoder architecture in Figure 7.9 (b), and it allows the transformer decoder block to directly incorporate the encoder's code. Transformer decoders are constructed by stacking several transformer decoder blocks in sequence, similarly to the stacking in the encoder, where each block gets the same $z^\star$. Similar to the encoder, the first block operates on positional embedded inputs. Further discussion of what these inputs are, is in the following subsection. On top of the final transformer decoder block (not illustrated in Figure 7.16), we add an additional layer transforming each output vector to a token. This is similar to the outputs of other encoder-decoder architectures. Often it is simply a linear layer with a softmax (i.e., a multinomial regression as in Section 3.3).

The cross attention layer inside each transformer decoder block is in fact a *multi-head cross attention* layer and follows equations similar to (7.25) and (7.26), with the difference being that the key and value inputs are the decoder output $z^\star$. Specifically, if we denote the input to the multi-head cross attention layer from earlier layers as $\tilde{u}^{\langle 1\rangle}, \ldots, \tilde{u}^{\langle T\rangle}$, then the self attention equation (7.25) is now modified to have cross attention (between $z^\star$ and the input to the layer $\tilde{u}$). This attention computation for head h is then,

$$u^{h,\langle t\rangle} = \sum_{\tau=1}^{T} \alpha_\tau^{h,\langle t\rangle}\, W_v^h\, z^{\star\langle\tau\rangle}, \qquad \text{with} \qquad \alpha_\tau^{h,\langle t\rangle} = \frac{e^{s(W_q^h\,\tilde{u}^{\langle t\rangle}, W_k^h\, z^{\star\langle\tau\rangle})}}{\sum_{t'=1}^{t} e^{s(W_q^h\,\tilde{u}^{\langle t\rangle}, W_k^h\, z^{\star\langle t'\rangle})}}, \tag{7.33}$$

followed by a combination of the heads using (7.26).

Observe that this pattern of cross attention is similar to the earlier encoder-decoder with an attention mechanism architecture presented in Figure 7.10. In the earlier architecture, the proxy vector $z_q^{\langle t\rangle}$ used the previous decoder state, somewhat similarly $W_q^h\,\tilde{u}^{\langle t\rangle}$ in (7.33). Further, in the earlier architecture, the proxy vector $z_k^{\langle t\rangle}$ is the same vector used as input to the attention mechanism. This agrees with using the decoder output, $z^\star$ as key and value inputs in (7.33).

Now after highlighting the differences between a transformer block as in Figure 7.15 (a), and the transformer decoder block in Figure 7.15 (b), we can briefly summarize the layers and steps of the transformer decoder block. The matrix of inputs to each block, with each input vector denoted $a_{\text{in}}^{\langle t\rangle}$, is first processed with causal multi-head self attention to yield $u_{[1]}^{\langle 1\rangle}, \ldots, u_{[1]}^{\langle T\rangle}$. Now exactly as in (7.31), layer normalization and residual connections yield $u_{[2]}^{\langle 1\rangle}, \ldots, u_{[2]}^{\langle T\rangle}$. Then this sequence of vectors (or matrix) is processed via the multi-head cross attention layer using (7.33) and (7.26), where $\tilde{u}^{\langle t \rangle}$ is $u_{[2]}^{\langle t \rangle}$, and the encoder output $z^{\star}$ is put to use. The result is $u_{[3]}^{\langle 1 \rangle}, \ldots, u_{[3]}^{\langle T \rangle}$. Then layer normalization and residual connections are applied again yielding $u_{[4]}^{\langle 1 \rangle}, \ldots, u_{[4]}^{\langle T \rangle}$. This sequence is now fed into the feedforward layer, to yield $u_{[5]}^{\langle 1 \rangle}, \ldots, u_{[5]}^{\langle T \rangle}$. Finally, each $u_{[5]}^{\langle t \rangle}$ is again applied with layer normalization and residual connections to yield the output $a_{\text{out}}^{\langle t \rangle}$. There are 6 steps here, in comparison to the 4 steps used in the transformer block of the encoder.[13]

The parameters of each transformer decoder block include those of the transformer block from Figure 7.15, as well as parameters resulting from the multi-head cross attention layer and its normalization. Specifically, for the decoder, the parameter count in (7.32) needs to be augmented with an additional $H \times (2mp + 2m_v p)$ term for the multi-head cross attention as well as $2p$ for the additional normalization parameters. As an example, using the same dimensions as above, we now have over 4 million parameters, or $4,199,936$ exactly, for a transformer decoder block.

If we now consider an encoder-decoder transformer architecture with 6 transformer blocks in the encoder and 6 transformer decoder blocks in the decoder, then the number of parameters in the encoder is about 19 million, and the number of parameters in the decoder is about 25 million. As mentioned above, we also require an additional layer on top of the decoder, transforming each output vector to a token (for example to a natural language word or part of it). This is simply a linear layer with a softmax (i.e., a multinomial regression). This type of layer is needed at the end of any pipeline that generates text. If we assume that the number of word tokens[14] is $d_{\mathcal{V}} \approx 37,000$, then the number of parameters of the final multinomial regression is $d_{\mathcal{V}} \times p + p$, which is about 19 million parameters in our case. Hence putting the pieces together we have about 63 million parameters in the whole model.[15]

The encoder-decoder transformer model is the cornerstone of *large language models*, and indeed by 2023, models with up to half a trillion parameters are already in use. Such parameter count is about 100 times more parameters than the size of the transformer discussed above.

Using the Encoder-Decoder in Production and Training

In our description of the encoder-decoder architecture above, except for indicating that positional embedding is applied, we did not specify the decoder inputs. The way that decoder inputs are used depends on the task at hand, and the form of the inputs varies between production and training. We now present the details.

First in production (inference or test-time), note that we use the decoder in an auto-regressive manner as in Figure 7.17. Specifically, the code from the encoder $z^{\star}$ is presented to the decoder and we iterate executions of the decoder until a `<stop>` token (or word) is realized. In the first iteration we set the input sequence to the decoder to only have a `<start>` token embedding, and then with every iteration we present the output sequence up to the previous iteration as input to the decoder. That is, at iteration t, the decoder computation can be

---

[13]We count layer normalization and residual connection as a single step.

[14]This is a common tokenization dimension and was used in the original transformers paper.

[15]The first introduced transformer architecture, [410], is estimated to have used about 65 million parameters. Such relatively small discrepancies are due to implementation.

represented as

$$\hat{y}^{\langle t \rangle} = f_{\text{decoder}}\big(z^{\star}, (\tilde{\hat{y}}^{\langle 1 \rangle}, \ldots, \tilde{\hat{y}}^{\langle t-1 \rangle})\big), \tag{7.34}$$

where $z^{\star}$ is the code vector from the encoder, and $\tilde{\hat{y}}^{\langle t \rangle}$ is an embedded and positionally embedded vector, resulting from the decoder output token $\widehat{\mathcal{Y}}^{\langle t \rangle}$. The transformation from the decoder output $\hat{y}^{\langle t \rangle}$ to the decoder output token $\widehat{\mathcal{Y}}^{\langle t \rangle}$, can naively be done via an argmax as in (3.34) in Chapter 3, or by sampling tokens according to the probability output $\hat{y}^{\langle t \rangle}$. More advanced multi-step techniques such as *beam search* can also be employed, where several consecutive tokens are considered together; we omit the details. In summary, as we see in (7.34), the transformer decoder output at time t is a function of its previous outputs while the first input $\tilde{\hat{y}}^{\langle 1 \rangle}$ is an embedded and positionally embedded representation of `<start>`.

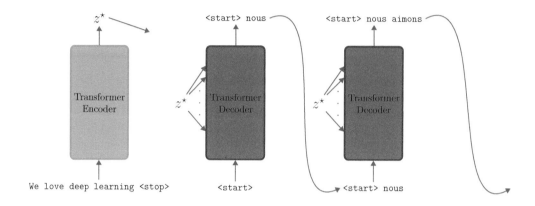

Figure 7.17: Auto-regressive application of a transformer decoder block for machine translation. The transformer encoder runs once on the whole input English sentence, creating the code $z^{\star}$. This code is then used with every iteration of the transformer decoder block, where with each iteration an additional word (or token) is created, and the previous output is fed as input in an auto-regressive manner. Generation stops (not illustrated in figure) when a `<stop>` token appears at the output.

Now considering training, a natural naive approach is to use (7.34) directly in each forward pass and backward pass iteration, when computing gradients. Namely, to use backward propagation through time. However, (7.34) naturally lends itself to use teacher forcing, similarly to the use of teacher forcing when training other encoder-decoder models, as discussed at the end of Section 7.4. With this approach, (7.34) is converted to

$$\hat{y}^{\langle t \rangle} = f_{\text{decoder}}\big(z^{\star}, (\tilde{y}^{\langle 1 \rangle}, \ldots, \tilde{y}^{\langle t-1 \rangle})\big),$$

where now the training data one-hot encoded labels $y^{\langle 1 \rangle}, \ldots, y^{\langle t \rangle}$, in their embedded and positionally embedded form, $\tilde{y}^{\langle 1 \rangle}, \ldots, \tilde{y}^{\langle t \rangle}$, are used as input to the transformer decoder instead of the predictions. This teacher forcing technique accelerates training by removing error during early phases of the process. Further, an important difference in teacher forcing of transformers vs. teacher forcing of recurrent encoder-decoder frameworks, is that with transformers we can exploit parallelization. Specifically, for the forward pass, we may compute each of these in parallel:

$$
\begin{cases}
\hat{y}^{\langle 2 \rangle} = f_{\text{decoder}}\left(z^{\star}, (\tilde{y}^{\langle 1 \rangle})\right) \\
\hat{y}^{\langle 3 \rangle} = f_{\text{decoder}}\left(z^{\star}, (\tilde{y}^{\langle 1 \rangle}, \tilde{y}^{\langle 2 \rangle})\right) \\
\hat{y}^{\langle 4 \rangle} = f_{\text{decoder}}\left(z^{\star}, (\tilde{y}^{\langle 1 \rangle}, \tilde{y}^{\langle 2 \rangle}, \tilde{y}^{\langle 3 \rangle})\right) \\
\quad \vdots \\
\hat{y}^{\langle T \rangle} = f_{\text{decoder}}\left(z^{\star}, (\tilde{y}^{\langle 1 \rangle}, \tilde{y}^{\langle 2 \rangle}, \ldots, \tilde{y}^{\langle T-1 \rangle})\right).
\end{cases}
$$

Note that while transformers were introduced for machine translation and are currently the power house of large language models for generative text modeling, adaptations of transformers have also been successful for other non-language tasks, including image tasks. In fact, transformer models compete with convolutional models, and in certain cases outperform convolutional models on images, especially in the presence of huge training datasets.

Notes and References

A useful applied introductory text about *time-series* sequence data analysis is [198] and a more theoretical book is [64]. Yet, while these are texts about sequence models, the traditional statistical and forecasting time-series focus is on cases where each $x^{\langle t \rangle}$ is a scalar or a low-dimensional vector. Neural network models, the topic of this chapter, are very different, and for an early review of recurrent neural networks and generalizations see chapter 10 of [142] and the many references there-in, where key references are also listed below.

As the most common application of sequence models is textual data, let us mention early texts on *natural language processing (NLP)*. General early approaches to NLP are summarized in [280] and [216] where the topic is tackled via rule-based approaches based on the statistics of grammar. A much more modern summary of applications is [228] and a review of applications of neural networks for NLP is in [138], yet this field is quickly advancing at the time of publishing of this current book. See also chapter 7 of the book [4] for a comprehensive discussion of RNNs as well as their *long short term memory (LSTM)* and *gated recurrent units* (GRU) generalizations.

Recurrent neural networks (RNN) are useful for broad applications such as DNA sequencing, see for example [375], image captioning as in [188], time series prediction as in [22], sentiment analysis as in [274], speech recognition as in [146], and many other applications. Possibly one of the first constructions of RNN in their modern form appeared in [117] and is sometimes referred to as an *Elman network*. Yet this was not the inception of ideas for RNNs and earlier ideas appeared in several influential works over the previous decades. See [367] for an historical account with notable earlier publications including [13] in 1972, and [185], and [357] in the 1980s.

The introduction of *bidirectional RNN* is in [369]. The introduction of long short term memory (LSTM) models in the late 1990s was in [184]. Gated recurrent units (GRUs) are much more recent concepts and were introduced in [80] and [85] after the big spark of interest in deep learning occurred. An empirical comparison of these various approaches is in [214]. A more contemporary review of LSTM is in [438]. These days, for advanced NLP tasks LSTMs and GRUs are generally outperformed by *transformer models*, yet in non-NLP applications we expect to see LSTMs remain a useful tool for many years to come. Some recent example application papers include [220], [295], [351], and [446], among many others.

Moving onto textual data, the idea of *word embeddings* is now standard in the world of NLP. The key principle originates with the *word2vec* work in [288]. Word embedding was further developed with *GloVe* in [327]. These days when considering dictionaries, lexicons, tokenizations, and word embeddings, one may often use dedicated libraries such as for example those supplied (and contributed to) with *HuggingFace*.[16] An applied book in this domain is [403] and since the field is moving quickly, many others are to appear as well.

The modern neural encoder-decoder approach was pioneered by [218] and then in the context of machine translation, influential works are [81] and [392]. The idea of using attention in recent times, first for *handwriting recognition*, was proposed in [145] and then the work in [20] extended the idea, and applied it to machine translation as we illustrate in our Figure 7.10. A massive advance was with the 2017 paper, "Attention is all you need", [410], which introduced the transformer architecture, the backbone of almost all of today's leading large language models. Ideas of *layer normalization* are from [19]. Further details of transformers can be found in [331], and a survey of variants of transformers as well as non-NLP applications can be found in [262].

At the time of publishing of this book the hottest topics in the world of deep learning are large language models and their multi-modal counterparts. A recent comprehensive survey is in [449], and other surveys are [73] and [155]. We should note that as this particular field is moving very rapidly at the time of publication of the book, there will surely be significant advances in the years coming. *Multimodal models* are being developed and deployed as well, and these models have images as input and output in addition to text; see [428] for a survey. Indeed beyond the initial task of machine translation, transformers have also been applied to images with incredible success. A first landmark paper on this avenue is [107]. See also the survey papers [227] and [269].

---

[16]https://huggingface.co.

While this list is certainly non-exhaustive, we also mention some of the key large language model architectures that emerged following the "Attention is all you need" paper [410]. Some of these are *BERT* [103], *Roberta* [268], *XLNET* [433], *GPT-2*[340], and *GPT-3* [66]. Other notable LLMs are *GLaM* [110], *Gopher* [341], *Chinchilla* [341], *Megatron-Turing NLG* [382], and *LaMDa* [400]. The topic of training and using these models is beyond our scope.

8 Specialized Architectures and Paradigms

In each of Chapters 5, 6, and 7, we presented one concrete deep learning paradigm, namely feedforward networks, convolutional neural networks, and sequence models, respectively. Such models are useful in their own right, yet in the world of deep learning one often integrates them within more complex architectures for specific activities. For example the convolutional neural networks of Chapter 6 may be inter-connected with sequence models of Chapter 7 for applications that involve both images and text. In addition, other specialized architectures and paradigms have also emerged where in each case, non-trivial ideas are employed to create powerful models. In the current chapter we present such ideas emerging from different domains, yet all using deep neural networks. Some of these domains include generative modeling, where we focus on diffusion models and generative adversarial networks, after an overview of variational autoencoders. Other domains are in the area of automatic control and decision making where we present concepts of reinforcement learning. Finally, we explore the domain of graph neural networks, an area that is proving to be ever so useful for complex problems that can be represented with graph structures. Without space constraints, each of these topics deserves its own chapter or a sequence of chapters, yet within this single chapter we hope that the reader gains an overarching view.

In Section 8.1 we introduce generative modeling principles. For this we introduce principles of variational autoencoders which are the basis for diffusion models, the topic of Section 8.2. In Section 8.3, we describe the ideas of generative adversarial networks, which also have some pinnings in game theory. These two variants, namely diffusion models of Section 8.2 and generative adversarial networks of Section 8.3, have become the most popular means of generative modeling to date. We continue in Section 8.4 where we outline principles of reinforcement learning. Toward that end we first define Markov decision processes and discuss principles of optimal control, and then tie the ideas to deep reinforcement learning. Note that while the application domain of reinforcement learning differs from the generative modeling domain of the earlier sections, ideas of Markov chains used in diffusion models, reappear in reinforcement learning of Section 8.4. Finally, graph neural networks are introduced in Section 8.5. As we see, graph neural networks generalize the convolutional neural networks of Chapter 6 while allowing us to have general graph structures within the data in contrast to simple spatial connections.

8.1 Generative Modeling Principles

The field of *generative modeling* deals with algorithms and models for creating (generating) data such as fake images, generated text, or similar. In this space we often think probabilistically and assume that the data has an underlying probability distribution. Our goal is then to train models that generate random yet realistic data from that distribution, with or without explicitly capturing the form of distribution. Generative modeling can be applied

DOI: 10.1201/9781003298687-8

both in the supervised case (features x and labels y) and the unsupervised case (no labels y). In Chapter 2, toward the end of Section 2.2, we mentioned a few names of supervised learning generative modeling approaches such as naive Bayes and others. In contrast, now we only focus on unsupervised learning. In such a case observed data $\mathcal{D}$ is composed of $\{x^{(1)}, \ldots, x^{(n)}\}$ and we assume that all $x^{(i)}$ are distributed according to a probability distribution, which we simply denote as $p(x)$. Note that in general $x^{(i)}$ is a high-dimensional object such as a high-resolution color image, and hence $p(x)$ is a complicated distribution.

There are many generative approaches, yet our focus in this chapter is on two approaches that have become very popular due to their ability to generate data that appears realistic. One approach that has recently become popular is the *diffusion model* approach, and this is the focus of Section 8.2. Another approach is the *generative adversarial network* (GAN) approach which is the focus of Section 8.3. The ideas of the diffusion approach require understanding of *variational autoencoders*. For this purpose, in the bulk of this section, we introduce variational autoencoders, a class of generative models that is also interesting and useful in their own right.

In Figure 8.1 we present a schematic of both the GAN approach and the diffusion approach. In both cases, an underlying idea is to first generate random noise. The space of this noise is typically called the *latent space*, and it has a very simple distribution, denoted[1] as $p(z)$, for example a multivariate standard normal. In both approaches, the sample z is processed as input to a model whose output is a point x, which is approximately distributed according to $p(x)$ and hence "looks realistic". In the GAN case, one often uses z which is of lower dimension than x; for example x may be a $3 \times 200 \times 200$ dimensional color image while z may be a vector of length 100. In contrast, in the diffusion models case, the latent variable z has the same dimension as the target sample x.

Both the diffusion approach and the GAN approach embody the conditional distribution of the data given the noise, denoted as $p(x \,|\, z)$. The GAN approach learns an approximate algorithm for sampling from $p(x \,|\, z)$ with a so-called *generator network*, without explicitly learning $p(x \,|\, z)$. In contrast, diffusion models are probabilistic models which approximately learn $p(x \,|\, z)$ within a so-called *decoder*. One additional difference between the approaches is that GANs generate x from z in one shot, i.e., via one application of the generator network. In contrast diffusion models iterate over multiple steps of the decoder, starting with the latent noise variable and eventually reaching the target output; see also Figure 8.4.

Both the diffusion decoder and the GAN generator use deep neural networks with learned parameters. As illustrated in Figure 8.1, these parameters are learned by training the models using the training data $\mathcal{D}$. In both cases an auxiliary network (not illustrated in Figure 8.1) also plays a part in training. In GANs this auxiliary network is called the *discriminator*, and to train the generator we train the discriminator network in parallel. For diffusion models the auxiliary network is called an *encoder* and in this case, it is fixed in advance and has no learned parameters. More details are in Section 8.3 for GANs and Section 8.2 for diffusion models. We begin the study of generative models with variational autoencoders which lay down the foundations for understanding diffusion models.

---

[1] Observe that in the context of this chapter we use $p(\cdot)$ in multiple ways, where the actual distribution used should be inferred from the argument of the function. For example $p(x)$ is the distribution of the data while $p(z)$ is the distribution of the latent space.

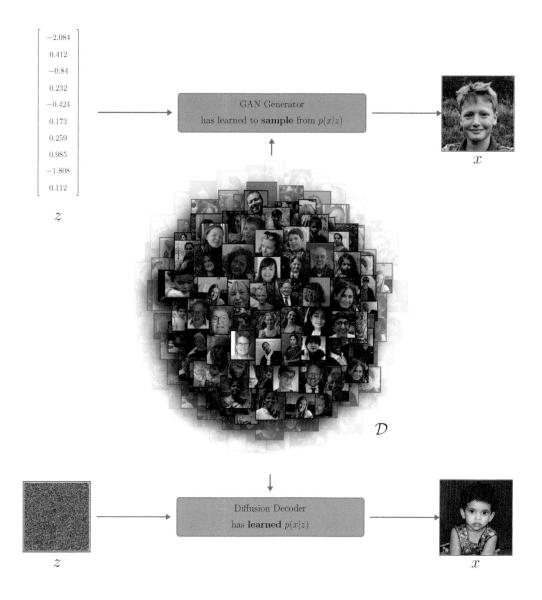

$$z$$

$$\begin{bmatrix} -2.084 \\ 0.412 \\ -0.84 \\ 0.232 \\ -0.424 \\ 0.173 \\ 0.259 \\ 0.985 \\ -1.808 \\ 0.112 \end{bmatrix}$$

GAN Generator
has learned to **sample** from $p(x|z)$

x

$\mathcal{D}$

Diffusion Decoder
has **learned** $p(x|z)$

z x

Figure 8.1: Generative adversarial networks (GANs) and diffusion models are different approaches for generative models. Both use a latent space z and are able to sample from the conditional distribution $p(x \,|\, z)$ to generate x. For image generation, GANs typically have a smaller-dimensional z while diffusion models use z of the same dimension as the image. Both cases are trained using data $\mathcal{D}$ where the resulting model in GANs is called a generator, and the resulting model in diffusion models is called a decoder.

Variational Autoencoders

In Section 3.5 we introduced basics of autoencoders within the context of shallow autoencoders. We now focus on variational autoencoders which are probabilistic enhancements of autoencoders. We first focus on a statistical perspective of variational autoencoders and then see a general framework for generating data from $p(x)$ after learning it.

Like many statistical models, a variational autoencoder is a *parametric model* where we assume some parameters θ determine $p(x)$ and hence denote the distribution as $p_\theta(x)$. By learning θ we can learn the distribution and then also generate samples from $p_\theta(x)$. A key principle for estimation of θ is maximum likelihood estimation (MLE). In Section 3.1, we briefly introduced this commonly used approach in the context of logistic regression.

With the maximum likelihood approach, similarly to (3.10) of Chapter 3, our estimate of θ for the data $\mathcal{D}$ is

$$\widehat{\theta}_{MLE} := \underset{\theta}{\mathrm{argmax}} \; \frac{1}{n} \sum_{i=1}^{n} \log p_\theta(x^{(i)}), \tag{8.1}$$

where the optimized quantity (barring the $1/n$ term) is the log-likelihood under the assumption of independent and identically distributed elements of $\mathcal{D}$. In practice, solving (8.1) requires information about the expression or form of $p_\theta(x)$.

In variational autoencoders, as alluded to at the start of the section, an unseen latent variable z plays a key role. More precisely we suppose that $p_\theta(x)$ is coupled with z which has distribution $p(z)$. The distribution $p(z)$ is assumed to be simple and does not depend on the parameters θ. The relationship involving z and x can then be represented as

$$p_\theta(x, z) = p_\theta(x \,|\, z) \, p(z). \tag{8.2}$$

Hence in summary, both the joint distribution $p_\theta(x, z)$ and the conditional distribution $p_\theta(x \,|\, z)$ are parameterized by θ, but $p(z)$ is not parameterized by θ.

We can use (8.2) to obtain $p_\theta(x)$ by marginalizing the joint distribution over z as

$$p_\theta(x) = \int p_\theta(x \,|\, z) \, p(z) dz. \tag{8.3}$$

Such a representation is helpful in expressing complex $p_\theta(x)$ using relatively simple expressions for $p_\theta(x \,|\, z)$ and $p(z)$. However, even in this setting, the integral (8.3) is intractable and hence indirect optimization using ELBO, defined and described in the sequel, is used.

To parameterize $p_\theta(x \,|\, z)$ and describe $p(z)$ we make use of *multivariate normal distributions* where the probability density for a random vector u is represented via $\mathcal{N}(u \,;\, \mu, \Sigma)$ with μ denoting the *mean vector* and Σ denoting the *covariance matrix*. A particular simple case is the *standard multivariate normal* where $\mu = 0$ (zero vector) and $\Sigma = I$ (identity matrix). A slightly more general case is setting $\Sigma = \sigma^2 I$, where the positive scalar σ^2 determines the variance shared among all coordinates.

We sometimes reparametrize[2] a covariance matrix $\Sigma \in \mathbb{R}^{p \times p}$ via

$$\Sigma = \Gamma \Gamma^\top, \quad \text{where} \quad \Gamma \in \mathbb{R}^{p \times p}. \tag{8.4}$$

Covariance matrices are always positive semi definite, and characterizing this constraint on the individual entries of the matrix can be difficult. In contrast, with reparameterizations to Γ, one may end up with an unconstrained matrix Γ which is easier to work with.

---

[2]For example Γ in $\Sigma = \Gamma \Gamma^\top$ can be obtained via a Cholesky factorization in which case Γ is an unconstrained lower triangular matrix. Other factorizations such as the spectral decomposition (or singular value decomposition) can also be used.

Starting with $p_\theta(x \mid z)$ we assume that this conditional distribution is multivariate normal where the mean vector and covariance matrix are functions of z that are both parameterized by θ. In particular, keeping in mind that m is the dimension of z, and p is the dimension[3] of x, we assume that we have learned functions,

$$\mu_\theta : \mathbb{R}^m \to \mathbb{R}^p \quad \text{and} \quad \Sigma_\theta : \mathbb{R}^m \to \mathbb{R}^{p \times p}, \tag{8.5}$$

each parameterized by θ. With these, we have the conditional distribution set as

$$p_\theta(x \mid z) = \mathcal{N}\big(x \,;\, \mu_\theta(z), \Sigma_\theta(z)\big). \tag{8.6}$$

In this case with (8.3) and (8.6), $p_\theta(x)$ is a *Gaussian mixture model*[4] with *mixture components* $p_\theta(x \mid z)$ indexed by the values of z, and corresponding *mixture weights* provided by $p(z)$. It is well known that any distribution can essentially be closely approximated by a Gaussian mixture model if the mixture components and mixture weights are properly selected.[5]

While the general form of Gaussian mixture models is very versatile, in variational autoencoders we make some simplifying assumptions. First and most importantly we assume that the distribution of the latent variable z is multivariate standard normal. Namely, the mixture weights are

$$p(z) = \mathcal{N}(z \,;\, 0, I). \tag{8.7}$$

Further, it is common to reduce the complexity of the mixture components (8.6) and assume that the covariance function is simply $\sigma^2 I$ where σ^2 is a pre-determined (not learned) hyper-parameter. Thus, (8.6) is reduced to

$$p_\theta(x \mid z) = \mathcal{N}\big(x \,;\, \mu_\theta(z), \sigma^2 I\big), \tag{8.8}$$

and now (8.8) together with (8.3) implies that θ parameterizes the distribution of x only via the mean function $\mu_\theta(\cdot)$. In the context of deep learning, this mean function is taken as a neural network with θ representing the weights and biases. However in the general framework of variational autoencoders, the mean function $\mu_\theta(\cdot)$ could be any model and not necessarily a deep neural network.

With the model defined, suppose now that based on data $\mathcal{D}$ we manage to approximate the maximum likelihood estimate (8.1) and learn θ for $\mu_\theta(\cdot)$. We can now use the model as a generative model similar to the spirit of Figure 8.1. In particular, to generate a new random data sample x^*, we first generate a random latent sample z^* from the standard multivariate normal distribution. We then compute $\mu_\theta(z^*)$ using the learned model, and then generate x^* using (8.8) where the mean taken is $\mu_\theta(z^*)$. In this sense, every random z^* yields its own $\mu_\theta(z^*)$ which in turn yields a random x^*. This generative process can be illustrated as follows:

$$z^* \quad \xrightarrow{\mu_\theta(\cdot)} \quad \mu_\theta(z^*) \quad \xrightarrow{p_\theta(x \mid z)} \quad x^*. \tag{8.9}$$

Note that the diffusion models alluded to in Figure 8.1 are a variant of variational autoencoders and we cover the details of this powerful class of generative models in Section 8.2.

---

[3] In case x is an image we vectorize it and assume the dimension is p.

[4] A Gaussian mixture model is a probability mixture model where a collection Gaussian distributions are "weighted" to create a new probability distribution. As with any mixture model, it is composed of *mixture components*, and *mixture weights*, both of which appear under the integral sign as in (8.3).

[5] Mathematically this property can be phrased as the fact that Gaussian mixture models are dense in the space of probability distributions.

As already mentioned, variational autoencoders are related to the standard autoencoders presented in Section 3.5, yet standard autoencoders are deterministic while variational autoencoders are probabilistic. It is this difference that allows variational autoencoders to be used as generative models while standard autoencoders on their own cannot. To see this distinction, assume we are going to try and carry out a generative procedure similar to (8.9) with a decoder trained on a standard autoencoder. This can be represented as follows:

$$z^* \xrightarrow{\text{Decoder}} x^*,$$

where again we would sample z^* from some predetermined distribution. However the standard autoencoder on its own does not capture the distribution in the latent space where meaningful samples of z^* are present. To see this, consider for example Figure 3.14 in Chapter 3, and observe that z^* would be some random point in the latent space and then mapped to a random output x^* via the decoder. With this, there is no guarantee to get any realistic x^* because the model does not capture the region of interest in the latent space. Hence standard autoencoders, while useful for other purposes as discussed in Section 3.5, are not useful as generative models on their own right.

The Encoder-Decoder Architecture for Variational Autoencoders

As alluded to above, a variational autoencoder has both an *encoder* and a *decoder*. The main object used for generative models, $p_\theta(x \mid z)$, is represented via the decoder and as described above this distribution is parameterized by the mean function $\mu_\theta(\cdot)$ as well as sometimes by the covariance function $\Sigma_\theta(\cdot)$. However, for learning the parameters θ, we require the full (variational) encoder-decoder architecture as illustrated in Figure 8.2.

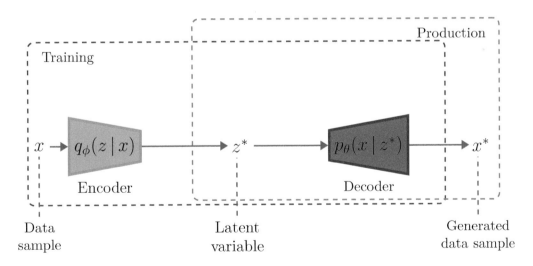

Figure 8.2: The variational autoencoder architecture comprised of an encoder $q_\phi(z \mid x)$ and a decoder $p_\theta(x \mid z)$.

Like the decoder, the encoder can be a neural network parameterized by weights and biases denoted as ϕ. Also like the decoder, the encoder is probabilistic in the sense that it describes a probability distribution, denoted as $q_\phi(z \mid x)$ and known as the *variational posterior* of the latent variable, that provides a distribution of the latent variable z given the data sample x. That is, the encoder transforms an input data sample $x \in \mathbb{R}^p$ from $\mathcal{D}$ to a latent variable

$z \in \mathbb{R}^m$. Like the decoder, the encoder has a mean function and a covariance function which we denote respectively as

$$\mu_\phi : \mathbb{R}^p \to \mathbb{R}^m \quad \text{and} \quad \Sigma_\phi : \mathbb{R}^p \to \mathbb{R}^{m \times m}, \tag{8.10}$$

and the conditional distribution of the latent variable z given a data point x is assumed to be normally distributed. Namely,

$$q_\phi(z \mid x) = \mathcal{N}\big(z \,;\, \mu_\phi(x), \Sigma_\phi(x)\big). \tag{8.11}$$

Hence, compare the encoder's (8.10) and (8.11) with the decoder's (8.5) and (8.6), respectively.

Taking all $x \in \mathcal{D}$ under consideration, the *latent variable sample marginal distribution* is given by

$$q_{\phi,\mathcal{D}}(z) = \frac{1}{n} \sum_{x \in \mathcal{D}} q_\phi(z \mid x). \tag{8.12}$$

Ideally, like (8.7), we would like $q_{\phi,\mathcal{D}}(z)$ to be a standard normal distribution in which case it no longer depends on ϕ and $\mathcal{D}$. This is then set as a training goal when jointly training the encoder (learning ϕ) and decoder (learning θ) in a variational autoencoder. To achieve such a goal, we aim to minimize a loss function of the general form

$$C(\phi, \theta \,;\, \mathcal{D}) = C_{\text{PriorMatching}}(\phi) + C_{\text{Reconstruction}}(\phi, \theta). \tag{8.13}$$

In general this loss depends on the data $\mathcal{D}$, yet for simplicity we omit this relationship for the two terms on the right hand side. Minimization of the first term, $C_{\text{PriorMatching}}(\phi)$, aims to set the latent distribution (8.12) to be as close as possible to a standard normal. It is called *prior matching* because when treating the variational autoencoder as a Bayesian inference model, the latent distribution can be viewed as a prior distribution. Minimization of the second term, $C_{\text{Reconstruction}}(\phi, \theta)$, aims to capture maximum likelihood estimation with respect to both encoder and decoder parameters and hence achieve optimal *reconstruction* of the output distribution. Details of implementing these loss functions are below, after we introduce the concept of the evidence lower bound.

Relations to Maximal Likelihood and ELBO

Our aim with training variational autoencoders is that minimization of (8.13) will act as a means for maximum likelihood estimation as in (8.1). In our exposition here we focus on a simple stochastic gradient descent setting and thus we consider a single datapoint $x \in \mathcal{D}$. Extending to multiple datapoints, or mini-batches is straightforward. Our goal is to optimize

$$\max_{\theta \in \mathbb{R}^p} \log p_\theta(x). \tag{8.14}$$

For this goal, let us decompose $p_\theta(x)$ and in our decomposition, we use the encoder distribution $q_\phi(z \mid x)$ as well as distributions associated with the joint distribution of the decoder from (8.2). Namely we use the joint distribution $p_\theta(x, z)$, as well as $p_\theta(z \mid x)$ which uses the decoder to describe the distribution of the latent variable z given x. This distribution can be obtained via

$$p_\theta(z \mid x) = \frac{p_\theta(x, z)}{p_\theta(x)}. \tag{8.15}$$

Note that $p_\theta(x,z)$ and $p_\theta(z\,|\,x)$ are not operationally used (in a learning or inference algorithms) but rather only appear theoretically for the decomposition. A key quantity in the decomposition is called the *evidence lower bound (ELBO)*, and is defined as

$$\mathsf{ELBO}(\theta,\phi\,;\,x) = \int q_\phi(z\,|\,x) \log \frac{p_\theta(x,z)}{q_\phi(z\,|\,x)} dz. \tag{8.16}$$

Note that ELBO is an expectation of $\log \frac{p_\theta(x,Z)}{q_\phi(Z\,|\,x)}$ where Z is distributed according to $q_\phi(\cdot\,|\,x)$. With ELBO defined, the decomposition of the log-likelihood $\log p_\theta(x)$ is

$$\log p_\theta(x) = \mathsf{ELBO}(\theta,\phi\,;\,x) + D_{\mathsf{KL}}\left(q_\phi(z|x)\,\|\,p_\theta(z|x)\right), \tag{8.17}$$

where $D_{\mathsf{KL}}\left(q_\phi(z|x)\,\|\,p_\theta(z|x)\right)$ denotes the KL-divergence (see (B.6) in Appendix B) which is a measure of how $q_\phi(z\,|\,x)$ is different from $p_\theta(z\,|\,x)$. Note that the KL-divergence is always non-negative, and equal to zero if and only if both distributions are identical. To see (8.17), observe that

$$\underbrace{\int q_\phi(z\,|\,x) \log \frac{p_\theta(x,z)}{q_\phi(z\,|\,x)} dz}_{\mathsf{ELBO}(\theta,\phi\,;\,x)} + \underbrace{\int q_\phi(z\,|\,x) \log \frac{q_\phi(z\,|\,x)}{p_\theta(z\,|\,x)} dz}_{D_{\mathsf{KL}}(q_\phi(z|x)\,\|\,p_\theta(z|x))} = \int q_\phi(z\,|\,x) \log \frac{p_\theta(x,z)}{p_\theta(z\,|\,x)} dz = \log p_\theta(x),$$

where moving from left to right, in the first step we combine the log terms and in the second step we use (8.15).

A consequence of the decomposition (8.17) as well as the non-negativity of the KL-divergence is that ELBO is a lower bound on the log-likelihood. Namely,

$$\log p_\theta(x) \geq \mathsf{ELBO}(\theta,\phi\,;\,x), \tag{8.18}$$

hence the name "lower bound" in ELBO. Note that the log-likelihood is sometimes called the *evidence*, and hence the term "evidence" in ELBO. Equality holds in (8.18) if and only if $q_\phi(z\,|\,x)$ and $p_\theta(z\,|\,x)$ are the same.

Now let us assume that the encoder model is flexible enough to yield any mean and covariance function as in (8.10). With this assumption on the flexibility of the encoder, there exists some ϕ such that $q_\phi(z\,|\,x)$ is equal to $p_\theta(z\,|\,x)$. Since $\log p_\theta(x)$ does not depend on ϕ we have from (8.18) and (8.17) that such a ϕ maximizes the evidence lower bound. Namely,

$$\log p_\theta(x) = \max_{\phi\in\mathbb{R}^m} \mathsf{ELBO}(\theta,\phi\,;\,x). \tag{8.19}$$

With such a representation of $\log p_\theta(x)$ for any θ, we can also maximize (8.19) over θ to obtain

$$\max_{\theta\in\mathbb{R}^p} \log p_\theta(x) = \max_{\theta\in\mathbb{R}^p,\,\phi\in\mathbb{R}^m} \mathsf{ELBO}(\theta,\phi\,;\,x). \tag{8.20}$$

We thus see that the maximum likelihood estimate (8.1), when constrained to a single $x \in \mathcal{D}$, can be obtained by maximization of ELBO in terms of both the encoder parameters ϕ and the decoder parameters θ. This general idea of indirect likelihood optimization via maximization of ELBO (with the additional ϕ set of parameters for the encoder) is the crux of training variational autoencoders. The importance of this approach is that approximate maximization of ELBO is computationally feasible in contrast to direct maximum likelihood estimation, where the integration in (8.3) is a computational barrier.

Details of the Loss Function

We now see how minimization of the encoder-decoder loss function (8.13) can be constructed for approximate maximization of ELBO. For this let us expand the expression in (8.16) as

$$
\begin{aligned}
\mathsf{ELBO}(\theta, \phi\,;\,x) &= \int q_\phi(z\,|\,x) \log \frac{p_\theta(x\,|\,z)\,p(z)}{q_\phi(z\,|\,x)} dz \\
&= \underbrace{\int q_\phi(z\,|\,x) \log p_\theta(x\,|\,z) dz}_{\mathbb{E}\log p_\theta(x\,|\,Z)} - \underbrace{\int q_\phi(z\,|\,x) \log \frac{q_\phi(z\,|\,x)}{p(z)} dz}_{D_{\mathsf{KL}}(q_\phi(z|x)\,\|\,p(z))}.
\end{aligned}
\tag{8.21}
$$

In the first equality we used the representation of the joint distribution $p_\theta(x, z)$ from (8.2). Then in the second equality we expand the logarithm. The resulting first term is an expectation of the function $\log p_\theta(x\,|\,Z)$ when Z is distributed according to $q_\phi(\cdot\,|\,x)$. For the second term, keeping in mind that $p(z)$ as in (8.7) is a parameter-free multivariate standard normal, we have a KL-divergence that measures how close the encoder distribution is to the standard normal. Keep in mind that this a different application of the KL-divergence to the one used for lower bounding the evidence in (8.17).

With the ELBO expression present, we can now fill in the details for the terms in the loss expression (8.13), such that minimization of this loss works toward maximization of ELBO. Our focus is on a stochastic gradient descent approach as in Section 4.2 where at any gradient descent step, instead of considering the whole data $\mathcal{D}$, we consider a single arbitrary $x \in \mathcal{D}$. In such a case, the loss components on the right hand side of (8.13) can be defined based on a single data sample $x \in \mathcal{D}$ where, as shown below, for the second component we also generate a random latent variable $z^* \in \mathbb{R}^m$ from (8.11).

The idea of minimizing the first term of (8.13), $C_{\mathrm{PriorMatching}}(\phi)$, is to drive the parameters such that $q_{\phi,\mathcal{D}}(z)$ of (8.12) is approximately a multivariate standard normal as in (8.7). For this we set

$$
C_{\mathrm{PriorMatching}}(\phi) = \mu_\phi(x)^\top \mu_\phi(x) - \log \det\left(\Sigma_\phi(x)\right) + \mathrm{tr}\left(\Sigma_\phi(x)\right),
\tag{8.22}
$$

where $\det(\cdot)$ is the determinant of a matrix and $\mathrm{tr}(\cdot)$ is the trace of a matrix. Minimization of this expression is equivalent to minimization of the KL-divergence where the first argument is a multivariate normal distribution with mean vector $\mu_\phi(x)$ and covariance matrix $\Sigma_\phi(x)$, and the second argument is an m-dimensional standard multivariate normal distribution. This is because

$$
D_{\mathsf{KL}}\left(q_\phi(z|x)\,\|\,p(z)\right) = \frac{1}{2} C_{\mathrm{PriorMatching}}(\phi) - \frac{m}{2}
$$

which can be verified with (B.11) of Appendix B.

Moving onto the second term in (8.13), using a single x and a single z^*, we construct this term as

$$
C_{\mathrm{Reconstruction}}(\phi, \theta) = \left(x - \mu_\theta(z^*)\right)^\top \Sigma_\theta(z^*)^{-1}\left(x - \mu_\theta(z^*)\right) + \log \det\left(\Sigma_\theta(z^*)\right).
\tag{8.23}
$$

This term aims to capture the $\mathbb{E}\log p_\theta(x\,|\,Z)$ term in (8.21). Observe that $C_{\mathrm{Reconstruction}}(\phi, \theta)$ is a random variable with distribution influenced by ϕ through the distribution of z^*. The form of (8.23) arises from the log-density expression in (B.8) of Appendix B. We do not have an explicit expression for the expectation $\mathbb{E}\log p_\theta(x\,|\,Z)$ term of (8.21), therefore, as a simple estimate we use a single sample z^* of the latent variable from the distribution $q_\phi(z \mid x)$ and rely on the approximate relationship,

$$\mathbb{E} \log p_\theta(x \mid Z) \approx \log p_\theta(x \mid z^*), \qquad (8.24)$$

as a proxy for the optimization. Now considering (8.23) we have

$$\log p_\theta(x \mid z^*) = -\frac{1}{2} C_{\text{Reconstruction}}(\phi, \theta) + \frac{1}{2} \log(2\pi).$$

Hence minimization of $C_{\text{Reconstruction}}(\phi, \theta)$ maximizes $\log p_\theta(x \mid z^*)$, which approximates maximization of $\mathbb{E} \log p_\theta(x \mid Z)$ via (8.24).

We have thus seen that the encoder-decoder architecture, learned via stochastic gradient descent with loss function (8.13) and individual terms (8.22) and (8.23) approximates maximization of ELBO, and thus facilitates maximum likelihood estimation of θ.

The Reparameterization Trick

During training of the variational autoencoder, we need to compute the gradient of the loss function (8.13). With standard backpropagation we can compute the gradient of the $C_{\text{PriorMatching}}(\phi)$ component in a straight forward manner. However, since the $C_{\text{Reconstruction}}(\phi, \theta)$ component is based on a random sample z^*, it is not obvious how to use backpropagation through the number random generator. Luckily, via a reparametrization of the random vector z^* as an affine function of a standard random vector ϵ^*, we can overcome this difficulty. Specifically, if ϵ^* is a standard m-dimensional multivariate normal random vector, then we may set

$$z^* = \mu_\phi(x) + \Gamma_\phi(x)\, \epsilon^*, \qquad (8.25)$$

where as before $\mu_\phi(x)$ is the learned mean function of the encoder. With the reparameterization (8.25), the desired learned covariance function $\Sigma_\phi(x)$ is decomposed using $\Sigma_\phi(x) = \Gamma_\phi(x)\, \Gamma_\phi(x)^\top$, as in (8.4). Now $\Gamma_\phi(\cdot)$ is learned in the encoder neural network in place of $\Sigma_\phi(\cdot)$ of (8.10). With this, the distribution of z^* is as required, and further there are no longer issues with enforcing the positive definite constraint that is required for the covariance matrix.

Importantly, backpropagation can now be carried out, because the random numbers generated in ϵ^* are now generated independently of the parameters whose gradients we seek. In some cases, the desired covariance matrix is diagonal. In such cases, the reparameterization is especially simple since $\Gamma_\phi(x)$ is also diagonal with each entry being the square root of the associated diagonal entry of $\Sigma_\phi(x)$.

8.2 Diffusion Models

Diffusion models are a class of generative models that have shown great promise in creating both realistic looking images, as well as highly impressive surreal artwork. Figure 8.3 presents a few examples with several paradigms of application including image generation, colorization, style transfer, and text to image creation. The ease of use from a user's perspective of such platforms, and the quality of the images created is impressive.

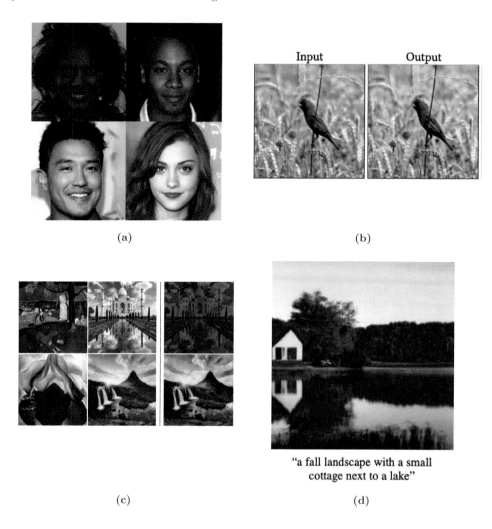

Input   Output

"a fall landscape with a small
cottage next to a lake"

(a)   (b)

(c)   (d)

Figure 8.3: Images generated via various diffusion architectures[6] with various types of paradigms. (a) A 256×256 generated images based on a label. (b) Image-to-image generation (colorization). (c) Applying a style (left column) to an image (middle column). (d) A text-to-image application.

The principles underlying diffusion models hinge on variational autoencoders presented in the previous section, as well as on a generalization of such models, called *hierarchical variational autoencoders* and in particular *Markovian hierarchical variational autoencoders*. Diffusion models are a special case of such models, hence in exploring diffusion models, we first study Markovian hierarchical variational autoencoders. We then construct diffusion models as a special case of Markovian hierarchical variational autoencoders, based on auto-regressive Gaussian processes.

Hierarchical Variational Autoencoders

In the autoencoders of Section 8.1 the encoder acts as a *noising mechanism* that converts the input x into a noisy latent variable z. On the other hand the decoder can be viewed as

---

[6]Image (a) is thanks to Ho, et al. [183]. Image (b) is thanks to Saharia, et. al. [358]. Image (c) is thanks to Wang, et. al. [418]. Image (d) is thanks to Nichol, et. al. [307].

a *denoising mechanism* for converting a noisy latent variable z to meaningful output. The main idea of *hierarchical variational autoencoders* is to break up this process into multiple levels. In the encoder, the input x is noised slightly to create z_1, then z_1 is further noised to create z_2 up until some final level T with a fully noisy latent variable z_T. In the decoder the reverse process takes place where each step from z_t to z_{t-1} enacts a slight denoising operation.

Hence in hierarchical variational autoencoders, instead of a single latent variable z, we have a sequence of T levels with latent variables, $z_1, \ldots, z_T$. Such models also enforce a specific dependence structure within this sequence and are called "hierarchical", since in the encoder each z_t can be generated based on the values of the lower-level latent variables $z_1, \ldots, z_{t-1}$ and in the decoder each latent variable z_t can be generated based on the values of the higher latent variables $z_{t+1}, \ldots, z_T$. In such a model, one can view the input $x \in \mathcal{D}$ as the 0-th level, namely the complete sequence of levels is $x, z_1, \ldots, z_T$, where we can treat x as z_0.

The most common hierarchical variational autoencoders are *Markovian* in which case the sequence $x, z_1, \ldots, z_T$ is a *Markov chain*, and the model is called a *Markovian hierarchical variational autoencoder*. That is, the sequence follows the *Markov property* which can be defined in multiple ways. One way is that for any $t = 1, \ldots, T - 1$, given the value of z_t, the sequence $x, z_1, \ldots, z_{t-1}$ and the sequence $z_{t+1}, \ldots, z_T$ are independent.

Considering the decoder, a consequence of the Markov property is that the joint distribution $p_\theta(x, z_1, \ldots, z_T)$ can be described as a product of *one step transition probabilities*. We have

$$p_\theta(x, z_1, \ldots, z_T) = p_{\theta_1}(x \,|\, z_1) p_{\theta_2}(z_1 \,|\, z_2) \cdots p_{\theta_T}(z_{T-1} \,|\, z_T) p(z_T), \qquad (8.26)$$

where $\theta = (\theta_1, \ldots, \theta_T)$ are the decoder parameters, and each $p_{\theta_t}(z_{t-1} \,|\, z_t)$ is the decoder conditional one step transition probability from level t to level $t - 1$. Note that in this expression, $p(z_T)$, the distribution of level T, is parameter free. Since it is assumed to be fully noisy, it is often taken as a multivariate standard normal similar to (8.7) from the (non-hierarchical) variational autoencoder.

In such models, the one step transition probabilities $p_{\theta_t}(\cdot \,|\, z_t)$ are generally represented as multivariate Gaussian distributions where for level t, we learn neural networks for the mean function $\mu_{\theta_t}(z_t)$ and potentially for the covariance function $\Sigma_{\theta_t}(z_t)$. Hence the learned parameters θ_t describe the distribution at level $t - 1$ given the value at the previous level, z_t. Observe that the Markovian hierarchical variational autoencoder joint distribution in (8.26) is the parallel of the joint distribution (8.2) for the non-hierarchical case. A potential benefit of using (8.26) is that by breaking up the parameterization of the decoder into T smaller steps, each with its own learned parameters, more expressive models can be achieved.

A decoder parameterized by $\theta = (\theta_1, \ldots, \theta_T)$ implements a generative model, similarly to the non-hierarchical case. First z_T^* is drawn from the distribution $p(z_T)$, and then for each transition we compute $\mu_{\theta_t}(z_t^*)$ and $\Sigma_{\theta_t}(z_t^*)$ and then draw z_{t-1}^* from the transition probability $p_{\theta_t}(\cdot \,|\, z_t^*)$ parametrized by these values. Hence while a non-hierarchical variational autoencoder draws z^* once and then generates x^*, here we draw z_T^*, then z_{T-1}^*, and so forth until attaining a generated data sample $x^* = z_0^*$.

Like non-hierarchial variational autoencoders, in Markovian hierarchical variational autoencoders, the encoder serves as an aid for training the generative model (the decoder). The encoder also incorporates a Markovian structure within levels. The encoder parameters

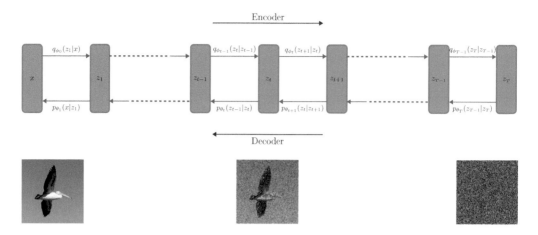

Figure 8.4: A Markovian hierarchical variational autoencoder has a latent space $z_1, z_2, \ldots, z_T$, with $x = z_0$. The encoder transforms from z_t to z_{t+1} (adding noise) and the decoder works in the opposite direction (removing noise).

are $\phi = (\phi_0, \ldots, \phi_{T-1})$, where this time for $t = 0, \ldots, T-1$, we have conditional one step transition probabilities, $q_{\phi_t}(z_{t+1} \mid z_t)$ describing the transition from level t to level $t+1$. Like the decoder, in the encoder we can assume Gaussian transition probabilities which are this time parameterized by $\mu_{\phi_t}(z_t)$ and $\Sigma_{\phi_t}(z_t)$. See Figure 8.4.

Considering the encoder, it is useful to represent the conditional distribution of $z_1, \ldots, z_t$ given x for any $t = 1, \ldots, T$, denoted as $q_\phi(z_1, \ldots, z_t \mid x)$. Due to the Markovian assumption we have

$$q_\phi(z_1, \ldots, z_t \mid x) = q_{\phi_0}(z_1 \mid x) q_{\phi_1}(z_2 \mid z_1) \cdots q_{\phi_{t-1}}(z_t \mid z_{t-1}), \quad \text{for} \quad t = 1, \ldots, T. \quad (8.27)$$

Further, the distribution of z_t for a given x is denoted as $q_\phi(z_t \mid x)$ which may formally be represented as a marginal distribution from $q_\phi(z_1, \ldots, z_t \mid x)$, and obtained via

$$q_\phi(z_t \mid x) = \int \cdots \int q_\phi(z_1, \ldots, z_t \mid x) \, dz_1 \cdots dz_{t-1}. \quad (8.28)$$

In the specific diffusion models below this distribution is constructed with an explicit form, yet for now the general expression (8.28) is appropriate.

With this notation and the construction of the encoder and decoder in place, we can seek to approximate maximum likelihood estimation of θ. This is done in a similar manner to the non-hierarchical variational autoencoder case outlined in Section 8.1. Specifically we maximize an ELBO term over both ϕ and θ.

The ELBO term in this case is similar to (8.16) yet can be represented via $z_1, \ldots, z_T$ in place of z. Namely,

$$\text{ELBO}(\theta, \phi \, ; x) = \int q_\phi(z_1, \ldots, z_T \mid x) \log \frac{p_\theta(x, z_1, \ldots, z_T)}{q_\phi(z_1, \ldots, z_T \mid x)} dz_1 \cdots dz_T. \quad (8.29)$$

Now using the Markovian structure of (8.26), (8.27), and after some extensive manipulation, the following expansion of $\mathsf{ELBO}(\theta, \phi \,;\, x)$ arises:

$$
\underbrace{\int q_{\phi_0}(z_1 \mid x) \log p_{\theta_1}(x \mid z_1) dz_1}_{\mathbb{E} \log p_{\theta_1}(x \mid Z_1)} \quad \text{(Reconstruction)}
$$

$$
- \underbrace{\int_{z_T} q_\phi(z_T \mid x) \log \frac{q_\phi(z_T \mid x)}{p(z_T)} dz_T}_{D_{\mathsf{KL}}(q_\phi(z_T \mid x) \,\|\, p(z_T))} \quad \text{(Prior matching)} \qquad (8.30)
$$

$$
- \underbrace{\sum_{t=2}^{T} \int_{z_t} q_\phi(z_t \mid x) \int q_\phi(z_{t-1} \mid z_t, x) \log \frac{q_\phi(z_{t-1} \mid z_t, x)}{p_{\theta_t}(z_{t-1} \mid z_t)} dz_t.}_{\mathbb{E} \, D_{\mathsf{KL}}\left(q_\phi(z_{t-1} \mid Z_t, x) \,\|\, p_{\theta_t}(z_{t-1} \mid Z_t)\right)} \quad \text{(Denoising matching)}
$$

With this expansion, ELBO is now represented in terms of three types of terms, namely a *reconstruction term*, a *prior matching term*, and *denoising matching terms*. The first two terms are also present in non-hierarchical variational autoencoders whereas the denoising matching terms arise due to the multi-level structure of the latent space.

The reconstruction term in (8.30) is an expectation of the conditional log-likelihood with respect to Z_1, distributed according to $q_{\phi_0}(z_1 \mid x)$. Maximization of this term drives maximum likelihood estimation. The prior matching term in (8.30) is the KL-divergence between the standard normal distribution $p(z_T)$ and the encoder based last step distribution $q_\phi(z_T \mid x)$. Minimization of this KL-divergence attempts to enforce that z_T at the exit of the encoder is distributed approximately as a standard normal.

In the additional $T - 1$ terms, we use $q_\phi(z_{t-1} \mid z_t, x)$ which can be viewed as a denoising distribution in the encoder. Note that in contrast to the direction of the encoder (z_{t-1} to z_t), this distribution is in the opposite direction. The expected KL-divergence in each denoising matching term is between $q_\phi(z_{t-1} \mid z_t, x)$ and the decoder's one step transition $p_{\theta_t}(z_{t-1} \mid z_t)$. Ideally both the encoder and the decoder are similar at level t and minimization of the expected KL-divergence enforces the denoising to be as close as possible. Note that when represented as an expectation, the expectation is with respect to the random variable Z_t distributed as $q_\theta(z_t \mid x)$.

As is evident, the $\mathsf{ELBO}(\theta, \phi \,;\, x)$ expressions rely on $q_\phi(z_t \mid x)$ which in general needs to be computed via (8.27) and (8.28). It also relies on $q_\phi(z_{t-1} \mid z_t, x)$, which is generally difficult to compute. We now see, that in the special case of a diffusion model, with the assumptions imposed, the expressions for $q_\phi(z_t \mid x)$ and $q_\phi(z_{t-1} \mid z_t, x)$ simplify and are efficient to compute.

The process of learning parameters θ for Markovian hierarchical variational autoencoders is in principle similar to learning variational autoencoders as presented in Section 8.1. Specifically, we approximately maximize $\mathsf{ELBO}(\theta, \phi \,;\, x)$ over both θ and ϕ using estimates obtained via samples of the latent variables. The resulting decoder with parameters θ emerges and approximates maximum likelihood estimation as in the non-hierarchical case.

The Diffusion Model Assumptions

A *diffusion model* is a Markovian hierarchical variational autoencoder with a few specific assumptions. All latent variables are of the same dimension of the data samples. Further, recalling that $x = z_0$, the one-step encoder and decoder transition probabilities are respectively,

$$q(z_{t+1} \mid z_t) = \mathcal{N}(z_{t+1}; \sqrt{1 - \beta_t}\, z_t, \beta_t I) \qquad \text{for} \qquad t = 0, \ldots, T-1, \qquad (8.31)$$

$$p_{\theta_t}(z_{t-1} \mid z_t) = \mathcal{N}(z_{t-1}; \mu_{\theta_t}(z_t), \sigma_t^2 I) \qquad \text{for} \qquad t = T, \ldots, 1. \qquad (8.32)$$

Here we have two sequences of scalar hyper-parameters. The encoder hyper-parameters are $\beta_0, \ldots, \beta_{T-1}$, each in the range $[0, 1]$. The decoder hyper-parameters are $\sigma_1^2, \ldots, \sigma_T^2$, each a positive number. Note that the encoder has no learned parameters and thus there are no ϕ_t subscripts for $q(z_{t+1} \mid z_t)$. The decoder learned parameters, $\theta_1, \ldots, \theta_T$, are used for the mean functions $\mu_{\theta_1}(\cdot), \ldots, \mu_{\theta_T}(\cdot)$, but not for the one step covariance matrices that are of the form $\sigma_1^2 I, \ldots, \sigma_T^2 I$. Each of these mean functions, $\mu_{\theta_t}(\cdot)$, is a neural network, and hence the model has T learned neural networks.

A key aspect of the diffusion model is the structure of the one step transition probabilities of the encoder (8.31). Since the encoder steps have a conditional mean vector and covariance matrix of $\sqrt{1 - \beta_t}\, z_t$ and $\beta_t I$, respectively, they describe an *auto-regressive stochastic sequence* of the form,

$$z_{t+1} = \sqrt{1 - \beta_t}\, z_t + \sqrt{\beta_t}\, \epsilon_t, \qquad (8.33)$$

where ϵ_t is a multivariate standard normal vector. Conditioned on the value of z_t, the expected value of z_{t+1} resulting from (8.33) is $\sqrt{1 - \beta_t}\, z_t$ and the covariance matrix is $\beta_t I$. Hence, this stochastic recursion then follows the next step distribution (8.31). The name "diffusion model" can be attributed to the fact that if (8.33) was in continuous time, then the process is a type of a stochastic diffusion process.

A strength of the recursion (8.33) is that it also enables closed-form expressions of the multi-step transition probabilities $q_\phi(z_t \mid x)$. Such probabilities are heavily used in the ELBO expression (8.30). For this purpose, it is useful to define the products γ_t for $t = 1, \ldots, T$, with

$$\gamma_t = (1 - \beta_0) \cdots (1 - \beta_{t-1}).$$

With this notation, standard recursive computations of means and variances involving geometric sums yield,

$$q(z_t \mid x) = \mathcal{N}(z_t; \sqrt{\gamma_t}\, x, (1 - \gamma_t)I) \qquad \text{for} \quad t = 1, \ldots, T. \qquad (8.34)$$

Hence for such an auto-regressive stochastic sequence, the distribution of z_t given the initial value x has a mean vector $\sqrt{\gamma_t}x$ and a covariance matrix $(1 - \gamma_t)I$.

It is of further interest to have explicit expressions for $q(z_{t-1} \mid z_t, x)$, a probability appearing in the denoising matching term in (8.30). The diffusion model assumptions enable us to obtain this conditional probability as well. Namely,

$$
\begin{aligned}
q(z_{t-1} \mid z_t, x) &= \frac{q(z_t \mid z_{t-1}, x)\, q(z_{t-1} \mid x)}{q(z_t \mid x)} \\
&= \frac{q(z_t \mid z_{t-1})\, q(z_{t-1} \mid x)}{q(z_t \mid x)} \\
&= \mathcal{N}\left(z_{t-1};\ \underbrace{\frac{(1-\gamma_{t-1})\sqrt{1-\beta_{t-1}}}{1-\gamma_t} z_t + \frac{\sqrt{\gamma_{t-1}}\,\beta_{t-1}}{1-\gamma_t} x}_{\text{Mean vector}},\ \underbrace{\frac{\beta_{t-1}(1-\gamma_{t-1})}{1-\gamma_t} I}_{\text{Covariance Matrix}} \right).
\end{aligned}
$$

$$(8.35)$$

The first equality follows Bayes' rule of conditional probability, maintaining the condition on x. In the second equality, we use the Markovian structure which implies that $q(z_t \mid z_{t-1}, x) = q(z_t \mid z_{t-1})$. Then to obtain the final expression we manipulate normal densities as given by (8.31) and (8.34). Note that this final manipulation requires a few algebraic steps involving completion of the squares; we omit the details. Hence we have explicit expressions for the conditional (on z_t and x) mean vector and covariance matrix of z_{t-1}.

Loss Function

Training a diffusion model involves learning the parameter vectors $\theta_1, \ldots, \theta_T$. This is the learning of T distinct neural networks together. While it is a special case of training general Markovian hierarchical variational autoencoders, the fact that there are no learned parameters in the encoder, and the fact that the decoder covariance functions are also without learned parameters, eases training.

With any Markovian hierarchical variational autoencoder, we would ideally like to maximize ELBO represented in (8.30), yet in the special case of a diffusion model, the closed-form diffusion expressions (8.31), (8.32), (8.34), and (8.35) simplify the training process. Some of the general parameters and associated probabilities in (8.30) are now captured by closed-form expressions and hyper-parameters in the diffusion model case.

Considering the ELBO expression (8.30) we see that for diffusion models we do not need a prior matching term, since this term only depends on the encoder parameters ϕ, and diffusion models do not have learned parameters for the encoder. With this, a single observation loss function for the diffusion model can then be represented as

$$
C(\theta_1 \ldots, \theta_T;\ x) = C_{\text{Reconstruction}}(\theta_1) + \sum_{t=2}^{T} C_{\text{DenoisingMatching}}(\theta_t), \tag{8.36}
$$

where $x \in \mathcal{D}$. The loss component $C_{\text{Reconstruction}}(\theta_1)$ relates to the reconstruction term in (8.30), and the loss components $C_{\text{DenoisingMatching}}(\theta_t)$ relates to the denoising matching terms in (8.30). Note that for brevity we omit the dependance on x in the terms on the right hand side. These loss components are implemented as

$$C_{\text{Reconstruction}}(\theta_1) = \frac{1}{\sigma_1^2} \left\| x - \mu_{\theta_1}(z_1^*) \right\|^2,\tag{8.37}$$

$$C_{\text{DenoisingMatching}}(\theta_t) = \frac{1}{\sigma_t^2} \left\| \frac{(1-\gamma_{t-1})\sqrt{1-\beta_{t-1}}}{1-\gamma_t} z_t^* + \frac{\sqrt{\gamma_{t-1}}\,\beta_{t-1}}{1-\gamma_t} x - \mu_{\theta_t}(z_t^*) \right\|^2,\tag{8.38}$$

for $t = 2, \ldots, T$. Here each z_t^* is generated randomly using (8.34) for given data sample x.

Minimization of the terms (8.37) and (8.38) is approximately equivalent to ELBO maximization, in a similar nature to the variational autoencoder. The reconstruction loss (8.37) is similar to the standard variational autoencoder reconstruction loss (8.23), based on the log-density expression (B.8) of Appendix B. Here we exploit the fact that there are no learned covariance parameters. Similarly to the standard variational autoencoder of Section 8.1, minimization of this reconstruction loss serves to approximately maximize $\mathbb{E}\log p_{\theta_1}(x\mid Z_1)$ in (8.30).

The denoising matching loss terms (8.38) are based on the KL-divergence expression (B.10) of Appendix B and in fact for every level t,

$$D_{KL}\left(q(z_{t-1}\mid z_t,x)\,\middle\|\,p_{\theta_t}(z_{t-1}\mid z_t)\right) = \frac{1}{2}C_{\text{DenoisingMatching}}(\theta_t) + \text{constant},$$

where the constant on the right hand side depends only on the (non-learnable) hyperparameters β_{t-1}, γ_{t-1}, and σ_t^2. From this expression, it is clear that minimization of the KL-divergence is equivalent to minimization of the Euclidean distance between the mean vectors of the distributions $q(z_{t-1}\mid z_t,x)$ and $p_{\theta_t}(z_{t-1}\mid z_t)$. Further, this term serves as an estimate of $\mathbb{E}\,D_{KL}\left(q(z_{t-1}|Z_t,x)\,\|\,p_{\theta_t}(z_{t-1}|Z_t)\right)$ from (8.30), where the mean and covariance expressions of (8.35) are used for the inner integral, and the single z_t^* drawn from $q(z_t\mid x)$ serves as an approximation of the outer integral.

This loss formulation in (8.36), (8.37), and (8.38) presents us with a simple mechanism for learning the parameters of the neural networks $\mu_{\theta_1}(\cdot), \ldots, \mu_{\theta_T}(\cdot)$ using gradient-based optimization where at each epoch we draw random $z_1^*, \ldots, z_T^*$ according to (8.34). Similar to other deep learning cases, we can also integrate mini-batch-based learning and other gradient descent variations. As we show now, these loss expressions can even be further simplified using the reparameterization trick.

The Reparameterization Trick and Simplified Loss

In (8.25) of Section 8.1, we saw the *reparameterization trick* in the context of variational autoencoders. This allows us to carry out model training using backpropagation. A similar reparameterization trick can be applied to diffusion models. In the context of diffusion models, this trick simplifies the loss function (8.36) and often yields better numerical stability during training. Such simplification is achieved by replacing the neural networks $\mu_{\theta_1}(z_1), \ldots, \mu_{\theta_T}(z_T)$, which are used for sequentially denoising the latent variables to obtain data sample x, with a different collection of neural networks $\hat{\mu}_{\theta_1}(z_1), \ldots, \hat{\mu}_{\theta_T}(z_T)$. Each such $\hat{\mu}_{\theta_t}(z_t)$ is trained to predict a standard noise component as constructed below.

Similar to the variational autoencoder case, the reparameterization trick we use is applied on the samples of the latent variables. In particular, based on (8.34) we can represent z_t^* as

$$z_t^* = \sqrt{\gamma_t}\, x + \sqrt{1 - \gamma_t}\, \epsilon_t^*, \tag{8.39}$$

where ϵ_t^* denotes an independent multivariate standard normal vector with dimension equal to the dimension of x. Using this reparameterization trick, as we show below, at a given data sample x, the loss function can be of the form

$$C(\theta_1 \ldots, \theta_T\,;\,x) = \sum_{t=1}^{T} C_{\text{Reparameterized}}(\theta_t), \tag{8.40}$$

where each $C_{\text{Reparameterized}}(\theta_t)$ is defined as

$$C_{\text{Reparameterized}}(\theta_t) = \frac{\beta_{t-1}^2}{\sigma_t^2\,(1 - \gamma_t)(1 - \beta_{t-1})} \left\| \epsilon_t^* - \hat{\mu}_{\theta_t}\Big(\underbrace{\sqrt{\gamma_t}\, x + \sqrt{1 - \gamma_t}\, \epsilon_t^*}_{z_t^*} \Big) \right\|^2, \tag{8.41}$$

and the dependence on x is notationally suppressed. Before delving into the mathematical formulation of obtaining the loss component (8.41), let us focus on how the diffusion model is trained.

When using (8.40) and (8.41) we train the neural networks $\hat{\mu}_{\theta_t}(\cdot)$ either with a single $x \in \mathcal{D}$ or with a mini-batch approach. Importantly, in each gradient evaluation during training, we generate a standard multivariate normal ϵ_t^* which is mapped to z_t^* using (8.39). We then obtain the gradient of $\hat{\mu}_{\theta_t}(z_t)$ at $z_t = z_t^*$ and this allows us to compute the gradient of $C_{\text{Reparameterized}}(\theta_t)$.

In the remaining part of the section, we provide a derivation of the loss expressions and then discuss how the trained diffusion model is used in the production. With (8.39), a sample of z_t^* can be obtained by first generating a sample of ϵ_t^* and then using a data sample x to get z_t^*. Thus, if the values of ϵ_t^* and the corresponding z_t^* are given, we can represent x as

$$x = \sqrt{\frac{1}{\gamma_t}}\, z_t^* - \sqrt{\frac{1 - \gamma_t}{\gamma_t}}\, \epsilon_t^*.$$

Using this, we can now manipulate the mean vector expression in (8.35). After some simplification, based on the fact that $\gamma_t/\gamma_{t-1} = 1 - \beta_{t-1}$, we obtain

$$\frac{(1 - \gamma_{t-1})\sqrt{1 - \beta_{t-1}}}{1 - \gamma_t}\, z_t^* + \frac{\sqrt{\gamma_{t-1}}\,\beta_{t-1}}{1 - \gamma_t}\, x = \frac{1}{\sqrt{1 - \beta_{t-1}}}\, z_t^* - \frac{\beta_{t-1}}{\sqrt{(1 - \gamma_t)(1 - \beta_{t-1})}}\, \epsilon_t^*.$$

Consequently, the denoising loss component in (8.38) can be expressed as

$$C_{\text{DenoisingMatching}}(\theta_t) = \frac{1}{\sigma_t^2} \left\| \frac{1}{\sqrt{1 - \beta_{t-1}}}\, z_t^* - \frac{\beta_{t-1}}{\sqrt{(1 - \gamma_t)(1 - \beta_{t-1})}}\, \epsilon_t^* - \mu_{\theta_t}(z_t^*) \right\|^2. \tag{8.42}$$

Hence the neural network $\mu_{\theta_t}(z_t^*)$ attempts to predict $\dfrac{1}{\sqrt{1 - \beta_{t-1}}}\, z_t^* - \dfrac{\beta_{t-1}}{\sqrt{(1 - \gamma_t)(1 - \beta_{t-1})}}\, \epsilon_t^*$. Now the essence of the reparameterization trick is to use a different neural network, denoted as $\hat{\mu}_{\theta_t}(z_t^*)$, that takes z_t^* to predict the noise ϵ_t^*. Note that only for the notational convenience we use the same notation θ_t to denote the parameters of new neural network as well.

With the new neural network, $\hat{\mu}_{\theta_t}(z_t^*)$, we replace $\mu_{\theta_t}(z_t^*)$ in (8.42) with a new random latent variable,

$$\frac{1}{\sqrt{1-\beta_{t-1}}} z_t^* - \frac{\beta_{t-1}}{\sqrt{(1-\gamma_t)(1-\beta_{t-1})}} \hat{\mu}_{\theta_t}(z_t^*), \tag{8.43}$$

where z_t^* is obtained from ϵ_t^* using (8.39). Now with basic manipulation the $C_{\text{DenoisingMatching}}(\theta_t)$ expression in (8.42) can be replaced by $C_{\text{Reparameterized}}(\theta_t)$ of (8.41).

With a similar argument, we can show that for $t = 1$, $C_{\text{Reconstruction}}(\theta_1)$ in (8.37) can be replaced with

$$C_{\text{Reparameterized}}(\theta_1) = \frac{\beta_0}{\sigma_1^2(1-\beta_0)} \|\epsilon_1^* - \hat{\mu}_{\theta_1}(z_1^*)\|^2.$$

As a consequence of these neural network replacements and the fact that $\gamma_1 = (1 - \beta_0)$, the loss component for each t must be of the form (8.41) and hence the final loss function is given by (8.40).

In production, once the diffusion model is trained, to generate a realistic looking sample x^*, we start at $t = T$ with a sample z_T^* from a multivariate standard normal. Iteratively, as we decrement t, for each step we generate a different standard multivariate normal ϵ^* and compute $\hat{\mu}_{\theta_t}(z_t^*)$. By using

$$z_{t-1}^* = \underbrace{\frac{1}{\sqrt{1-\beta_{t-1}}} z_t^* - \frac{\beta_{t-1}}{\sqrt{(1-\gamma_t)(1-\beta_{t-1})}} \hat{\mu}_{\theta_t}(z_t^*)}_{\hat{z}_t^*} + \sigma_t \epsilon^*, \tag{8.44}$$

we obtain the (denoised) input to the next iteration, where $\hat{z}_t^*$ is given by (8.43). We repeat until $x^* = z_0^*$ yields a realistic looking generated data sample. Observe that randomness of this sample is due to the initial random z_T^* as well as to the ϵ^* that was generated in each iteration.

Notice from (8.44) that z_{t-1}^* follows a multivariate normal distribution with the conditional (on z_t^*) mean vector being $\hat{z}_t^*$ and the covariance matrix being $\sigma_t^2 I$. Since with the reparameterization trick, $\hat{z}_t^*$ is a replacement of $\mu_{\theta_t}(z_t^*)$, in production, the distribution of z_{t-1}^* is approximately equal to $p_{\theta_t}(\cdot \mid z_t^*)$ which is what we aim to achieve.

8.3 Generative Adversarial Networks

Generative adversarial networks or GANs, are generative deep learning architectures that have made great impact on the field of synthetic data generation. For example in Figure 8.5 we can see synthetic images, all created via various GAN architectures, with several paradigms of application. In this section we aim to highlight the key ideas of GANs, focusing on mathematical principles. The basic setup of using such a generative model to approximately create samples from the distribution $p(x \mid z)$ as is illustrated in Figure 8.1. We now present details of GAN architectures.

The GAN Approach to Generative Modeling

The general idea of GAN modeling is to train a neural network, called the *generator* and denoted here as $f_\theta^G(\cdot)$. This model applied to a noise vector z^*, generates a data point

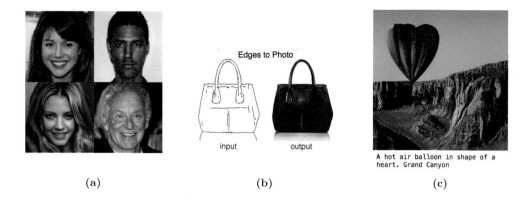

(a) **(b)** **(c)**

Figure 8.5: Images generated via various GAN architectures[7] with various types of paradigms. (a) Faces with 1024×1024 resolution. (b) Picture to picture generation. (c) A text to image application.

x^* that ideally appears similar to other data points $x \in \mathcal{D}$. Training is motivated by an *adversarial* game like approach, where we simultaneously train both the generator $f_\theta^G(\cdot)$, and another network called the *discriminator*, and denoted here as $f_\phi^D(\cdot)$. Hence training implies learning the generator parameters θ while in parallel also learning the discriminator's parameters ϕ.

The generator's purpose is to create "fake" data samples and the discriminator's purpose is to try and distinguish between fake and real samples. As such, the discriminator is a binary classifier, which when presented with some data point $\tilde{x}$, would ideally be able to output a probability value near 1 if $\tilde{x} \in \mathcal{D}$ (real) and a probability value near 0 if $\tilde{x} = x^*$ (fake).

While in a normal classification setting, having a well-performing disciriminator $f_\phi^D(\cdot)$ is desirable, in the GAN setting, the discriminator is actually used to help train the generator $f_\theta^G(\cdot)$. Once training is complete, we would generally have that $f_\phi^D(\tilde{x})$ is on average near $1/2$ for both real $\tilde{x} \in \mathcal{D}$ and fake $\tilde{x} = x^*$, with $x^* = f_\phi^G(z^*)$. As such, during training of a GAN, we can expect to see a sequence of network parameters,

$$\underbrace{(\phi^{(1)}, \theta^{(1)})}_{\text{iteration 1}}, \underbrace{(\phi^{(2)}, \theta^{(2)})}_{\text{iteration 2}}, \dots, \tag{8.45}$$

such that as training progresses to high iterations t, $\theta^{(t)}$ are the parameters of a generator that creates "sensible fake samples", while the discriminator parameters $\phi^{(t)}$, are no longer able to discriminate between "fake" and "real", and then yield performance near $1/2$.

Empirically it has been experienced that one needs to train both the generator and discriminator together, in order to achieve the desired behavior. The alternative of starting with an already trained discriminator, and then training a generator is not feasible, first because such a trained discriminator is not available without the generator trained, and secondly because the joint training provides a loss landscape for the generator parameters θ that is suitable for gradient-based learning. In this sense, training a GAN, follows a *dynamic equilibrium* approach which can often be posed as a *mathematical game* between two players.

---

[7]Image (a) is thanks to Karras, et. al. [222]. Image (b) is thanks to Isola, et. al. [204]. Image (c) is thanks to Kang, et. al. [219].

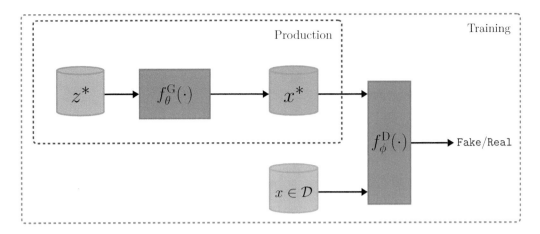

Figure 8.6: The GAN architecture for learning the generator $f_\theta^G(\cdot)$. A random z^* passes through the generator to produce x^*. The generator is trained simultaneously with the discriminator $f_\phi^D(\cdot)$, whose role is to determine if its input is fake, x^*, or real, $x \in \mathcal{D}$.

In fact, as we present below, at the equilibrium point of this game, one can properly use mini-max game analysis to analyze some properties of the optimization.

Figure 8.6 illustrates the general GAN architecture where in training we use both the generator and discriminator networks, while in production only the generator network is used for creating samples from $p(x \mid z)$. Note that the latent space size of z, denoted m, is typically in the order of several dozen to several hundred dimensional and it is typically taken to be a multivariate standard normal distribution as in the previous generative models of this chapter.

Training a GAN

The key to training a GAN with steps of a sequence such as (8.45) is to alternate between learning ϕ for the discriminator, and learning θ for the generator. For the discriminator, training involves improving the binary classifier $f_\phi^D(\cdot)$ and for the generator, training is *adversarial* since it involves seeking parameters θ that yield the "worst possible" generator $f_\theta^G(\cdot)$ for a given discriminator $f_\phi^D(\cdot)$. We cast these alternative goals as a *minimax objective*, a notion that we now describe.

First note that as in Section 8.1 we use $\mathcal{D}$ to denote the set of data points and assume that there are n such data points. In addition we assume some set of random latent variable samples which we denote as $\mathcal{Z}^*$ with each $z^* \in \mathcal{Z}^*$ being an m-dimensional random vector. In practice, one does not need to randomly sample all of $\mathcal{Z}^*$ in one go, but can rather resample from a standard multivariate normal distribution every time we seek an arbitrary element of $\mathcal{Z}^*$. Yet for simplicity of the exposition and analysis we assume that the number of elements in $\mathcal{Z}^*$ is the same as in $\mathcal{D}$, i.e., n.

With this notation, based on $\mathcal{D}$ and $\mathcal{Z}^*$, for fixed generator parameters θ, the objective for the discriminator parameters ϕ can be taken as the classic binary cross entropy objective; recall (3.11) of Chapter 3. Specifically, from the discriminator's perspective we want $x \in \mathcal{D}$ to be considered as a positive example and we want $f_\theta^G(z^*)$ for $z^* \in \mathcal{Z}^*$ to be considered as a negative example. That is, the discriminator applied to x should be as close to 1 as possible, and the discriminator applied to $f_\theta^G(z^*)$ should be as close to 0 as possible. Now given the generator's parameters θ, together with $\mathcal{D}$ and $\mathcal{Z}^*$, a binary cross entropy loss function (scaled by a factor of 2) can be formulated as

$$C_\theta(\phi\,;\mathcal{D},\mathcal{Z}^*) = -\frac{1}{n}\Big(\sum_{x\in\mathcal{D}}\log\big(f_\phi^D(x)\big) + \sum_{z^*\in\mathcal{Z}^*}\log\big(1 - f_\phi^D\big(f_\theta^G(z^*)\big)\big)\Big). \tag{8.46}$$

Hence the objective of learning the discriminator parameters is

$$\min_\phi C_\theta(\phi\,;\mathcal{D},\mathcal{Z}^*). \tag{8.47}$$

The generator's learning process is adversarial as the generator wishes for $C_\theta(\phi\,;\mathcal{D},\mathcal{Z}^*)$ to be as high as possible. Specifically, from the generator's point of view, we seek to find parameters θ that are the worst possible parameters (maximization of C_θ) for the optimal choice of the discriminator. That is, we seek to solve

$$\max_\theta\Big(\min_\phi C_\theta(\phi\,;\mathcal{D},\mathcal{Z}^*)\Big). \tag{8.48}$$

The joint objective (8.48) is generally called a minimax objective as it captures the competing goals of both players in the game.

To implement (8.48) we use an iterative algorithm that goes through iterates as in (8.45) and in each iteration alternates between ϕ and θ, with multiple steps within the iteration for each. Specifically, at iteration t we first carry out mini-batch-based gradient descent steps on the loss function (8.46) where $\theta = \theta^{(t-1)}$ is fixed and we improve ϕ, starting with $\phi^{(t-1)}$ in the iteration. The result of these mini-batch gradient descent steps for iteration t is $\phi^{(t)}$.

In the second part of iteration t, we seek to maximize (8.46) over θ while keeping $\phi^{(t)}$ fixed. Now since the first term in (8.46) does not depend on θ, this maximization is equivalent to gradient descent steps (minimization) for

$$\min_\theta\frac{1}{n}\sum_{z^*\in\mathcal{Z}^*}\log\big(1 - f_{\phi^{(t)}}^D\big(f_\theta^G(z^*)\big)\big). \tag{8.49}$$

Such iterative gradient-based learning, ideally approximates a solution of (8.48). In practice one often tunes the number of mini-batch discriminator steps and the number of z^* samples for the generator steps, carried out per iteration t. One problem that sometimes arises during training is called *mode collapse* and it is a situation where the generator is stuck at a point in the parameter space of θ where it generates samples that do not cover the breadth of the data $\mathcal{D}$. In general, when carrying out diagnostics of GAN training, we expect the performance of the discriminator to converge to about $1/2$.

Minimization of the Jensen-Shannon Divergence

The training procedure outlined above can be cast in a theoretical setting. Motivated by the discriminator loss (8.46) we can describe a theoretical construct which abstracts what a GAN training procedure actually optimizes. Specifically, we now shift to probabilistic thinking, similar to the notions of Section 8.1 in the context of variational autoencoders and Section 8.2 in the context of diffusion models.

For this let us define three probability distributions used in the analysis. First, the distribution of the data $\mathcal{D}$ is assumed to be captured by $p_{\mathcal{D}}(\cdot)$. Let us assume this is a probability distribution covering all of $\mathbb{R}^p$. Then the distribution of each of the latent variables $\mathcal{Z}^*$ is assumed be captured by $p_{\mathcal{Z}*}(\cdot)$. Let us assume that this a probability distribution over $\mathbb{R}^m$ which typically is multivariate standard Gaussian and hence covers the whole latent space. Finally, the distribution of the output of the generator is captured by $p_G(\cdot)$. This last distribution can in theory be obtained by applying the generator $f_\theta^G(\cdot)$ as a transformation on the random variables in $\mathcal{Z}^*$. Like the distribution of the data, this is a distribution over $\mathbb{R}^p$ and we assume it also covers the whole space.

Ideally we would like a GAN to learn the generator parameters θ such that $p_G = p_{\mathcal{D}}$, i.e., that the distribution of the generator is the same as the distribution of the data. To capture such a goal, it turns out that in the context of GANs, a good measure of the distance between the two distributions is the Jenson-Shannon divergence, defined in (B.7) of Appendix B. Specifically, let us denote

$$J_\theta = \mathrm{JSD}(p_G \parallel p_{\mathcal{D}}) = \frac{D_{\mathrm{KL}}(p_{\mathcal{D}} \parallel p_{\mathcal{M}}) + D_{\mathrm{KL}}(p_G \parallel p_{\mathcal{M}})}{2}, \quad \text{where } p_{\mathcal{M}}(u) = \frac{1}{2}\left(p_G(u) + p_{\mathcal{D}}(u)\right),$$

and $D_{\mathrm{KL}}(\cdot \parallel \cdot)$ is the KL-divergence. Hence for generator parameters θ, the divergence value J_θ captures how far off our generator is from the actual data. In the ideal situation of $p_G = p_{\mathcal{D}}$ we have that $J_\theta = 0$, otherwise it is greater than 0. Note that upon representing the KL-divergences as integrals and manipulating the 2 constant we have

$$J_\theta = \frac{1}{2}\int_{\mathbb{R}^p} p_{\mathcal{D}}(u) \log\left(\frac{p_{\mathcal{D}}(u)}{p_{\mathcal{D}}(u) + p_G(u)}\right)du + p_G(u) \log\left(\frac{p_G(u)}{p_{\mathcal{D}}(u) + p_G(u)}\right)du + \log 2. \quad (8.50)$$

Consider now the discriminator loss (8.46) and assume that $n \to \infty$. In this case, due to laws of large numbers, the loss can be rephrased in terms of expectations and expressed as

$$\overline{C}(\phi, \theta\,; p_{\mathcal{D}}, p_{\mathcal{Z}*}) = -\mathbb{E}_{p_{\mathcal{D}}} \log\left(f_\phi^D(X)\right) - \mathbb{E}_{p_{\mathcal{Z}*}} \log\left(1 - f_\phi^D(f_\theta^G(Z))\right), \quad (8.51)$$

where the first expectation is of the random variable X distributed according to $p_{\mathcal{D}}(\cdot)$ and the second expectation is with respect to the latent random variable Z distributed according to $p_{\mathcal{Z}*}(\cdot)$. This expected loss can further be manipulated as

$$\overline{C}(\phi, \theta\,; p_{\mathcal{D}}, p_{\mathcal{Z}*}) = -\mathbb{E}_{p_{\mathcal{D}}} \log\left(f_\phi^D(X)\right) - \mathbb{E}_{p_G} \log\left(1 - f_\phi^D(W)\right)$$

$$= -\int_{\mathbb{R}^p} p_{\mathcal{D}}(u) \log\left(f_\phi^D(u)\right)du - \int_{\mathbb{R}^p} p_G(u) \log\left(1 - f_\phi^D(u)\right)du$$

$$= -\int_{\mathbb{R}^p} p_{\mathcal{D}}(u) \log\left(f_\phi^D(u)\right)du + p_G(u) \log\left(1 - f_\phi^D(u)\right)du,$$

where in the first equation we set $W = f_\theta^G(Z)$ which is a random variable distributed according to $p_G(\cdot)$. Hence the second expectation in the first equation is with respect to the distribution $p_G(\cdot)$. Moving from the first equation to the second step we simply represent the expectations as integrals and since both integrations are on the same domain we can move from the second step to the third step.

Now let us identify a pattern inside the final integral of the form $a\log(y) + b\log(1-y)$, where $a = p_{\mathcal{D}}(u)$, $b = p_G(u)$, and $y = f_\phi^D(u)$. It is easy to check that for $a, b > 0$, the function

$g(y) = a \log(y) + b \log(1 - y)$ is maximized at $y = a/(a + b)$. Hence, for any y,

$$a \log(y) + b \log(1 - y) \leq a \log(\frac{a}{a + b}) + b \log(\frac{b}{a + b}) \tag{8.52}$$

with the inequality being an equality at $y = a/(a + b)$.

We can now use (8.52) to bound the integrand in the final expression for $\overline{C}(\phi, \theta \, ; p_{\mathcal{D}}, p_{\mathcal{Z}^*})$ as follows:

$$- \int_{\mathbb{R}^p} p_{\mathcal{D}}(u) \log \left(\frac{p_{\mathcal{D}}(u)}{p_{\mathcal{D}}(u) + p_G(u)} \right) du + p_G(u) \log \left(\frac{p_G(u)}{p_{\mathcal{D}}(u) + p_G(u)} \right) du \leq \overline{C}(\phi, \theta \, ; p_{\mathcal{D}}, p_{\mathcal{Z}^*}).$$

Or, using the expression for J_θ in (8.50) we have

$$2 \log 2 - 2 J_\theta \leq \overline{C}(\phi, \theta \, ; p_{\mathcal{D}}, p_{\mathcal{Z}^*}), \tag{8.53}$$

where at the best theoretical generator parameters, θ, we have that $J_\theta = 0$ and the inequality is an equality with $\overline{C} = 2 \log 2 \approx 1.386$.

Let us note some parallels between this Jensen-Shannon-based analysis for GANS and the ELBO analysis of Section 8.1 for variational autoencoders. In the variational autoencoder case, our inherent implicit goal is maximum likelihood estimation of θ and since we cannot achieve such estimation in a computationally efficient manner, we resort to (approximate) optimization of the lower bound, ELBO. Nevertheless, a theoretical inequality such as (8.18), and the additional optimization over the encoder's ϕ in (8.20) ensures that once we have optimized ELBO both in terms of θ and ϕ, then the desired maximum likelihood estimation is also achieved.

In our GAN case, the inequality (8.53) helps justify a similar idea. If we apply a minimax style optimization on it, as in (8.48) then we obtain

$$2 \log 2 = \max_\theta \left(\min_\phi \overline{C}(\phi, \theta \, ; p_{\mathcal{D}}, p_{\mathcal{Z}^*}) \right),$$

which is the best possible value.

Variations to the Objective Function

While the ideas of GANs presented up to now are useful in their own right, there is much room for architectural improvements. On the one hand, one may consider different structures for $f_\phi^D(\cdot)$ and $f_\theta^G(\cdot)$, such as for example setting the discriminator and generator neural networks to be convolutional networks. On the other hand, one may revise the objective function (8.46) as well as its expected value formulation (8.51), to yield better training performance. We now focus on this approach, where variations of the objective function are considered.

In general, problems may occur in initial training iterations when generator parameter values, $\theta^{(t)}$, are nearly arbitrary. In such a case there is often extreme separation between $p_{\mathcal{D}}(\cdot)$ and $p_G(\cdot)$, because the latter distribution is nearly arbitrary. Such a difficult situation requires a "signal" from the discriminator that will force learning to improve subsequent $\theta^{(t)}$ values. However, an objective such as (8.46), does not always cater for such learning. In particular,

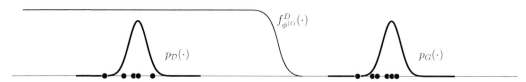

Figure 8.7: A schematic of a potential scenario during early phases of GAN training. The data distribution $p_{\mathcal{D}}(\cdot)$ and the generator distribution $p_G(\cdot)$ are far while the discriminator $f^D_{\phi^{(t)}}(\cdot)$ is already capable of separating the two distributions well. This reduces the "signal" for generator training.

see Figure 8.7, where for simplicity we assume the data is one-dimensional and we plot data points together with $p_{\mathcal{D}}(\cdot)$ as well as the generator distribution $p_G(\cdot)$ associated with some arbitrary $\theta^{(1)}$. In such a case, after only a few iterations, the discriminator parameters $\phi^{(t)}$ may quickly be trained to separate between the two distributions. The problem with this is that at that point, when considering gradients for minimization of the generator, as in (8.49), there will often be a "lack of signal" (or nearly zero gradients) for learning the generator parameters $\theta^{(t)}$. This situation is quite common in practice, and often prohibits effective learning of GANs.

As a consequence of such scenarios, multiple alternative objective formulations have been proposed, some of which yield much better training performance. We now outline a few of these ideas where we first describe a simple modification called *non-saturating GAN* (NS-GAN). We then describe a deeper idea using *Wasserstein distances* which yields a framework called *Wasserstein GAN* (W-GAN), and finally we discuss improvements to W-GAN, which involve regularization concepts.

The first idea of *non-saturating GAN* (NS-GAN) is simply to modify the generator objective (8.49) to use $-\log(u)$ instead of $\log(1-u)$. Namely, the NS-GAN generator objective is

$$\min_{\theta} \quad -\frac{1}{n} \sum_{z^* \in \mathcal{Z}^*} \log\left(f^D_{\phi^{(t)}}\left(f^G_{\theta}(z^*)\right)\right), \tag{8.54}$$

and NS-GAN retains the same discriminator objective (8.46). The motivation for such a modification is simply the fact that at $u \approx 0$, the derivative of $\log(1-u)$ (as in the original formulation) is approximately -1, while the derivative of derivative of $-\log(u)$ (as in NS-GAN) is nearly negative infinite. This may yield an improvement for initial phases of training and is particularly useful for cases when $p_{\mathcal{D}}(\cdot)$ is highly separated from $p_G(\cdot)$ and the discriminator parameters already yield good separation as in Figure 8.7. In this case, $f^D_{\phi^{(t)}}\left(f^G_{\theta^{(t)}}(z^*)\right) \approx 0$ and thus the NS-GAN generator objective (8.54) yields much higher magnitude gradients than the original generator objective (8.49), due to the nearly negative infinite derivative.

For both the original GAN and NS-GAN, as training progresses and the expected value of $f^D_{\phi^{(t)}}\left(f^G_{\theta}(z^*)\right)$ approaches $1/2$, the objectives (8.54) and (8.49) behave similarly. Hence in principle, the NS-GAN modification should not hamper with training. However, while the simple idea of NS-GAN may appear like a good improvement, in practice it has shown to yield unstable gradient updates. We chose to present NS-GAN here, merely as simple idea where modification of the objective can potentially yield training performance improvement.

A more fundamental way to improve GANs is based on the *Wasserstein distance*. This is a concept used to determine the "distance" between two probability distributions, with a similar goal of JS-divergence, yet with a completely different approach that yields different properties of the distance metric.

Formally when presented with two probability distributions, in our case, $p_D(\cdot)$, and $p_G(\cdot)$, the Wasserstein distance, $\mathcal{W}(\cdot, \cdot)$ may be represented as

$$\mathcal{W}(p_D, p_G) = \inf_{p_\Pi \in \Pi(p_D, p_G)} \mathbb{E}_{(x,\tilde{x}) \sim p_\Pi} \|x - \tilde{x}\|. \tag{8.55}$$

Let us unpack (8.55) by assuming that our data and generator are each in $\mathbb{R}^p$. Here $\Pi(p_D, p_G)$ is the space of all probability distributions over $\mathbb{R}^{2p}$ where the first p coordinates are for the data and the following p coordinates are for the generator. This space of distributions is defined such that each element distribution $p_\Pi \in \Pi(p_D, p_G)$ has a marginal distribution of the first p coordinates of p_Π that agrees with p_D, and likewise the marginal distribution of the remaining p coordinates of p_Π agrees with p_G. That is, for every distribution $p_\Pi \in \Pi(p_D, p_G)$,

$$
\begin{aligned}
p_D(x_1, \ldots, x_p) &= \int \cdots \int p_\Pi(x_1, \ldots, x_p, u_1, \ldots, u_p) \, du_1 \cdots du_p, \\
p_G(\tilde{x}_1, \ldots, \tilde{x}_p) &= \int \cdots \int p_\Pi(u_1, \ldots, u_p, \tilde{x}_1, \ldots, \tilde{x}_p) \, du_1 \cdots du_p.
\end{aligned}
\tag{8.56}
$$

The infimum in (8.55) acts "like a minimum" and formally seeks the greatest lower bound. This minimization is over the expectation of the Euclidean norm $\|x - \tilde{x}\|$ with x and $\tilde{x}$ each elements of $\mathbb{R}^p$. Thus the pair $(x, \tilde{x}) \in \mathbb{R}^{2p}$ is distributed according to p_Π and hence x is (marginally) distributed according to p_D and $\tilde{x}$ is (marginally) distributed according to p_G. For clarity, note that the expectation in (8.55) can be represented as

$$\mathbb{E}\|x - \tilde{x}\| = \int \cdots \cdots \int \sqrt{\sum_{i=1}^p (x_i - \tilde{x}_i)^2} \; p_\Pi(x_1, \ldots, x_p, \tilde{x}_1, \ldots, \tilde{x}_p) \, dx_1 \cdots dx_p \, d\tilde{x}_1 \cdots d\tilde{x}_p.$$

$$\tag{8.57}$$

As an extreme example, consider the case where $p_D = p_G$. Then one particular $p_\Pi^* \in \Pi(p_D, p_G)$ is a distribution that only has support on elements $u_1, \ldots, u_p, u_{p+1}, \ldots, u_{2p}$ where $(u_1, \ldots, u_p) = (u_{p+1}, \ldots, u_{2p})$, and the density over each such element is $p_D(u_1, \ldots, u_p)$ which is the same as $p_G(u_{p+1}, \ldots, u_{2p})$. In this case, using p_Π^* in (8.57) we get 0 because probability mass (or density) is only concentrated on points $(x, \tilde{x}) \in \mathbb{R}^{2p}$ where $x = \tilde{x}$. Hence this p_Π^* minimizes (8.57), and hence the Wasserstein distance is 0. However, when $p_D \neq p_G$, one can show that the Wasserstein distance is strictly positive.

In general, the joint distribution p_Π captures an "earth moving" plan which describes how to shift mass from one distribution p_D to the other p_G. To get a feel for this interpretation, let us grossly simplify the situation and assume that p_D and p_G are discrete one-dimensional ($p = 1$) distributions on a finite domain with values, such as 1, 2, and 3. In this case, the joint distribution p_Π is on the 3×3 grid, capturing the probability mass over the 9 possible joint locations. In this simplistic case, the marginal distribution assumptions (8.56) can be written as

$$
\begin{aligned}
p_D(x) &= \textstyle\sum_{\tilde{x}=1}^3 p_\Pi(x, \tilde{x}) \quad \text{for} \quad x = 1, 2, 3, \text{ and} \\
p_G(\tilde{x}) &= \textstyle\sum_{x=1}^3 p_\Pi(x, \tilde{x}) \quad \text{for} \quad \tilde{x} = 1, 2, 3,
\end{aligned}
\tag{8.58}
$$

where $p_\Pi(\cdot, \cdot)$, $p_D(\cdot)$, and $p_G(\cdot)$ are probability masses.

Now we may interpret $p_\Pi(x, \tilde{x})$ as a "plan" of how much mass (or earth) to shift from location x to location $\tilde{x}$. For this, the marginal distribution assumptions (8.58) serve as constraints where the first constraints imply that mass is shifted properly out of p_D and the second constraints imply that all resulting mass ends up creating p_G. The Wasserstein distance is then the expectation of $\|x - \tilde{x}\|$ for the optimal plan, or optimal joint distribution. Note that in such a discrete finite case, one may actually formulate and solve this optimization problem using *linear programming*. While such an approach is not important in deep learning practice, ideas of *duality theory* from linear programming play a role as described below.

One important attribute of the Wasserstein distance is that it is sensitive to differences in the "location" of the distributions and captures such differences in a much better way than the JS-divergence (based on the KL-divergence). As an illustration, let us focus on a case with $p = 1$ where we can easily plot p_D and p_G as marginals. Consider Figure 8.8 where we plot a red pair (p_D, p_G) with the distributions concentrated at quite far away locations, and a green pair (p_D, p_G) with the distributions closer together. We cannot exactly see the Wasserstein distance because it requires minimization over all possible $\Pi(p_D, p_G)$ distributions. However, in this example any possible $p_\Pi \in \Pi(p_D, p_G)$ is concentrated around the respective red or green cloud. Hence we roughly see the Wasserstein distance as marked in the figure, representing the difference between the center of the cloud and the diagonal $x = \tilde{x}$ line. Importantly, the distance for the red pair is much higher than that of the green pair. Now in contrast, if we were to consider the JS-divergence, then in both the red and the green pair cases, the divergence would be very close[8] to 0.

An important result for the Wasserstein distance, called the *Kantorovich-Rubinstein duality theorem*, provides us with a dual representation of the minimization problem in (8.55). This result implies the following representation of $\mathcal{W}(p_D, p_G)$, for any $K > 0$,

$$\mathcal{W}(p_D, p_G) = \frac{1}{K} \sup_{\|h\|_L \leq K} \left\{ \mathbb{E}_{x \sim p_D} h(x) - \mathbb{E}_{\tilde{x} \sim p_G} h(\tilde{x}) \right\}. \tag{8.59}$$

Let us unpack (8.59). The supremum in (8.59) acts "like a maximum" and formally seeks the lowest upper bound. Continuing to assume that p_D and p_G are distributions over $\mathbb{R}^p$, this maximization searches over functions $h : \mathbb{R}^p \to \mathbb{R}$ that also satisfy the K-Lipschitz property, denoted by $\|h\|_L \leq K$. A function $h(\cdot)$ is K-Lipschitz if

$$\|h(u) - h(v)\| \leq K \|u - v\|, \qquad \text{for all} \qquad u, v \in \mathbb{R}^p.$$

This implies that the function does not ascend or descend in any direction with steepness greater than K. Note that in stating the Kantorovich-Rubinstein duality theorem using $K = 1$ is common, yet for our GAN purposes keeping K free is preferable.

The elegance and applicability of (8.59) is that instead of searching over all possible joint probability distributions (earth moving plans) as in (8.55), we now only need to search over all K-Lipschitz functions. It turns out, that in the context of GANs, the latter formulation is much easier to implement. We omit further details of the duality derivation between (8.55) and (8.59), and now show how to use (8.59) in a GAN setting.

Since p_G is determined by the generator $f_\theta^G(\cdot)$, with a Wasserstein distance approach our overall goal in a generative setting is to find generator parameters θ that minimize $\mathcal{W}(p_D, p_G)$.

---

[8]In fact, one needs to assume that the support of the p_D and p_G distributions overlaps for the JS-divergence to be properly defined.

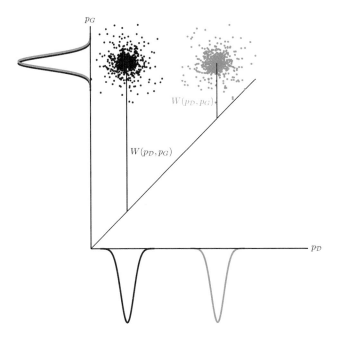

Figure 8.8: Two pairs of distributions where in the red pair, the distributions are farther away than in the green pair. While we cannot exactly see the Wasserstein distance in this case, we get an approximate feeling for it using the distances marked with vertical lines between the centers of the point clouds and the diagonal line.

Replacing the supremum in (8.59) by a maximum, and dropping the $1/K$ constant term, we have and overall objective

$$\min_{\theta} \ \max_{\|h\|_L \leq K} \left\{ \mathbb{E}_{x \sim p_\mathcal{D}} \, h(x) \ - \ \underbrace{\mathbb{E}_{\tilde{x} \sim p_G} \, h(\tilde{x})}_{\text{Depends on } \theta} \right\}. \tag{8.60}$$

The key idea of W-GAN is now to treat the function $h(\cdot)$ as a discriminator, even though it is no longer a binary classifier. Following the previous notation, we still denote this discriminator as $f_\phi^D(\cdot)$ with parameters ϕ, even though now the architecture of this discriminator allows it to output any potential real value, not just in $[0, 1]$. With this, (8.60) can be approximated with the objective

$$\min_{\theta} \ \max_{\phi \in \Phi_K} \left\{ \frac{1}{n} \sum_{x \in \mathcal{D}} f_\phi^D(x) \ - \ \frac{1}{n} \sum_{z^* \in \mathcal{Z}^*} f_\phi^D\big(f_\theta^G(z^*)\big) \right\}, \tag{8.61}$$

where now Φ_K is some space of discriminator parameters that approximately enforces a K-Lipstchitz condition on $f_\phi^D(\cdot)$. That is, with every $\phi \in \Phi_K$, we have that $f_\phi^D(\cdot)$ is K-Lipstchitz. Hence the discriminator objective, posed as a minimization problem constrained on Φ_k (and dropping the $1/n$ terms) is

$$\min_{\phi \in \Phi_K} \left\{ - \sum_{x \in \mathcal{D}} f_\phi^D(x) \ + \ \sum_{z^* \in \mathcal{Z}^*} f_\phi^D\big(f_\theta^G(z^*)\big) \right\}. \tag{8.62}$$

In training a W-GAN we iterate between discriminator and generator steps, and hence in the discriminator steps we carry out mini-batch gradient descent steps for (8.62). The constraint of keeping $\phi \in \Phi_K$ is discussed below. The outcome of such a discriminator iteration is $\phi^{(t)}$ and then in carrying out generator steps for (8.61) we carry out iterations for

$$\min_{\theta} \quad - \sum_{z^* \in \mathcal{Z}^*} f^D_{\phi^{(t)}}\left(f^G_{\theta}(z^*)\right). \tag{8.63}$$

Compare the W-GAN discriminator problem (8.62) with the original GAN discriminator problem (8.47), and similarly compare the W-GAN generator problem (8.63) with the original GAN generator problem (8.49) (similarly we may compare to NS-GAN with generator problem as in (8.54)). Quite surprisingly, if we ignore the Φ_K constraint, the differences are only very minor where the W-GAN does not use logarithms. In practice, W-GAN generally overcomes the problems illustrated in Figure 8.7 by enhancing the algorithm with a much better discrimination signal.

We still need to specify how to enforce the Φ_K, K-Lipstchitz constraint in (8.62). For this there are multiple techniques. The most basic technique is called *weight clipping* where we constrain the weights for the neural network $f^D_\phi(\cdot)$ to lie in some range, e.g., $[-0.05, 0.05]$. For almost all standard deep learning architectures, this will mean that $f^D_\phi(\cdot)$ will be K-Lipstchitz for some K that depends on the architecture of the network and on the clipping half-range 0.05.

A different approach which in practice has generally shown better performance is *gradient penalty*. This approach relies on the following property of (8.59). If a function $h^* : \mathbb{R}^p \to \mathbb{R}$ attains the supremum in (8.59), then this function has with probability one, a unit gradient norm on any random point x_η that is a convex combination between a point drawn from $p_\mathcal{D}$ and a point drawn from p_G. That is, if we take x as a random point from $p_\mathcal{D}$ and x^* as a random point from p_G (for some generator), then for any $\eta \in [0, 1]$, the point $x_\eta = \eta x + (1 - \eta)x^*$ satisfies,

$$\|\nabla h^*(x_\eta)\| = 1, \quad \text{with probability 1.} \tag{8.64}$$

We omit the proof, yet we use this property algorithmically to approximately enforce the constraint $\phi \in \Phi_K$. To do so, the gradient penalty approach uses (8.64) as an optimization constraint in place of $\phi \in \Phi_K$. This constraint is then integrated in the objective using an approximate Lagrange multiplier approach.

Specifically for some tunable $\lambda > 0$, we modify the discriminator objective (8.62) to

$$\min_{\phi} \left\{ - \sum_{x \in \mathcal{D}} f^D_\phi(x) + \sum_{z^* \in \mathcal{Z}^*} f^D_\phi\left(f^G_\theta(z^*)\right) + \lambda \sum_{x_\eta \in \mathcal{D}_\eta \mathcal{Z}^*} \left(\|\nabla f^D_\phi(x_\eta)\| - 1\right)^2 \right\}, \tag{8.65}$$

where the set of random points $\mathcal{D}_\eta \mathcal{Z}^*$ is a set of points based on pairs $x \in \mathcal{D}$ and $z^* \in \mathcal{Z}^*$, where each x has as single matching z^*, and for each pair we generate a uniformly random $\eta \in [0, 1]$ and set $x_\eta = \eta x + (1 - \eta)f^G_\theta(z^*)$.

In practice the gradient penalty approach works very well and out of all of the GAN objective variations that we presented, it is the most common approach used. In a typical algorithm that implements mini-batch gradient descent steps for (8.65), we would sample a subset of the data $\mathcal{D}$ and a matching sized sample of $\mathcal{Z}^*$ followed by a sample of $\mathcal{D}_\eta \mathcal{Z}^*$ with each element determined by a uniform η.

Beyond Data Generation with GANs

Up to now we considered the application of generative adversarial networks for generative modeling where an unlabeled dataset $\mathcal{D} = \{x^{(1)}, \ldots, x^{(n)}\}$ is used to train a generator $f_\theta^G(\cdot)$ and then this generator can create samples similar to those in $\mathcal{D}$. This is already useful for a variety of applications, one of which is *data augmentation*, where we can then use additional generated (fake) samples to train other models. Yet, there are many more tasks where GANs can be employed. We now outline a few paradigms that extend the basic GAN architecture and allow us to handle additional tasks. We outline the *conditional generation paradigm*, the *image to image paradigm*, and the *style transfer paradigm*.

With the *conditional generation paradigm* the resulting generator from training is not just a function of the latent space, but is also a function of additional variables. Namely we can describe the generator as $f_\theta^G(z, w)$ where z is still a random latent variable and the newly introduced w contains some additional information. One such example of conditional generation, called a *conditional generative adversarial network* (C-GAN) is where the data is labeled and is of the form $\mathcal{D} = \{(x^{(1)}, y^{(1)}) \ldots, (x^{(n)}, y^{(n)})\}$. Here we can use the additional information w as some attribute vector that potentially encodes the particular label. Hence for example in a case of the MNIST digits dataset, our GAN $f_\theta^G(z, w)$ will create random digits, where w may indicate which digit to create using one-hot encoding. The basic way in which such a GAN is trained, is by also supplying the attribute vector to the discriminator during training. We omit further details.

A slightly different form of conditional generation is also described in the *auxiliary classifier generative adversarial network* (AC-GAN) architecture. Here the generator is similar to the C-GAN case where w is specifically a class of the desired sample to generate, yet in training, the discriminator has two outputs, one of which is the binary classification fake/real as before, and another is a probability vector determining which class is detected. Such a training process, often results in a better architecture for conditional generation than the original C-GAN.

Ideas of conditional generation can be extended in multiple ways, where one notable way is the *interpretable representation learning generative adversarial network* (Info-GAN) approach. Here we are back to unlabeled data such as $\mathcal{D} = \{x^{(1)}, \ldots, x^{(n)}\}$, and we use the GAN training process to separate some elements of w "out of z", where typically w is of smaller dimension than z. The resulting generator, $f_\theta^G(z, w)$, then uses w for tweaking specific attributes of the image, where these attributes are not controlled apriori, but are rather discovered during the learning process. As a concrete example, again in the case of MNIST digits, we may have $w \in \mathbb{R}^4$ where after the training process it turns out that the first coordinate of w controls the stroke widths of the digits, the second coordinate controls the slant of the digit, and other two coordinates of w either have some interpretation or not. The beauty of Info-GAN is in the self discovery of these properties. The general idea of training in such a framework is that the discriminator outputs additional variables that are meant to match the inputs w. The analysis involves mutual information with basic ideas from information theory, and is beyond our scope.

With the *image to image paradigm*, or more generally *data to data paradigm* our goal is not to generate random general images or data examples, but rather to be able to *enhance* images or data. Assume a training dataset such as $\mathcal{D} = \{(x_{\text{in}}^{(1)}, x_{\text{out}}^{(1)}), \dots, (x_{\text{in}}^{(n)}, x_{\text{out}}^{(n)})\}$, where each $x_{\text{in}}^{(i)}$ has lower information content than the corresponding $x_{\text{out}}^{(i)}$. For example $x_{\text{in}}^{(i)}$ may be a black and white image of a portion a street map, while $x_{\text{out}}^{(i)}$ may be a color satellite image of the same geographic location.[9] The goal of an image to image generator, say $\tilde{G}$, is to operate on input images similar to $x_{\text{in}}^{(i)}$ in structure, and create the corresponding enhanced output image. Then when presented with a new input image $\tilde{x}_{\text{in}}$, we would have that the trained generator applied to it, $f_\theta^{\tilde{G}}(\tilde{x}_{\text{in}})$, is the enhanced image. Hence for example in the geographic case, we can generate fake satellite images, based on street map images. Other applications include the color enhancement of images or films, and almost any domain where datasets such as $\mathcal{D} = \{(x_{\text{in}}^{(1)}, x_{\text{out}}^{(1)}), \dots, (x_{\text{in}}^{(n)}, x_{\text{out}}^{(n)})\}$ appear.

The basic training architecture in the image to image paradigm is based on a discriminator that distinguishes between real pairs $(x_{\text{in}}^{(i)}, x_{\text{out}}^{(i)})$, and fake pairs $(x_{\text{in}}^{(i)}, x_{\text{out}}^{*(i)})$, where in the first pair, $x_{\text{out}}^{(i)}$ matches $x_{\text{in}}^{(i)}$ from the data, while the second pair, $x_{\text{out}}^{*(i)}$ is generated. That is, the discriminator, is the function $f_\phi^{\tilde{D}}(\tilde{x}_{\text{in}}, \tilde{x}_{\text{out}})$ that tries to determine if the pair it is fed with is completely real or if the second image fed to it is generated. In this sense, the image to image paradigm is similar to the C-GAN paradigm. Key differences include the fact, that in the image to image case, one would often want a complicated generator network such as a U-Net.[10]

In deep learning, *style transfer*, sometimes called *neural style transfer*, is an area broader than GANs, mostly focusing on image data. The general theme of style transfer is to modify input images such that they appear to resemble a certain style. Examples that have been popularized include the recreation of arbitrary images using the distinctive drawing style of Vincent van Gogh. Within the context of GANs, the *style transfer paradigm* uses GANs for style transfer and similar image modifications.

A notable GAN architecture specifically focused on style transfer is the *style-GAN*. Here, multiple new ideas are introduced in the context of the generator network, specifically for image data using convolutional architectures. One of the notable features of StyleGAN is its ability to generate high-resolution images with remarkable detail and diversity. Style-GAN has been widely used in applications such as image synthesis, artistic expression, and creating realistic faces with customizable attributes.

Several of the key ideas of Style-GAN are specific to convolutional architectures and include *upsampling*, as well as *adaptive instance normalization* which is somewhat similar to batch normalization of Section 5.6 and layer normalization, used in Section 7.5. We do not focus on these ideas here but rather comment on the general structure of the generator which differs from generators used in other paradigms. The Style-GAN generator is composed of two distinct functions, a *mapping network* and a *synthesis network*, which we denote as $f_{\theta_m}^G(z_1)$ and $f_{\theta_s}^G(z_2, w)$, respectively.

The mapping network's role is to convert a latent space noise vector z_1 into a more meaningful representation $w = f_{\theta_m}^G(z_1)$ which is an intermediate variable that is later amenable to manipulation. The synthesis network's role is somewhat similar to the Info-GAN generator

---

[9]For this specific example and other examples of image to image generation, it is not hard to create such training data.

[10]See the notes and references in Chapter 6 for background on the U-Net architecture.

where z_2 serves as noise, while w captures style information. One way to use a trained generator without any style modifications is via,

$$x^* = f_{\theta_s}^G\left(z_2^*, \underbrace{f_{\theta_m}^G(z_1^*)}_{w}\right),$$

where z_1^* and z_2^* are noise vectors, and x^* is the resulting random output image. However, in more advanced applications we may keep one latent noise vector fixed, while perturbing the other noise vector, and/or varying the intermediate variable w. Variations of this form enable Style-GAN to create meaningful modifications of images.

8.4 Reinforcement Learning

The topic of *reinforcement learning* deals with controlling dynamic processes over time. This is different than most tasks presented in the book, which on their own are typically applied at one instant of time. For example, classification is based on a static input x at some time, to determine an output $\hat{y}$ relevant for that time. Yet the classification task does not care about previous or future classification decisions or the time at which it is carried out. In contrast, with reinforcement learning, we have the task of making decisions as time evolves, where our decisions often affect the evolution of the system and thus our decisions need to take planning ahead into consideration.

Typical applications of reinforcement learning include decisions for automatic pilots, robot control, playing strategy games, financial management, and other applications that involve decisions over a time horizon. Indeed, in the second decade of this century, the deep learning variant of reinforcement learning, namely *deep reinforcement learning*, made big headlines with the game of Go, where the world's best Go players were eventually beaten by an engineered deep reinforcement learning platform. Earlier, it was shown that deep reinforcement learning systems can be trained to automatically play classic arcade Atari video games.

In general, the field of *control theory* is the engineering field where *automatic control* systems are designed and analyzed. This field has a rich history, including many advances made during the space programs of the 1950s and the 1960s, including concepts such as *Kalman filtering* and many related ideas. The area of reinforcement learning can be cast as part of the control theory world, yet it has much more of a computer science flavor to it, and with the advent of deep learning, has essentially become part of the deep learning toolkit. Nevertheless, today, ideas of reinforcement learning are part of the control theory world.

A schematic representation of reinforcement learning is in Figure 8.9. The basic setup is that a system, or *environment*, is controlled by an *agent*. Other processes may be taking place in the environment as well, perhaps not directly controlled by the agent, these are modeled via randomness. As time progresses the agent sends their control decisions in the form of *actions* to the environment, and in turn it receives *reward* from the environment as well as *observations*.

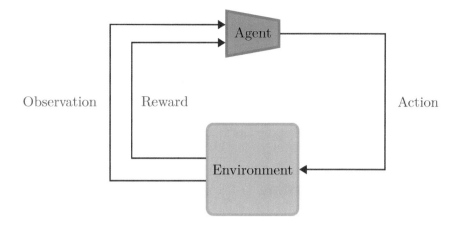

Figure 8.9: The setup in reinforcement learning is that an agent (or controller) applies actions to an environment (also known as system). The agent's decisions of which actions to take, are based on reward obtained together with observations. One illustrative example of the environment is a video game where multiple random things happen and the agent is a player.

Mathematically, we cast reinforcement learning in terms of an area called *Markov decision processes*.[11] Here, as we explain in this section, the evolution of the environment being controlled is modeled as a variant of a Markov chain, where the state evolution of the chain also depends on control decisions made. A key goal in the area of Markov decision processes is the study and application of algorithms for controlling the system in some optimal manner. Reinforcement learning, aims to solve the same problem, with the difference that in the Markov decision processes case we assume to have full knowledge of the system evolution model, whereas in the reinforcement learning case, the system model is unknown and hence needs to either be learned, or optimal control strategies need to be learned. With reinforcement learning, the agent often learn incrementally while controlling actual systems. In some cases learning is based on simulations of the system before deployment into the real world, while in other cases, operational systems are controlled and learned simultaneously.

Our focus of this section is to first present the Markov decision processes framework and to further discuss optimal strategies using Bellman equations. Then, after briefly discussing solution methods for such equations we move onto the reinforcement learning case, where the system model is unknown. There are multiple types of algorithms for reinforcement learning, and we focus only on the Q-learning approach. After understanding the basic Q-learning algorithm, we finally see how concepts of deep neural networks can be integrated by replacing the so-called Q-function with a deep neural network. This presentation here is merely an introduction to the field, and more readings are suggested at the end of the chapter.

Markov Decision Processes

Assume some system evolving over discrete time $t = 0, 1, 2, \ldots$, where at any time t, the system state is z_t. This state may describe the location of a robot, or the vector of velocities

---

[11]In certain cases the observations are only a partial description of the environment state, in which case the formal framework is that of *partially observable Markov decision processes*. Yet in our brief exposition, we only consider full state observations.

in multiple directions together with accelerations, or a discrete state of a game, or one of many other possibilities depending on the application.

Let us first ignore decisions and system control. In this case we can model the evolution of $z_0, z_1, z_2, \ldots$ as a *Markov chain*, where some transition kernel $p_t(z_{t+1} \mid z_t)$ determines the probability of moving from a given state z_t to state z_{t+1}, between time t and time $t+1$. In this form, the transitions appear to depend on the time t, yet we may also assume that the evolution is *time homogenous* and does not depend on time t. In this case, common to this section, the transition kernel can be presented with the notation $p(z_{t+1} \mid z_t)$, i.e., without a subscript t. As already stated in Section 8.2 in the context of Markovian hierarchical variational autoencoders, a Markov chain satisfies the Markov property. In our case here, this property is best interpreted as implying that given any history prior to time t, if we are given z_t, that history does not have an effect on the future. This then means that the *state information* at time t, namely z_t, is enough to determine probabilities of the future.

As a simple example, consider a scenario focusing on the engagement level of a student. Assume that there are 10 levels of engagement, $1, 2, \ldots, 10$, where at level 1 the student is not engaged at all and at the other extreme, at level 10, the student is maximally engaged. One may model the probability transitions of engagement as

$$p(j \mid 1) = \begin{cases} 0.6 & j = 1, \\ 0.4 & j = 2, \\ 0 & \text{otherwise}, \end{cases}$$

$$p(j \mid i) = \begin{cases} 0.6 & j = i - 1, \\ 0.2 & j = i, \\ 0.2 & j = i + 1, \\ 0 & \text{otherwise}, \end{cases} \quad \text{for} \quad i = 2, \ldots, 9,$$

$$p(j \mid 10) = \begin{cases} 0.6 & j = 9, \\ 0.4 & j = 10, \\ 0 & \text{otherwise}. \end{cases} \tag{8.66}$$

In this case, it is evident that at any time step, the student's engagement can either improve by one level, decrease by one level, or stay unchanged, with the probabilities specified by $p(\cdot \mid \cdot)$ as above. If at onset at time 0 we start at some given level $z_0 \in \{1, \ldots, 10\}$, as time progresses, the level will fluctuate according to the Markov chain stochastic process. The level of engagement is then the system or environment in this very simple case.

A Markov decision process is a generalization of a Markov chain where now some decision (also known as or control) is used by the agent. In this simple student engagement example, assume that at any time t we can choose to "stimulate" the student, for example by suggesting a prize, or "not to stimulate". Naturally, it is sensible to model the fact that stimulation will improve transitions up toward higher levels. For example, one model for these transitions under stimulation captures transition probabilities via $p^{(1)}(z_{t+1} \mid z_t)$, as,

$$p^{(1)}(j \mid 1) = \begin{cases} 0.2 & j = 1, \\ 0.8 & j = 2, \\ 0 & \text{otherwise}, \end{cases}$$

$$p^{(1)}(j \mid i) = \begin{cases} 0.1 & j = i - 1, \\ 0.1 & j = i, \\ 0.8 & j = i + 1, \\ 0 & \text{otherwise}, \end{cases} \qquad \text{for} \qquad i = 2, \dots, 9,$$

$$p^{(1)}(j \mid 10) = \begin{cases} 0.2 & j = 9, \\ 0.8 & j = 10, \\ 0 & \text{otherwise}. \end{cases} \tag{8.67}$$

Compare now the original transition probabilities (8.66) with the revised transition probabilities under "stimulation" (8.67). With these particular values in the model that we chose,
transitions with "stimulation" generally have a higher probability of pushing engagement
level up. In a Markov decision process, in addition to the state evolution $z_0, z_1, z_2, \dots$, we
also have an *action* chosen by the *controller* or *agent* for any time t. Specifically the sequence
$a_0, a_1, a_2, \dots$ denotes the actions, where in this example we can set that $a_t = 0$ implies "not
to stimulate" at time t and that $a_t = 1$ implies to "stimulate" at time t.

If we now define $p^{(0)}(j \mid i)$ as the original probabilities (8.66), then each of the probabilities,

$$p^{(a)}(j \mid i), \qquad \text{for} \quad a \in \{0, 1\}, \quad i, j \in \{1, \dots, 10\}, \tag{8.68}$$

constitutes the transition probability from $z_t = i$ to $z_{t+1} = j$ if action $a_t = a$ is chosen at
time t. Hence (8.68) is a specification of the environment (or system) and of how the agent's
actions affect that system.

Now with the action sequence, $a_0, a_1, a_2, \dots$, the state sequence $z_0, z_1, z_2, \dots$ does not evolve
autonomously, but rather depends on the decisions made $a_0, a_1, a_2, \dots$. So if for example
the first decision is $a_0 - 0$, the second is $a_1 = 0$, and the third is $a_2 = 1$, then given some
initial z_0, the first probabilities are $p^{(0)}(\cdot \mid z_0)$ which determine a random z_1; then given this
value, the second transition probabilities are $p^{(0)}(\cdot \mid z_1)$ which determine a random z_2; and
the third transition probabilities are $p^{(1)}(\cdot \mid z_2)$ which determine a random z_3, and so forth.

Having the sequence of decisions $a_0, a_1, a_2, \dots$ fixed a-priori is called an *open loop control*
and in our context this is not common. Instead, we typically wish to determine the action
a_t based on the state z_t. This is called *closed loop control* or *feedback control*, because the
control decision is based on the current state. A *control policy*, or just a *policy*[12] is a function
of the form,

$$g_{\text{policy}} : \underbrace{\{1, \dots, 10\}}_{\text{State space}} \;\to\; \underbrace{\{0, 1\}}_{\text{Action space}},$$

where in more general examples the *state space* and *action space* may be much more complex.

Any policy function, $g_{\text{policy}}(\cdot)$, induces a Markov chain specific for that policy. This is because
with $a_t = g_{\text{policy}}(z_t)$, the transitions $p^{(a_t)}(\cdot \mid z_t)$ are used for that Markov chain. Hence in

---

[12]This is in fact called a deterministic Markovian policy.

this example a policy can be seen as a rule for mixing (8.66) and (8.67). With this specific example there are only $2^{10} = 1,024$ possible policies, yet with more complex state spaces or action spaces, the number of possible policies can be huge.

The activity of finding the best feedback control policy for a Markov decision process is the act of solving a Markov decision process. To do so, our model requires a *reward function* to be specified. This reward function applies to every time step t, and captures the benefit of being in a given state, and choosing a given action at that time. The reward function[13] can be viewed as part of the Markov decision process model specification, and is of the form

$$r : \underbrace{\{1, \ldots, 10\}}_{\text{State space}} \times \underbrace{\{0, 1\}}_{\text{Action space}} \rightarrow \mathbb{R}.$$

Hence for example, $r(4, 0)$ is the reward obtained at a time where the state is $z_t = 4$ and the chosen action is $a_t = 0$ (no stimulation).

For our specific example, let us assume a reward function,

$$r(z, a) = z - 1.5a, \tag{8.69}$$

where higher student engagement levels have higher reward with a linear increase via the z component, and further, we pay a price of 1.5 reward units every time we set $a_t = 1$. Hence for example $r(4, 0) = 4$ and $r(4, 1) = 2.5$. From a modeling perspective, this latter reward being lower can be viewed as due to the cost of the prize for stimulation. Observe that the reward is always from the viewpoint of the agent controlling the system (the student in this particular example is the environment).

We now want to find a policy $g_{\text{policy}}(\cdot)$ that is best in terms of this reward over the whole time horizon. There are multiple ways to accumulate reward and in our exposition we consider only the *infinite horizon expected discounted reward* objective. Here, we first set $\gamma \in (0, 1)$ which a fixed hyper-parameter called a *discount factor*. Then the contribution at time t is taken to be

$$\gamma^t \, r(z_t, \underbrace{g_{\text{policy}}(z_t)}_{a_t}).$$

These contributions are accumulated to form the infinite horizon expected discounted reward objective, which depends on the initial state $z_0 = z$ and is

$$V_{g_{\text{policy}}}(z) = \mathbb{E}\left[\sum_{t=0}^{\infty} \gamma^t \, r(z_t, \underbrace{g_{\text{policy}}(z_t)}_{a_t}) \,\middle|\, z_0 = z \right]. \tag{8.70}$$

The role of the discount factor, γ, is to capture the importance of near present times vs. far future times. With γ low (near 0), far future times t have little effect on this contribution. Similarly with γ high (near 1), these far future times have much more of an affect on the contribution.

Observe that this objective function, (8.70), is parameterized by the policy, $g_{\text{policy}}(\cdot)$, because the policy determines how actions a_t are chosen, given values of z_t. Our goal is to find an optimal policy that maximizes (8.70), yet it may appear that there are multiple objectives

---

[13] Here we focus on the time-homogenous reward function which does not depend on the current time.

here because (8.70) depends for every initial state $z_0 = z$. Nevertheless, a property of the type of Markov decision processes that we are using is that there exists a policy that can maximize $V_{g_{\text{policy}}}(z)$ for all initial states $z_0 = z$. Hence we seek,

$$g^*_{\text{policy}} = \operatorname*{argmax}_{g_{\text{policy}}} V_{g_{\text{policy}}}(z), \qquad \text{for all initial states} \quad z. \tag{8.71}$$

In our simple student engagement example, one may even find such a policy by enumerating all possible policies and for each policy evaluating the expectation in (8.70) either via Monte-Carlo simulation or via analytic properties of Markov chains. For example, if the discount factor is at $\gamma = 0.6$, then an optimal policy turns out to be

$$g^*_{\text{policy}}(z) = \begin{cases} 0 & z \in \{1,2\}, \\ 1 & z \in \{3,4,5,6,7\}, \\ 0 & z \in \{8,9,10\}. \end{cases} \tag{8.72}$$

It is not obvious a-priori why this is the best policy, but considering it we see that for low and high engagement levels (1, 2, 8, 9, and 10) it is not worth to pay the price for stimulation, whereas otherwise for intermediate engagement levels (3, 4, 5, 6, and 7), it is worth stimulating the student. The strength of Markov decision processes is that they expose such policies, where the optimal control for any time t and current state z_t implicitly takes the future evolution into account via (8.70).

While such a Markov decision processes framework is in principle very powerful, we are faced with two problems. The first problem is to have better means than exhaustive search for solving (8.71) to find optimal policies, and we discuss such means shortly. The second problem is the fact that models can seldom be specified as we did here, with probabilities that correctly capture reality. That is, realistic scenarios would involve very complex transition kernels $p^{(a)}(j \mid i)$ in contrast to the simplistic specification of probabilities in (8.66) and (8.67). This second problem is handled by reinforcement learning methods.

Bellman Equations, the Value Function, and the Q-function

In characterizing the solution of (8.71) an important object is the *value function*. This real valued function, denoted as $V^*(z)$, where z is any element of the state space, is defined as

$$V^*(z) = \max_{g_{\text{policy}}} V_{g_{\text{policy}}}(z), \tag{8.73}$$

where $V_{g_{\text{policy}}}(z)$ is from (8.70). Hence the value function determines the optimal infinite horizon expected discounted reward, when starting at a state $z_0 = z$.

A result in the study of Markov decision processes is that we can characterize the value function via a non-linear system of equations called the *Bellman equations*. For the case of finite state and action spaces, these equations are,

$$V^*(z) = \max_a \underbrace{\left\{ r(z,a) + \gamma \sum_{z'} p^{(a)}(z' \mid z) V^*(z') \right\}}_{Q^*(z,a)}, \qquad \text{for all states} \quad z. \tag{8.74}$$

Here the maximum is over all actions a (e.g., $\{0,1\}$ in the student engagement example). Further, inside the maximum on the right hand side, we have the *Q-function*, $Q^*(z,a)$, which captures the "quality" of choosing action a on state z. Note that the summation in (8.74) is taken over all states z' (e.g., $\{1, \ldots, 10\}$ in the student engagement example).

One can informally derive the Bellman equations (8.74) via what is known as the *dynamic programming principle* where the first term of $Q^*(z, a)$ is the immediate reward, $r(z, a)$ and the second part is the expected next step reward, discounted by one time step and hence multiplied by γ. Theoretically it can be shown that any function $V^*(\cdot)$ that satisfies the Bellman equations is the value function as in (8.73).

With more complex state spaces that are not necessarily discrete, we can rewrite the Q-function as

$$Q^*(z, a) = r(z, a) + \gamma \, \mathbb{E}\big[V^*(z_{t+1}) \,|\, z_t = z, a_t = a \big], \qquad (8.75)$$

while keeping in mind that the time-homogenous assumptions imply that $Q^*(z, a)$ is the same for every time t and hence we can use z_0, a_0, and z_1 in place of z_t, a_t, and z_{t+1}, respectively. With the formulation of the Q-function as in (8.75), we have a more general expression for the Bellman equation (8.74), where $\sum_{z'} p^{(a)}(z' \,|\, z) \, V^*(z')$ is replaced by $\mathbb{E}[V^*(z_1) \,|\, z_0 = z, a_0 = a]$. This formulation encompasses states spaces that are not discrete.

Importantly, knowing either the Q-function or the value function presents us with an optimal policy. If we know the Q-function, $Q^*(\cdot, \cdot)$, then we can determine an optimal policy $g_{\text{policy}}^*(\cdot)$ as in (8.71) via

$$g_{\text{policy}}^*(z) = \underset{a}{\text{argmax}} \, Q^*(z, a). \qquad (8.76)$$

If we know the value function, $V^*(\cdot)$, then we can first evaluate the Q-function via (8.75) where we compute the expectation using the explicit model (sometimes using Monte Carlo simulation if needed) and then with this computed Q-function we can evaluate (8.76).

The classic study of Markov decision processes deals with the existence and optimality of solutions to the Bellman equations. It also deals with algorithms for solving these equations to find the value function and hence to find an optimal policy via (8.75) and (8.76). We briefly discuss such solution methods now.

Solving Bellman Equations

The two most common algorithms for solving Bellman equations are *value iteration* and *policy iteration*. We focus on value iteration on discrete state spaces. The algorithm is based on the recursion[14]

$$Q^{(t+1)}(z, a) = r(z, a) + \gamma \sum_{z'} p^{(a)}(z' \,|\, z) \left(\max_{a'} Q^{(t)}(z', a') \right), \qquad \text{for all } z \text{ and } a. \quad (8.77)$$

Value iteration starts with some initial or arbitrary guess $Q^{(0)}(\cdot, \cdot)$ and then with each step t we apply (8.77) on all states z and all actions a to get from $Q^{(t)}(\cdot, \cdot)$ to $Q^{(t+1)}(\cdot, \cdot)$. The algorithm terminates when some distance measure applied to $Q^{(t)}(\cdot, \cdot)$ and $Q^{(t+1)}(\cdot, \cdot)$ is below a specified small threshold.

---

[14] Note that one often writes the recursion in terms of $V^*(\cdot)$ and not in terms of Q-functions as we did here. However, the two formulations are equivalent and our representation is preferable for understanding Q-learning in the sequel.

In quite general settings, convergence of repeated applications of (8.77) to the optimal Q-function is guaranteed and a proof of this relies on the fact that the right hand side of (8.77) is a *contraction mapping*. We omit the details. Yet with value iteration we do not have an indication at what iteration the optimal policy has been discovered. Hence one may need to apply (8.77) many times.

Note that policy iteration, the other algorithm we mentioned above, remedies the situation. On finite state and action spaces, policy iteration needs to execute for only a finite number of steps before discovering an optimal policy. We do not discuss the policy iteration algorithm further here because Q-learning is based on value iteration.

Back to the simple student engagement example presented earlier, each $Q^{(t)}(\cdot,\cdot)$ can simply be implemented as a table with $10 \times 2 = 20$ entries since the state space is $\{1,\ldots,10\}$ and the action space is $\{0,1\}$. One sometimes informally calls this a *Q-table*. If we were to use value iteration for finding the optimal policy, we would first initialize $Q^{(0)}(\cdot,\cdot)$ with some 20 arbitrary values. We would then apply (8.77) for each $z \in \{1,\ldots,10\}$ and $a \in \{0,1\}$. This application would directly use the probabilities specified in (8.66) and (8.67), and the reward function specified in (8.69). Applying such value iteration steps is thus straightforward to execute recursively. After multiple steps, we can then determine the policy using (8.76) and this is in fact how we obtained the example optimal policy (8.72). However, more complex examples are harder to implement and importantly in realistic scenarios we often do not know the exact transition probabilities and reward function. Instead, we take the approach of learning the Q-function, which we describe now.

Q-learning

The idea with *Q-learning* is to learn the Q-function (8.75) and obtain some estimate $\hat{Q}(z,a)$ for all states z and all actions a. Learning the Q-function does not mean learning, the individual components, $r(\cdot,\cdot)$, $p^{(\cdot)}(\cdot\,|\,\cdot)$, and $V^*(\cdot)$. It rather means learning $\hat{Q}(\cdot,\cdot)$ as a whole. One can then use this estimate in (8.76) in place of $Q^*(z,a)$ to obtain a policy that is approximately optimal via

$$\hat{g}_{\text{policy}}(z) = \underset{a}{\text{argmax}}\ \hat{Q}(z,a). \tag{8.78}$$

Before seeing how Q-learning learns $\hat{Q}(\cdot,\cdot)$, we mention that as a reinforcement learning algorithm, one sometimes applies Q-learning in parallel to controlling a system, or controlling a simulation of the system. This is different from other learning paradigms in this book where the learning and production activities are often separate. Such a mix of controlling and learning involves ongoing estimates of the Q-function, $\hat{Q}^{(t)}(\cdot,\cdot)$ where at any time t we use $\hat{Q}^{(t)}(\cdot,\cdot)$ in place of $\hat{Q}(\cdot,\cdot)$ for (8.78).

When carrying out such a mix of learning and controlling, we know that at time t, the latest $\hat{Q}^{(t)}(\cdot,\cdot)$ is only an estimate. Hence we also employ *state exploration* as part of the policy. For this one typical approach known as the *epsilon greedy* approach, uses some pre-specified decreasing sequence of probabilities, $\varepsilon_0, \varepsilon_1, \varepsilon_2, \ldots$, and at any time t, we control the system with a randomized policy,

$$\hat{g}_{\text{policy}}^{(t)}(z) = \begin{cases} \underset{a}{\text{argmax}}\ \hat{Q}^{(t)}(z,a), & \text{with probability } 1 - \varepsilon_t, \\ \text{random action } a, & \text{with probability } \varepsilon_t. \end{cases} \tag{8.79}$$

With this epsilon greedy approach we know that at time t there is a chance of ε_t of selecting an arbitrary action that is most likely sub-optimal. Yet it allows us to potentially navigate the system to parts of the state space that would otherwise remain unexplored.

Now let us consider the Q-learning algorithm. Here the idea is to update some part of $\hat{Q}^{(t)}(\cdot, \cdot)$ at any time step based on the following available information: (i) the previous state z_t; (ii) the new state z_{t+1}; (iii) the previous action chosen a_t; (iv) the observed reward denoted as r_t which is the reward after applying the action at time t. For this, Q-learning relies on hyper-parameters, $\alpha_0, \alpha_1, \alpha_2, \ldots$, a pre-specified decreasing sequence of probabilities. The recursion of Q-learning is

$$\hat{Q}^{(t+1)}(z,a) = \begin{cases} (1-\alpha_t)\,\hat{Q}^{(t)}(z,a) + \alpha_t\Big(\ \underbrace{r_t + \gamma \max_{a'} \hat{Q}^{(t)}(z_{t+1},a')}_{\text{Single sample Bellman estimate}}\ \Big), & \text{if } z_t = z, a_t = a \\ \hat{Q}^{(t)}(z,a), & \text{otherwise.} \end{cases}$$

(8.80)

Key to (8.80) is that we only update the Q-function estimate for one specific z_t, a_t pair at any time step. Further observe that this update is a weighted average of the previous entry $\hat{Q}^{(t)}(z,a)$ and a new component which we denote as the *single sample Bellman estimate*. This term is directly motivated by the value iteration recursion (8.77) and we can observe that it agrees with the right hand side of (8.77) if we were to ignore the probabilities $p^{(a)}(z' \mid z)$. This general form of approximation is called a *stochastic approximation* and its theoretical analysis allows one to prove certain convergence properties of Q-learning. We omit these details.

From an operational point of view, we can now integrate the epsilon-greedy control of (8.79) with the Q-learning recursion of (8.80) to develop a learning scheme that both controls the system and learns how to control it as time progresses. With such a scheme, calibration of the hyper-parameter sequences $\varepsilon_0, \varepsilon_1, \varepsilon_2, \ldots$ and $\alpha_0, \alpha_1, \alpha_2, \ldots$ is often delicate, and one often has to experiment with various hyper-parameter settings in order to get useful results. A theoretical result for Q-learning is that under certain conditions we need the sequence $\alpha_0, \alpha_1, \alpha_2, \ldots$ to satisfy,

$$\sum_{t=0}^{\infty} \alpha_t = \infty, \quad \text{and} \quad \sum_{t=0}^{\infty} \alpha_t^2 < \infty.$$

(8.81)

Hence the probabilities need to decay quickly enough, but not too quickly. So for example probabilities such as $\alpha_t = 1/(t+1)$ suffice since with these the first (harmonic) series diverges while the second series converges.

While theoretical results based on the condition (8.81) help to place Q-learning on a rigorous footing, from a practical perspective Q-learning on its own is difficult to use effectively. Even in our simple student engagement example where we try to learn the 10×2 Q-table, Q-learning can be challenging. At any time step we only update a single entry of this table. This is already slow, and is further slowed down due to the fact that if for non-small times t, α_t is a low probability, then the averaging in (8.80) would keep the new entry $\hat{Q}^{(t+1)}(z,a)$ not far from the previous entry $\hat{Q}^{(t)}(z,a)$. In more complex and realistic scenarios where the state and action spaces are big and sometimes non-discrete, we cannot even tabulate the Q-function in a naive Q-table. Hence approximate Q-learning needs to be carried out. This brings us to *deep reinforcement learning*.

Deep Reinforcement Learning

The key idea with deep reinforcement learning is to approximate the Q-function, $Q^*(z,a)$, with a neural network, $f_\theta^Q(z,a)$. Such a setup allows us to deal with highly complex state spaces and action spaces. The parameters of this network are learned during a *deep Q-learning* algorithm where as the algorithm evolves, we have a sequence of learned parameters $\theta^{(1)}, \theta^{(2)}, \ldots$. With each such $\theta^{(t)}$, we can still use an epsilon greedy policy similar to (8.79) where the control decision is taken as

$$g_{\text{policy}}(z, \theta^{(t)}) = \begin{cases} \underset{a}{\operatorname{argmax}}\ f_{\theta^{(t)}}^Q(z,a), & \text{with probability } 1 - \varepsilon_t, \\ \text{random action } a, & \text{with probability } \varepsilon_t. \end{cases} \quad (8.82)$$

The key is now how to learn $f_\theta^Q(\cdot,\cdot)$ and for this there are many variants of the deep Q-learning algorithm of which we only outline the simplest form. To create a loss function that will help to learn the parameters θ, we use a reinforcement learning concept called *temporal difference learning*. Here the loss at time t is given by

$$C_t(\theta\,;z_t,a_t,r_t,z_{t+1}) = \Big(\underbrace{r_t + \gamma \max_a f_\theta^Q(z_{t+1},a)}_{\text{Single sample Bellman estimate}} - f_\theta^Q(z_t,a_t)\Big)^2, \quad (8.83)$$

where the observed state is z_t, the system is controlled via the action a_t, a reward of r_t is obtained, and the resulting next state is z_{t+1}. With such a temporal difference loss, if the neural network parameters θ determine a good approximation for the actual Q-function, then the loss would be generally low, whereas if θ does not describe the Q-function well, then the loss is higher. To see this, revisit the value iteration equation (8.77).

Now with the neural network loss (8.83) defined, we have a very basic deep Q-learning algorithm. We control the system using the control policy of (8.82) and at any time step t we apply a single gradient descent update for the parameters θ based on the loss $C_t(\cdot\,;z_t,a_t,r_t,z_{t+1})$. The gradient descent update then modifies the Q-function estimate from $f_{\theta^{(t)}}^Q(\cdot,\cdot)$ to $f_{\theta^{(t+1)}}^Q(\cdot,\cdot)$. Note that like the Q-learning algorithm (without a neural network approximation) described above, in this framework every time involves both a control decision and learning.

In practice, this basic deep Q-learning algorithm typically does not perform well. One problem is the coupling of the gradient descent step and the control decision, both happening only once at any time t. In practice we often seek an algorithm that can on the one hand learn the Q-function approximation using multiple control steps, and on the other hand perform multiple gradient descent steps. For this, one popular variant is to maintain two copies of the network approximating the Q-function, $f_\theta^Q(z,a)$. One copy is updated only every several time steps and is used for the control decisions (8.82), and the other copy is updated with every gradient descent step. Another concept often used is *reply memory* where we control the system for some multiple time steps and use the combination of these time steps for gradient-based learning. We do not dive into these technical details here.

8.5 Graph Neural Networks

In this section we introduce and explore neural networks for graph objects, called *graph neural networks* (GNNs). In a similar way to how convolutional neural networks are primarily used for image data, and RNNs are primarily used for sequence data, GNNs are applied on data organized as a (combinatorial) graph. The input data is typically of the form $G = (V, E)$ together with related feature data, where V is some vertex set of the graph, and E is the edge set of that graph. We introduce basic concepts of graphs in the sequel.

Abstractly, a GNN can be viewed as a function

$$f_\theta : \mathcal{G} \times \mathcal{F} \to \text{output}, \tag{8.84}$$

where $\mathcal{G}$ is the set of possible input graphs G, $\mathcal{F}$ is the set of possible input features, and the output may be a graph, a classification vector, a scalar value, or similar. Specifically, the features in $\mathcal{F}$ may include weights on edges, or much more complex features associated with edges and nodes. In similar spirit to other deep learning models, GNNs often implement $f_\theta(\cdot)$ via a composition of a sequence of steps or layers, each denoted $f_{\theta^{[\ell]}}^{[\ell]}(\cdot)$ as in (5.1). At each such layer ℓ, the input graph and features are transformed.

Applications of Graph Neural Networks

In contrast to domains such as image classification using convolutional neural networks, or natural language translation using sequence models, the applications associated with graph neural networks often require some transformation of the given problem into a graphical representation. Our focus in this section is not on such transformations for applications; see the notes and references at the end of this chapter for specific reading suggestions. An important point to highlight is that most applications of GNNs are not for mimicking human level performance (e.g., image recognition or text understanding), but are rather for gaining insights from complex data, that is otherwise hard to process. We now mention a few general application domains. See Figure 8.10 for an illustration.

One key area of application is social network analysis, where GNNs assist in identifying influential users, detecting communities, and predicting connections between individuals. Additionally, in recommendation systems, GNNs enhance the accuracy and personalization of suggestions by modeling user-item interactions within collaborative filtering settings. Another prominent application is in fraud detection, particularly in financial and e-commerce contexts, where GNNs uncover hidden relationships and anomalous patterns within transaction graphs. In the fields of biology and chemistry, GNNs excel at modeling molecular structures and interactions, offering valuable insights into drug discovery, protein-protein interaction prediction, and chemical property estimation. GNNs are also widely used in traffic analysis for optimizing transportation networks, predicting congestion, and enhancing routing and scheduling.

All of the above applications are generally cases where the tasks are not supposed to replace human level tasks, yet there are also situations where GNNs can enhance or replace human level tasks. Specifically, GNNs have applications in image analysis, enabling tasks like image segmentation, object tracking, and scene understanding, particularly when graphs represent relationships between image regions or structures. Further, in the natural language processing

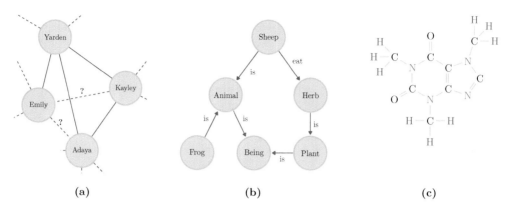

Figure 8.10: An illustration of graph structures arising in several application domains. (a) Connections in social networks described via graphs. The presence of edges with question marks is not known and can be predicted via a graph neural network. (b) Knowledge graphs capturing relationships between entities. Learning about such relationships can be assisted with graph neural networks. (c) Molecular bonds can be described via graphs. Thus classification of molecular structures and the design and discovery of new structures, is also aided by GNN.

domain, GNNs enhance tasks such as named entity recognition, sentiment analysis, and text classification by analyzing text data represented as dependency trees or semantic graphs.

Graph Structures

As alluded to in (8.84) the input to a GNN consists of a graph and features which are elements of $\mathcal{G}$ and $\mathcal{F}$ respectively. We now discuss possible representations of such objects. We begin with a brief outline of graph-theoretic terminology. See Figure 8.11 as a guide.

Recall that a *graph* $G \in \mathcal{G}$, is often denoted $G = (V, E)$. The graph is composed of a *node set* V and an *edge set* E. The node set can be denoted as $V = \{v_1, v_2, \ldots, v_r\}$ where each v_i is a node (also known as a *vertex*), and each of the r nodes can represent a distinct entity in the application domain (e.g., a single person, an atom, etc.). The edge set E is a subset of $V \times V$, or in particular is composed of tuples of the form (v_i, v_j) where each such tuple represents an *edge* connecting v_i and v_j. In our terminology we do not allow the elements (v_i, v_i) to be in E, that is, there are no *self loops*.

In some cases the graph is a *directed graph*, where (v_i, v_j) is different from (v_j, v_i) for $i \neq j$, and the former represents an edge (arrow) from v_i to v_j, while the latter is in the opposite direction. In other cases, the graph is an *undirected graph* in which case we can either treat (v_i, v_j) as unordered, or more formally require that if $(v_i, v_j) \in E$ then also $(v_j, v_i) \in E$. In the undirected case, edges are not represented as arrows but rather as links between nodes.

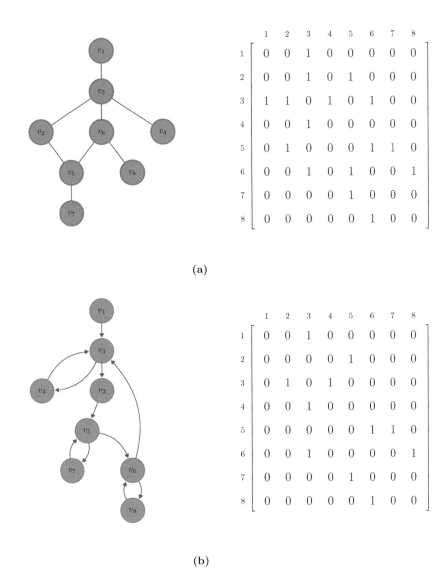

(a)

(b)

Figure 8.11: Directed and undirected graphs each with their associated adjacency matrix. (a) An undirected graph has a symmetric adjacency matrix. (b) A directed graph has an adjacency matrix that is typically not symmetric.

For undirected graphs, the *degree* of a node v_i is the number of edges connected to it, or more formally the number of nodes v_j such that $(v_i, v_j) \in E$. For directed graphs we differentiate between the *out-degree* of node v_i and the *in-degree*. The former is the number of nodes v_j such that $(v_i, v_j) \in E$, while the latter is the number of nodes v_j such that $(v_j, v_i) \in E$. In the undirected case, both out-degree and in-degree are the same at each node.

In the context of undirected graphs, the *neighbors* of a node v_i are the nodes v_j that are connected to v_i with an edge. We denote the set of indices of these neighbors via $\mathcal{N}(v_i)$. Hence the degree is the number of elements in this set. See for example, the undirected graph illustrated in Figure 8.11 (a) where $\mathcal{N}(v_1) = 1$, $\mathcal{N}(v_2) = 2$, $\mathcal{N}(v_3) = 4$, $\mathcal{N}(v_4) = 1$, $\mathcal{N}(v_5) = 3$, etc.

A *path* between two nodes v_i and v_j is a sequence of nodes $(v_{k_1}, v_{k_2}, \ldots, v_{k_m})$ where $v_{k_1} = v_i$, $v_{k_m} = v_j$ and $(v_{k_\ell}, v_{k_{\ell+1}}) \in E$ for $\ell = 1, \ldots, m-1$. A graph is said to be *connected* if there is a path between any two nodes $v_i, v_j \in V$. One trivial graph (which is also connected) is the *complete graph* (also known as the *fully connected graph*) where $(v_i, v_j) \in E$ for all $v_i, v_j \in V$.

Mathematically, one way to represent a graph is via an *adjacency matrix* where we number the nodes as $1, 2, \ldots, r$. Such an $r \times r$ matrix A has entries that are either 0 or 1 where,

$$A_{ij} = \begin{cases} 1 & \text{if } (v_i, v_j) \in E, \\ 0 & \text{otherwise.} \end{cases} \tag{8.85}$$

For undirected graphs, the adjacency matrix is symmetric ($A_{ij} = A_{ji}$) whereas for directed graphs it is not necessarily symmetric. In either case, since we do not allow self loops, $A_{ii} = 0$ for all i. When representing a graph in computer memory, sometimes an adjacency matrix is suitable, yet at other times, when the graph has fewer edges, sparser representations called *adjacency lists* are more suitable.

Since the node numbering is typically arbitrary, it is often useful to allow all permutations of the node numbering to be valid. Mathematically, we can express this property with the aid of an $r \times r$ *permutation matrix* P with,

$$P_{ij} = \begin{cases} 1 & \text{if } j \text{ replaces } i \text{ in the permutation,} \\ 0 & \text{otherwise.} \end{cases} \tag{8.86}$$

When P is applied to a vector x, namely $\tilde{x} = Px$, the result $\tilde{x}$ has x_j in at index i. Similarly when P is left multiplied to another matrix, the matrix's rows are permuted according to the permutation encoded in P. Also when $P^\top$ is right multiplied to another matrix, the matrix's columns are permuted as such. Finally, applying the permutation to the adjacency matrix we have, the obtained new matrix $PAP^\top$ represents the adjacency matrix after transforming the numbering of the nodes according to the permutation described by P.

A graph is called a *weighted graph* if for every edge $(v_i, v_j) \in E$ there is also an associated weight $w_{ij} \in \mathbb{R}$ with the edge. One can also associate weights with all elements of $V \times V$ and consider a weight of 0 as the absence of the edge. In this case the adjacency matrix can be enhanced to an *adjacency weight matrix* which we still denote as A, and have $A_{ij} = w_{ij}$. With this representation,

$$A_{ij} = \begin{cases} w_{ij} & \text{if } (v_i, v_j) \in E, \\ 0 & \text{otherwise.} \end{cases}$$

One example application of a weighted graph is a map of cities where edges between cities are roads and the weights are the distances in kilometers. Note that this is also a *planar graph* since it can be "drawn" on a plane. Not all graphs are planar. We also mention that a more general object than a graph is a *multi-graph* which permits multiple distinct edges (e.g., roads) between vertices. Some graph neural networks deal with multi-graphs, yet we do not go into this specialization in this section.

The definitions above associated with $G \in \mathcal{G}$ are general graph-theoretic concepts, not specific to graph neural networks. Yet in graph neural networks, there are also additional features, which we denote as elements of $\mathcal{F}$. Specifically, each individual node v_i can carry an associated feature vector $x_{(i)}$ which we generally consider an element of $\mathbb{R}^{p_V}$, for some $p_V \geq 1$. For example, in social networks applications where each individual node is a person, $x_{(i)}$ denotes the p_V features associated with the person such as age, marital status, etc. We call these features *node level features*.

Similarly to node level features, in certain cases we can also consider *edge level features*. Here we have the features $x_{(i \to j)}$ for edge (v_i, v_j), and we assume $x_{(i \to j)} \in \mathbb{R}^{p_E}$, for some $p_E \geq 1$. One can treat a weighted graph as a very simple case where $x_{(i \to j)} = w_{ij}$ and $p_E = 1$. However, in most cases that involve edge level features, $p_E > 1$. For example, in a transportation case where edges are roads (or transportation links) we may have each $x_{(i \to j)}$ represent multiple characteristics of the road such as the number of lanes, the speed limit, toll information, etc.

In this section's exposition of GNNs we generally ignore edge level features and focus on data with node level features. In such a case, the features can be organized in a $r \times p_V$ matrix X where each row represents the features of node v_i, namely $x_{(i)}$. That is, $\mathcal{F}$ is the set of all possible feature matrices[15] X. Note also that if we wish to apply a permutation to the node numbering as encoded in a permutation matrix P, then the permuted feature matrix is PX.

The Structure of Input Data and Tasks

Similarly to the rest of the book, we generally denote the data via $\mathcal{D}$. For GNNs this data can come in various forms:

$$
\mathcal{D} = \begin{cases} (G, X), & \text{for transductive learning (i)}, \\ \left\{ \left((G^{(1)}, X^{(1)}), y^{(1)} \right), \ldots, \left((G^{(n)}, X^{(n)}), y^{(n)} \right) \right\}, & \text{for inductive learning (ii)}, \\ \widetilde{X}_{(G,X)}, & \text{for graph embedding (iii)}. \end{cases}
$$

$$(8.87)$$

The forms (i) and (ii) in (8.87) are for the two overarching graph neural network learning paradigms, namely, *transductive learning* and *inductive learning*. In the case of transductive learning the data is one (big) input graph together with the features. In the case of inductive learning, the data consists of n different graphs (potentially each of them smaller), each with its own features as well, and also potential labels, $y^{(i)}$. See Figure 8.12 for an illustration of the difference between transductive and inductive learning.

To further understand the difference between transductive and inductive learning consider the following illustrative examples. For transductive learning, consider the social networks domain where we treat a big social network with some missing data as an input and then learn a model for predicting potential connections between individuals (predicting missing edges). As an inductive learning example consider classification of molecules where for example $y^{(i)} = 0$ implies a non-toxic molecule and $y^{(i)} = 1$ implies that molecule is toxic. The training data is then composed of many input graphs, some of which have $y^{(i)} = 0$ and some of which have $y^{(i)} = 1$. We then train a model that operates on an input graph

---

[15] Note that if one was to organize edge level features in such an object then a tensor of dimension $r \times r \times p_E$ would be appropriate.

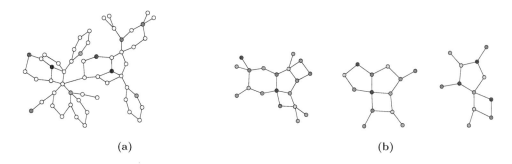

(a)   (b)

Figure 8.12: Transductive vs. inductive learning. (a) Transductive learning involves a single large input graph. (b) Inductive learning involves multiple separate input graphs.

(molecule structure) and outputs $\hat{y}$ which is the estimated probability of being toxic. It can then be transformed into a decision rule, as with any binary classifier.

Form (iii) of the data in (8.87) is different. In this case the data is no longer a graph but rather a transformation of graphical data into vector form using techniques called *graph embeddings*. Here, in a similar nature to word embeddings in the context of natural language processing (see Section 7.1), input graph data and features (G, X) are pre-processed to create a matrix $\widetilde{X}_{(G,X)}$ which summarizes the graph and the features via real-valued vectors. In contrast to the first two forms of data (i) and (ii) in (8.87), which are used as input to GNNs, the graph embedding form (iii) can be used as input to feedforward networks or other (non graphical) models trained on $\widetilde{X}_{(G,X)}$. Interestingly, some graph embedding techniques themselves are based on GNNs; we omit the details.[16]

Given graphical data of the form (8.87), one can train a model as in (8.84) for various different tasks. These can be dichotomized as *tasks on nodes*, *tasks on edges*, or *tasks on graphs*. In the first case the output is associated with nodes and our main goal is to predict outcomes, or impute missing features, for specific nodes in the graph. In the second case the output is associated with edges and our goal is to determine the presence of certain edges that were not originally available in the data, or similarly predict outcomes associated with available edges. In the third case the output is associated with the whole graph and in this case we predict properties of the graph, including classification of graphs, regression, and similar. See Figure 8.13 for an illustration.

Generally, tasks on graphs are carried out in an inductive setting as we require form (ii) of the training data from (8.87). In contrast, tasks on nodes or tasks on edges can be carried out both in an inductive and a transductive level.

The General Structure of a Graph Neural Network Model

In general, a GNN model is a function of graph data x from $\mathcal{D}$ of (8.87). Like (5.1) of Chapter 5 we construct the neural network $f_\theta(x)$ of (8.84) with a recursive computation of layers, $f_\theta(x) = f_{\theta^{[L]}}^{[L]}(f_{\theta^{[L-1]}}^{[L-1]}(\ldots(f_{\theta^{[1]}}^{[1]}(x))\ldots))$. Here $f_{\theta^{[\ell]}}^{[\ell]}(\cdot)$ is the ℓ-th layer, and $\theta^{[\ell]}$ are the parameters of that layer. As with other deep learning models, we denote $a^{[\ell]}$ as the result

---

[16]Various techniques for graph embedding include DeepWalk, node2vec, GraphSAGE, LINE (Large-scale Information Network Embedding), and HOPE (High-Order Proximity preserved Embedding).

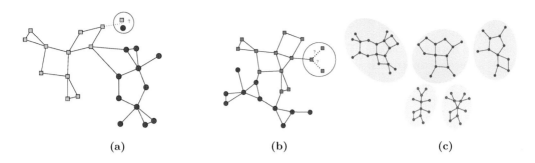

(a)	(b)	(c)

Figure 8.13: Various forms of tasks for graph neural networks. (a) Tasks on nodes deal with inference about individual nodes. For example classification of the type of node. (b) Tasks on edges deal with inference about individual edges. For example the existence of an edge or not. (c) Tasks on graphs deal with inference about the whole graph. For example classification of the graph.

of $f_{\theta^{[\ell]}}^{[\ell]}(a^{[\ell-1]})$ and $a^{[0]} = x$, keeping in mind that x contains both the graph structure and features.

Some models are *dynamic graph neural networks* in which case the graph structure is modified as it is processed by the neural network layers. Other models, are *static graph neural networks* for which $G = (V, E)$ is not modified across layers, and thus G has the same value within each layer. In such a case, it is convenient to represent $a^{[\ell]} = (h^{[\ell]}, G)$ where $h^{[\ell]}$ is sometimes called the *hidden state*. We set $h^{[0]}$ as the input features from x and denote the neural network function as $f_\theta(h^{[0]}, G)$. With this representation, the function of each layer can be denoted as $f_{\theta^{[\ell]}}^{[\ell]}(h^{[\ell-1]}; G)$. Note that G is fixed for all layers, but with each application of the complete $f_\theta(h^{[0]}, G)$, a different graph structure G is possible. As with feedforward neural network models, the parameters of the model are $\theta = (\theta^{[1]}, \ldots, \theta^{[L]})$.

Our focus in this exposition is only on static GNNs for which the forward pass[17] is

$$h^{[1]} = f_{\theta^{[1]}}^{[1]}(\overbrace{h^{[0]}}^{\text{Input features}}; G)$$

$$h^{[2]} = f_{\theta^{[2]}}^{[2]}(h^{[1]}; G)$$

$$\vdots$$

$$h^{[L-1]} = f_{\theta^{[L-1]}}^{[L-1]}(h^{[L-2]}; G)$$

$$\underbrace{\hat{y}}_{\text{output}} = f_{\theta^{[L]}}^{[L]}(h^{[L-1]}; G).$$

The specific structure of $h^{[\ell]}$ can vary between applications and may be a matrix, or a higher-dimensional tensor. In our case, for simplicity, let us assume it is a matrix of dimension $r \times p_V$ where the i-th row represents the node level features for node v_i in the graph, and there are r nodes in total. In general one may have different column dimensions (number of features) per layer ℓ, yet for simplicity let us assume that this is fixed as p_V throughout the layers.

---

[17]Compare with the feedforward network forward pass in (5.4).

An important requirement of the layer function $f_{\theta^{[\ell]}}^{[\ell]}(h^{[\ell]}\,;\,G)$ is *permutation invariance* where the order of nodes (and consequently, the order of entries in the adjacency matrix) does not affect the network's output. To understand this requirement, assume that we represent G via an adjacency matrix A, and hence the layer's operation can be represented with A in place of G, namely we can denote the layer's function as $f_{\theta^{[\ell]}}^{[\ell]}(h^{[\ell-1]}\,;\,A)$. Now for any permutation on the nodes represented via P as in (8.86), we require that,

$$f_{\theta^{[\ell]}}^{[\ell]}(\;\underbrace{Ph^{[\ell]}}_{\text{Permuted hidden state}}\;;\;\overbrace{PAP^{\top}}^{\text{Permuted adjacency matrix}}\;)\;=\;\underbrace{P\,f_{\theta^{[\ell]}}^{[\ell]}(h^{[\ell]}\,;\,A)}_{\text{Permuted hidden state at layer }\ell+1}. \qquad (8.88)$$

The permutation invariance requirement in (8.88) then ensures that node numbering is indeed arbitrary and does not does not affect the operation of the model.

Let us now dive into the structure of a single layer except for the final layer. Namely, let us see the structure of $f_{\theta^{[\ell]}}^{[\ell]}(h^{[\ell-1]}\,;\,G)$ for $\ell = 1, \ldots, L-1$. Here, similarly to the convolutional neural networks of Chapter 6, GNNs try to enforce *locality* and *translation invariance*; see Section 6.1. The translation invariance property is analogous to the permutation invariance property of (8.88). Locality is enforced by requiring that the output of $f_{\theta^{[\ell]}}^{[\ell]}(h^{[\ell-1]}\,;\,G)$ for node v_i only depends on the neighbors of v_i. Specifically in our data representation it means that only the rows with indices of neighbors $\mathcal{N}(v_i)$ as well as v_i itself are used to compute the i-th row of $h^{[\ell]}$. Permutation invariance is enforced by using the same form of function (and same parameters) for each target output node, and not considering the actual index of a node but rather only the graphical structure.

Specifically, if we denote the i-th row of $h^{[\ell]}$ via $h_{(i)}^{[\ell]}$, then the computation of the layer associated with node v_i can be represented via

$$h_{(i)}^{[\ell]} = f_{\text{node},\theta^{[\ell]}}^{[\ell]}\left(h_{(j)}^{[\ell-1]} \text{ for } v_j \in \mathcal{N}(v_i) \cup \{v_i\}\right), \qquad (8.89)$$

where the function $f_{\text{node},\theta^{[\ell]}}^{[\ell]}(\cdot)$ determines how the hidden state of a node in the next layer is determined by the hidden state of the node and its neighbors in the previous layer.

The final layer L is typically different because the output $\hat{y}$ is often not of the same dimension as $h^{[\ell]}$. In this case, we just retain the final layer action via the general function $f_{\theta^{[L]}}^{[L]}(\cdot)$.

Message Passing Schemes

The operation of (8.89) is based on a so-called *message passing* scheme where two steps called *aggregate* and *update*[18] are executed one after the other. These steps break up the operation of the $f_{\text{node},\theta^{[\ell]}}^{[\ell]}(\cdot)$ function via, a function $f_{\text{aggregate}}(\cdot)$ and a function $f_{\text{update},\theta^{[\ell]}}^{[\ell]}(\cdot)$

---

[18]Sometimes this is called *combine*.

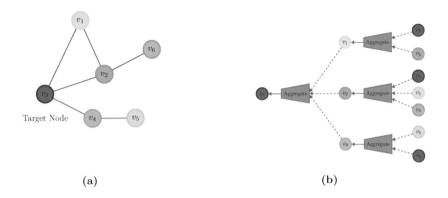

(a)   (b)

Figure 8.14: Aggregation in a message passing scheme. (a) An input graph where the messages for node v_3 are considered. (b) The application of the aggregation via multiple layers (2 layers in this case) yielding the message aggregated for the node. Observe the neighbor of v_3 in each layer.

and are executed as,

$$m_{(i)}^{[\ell]} = f_{\text{aggregate}} \left(h_{(j)}^{[\ell-1]} \text{ for } v_j \in \mathcal{N}(v_i) \cup \{v_i\} \right),$$

$$h_{(i)}^{[\ell]} = f_{\text{update},\theta^{[\ell]}}^{[\ell]} \left(h_{(i)}^{[\ell-1]}, m_{(i)}^{[\ell]} \right).$$

Observe that the aggregate function is the same for all layers and is not parameterized, while the parameters for layer ℓ, $\theta^{[\ell]}$ are associated only with the update function.

As is evident, the aggregation step utilizes all the hidden states of the neighbors of the node, say v_i, to achieve a summary, $m_{(i)}^{[\ell]}$, which we call the *message*. Note that in some cases it also uses node v_i itself (*self loops*) and in other cases not (*no self loops*). The message collects hidden state information of neighboring nodes.

In our exposition, we assume that the message for node i is an element of $\mathbb{R}^{p_V}$. Similarly to the hidden state, for simplicity, we assume this dimension, p_V, does not depend on the layer ℓ.

Typical aggregation functions are the sum, the mean, or the element-wise maximum. In our exposition we focus on the sum as an illustrative simple case where,

$$m_{(i)}^{[\ell]} = \begin{cases} \sum_{j\,:\,v_j \in \mathcal{N}(v_i)} h_{(j)}^{[\ell-1]}, & \text{for sum aggregation without self loops,} \\ \left(\sum_{j\,:\,v_j \in \mathcal{N}(v_i)} h_{(j)}^{[\ell-1]} \right) + h_{(i)}^{[\ell-1]}, & \text{for sum aggregation with self loops.} \end{cases} \quad (8.90)$$

The aggregate and update steps are carried out for all nodes in the graph and thus one may view this scheme as messages passing along neighbors within the graph. Note that this type of architecture is sometimes called a *message passing neural network (MPNN)*; see Figure 8.14 for an illustration.

Considering the whole graph, note that like the $r \times p_V$ hidden state matrix $h^{[\ell]}$, we can also consider an $r \times p_V$ message matrix $m^{[\ell]}$, where the i-th row is $m_{(i)}^{[\ell]}$. Now focusing on the unweighted case, using the graph's adjacency matrix A from (8.85), one can verify that the message matrix resulting from the sum aggregation (8.90) is

$$
m^{[\ell]} = \begin{cases} A\, h^{[\ell-1]}, & \text{for sum aggregation without self loops,} \\ \\ (A+I)\, h^{[\ell-1]}, & \text{for sum aggregation with self loops.} \end{cases} \tag{8.91}
$$

The update step is where a neural network approach is used. In this step, the hidden state of the node, $h_{(i)}^{[\ell-1]}$, and the message $m_{(i)}^{[\ell]}$ are used to determine the hidden state of the node in the output of the ℓ-th layer. The update function is typically composed of an affine transformation together with a non-linear activation. A simple typical form of $f_{\text{update},\theta^{[\ell]}}^{[\ell]}(\cdot)$ is

$$
h_{(i)}^{[\ell]} = S\left(m_{(i)}^{[\ell]} W_{\mathrm{m}}^{[\ell]} + h_{(i)}^{[\ell-1]} W_{\mathrm{u}}^{[\ell]} + b^{[\ell]} \right), \tag{8.92}
$$

where we treat the vectors as row vectors. Here $S(\cdot)$ is a vector activation function typically formed of scalar activation functions such as $\sigma_{\text{Sig}}(\cdot)$, similarly to other neural networks. With this form, the learned parameters $\theta^{[\ell]} = (W_{\mathrm{m}}^{[\ell]}, W_{\mathrm{u}}^{[\ell]}, b^{[\ell]})$ include weight matrices and a bias vector, and under our (simplifying) dimensionality assumptions, we have $W_{\mathrm{m}}^{[\ell]}, W_{\mathrm{u}}^{[\ell]} \in \mathbb{R}^{p_V \times p_V}$, and $b^{[\ell]} \in \mathbb{R}^{p_V}$ (considered as a row vector). Note that more generally, dimensions may vary across layers and hence the matrices may be non-square.

If we focus on sum aggregation without self loops, then (8.92) becomes

$$
h_{(i)}^{[\ell]} = S\left(\sum_{j:v_j \in \mathcal{N}(v_i)} \left(h_{(j)}^{[\ell-1]} W_{\mathrm{m}}^{[\ell]} \right) + h_{(i)}^{[\ell-1]} W_{\mathrm{u}}^{[\ell]} + b^{[\ell]} \right). \tag{8.93}
$$

Further, in this case, if we consider the whole graph, we can use (8.91) to represent (8.93) via a network wide equation,

$$
h^{[\ell]} = S\left(A h^{[\ell-1]} W_{\mathrm{m}}^{[\ell]} + h^{[\ell-1]} W_{\mathrm{u}}^{[\ell]} + B^{[\ell]} \right), \tag{8.94}
$$

where in similar nature to (5.10) of Chapter 5, $B^{[\ell]}$ is a $r \times p_V$ matrix with each row equal to the bias vector $b^{[\ell]}$. Note that here $S(\cdot)$ is taken as the activation function over the whole matrix, typically element-wise.

It is also of interest to note that in case where we restrict the parameters with $W_{\mathrm{m}}^{[\ell]} = W_{\mathrm{u}}^{[\ell]}$, both denoted as $W^{[\ell]}$, then (8.94) is reduced to

$$
h^{[\ell]} = S\left((A+I) h^{[\ell-1]} W^{[\ell]} + B^{[\ell]} \right). \tag{8.95}
$$

This case of $W^{[\ell]} = W_{\mathrm{m}}^{[\ell]} = W_{\mathrm{u}}^{[\ell]}$ yields a similar update rule to what we would have if we consider sum aggregation with self loops; see (8.91).

In practice, one can make a choice if to use the formulation with more parameters (8.94) or the less parameterized formulation (8.95). With the latter, there can be some restriction on the expressivity of the GNN as there is no separation of the information from the node and from its neighbors. Nevertheless in some cases, the less parameterized case, (8.95) suffices.

Model Variants

We close this section with a few model variants of GNNs. Each of the variants has its advantages and disadvantages, and the applicability of the variants for applications is beyond our scope. Our purpose here is simply to explore the various basic ideas and equations associated with each of the models. We consider graph convolutional networks, spectral approaches, and the use of the attention mechanism.

In a *graph convolutional network*, (8.95) is modified to

$$h^{[\ell]} = S\left(\tilde{D}^{-\frac{1}{2}} (A + I) \tilde{D}^{-\frac{1}{2}} h^{[\ell-1]} W^{[\ell]} + B^{[\ell]} \right), \tag{8.96}$$

where the matrix $\tilde{D}$ is a diagonal matrix and $\tilde{D}_{ii}$ is the degree of node v_i plus one. Thus $\tilde{D}^{-\frac{1}{2}}$ is a diagonal matrix with entries that are inverse of the square root of the degree plus one. At the node level, again representing vectors as row vectors, we can unpack (8.96) for node v_i as

$$h^{[\ell]}_{(i)} = S\left(\sum_{j:v_j \in \mathcal{N}(v_i)} \left(\frac{1}{\sqrt{\tilde{D}_{ii}\tilde{D}_{jj}}} h^{[\ell-1]}_{(j)} W^{[\ell]} \right) + \frac{1}{\tilde{D}_{ii}} h^{[\ell-1]}_{(i)} W^{[\ell]} + b^{[\ell]} \right). \tag{8.97}$$

If we compare (8.97) with (8.93), we see that a graph convolutional network is similar to sum aggregation, yet with weighting in the summation proportional to the degrees. Specifically, when considering the update for the hidden state $h^{[\ell]}_{(i)}$ of node v_i, the weight of the hidden state of neighboring nodes that have more neighbors than i is reduced, and conversely the weight for nodes that have less neighbors is increased. This scaling acts as a form of regularization in the network.

Graph convolutional networks can be extended with a *spectral* approach. Specifically let us now outline key ideas of *spectral graph neural networks*, also known as *spectral convolutional graph neural networks*. Generally, in the world of signal processing and mathematics, a spectral approach deals with analyzing a transform of a signal in place of the signal itself. For example in the context of time signals, as briefly discussed in Section 6.2, instead of the signal, one may sometimes consider the *Fourier transform* of the signal. Then based on the so-called *convolution theorem*, the Fourier transform of the convolution between two signals, as for example in equation (6.2) of Chapter 6, can be represented as the product of the Fourier transforms of the individual signals. [19]

In the context of graphs, an analogy to the convolution theorem can be considered using eigenvalue decompositions of matrices, which is the topic of study of an area called *spectral graph theory*. Let us focus on undirected graphs that are connected, and thus the degree of each node is at least one.

In spectral graph theory, an important matrix associated with a graph with r nodes is the $r \times r$ *Laplacian matrix*, appearing in either the unnormalized form, or the normalized form, and defined as

$$\mathcal{L} = \begin{cases} D - A & \text{(unnormalized)}, \\ I - D^{-\frac{1}{2}} A D^{-\frac{1}{2}} & \text{(normalized)}. \end{cases}$$

---

[19] Note that in Chapter 6 we actually do not discuss Fourier transforms, yet we point the reader there for general context of signals and systems.

Here, similarly to $\tilde{D}$ from above, D a diagonal matrix where this time D_{ii} is the degree of node v_i.

One may consider a vector $u^{[\ell]} \in \mathbb{R}^r$ which represents some state values for each node in the graph, and the operation

$$u^{[\ell+1]} = \mathcal{L} u^{[\ell]}, \tag{8.98}$$

can encompass the effect of applying the Laplacian matrix of the graph (either unnormalized or normalized) on the state $u^{[\ell]}$ to achieve the next state $u^{[\ell+1]}$. Interpretations of this operation for specific graph contexts are beyond our scope, yet we mention that in the context of random walks on graphs, electrical networks (Kirchhoff laws), and other contexts, this operation is common.

A key idea in spectral graph networks is to modify the operation (8.98) to

$$u^{[\ell+1]} = \tilde{\mathcal{L}} u^{[\ell]}, \tag{8.99}$$

where the modification from $\mathcal{L}$ to $\tilde{\mathcal{L}}$ is the essence of the learned parameters in the model and is further explained below. With this modification, it is useful to use the *spectral decomposition* of $\mathcal{L}$ and learn how to modify the eigenvalues of $\mathcal{L}$ effectively. This is analogous to learning how to modify filters, as is done in Chapter 6, yet unlike Chapter 6, learning is on the spectral domain (eigenvalues) and not on the time domain (convolutions).

Now looking at the spectral decomposition, since the graph is undirected, $\mathcal{L}$, either in the unnormalized or normalized form, is a symmetric matrix, and further it can be shown to be positive semidefinite. The spectral decomposition of $\mathcal{L}$ is

$$\mathcal{L} = U \Lambda U^\top, \tag{8.100}$$

where $U \in \mathbb{R}^{r \times r}$ has normalized eigenvectors of $\mathcal{L}$ as columns, and $\Lambda \in \mathbb{R}^{r \times r}$ is a diagonal matrix with corresponding eigenvalues, each of which is real and non-negative (due to being positive semidefinite). With this spectral decomposition, U is an orthogonal matrix, namely $U^\top U = I$ (the inverse of U is its transpose). Note that it is also customary to order the eigenvalues in the diagonal of Λ in descending order (and the associated eigenvectors in the columns of U are obviously ordered accordingly).

Now when we consider (8.98) and wish to modify it to (8.99), we do so by transforming the eigenvalues of $\mathcal{L}$, appearing in the diagonal of Λ. For example with so-called *high-pass filtering* we retain the high-valued eigenvalues (above some threshold), while shrinking or zeroing out the other eigenvalues. Similarly, *low-pass filtering* works in the other direction, retaining only low eigenvalues. In each such case, we may view the transformation of the eigenvalues as some function $F(\cdot)$ which applied to the diagonal matrix Λ is denoted as $F(\Lambda) \in \mathbb{R}^{r \times r}$. With this we have the modified Laplacian matrix as $\tilde{\mathcal{L}} = U F(\Lambda) U^\top$, and thus (8.99) is represented as

$$u^{[\ell+1]} = U F(\Lambda) U^\top u^{[\ell]}. \tag{8.101}$$

Note that we may view (8.101) as first projecting $u^{[\ell]}$ onto the orthogonal eigenvector space via the transformation $U^\top u^{[\ell]}$, then applying individual (adjusted via $F(\cdot)$) eigenvalues on each coordinate of the basis via the left multiplication by the diagonal matrix $F(\Lambda)$, and finally, transforming back to the original basis via another left multiplication by U.

Having understood how the spectral decomposition can be used, we now return to the general setup of GNNs where as before $h^{[\ell]} \in \mathbb{R}^{r \times p_V}$ is the hidden state matrix updated from layer ℓ to layer $\ell + 1$. Now also denote $h^{[\ell]}_{(|i)}$ as a vector in $\mathbb{R}^r$ which is the i-th column of this matrix. This vector represents the hidden state information for each node in the network, based on the i-th hidden state feature at layer ℓ. With this notation, the update for this hidden state vector per feature i depends on all features in the previous layer and is

$$h^{[\ell+1]}_{(|i)} = S\left(\left(\sum_{k=1}^{p_V} U \, F^{[\ell]}_{(k,i)} \, U^\top h^{[\ell]}_{(|k)}\right) + b^{[\ell]}_{(i)}\right), \qquad \text{for} \qquad i = 1, \ldots, p_V. \tag{8.102}$$

Here for each pair of features k and i, $F^{[\ell]}_{(k,i)}$ is an $r \times r$ diagonal matrix of learned parameters that we call *spectral weights*, and $b^{[\ell]}_{(i)}$ is a bias vector in $\mathbb{R}^r$. As always, $S(\cdot)$ is a vector activation function. For one such layer ℓ, we can observe that the total number of learned parameters for spectral weights is $r \times (p_V)^2$ and the total number of learned parameters for the bias vectors is $r \times p_V$.

To get a better feel for the update equation (8.102), compare it with (8.101). In (8.101) we see that $F(\Lambda)$ is a filter applied to the eigenvalues whereas in (8.102) we represent a learned $F(\Lambda)$ via $F^{[\ell]}_{(k,i)}$. Similarly to the concept of channels in convolutional neural networks of Chapter 6, the summation over all updates $U \, F^{[\ell]}_{(k,i)} \, U^\top h^{[\ell]}_{(|k)}$ integrates all the features from layer ℓ into the associated feature of layer $\ell + 1$.

The story of spectral GNNs does not stop here because ideally one would like to reduce the dimension of the learned parameters $F^{[\ell]}_{(k,i)}$ so that they do not depend on the whole graph and are independent of the number of nodes in the graph r. The study and application of spectral GNNs deals with such approaches and indeed in certain cases it has been shown that spectral graph neural networks generalize well and sometimes perform better than their convolutional graph neural networks counter parts. The details are beyond our scope.

A different variant is *graph attention networks* where the aggregation step uses an attention mechanism. Concepts of attention were heavily discussed in Chapter 7 in the context of sequence models. Specifically, in Section 7.4 we introduced the general concept of attention where attention weights are calculated as in (7.20) and are then used for linear combinations of inputs in (7.21). We then also used attention in the context of transformers as in Section 7.5.

In the graph context, attention can be incorporated via the aggregation step. Specifically, we can enhance the basic sum aggregation as in (8.90), focusing here only on the case without self loops, to be,

$$m^{[\ell]}_{(i)} = \sum_{j \, : \, v_j \in \mathcal{N}(v_i)} \alpha^{[\ell]}_{(i),j} \, h^{[\ell-1]}_{(j)}. \tag{8.103}$$

Here for node v_i, and each neighbor $v_j \in \mathcal{N}(v_i)$, we have attention weight $\alpha^{[\ell]}_{(i),j}$ where

$$\sum_{j \, : \, v_j \in \mathcal{N}(v_i)} \alpha^{[\ell]}_{(i),j} = 1.$$

Similarly to Chapter 7, attention weights are calculated using a scoring mechanism via an alignment function $s : \mathbb{R}^{p_V} \times \mathbb{R}^{p_V} \to \mathbb{R}$, where $s(h^{[\ell-1]}_{(i_1)}, h^{[\ell-1]}_{(i_2)})$ measures the proximity between the hidden states of nodes v_{i_1} and v_{i_2} at layer $\ell - 1$. The basic alignment function is the inner product between $h_{(i_1)}^{[\ell-1]}$ and $h_{(i_2)}^{[\ell-1]}$, yet more complex options with learned parameters are also possible. One option with linear re-weighting is

$$s(h_{(i_1)}^{[\ell-1]}, h_{(i_2)}^{[\ell-1]}) = \left(h_{(i_1)}^{[\ell-1]} W_a^{[\ell]}\right)\left(h_{(i_2)}^{[\ell-1]} W_a^{[\ell]}\right)^{\top}, \tag{8.104}$$

where $W_a^{[\ell]} \in \mathbb{R}^{p_V \times m'}$ is a matrix of learned parameters for layer ℓ with m' set as some dimension (recall here that our hidden state vectors are taken as rows).

Based on the alignment function, the attention weights are calculated for each node v_i. As with all attention mechanisms, we use a softmax function over neighbors indices j to obtain the attention weights

$$\alpha_{(i),j}^{[\ell]} = \frac{e^{s(h_{(i)}^{[\ell-1]}, h_{(j)}^{[\ell-1]})}}{\sum_{k \,:\, v_k \in \mathcal{N}(v_i)} e^{s(h_{(i)}^{[\ell-1]}, h_{(k)}^{[\ell-1]})}},$$

which are then used in the aggregation step (8.103). Once the message is computed, it can be used in an update function as in (8.92).

Note that with our description of graph attention networks here, the learned parameters per layer are $W_a^{[\ell]}$ for the alignment function (8.104) as well as $\theta^{[\ell]} = (W_{\mathrm{m}}^{[\ell]}, W_{\mathrm{u}}^{[\ell]}, b^{[\ell]})$ used in the update (8.92). However, other options, typically with reduced parameters, reusing weight matrices either across layers, or between the alignment function and the update equation, are also possible. Also, similar to the transformer architecture, multi-head attention has also been introduced and in certain GNN applications, such architectures are very popular.

Notes and References

This chapter covered a broad range of specialized architectures and paradigms where each section covers a major topic which could have in fact made a whole chapter. Hence in our notes and references about the topics of this chapter we only summarize key references and developments in each of the sub-fields. A further recent overarching text that we recommend is [336] with multiple chapters, one per each of the topics covered here.

The field of generative modeling has multiple origins. Early models include *hidden Markov models* and *Gaussian mixture models* with origins in the 1950s and 1960s; see chapters 11 and 17 of [298] for background. Somewhat more recently, some authors consider the study of *Boltzman machine* models introduced in the 1980s in [2], and deep Boltzman machines in [361], as the initial meaningful generative models in the context of deep learning. See also chapter 20 of [142] for an overview. A more recent survey of generative models in machine learning is [162] and a comparison of deep generative modeling approaches is in [46].

Up to 2014, while generative models were useful for some applications and certainly interesting, in terms of images, they lacked the ability to create real life looking data. The big advance came with the development of *generative adversarial networks* (GANs), in Goodfellow et al.'s work [143]. This opened up possibilities for creation of realistic looking images (and data) and is still a very active topic. *Variational autoencoders*, initially introduced in [234], grew into multiple directions and contemporary *diffusion models* such as [183], and those surveyed in [430], constitute the state of the art in image generative modeling. As of the time of publishing of this book, diffusion models and GANs still compete, with diffusion models generally able to produce more impressive images, while GANs are much faster in production since they do not require multiple neural networks.

Ideas of variational autoencoders are rooted in modern developments of *Bayesian statistics*. See [92] for an introductory general text on Bayesian statistics and [407] for an accessible review of the area. Specifically, the *variational Bayes* methods, a well-known optimization-based approach in the field of *approximate Bayesian computation*, captures the key ideas used in variational autoencoders. See [41] and [444] for reviews of variational Bayes. This approach also falls in the realm of *approximate Bayesian computation* and entails a method for approximating posterior distributions using simpler surrogate distributions. See [381] for a collection of approximate Bayesian computation methods. Specifically, for more details about variational autoencoders, see [235].

Our presentation of variational autoencoders was geared toward *hierarchical Markovian variational autoencoders* of which diffusion models are a special case. Nevertheless, variational autoencoders and their variants are interesting and useful in their own right. They have been applied to many fields. In image processing, prediction of the trajectory of pixels of an image is tackled in [414] and natural image modeling is in [152]. In the field of speech analysis, voice conversion is handled in [191] and speech synthesis in [8]. In the area of text processing as in [54], reccurent neural network-based variational autoencoders for generating sentences are put forward and in [193], controlled text generation is handled. Another field is graph-based data analysis as in [236] where learning on graph-structured data is handled, and [210] which deals with molecular graph generation. As we presented diffusion models as special cases of *hierarchical variational autoencoders*, the literature on these models is also relevant. In particular, see [343] for an application in black box variational inference and [385] for a variant called ladder variational autoencoder.

Diffusion models, initially introduced in [384], gained significant prominence following [183], which showcased exceptional image synthesis results. These models were further improved with [104], where for the first time the prolonged dominance of GANs was broken. For recent surveys of diffusion models, refer to [71], [96], and [430]. In terms of applications in the realm of image processing, diffusion models are utilized for tasks such as *colorization, inpainting, uncropping,* and *restoration* as in [358]. Other image processing applications include super-resolution as in [360], and image editing as in [176]. There is an extensive study on applications of diffusion models for *text to image generation* such as for example the work in [359] which introduced *Imagen*. This system utilizes a transformer-based large language model which is used for understanding text combined with a diffusion model used for image generation. The application of diffusion models extends to video data as well. Notable contributions include [163] where an approach for long-duration video completions is put forward, and [182] which introduced *Imagen Video*, a text-conditional video generation system based on a cascade of video diffusion models.

A related paradigm to variational autoencoders and diffusion models that we did not cover is *normalizing flows*. This paradigm was first introduced in [347] for representing the posterior in variational autoencoders. These generative models construct complex distributions by transforming a distribution through a series of invertible mappings. See chapter 16 of [336] for an overview and [322] for a recent survey.

As already mentioned, GANs were introduced in [143]. After the introduction of this paradigm, multiple generalizations appeared. Particular early variants included *C-GAN* [293], and convolutional versions of GANs as in [339]. The idea of *NS-GAN* was already introduced in [143]. The *W-GAN* concept first appeared in [15] and was later developed in [151] where the gradient penalty approach was introduced. See also [3] as well as the empirical comparison in [271] and the general GAN surveys [150] and [200]. The ideas of *AC-GAN* were developed in [316] and the ideas of *Info-GAN* were developed in [78]. The image to image paradigm in GANs is broad. A survey on this topic is [11] with initial ideas in [83], [204], and [417]. *Style-GAN* was developed in [224], and further developments are in [225] and [223]. Finally we mention a few general developments in GANs including *sequence GAN* from [437], and *EditGAN* from [264].

Modern approaches to deep reinforcement learning are surveyed extensively in the texts [125] and [393], whereas more dated accounts are [36] and [217]. At the more fundamental level, basics of Markov decision processes are presented in [33], [34], and [337]. An expository overview of engineering control theory is in [9]. The criterion for convergence of *Q-learning* as in our equation (8.81) is from [419]. The success of reinforcement learning in playing Atari video games is in [294], and later landmark results in the game of Go are documented in [377]. Apart from games, reinforcement learning is successfully used in several domains one of which addresses hard combinatorial problems; see [282] for a survey of such approaches. Nowadays, reinforcement learning is applied for fine tuning large language models through human feedback as documented in [318] and [429]. Reinforcement learning is also critical for some robotic tasks as described in the survey [448]. However, as one would have thought initially that self driving cars are best handled as a system via reinforcement learning, in practice other techniques have so far prevailed; see [79] for a general survey. A related approach often used is *imitation learning*, see [286] for a survey of this technique in the context of autonomous vehicles.

General texts about graph neural networks are [273], and [157], with recent survey papers in [413] and [427] as well as a notable chapter in [336]. Early ideas in graph neural networks arose in [387] where a concept called at the time the *generalized recursive neuron* was introduced. Further ideas of graph neural networks were developed in [144] and [365]. These days graph neural networks are used in multiple applications such as those discussed in [242], [445], and [451] among many others. There are various techniques for graph embedding including *DeepWalk* [329], *node2vec* [148], *GraphSAGE* [156], *LINE* (Large-scale Information Network Embedding) [397], and *HOPE* (High-order Proximity preserved Embedding), [317]. For an introductory computer science overview of traditional graph algorithms and data structures see [93].

General ideas of *message passing schemes* in graph neural networks were introduced in [287]. Ideas of *graph convolutional networks* were developed in [18], [134], and [309]. Ideas of *spectral convolutional graph neural networks* were developed in [67] and [174]. *Graph attention networks* were developed in [411], motivated by the effective application of the attention mechanism to sequence models as in [410]. Indeed many developments in deep learning cross architectures, domains, and sub-disciplines, and the development of graph attention networks serve as one such example. We also close by mentioning *spatial-temporal graph neural networks* which are designed to deal with *dynamic graphs*, sometimes also with a spatial component. These models were introduced in [257] and [371], and are further described in the recent review [209].

Epilogue

Our story was about the **mathematical engineering of deep learning**. Our goal was to describe deep learning ideas in simple mathematical terms. Our goal was not to study implementation of deep learning; it was not to discuss the history and evolution of deep learning; and it was not to dive into subtle mathematical properties of deep learning. We simply wanted to present a basic **mathematical** description, empowering the reader with an understanding of key concepts and terminology. Mathematics is a language of choice.

We focused on the most popular and successful **deep learning** architectures and ideas that emerged over recent years. Somewhat anti-climatically we claim that the popularity and success of these ideas is due to their practical applicability, and not so much due to mathematical elegance. There are many other variants that we did not present here which are interesting and elegant yet have not been as popular from a practical perspective. With this we note that the aspect of **engineering** focusing on the empirical evaluation of architectures was not discussed and studied in the book at all.

Take as an example the *transformer architecture* studied in Section 7.5. This architecture has been pivotal in *large language models*. Indeed, in the same years that we worked on writing this book, 2021–2023, large language models, almost exclusively powered by the transformer architecture, have risen in popularity. Yet it is fair to say that the transformer architecture is quite arbitrary. If a couple of years prior to the development of this architecture, published in 2017 with [410], we the authors would have been presented with a transformer without empirical trials and experimentation results, we would have no proof that transformers work so well.

It is also important to note that the pace and unpredictability of deep learning developments moves fast. By now, large language models have effectively beaten the Turing test, [38], a goal which seemed yet unattainable in the days when we conceived this book in late 2020. So our humble claim is that while **mathematical engineering** is important, in its own right, without computers, GPUs, software, data, and experimentation, it is void of substance. Nevertheless, we do believe that our presentation approach is succinct and unique, and given that the ideas that we present were previously shown to be winning ideas, the knowledge that you gained by reading this book will be beneficial.

Finally we close by mentioning that while this is a mathematical book, one cannot ignore the vast area of ethical issues associated with deep learning and artificial intelligence. Now, as we are in the third decade of the 21st century, artificial intelligence is at the center of discussions associated with politics, freedom, social justice, violence, equity, and many other domains. Since this book is not about applications, we as authors had the luxury of ignoring the many ethical issues associated with deep learning in our exposition. Nevertheless, any practitioner using deep learning should at onset make sure to consider what defines responsible use and what not. We certainly want the technology to be used for purposes that do good rather than bad.

A Some Multivariable Calculus

This appendix provides key results and notation from multivariable calculus. It is not an exhaustive summary of multi-variable calculus but rather contains the results needed for the contents of the book.

A.1 Vectors and Functions in $\mathbb{R}^n$

Denote the set of all the real numbers by $\mathbb{R}$ and the real coordinate space of dimension n by $\mathbb{R}^n$. Each element of $\mathbb{R}^n$ is an n-dimensional vector, interpreted as a column of the form

$$u = (u_1, \ldots, u_n) = [u_1 \ \cdots \ u_n]^\top = \begin{bmatrix} u_1 \\ \vdots \\ u_n \end{bmatrix}.$$

The *Euclidean norm* of $u \in \mathbb{R}^n$, measuring the geometric length of u and also known as the L_2 norm, is

$$\|u\|_2 = \sqrt{u^\top u} = \left(\sum_{i=1}^n u_i^2 \right)^{1/2}.$$

Here the scalar $u^\top v$ is the *inner product* between two vectors $u, v \in \mathbb{R}^n$. A normalized form of the inner product, called the *cosine of the angle* between the two vectors, sometimes simply denoted $\cos \theta$, is

$$\cos \theta = \frac{u^\top v}{\|u\|_2 \|v\|_2}. \tag{A.1}$$

The Euclidean norm is a special case of the L_p norm which is defined via

$$\|u\|_p = \left(\sum_{i=1}^n |u_i|^p \right)^{1/p},$$

for $p \geq 1$. When p in $\| \cdot \|_p$ is not specified, we interpret $\| \cdot \|$ as the L_2 norm.

Focusing on the L_2 norm and the inner product $u^\top v$, the *Cauchy-Schwartz* inequality is

$$|u^\top v| \leq \|u\|\|v\|, \tag{A.2}$$

where the two sides are equal if and only if u and v are linearly dependent (that is, $u = c\,v$ for some $c \in \mathbb{R}$). Also, the *Euclidean distance* (or, simply the *distance*) between u and v is

DOI: 10.1201/9781003298687-A

defined as

$$\|u - v\| = \left(\sum_{i=1}^{n} (u_i - v_i)^2 \right)^{1/2}.$$

An important consequence of the Cauchy-Schwartz inequality is that the Euclidean norm satisfies the *triangle inequality*: For any $u, v \in \mathbb{R}^n$,

$$\|u + v\| \leq \|u\| + \|v\|. \tag{A.3}$$

To see this, observe that

$$\begin{aligned}
\|u + v\|^2 &= \|u\|^2 + \|v\|^2 + 2u^\top v \\
&\leq \|u\|^2 + \|v\|^2 + 2\|u\|\|v\| \\
&= (\|u\| + \|v\|)^2.
\end{aligned}$$

Convergence of a sequence of vectors can be defined via scalar converge of the distance. That is, a sequence of vectors $u^{(1)}, u^{(2)}, \ldots$ in $\mathbb{R}^n$ is said to *converge* to a vector $u \in \mathbb{R}^n$, denoted via $\lim_{k \to \infty} u^{(k)} = u$, if

$$\lim_{k \to \infty} \|u^{(k)} - u\| = 0.$$

That is, if for every $\varepsilon > 0$, there exists an N_0 such that for all $k \geq N_0$,

$$\|u^{(k)} - u\| < \varepsilon.$$

Let $f : \mathbb{R}^n \to \mathbb{R}$ be an n-dimensional multivariate function that maps each vector $u = (u_1, \ldots, u_n)^\top \in \mathbb{R}^n$ to a real number. Then, the function is said to be *continuous* at $u \in \mathbb{R}^n$ if for any sequence $u^{(1)}, u^{(2)}, \ldots$ such that $\lim_{k \to \infty} u^{(k)} = u$, we have that

$$\lim_{k \to \infty} f(u^{(k)}) = f(u).$$

Alternatively, f is continuous at $u \in \mathbb{R}^n$ if for every $\varepsilon > 0$ there exists $\delta > 0$ such that

$$|f(u) - f(v)| < \varepsilon,$$

for every $v \in \mathbb{R}^n$ with $\|u - v\| < \delta$. Continuity of f at u implies that the values of f at u and at v can be made arbitrarily close by setting the point v to be arbitrarily close to u.

We can extend the above continuity definitions to multivariate vector valued functions of the form $f : \mathbb{R}^n \to \mathbb{R}^m$ that map every n-dimensional real-valued vector to an m-dimensional real-valued vector. Such functions can be written as

$$f(u) = [f_1(u) \quad \cdots \quad f_m(u)]^\top, \tag{A.4}$$

where $f_i : \mathbb{R}^n \to \mathbb{R}$ for each $i = 1, \ldots, m$. Then, the function f is called continuous at u if each f_i is continuous at u. We say that the function f is *continuous on* a set $\mathcal{U} \subseteq \mathbb{R}^n$ if f is continuous at *each point* in $\mathcal{U}$.

A.2 Derivatives

Consider an n-dimensional multivariate function $f : \mathbb{R}^n \rightarrow \mathbb{R}$. The *partial derivative* $\frac{\partial f(u)}{\partial u_i}$ of f with respect u_i is the derivative taken with respect to the variable u_i while keeping all other variables constant. That is

$$\frac{\partial f(u)}{\partial u_i} = \lim_{h \rightarrow 0} \frac{f(u_1, \ldots, u_{i-1}, u_i + h, u_{i+1}, \ldots, u_n) - f(u)}{h}. \tag{A.5}$$

Suppose that the partial derivative (A.5) exists for all $i = 1, \ldots, n$. Then the *gradient* of f at u, denoted by $\nabla f(u)$ or $\frac{\partial f(u)}{\partial u}$, is a concatenation of the partial derivatives of f with respect to all its variables, and it is expressed as a vector:

$$\nabla f(u) = \frac{\partial f(u)}{\partial u} = \left[\frac{\partial f(u)}{\partial u_1} \quad \cdots \quad \frac{\partial f(u)}{\partial u_n} \right]^{\top}. \tag{A.6}$$

The gradient $\nabla f(u)$ is a vector capturing the direction of the steepest ascent at u. Further, $h \|\nabla f(u)\|$ is the increase in f when moving in that direction for infinitesimal distance h.

In some situations, instead of a vector form, variables of the function are represented as a matrix. In that scenario, multivariate functions are of form $f : \mathbb{R}^n \times \mathbb{R}^m \rightarrow \mathbb{R}$, that is, f maps matrices $U = (u_{i,j})$ of dimension $n \times m$ to real values $f(U)$. If the partial derivative $\frac{\partial f(U)}{\partial u_{i,j}}$ exists for all $i = 1, \ldots, n$ and $j = 1, \ldots, m$, it is convenient to use the notation $\frac{\partial f(U)}{\partial U}$ to denote the collection of the partial derivatives of f with respect to all its variables as a matrix of the same dimension $n \times m$,

$$\frac{\partial f(U)}{\partial U} = \begin{bmatrix} \frac{\partial f(U)}{\partial u_{1,1}} & \cdots & \frac{\partial f(U)}{\partial u_{1,m}} \\ \vdots & \ddots & \vdots \\ \frac{\partial f(U)}{\partial u_{n,1}} & \cdots & \frac{\partial f(U)}{\partial u_{n,m}} \end{bmatrix}. \tag{A.7}$$

Directional Derivatives

The *directional derivative* of $f : \mathbb{R}^n \rightarrow \mathbb{R}$ at u in the direction $v \in \mathbb{R}^n$ is the scalar defined by

$$\nabla_v f(u) = \lim_{h \rightarrow 0} \frac{f(u + hv) - f(u)}{h}.$$

The directional derivative generalizes the notion of the partial derivative. In fact, the partial derivative $\frac{\partial f(u)}{\partial u_i}$ is the directional derivative at u in the direction of the vector e_i which consists of 1 at the i-th coordinate and zeros everywhere else. This simply follows from the observation that

$$\nabla_{e_i} f(u) = \lim_{h \rightarrow 0} \frac{f(u_1, \ldots, u_{i-1}, u_i + h, u_{i+1}, \ldots, u_n) - f(u)}{h} = \frac{\partial f(u)}{\partial u_i}.$$

As consequence, if the gradient of f exists at u, the directional derivative exists in every direction v and we have

$$\nabla_v f(u) = v^\top \nabla f(u). \tag{A.8}$$

One way to see (A.8) in the case of continuity of the partial derivatives is via a Taylor's theorem-based first-order approximation (see Theorem A.1):

$$f(u + hv) = f(u) + (h\,v)^\top \nabla f(u) + O(h^2),$$

where $O(h^k)$ denotes a function such that $O(h^k)/h^k$ goes to a constant as $h \to 0$. Thus,

$$\frac{f(u + hv) - f(u)}{h} = v^\top \nabla f(u) + O(h).$$

Now take the limit $h \to 0$ on both the sides to get (A.8).

It is useful to note that the directional derivative $\nabla_v f(u)$ is maximum in the direction of the gradient in the sense that for all unit length vectors v, the choice $v = \nabla f(u)/\|\nabla f(u)\|$ maximizes $\|\nabla_v f(u)\|$. This is a consequence of the Cauchy-Schwartz inequality (A.2):

$$|\nabla_v f(u)| = |v^\top \nabla f(u)| \le \|v\| \|\nabla f(u)\| = \|\nabla f(u)\|.$$

Setting $v = \nabla f(u)/\|\nabla f(u)\|$ achieves the equality.

Jacobians

The Jacobian is useful for functions of the form $f : \mathbb{R}^n \to \mathbb{R}^m$ as in (A.4) where each f_i is a real-valued function of u. The *Jacobian* of f at u, denoted by J_f, is the $m \times n$ matrix defined via

$$J_f(u) = \begin{bmatrix} \frac{\partial f_1(u)}{\partial u_1} & \cdots & \frac{\partial f_1(u)}{\partial u_n} \\ \vdots & \ddots & \vdots \\ \frac{\partial f_m(u)}{\partial u_1} & \cdots & \frac{\partial f_m(u)}{\partial u_n} \end{bmatrix}. \tag{A.9}$$

In other words, the i-th row of the Jacobian is the gradient $\nabla f_i(u)$. In some situations, it is convenient to use the notation $\frac{\partial f(u)}{\partial u}$ to denote the transpose of the Jacobian of f at u. That is

$$\frac{\partial f(u)}{\partial u} = (J_f(u))^\top. \tag{A.10}$$

Hessians

Returning to functions of the form $f : \mathbb{R}^n \to \mathbb{R}$, to describe the curvature of the function f at a given $u \in \mathbb{R}^n$, it is important to consider the second-order partial derivatives at u.

These partial derivatives are arranged as an $n \times n$ matrix, called the *Hessian* and defined by

$$\nabla^2 f(u) = \frac{\partial \nabla f(u)}{\partial u} = \begin{bmatrix} \frac{\partial^2 f}{\partial u_1^2} & \frac{\partial^2 f}{\partial u_1 \partial u_2} & \cdots & \frac{\partial^2 f}{\partial u_1 \partial u_n} \\ \frac{\partial^2 f}{\partial u_2 \partial u_1} & \frac{\partial^2 f}{\partial u_2^2} & \cdots & \frac{\partial^2 f}{\partial u_2 \partial u_n} \\ \vdots & \vdots & \ddots & \vdots \\ \frac{\partial^2 f}{\partial u_n \partial u_1} & \frac{\partial^2 f}{\partial u_n \partial u_2} & \cdots & \frac{\partial^2 f}{\partial u_n^2} \end{bmatrix}, \tag{A.11}$$

where $\frac{\partial^2 f}{\partial u_i \partial u_j} = \frac{\partial f}{\partial u_i}\left(\frac{\partial f}{\partial u_j}\right)$. Note that if all the second-order partial derivatives are continuous at u, then the Hessian $\nabla^2 f(u)$ is a symmetric matrix. That is, for all $i, j \in \{1, \dots, n\}$,

$$\frac{\partial^2 f}{\partial u_i \partial u_j} = \frac{\partial^2 f}{\partial u_j \partial u_i}.$$

This result is known as *Schwarz's theorem* or *Clairaut's theorem*. Observe that using the Jacobian, we can treat the Hessian as the Jacobian of the gradient vector. That is,

$$\nabla^2 f(u) = J_{\nabla f}(u).$$

Certain attributes of optimization problems are often defined via *positive (semi) definiteness* of the Hessian $\nabla^2 f(\theta)$ at θ. In particular, a symmetric matrix A is said to be *positive semidefinite* if for all $\phi \in \mathbb{R}^d$,

$$\phi^\top A \phi \geq 0. \tag{A.12}$$

Furthermore, A is said to be *positive definite* if the inequality in (A.12) is strict for all $\phi \in \mathbb{R}^d \setminus \{0\}$. Note that the matrix A is called is *negative semidefinite* (respectively, *negative definite*) when $-A$ is positive semidefinite (respectively, positive definite).

Differentiability

A multivariate vector valued function $f : \mathbb{R}^n \to \mathbb{R}^m$ is said to be *differentiable* at $u \in \mathbb{R}^n$ if there is an $m \times n$ dimensional matrix A such that

$$\lim_{v \to u} \left(\frac{\|f(u) - f(v) - A(u - v)\|}{\|u - v\|} \right) = 0.$$

Here the limit notation $v \to u$ implies that the limit exists for every sequence $\{v^{(k)} : k \geq 1\}$ such that $\lim_{k \to \infty} v^{(k)} = u$. The matrix A is called the *derivative*. If the function f is differentiable at u, then the derivative at u is equal to the Jacobian $J_f(u)$. In particular, if f is a real-valued function (that is, $m = 1$) and differentiable at u, then the derivative at u is $\nabla f(u)^\top$. If the derivative is continuous on a set $\mathcal{U} \subseteq \mathbb{R}^n$, we say that f is *continuously differentiable* on $\mathcal{U}$, and in that case all the partial derivatives $\frac{\partial f(u)}{\partial u_i}$ are continuous on $\mathcal{U}$.

A.3 The Multivariable Chain Rule

Consider a multivariate vector valued function $h : \mathbb{R}^n \to \mathbb{R}^k$ and a multivariate real-valued function $g : \mathbb{R}^k \to \mathbb{R}$. Suppose that h is differentiable at $u \in \mathbb{R}^n$ and g is differentiable at $h(u) = [h_1(u) \cdots h_k(u)]^\top$. Let $f : \mathbb{R}^n \to \mathbb{R}$ be the composition $f = g \circ h$ or $f(u) = g(h(u))$. For each $i = 1, \ldots, n$, the *multivariate chain rule* is

$$\frac{\partial f(u)}{\partial u_i} = \frac{\partial g(h(u))}{\partial v_1} \frac{\partial h_1(u)}{\partial u_i} + \cdots + \frac{\partial g(h(u))}{\partial v_k} \frac{\partial h_k(u)}{\partial u_i},$$

where $\frac{\partial g}{\partial v_i}$ denotes the partial derivative of g with respect to the i-th coordinate. Thus,

$$\frac{\partial f(u)}{\partial u_i} = \begin{bmatrix} \frac{\partial h_1(u)}{\partial u_i} & \cdots & \frac{\partial h_k(u)}{\partial u_i} \end{bmatrix} \nabla g(h(u)),$$

and combining for all $i = 1, \ldots, n$,

$$\nabla f(u) = J_h(u)^\top \nabla g(h(u)).$$

Now consider the case where g is also a multivariate vector valued function. That is, suppose $h : \mathbb{R}^n \to \mathbb{R}^k$ is differentiable at u and $g : \mathbb{R}^k \to \mathbb{R}^m$ is differentiable at $h(u)$. Then the composition $f = g \circ h : \mathbb{R}^n \to \mathbb{R}^m$ is a vector valued function with Jacobian,

$$J_f(u) = J_g(h(u)) J_h(u). \tag{A.13}$$

The expression in (A.13) is called the multivariable chain rule. In terms of the notation (A.10), we may represent the multivariable chain rule as

$$\left[\frac{\partial f}{\partial u}\right]^\top = \left[\frac{\partial g}{\partial h}\right]^\top \left[\frac{\partial h}{\partial u}\right]^\top, \quad \text{or} \quad \frac{\partial f}{\partial u} = \frac{\partial h}{\partial u} \frac{\partial g}{\partial h}. \tag{A.14}$$

The Chain Rule for a Matrix Derivative of an Affine Transformation

Let us focus on the case $y = g(h(u))$ where $h : \mathbb{R}^n \to \mathbb{R}^k$ and $g : \mathbb{R}^k \to \mathbb{R}$. Specifically let us assume that $h(\cdot)$ is the affine function $h(u) = Wu + b$ where $W \in \mathbb{R}^{k \times n}$ and $b \in \mathbb{R}^k$. That is,

$$y = g(z), \quad \text{with} \quad z = Wu + b.$$

We are often interested in the derivative of the scalar output y with respect to the matrix $W = [w_{i,j}]$. This is denoted via $\frac{\partial y}{\partial W}$ as in (A.7).

It turns out that we can represent this matrix derivative as the outer product

$$\frac{\partial y}{\partial W} = \frac{\partial y}{\partial z} u^\top, \tag{A.15}$$

where $\frac{\partial y}{\partial z}$ is the gradient of $g(\cdot)$ evaluated at z.

To see (A.15) denote the columns of W via $w_{(1)}, \ldots, w_{(n)}$, each an element of $\mathbb{R}^k$, and observe that

$$z = b + \sum_{i=1}^{n} u_i w_{(i)}.$$

We may now observe that Jacobian transposed $\partial z/\partial w_{(i)}$ is $u_i I$, where I is the $k \times k$ identity matrix. Hence now, using (A.13), we have

$$\frac{\partial y}{\partial w_{(i)}} = \frac{\partial z}{\partial w_{(i)}}\frac{\partial y}{\partial z} = u_i\frac{\partial y}{\partial z}.$$

Now we can construct $\frac{\partial y}{\partial W}$, column by column as

$$\frac{\partial y}{\partial W} = \left[\frac{\partial y}{\partial w_{(1)}} \cdots \frac{\partial y}{\partial w_{(n)}}\right] = \left[u_1\frac{\partial y}{\partial z} \cdots u_n\frac{\partial y}{\partial z}\right] = \frac{\partial y}{\partial z}u^\top.$$

Jacobian Vector Products and Vector Jacobian Products

Let $f = (f_1,\ldots,f_m) = h_L \circ h_{L-1} \circ \cdots \circ h_1$ be a composition of L differentiable functions $h_1, h_2,\ldots,h_L$ such that $h_\ell : \mathbb{R}^{m_{\ell-1}} \to \mathbb{R}^{m_\ell}$ where $m_0, m_1,\ldots,m_L$ are positive integers with $m_0 = n$ and $m_L = m$.

Further, to simplify the notation, for each $\ell = 1,\ldots,L$, let

$$g_\ell(u) = h_\ell\left(h_{\ell-1}\left(\cdots(h_1(u))\cdots\right)\right).$$

Then, $g_L(u) = f(u)$ and by recursive application of (A.13), we obtain

$$J_f(u) = J_{h_L}\left(g_{L-1}(u)\right)J_{h_{L-1}}\left(g_{L-2}(u)\right)\cdots J_{h_1}(u). \tag{A.16}$$

Note that from the definition of the Jacobian, the j-th column of $J_f(u)$ is the m-dimensional vector

$$\frac{\partial f(u)}{\partial u_j} = \left(\frac{\partial f_1(u)}{\partial u_j},\ldots,\frac{\partial f_m(u)}{\partial u_j}\right) = J_f(u)e_j,$$

where e_j is the j-th unit vector of appropriate dimension. Therefore, using (A.16), for each $j = 1,\ldots,n$,

$$\frac{\partial f(u)}{\partial u_j} = J_{h_L}\left(g_{L-1}(u)\right)\left[J_{h_{L-1}}\left(g_{L-2}(u)\right)\left[\cdots[J_{h_1}(u)e_j]\cdots\right]\right]. \tag{A.17}$$

That is, for each $j = 1,\ldots,n$, $\frac{\partial f(u)}{\partial u_j}$ can be obtained by recursively computing the *Jacobian vector product* given by

$$v_\ell := J_{h_\ell}\left(g_{\ell-1}(u)\right)v_{\ell-1},$$

for $\ell = 1,\ldots,L$, starting with $v_0 = e_j$ and $g_0(u) = u$.

On the other hand, since the i-th row of $J_f(u)$ is the gradient $\nabla f_i(u)$, we have

$$\nabla f_i(u) = e_i^\top J_f(u)$$
$$= \left[\cdots\left[\left[e_i^\top J_{h_L}\left(g_{L-1}(u)\right)\right]J_{h_{L-1}}\left(g_{L-2}(u)\right)\right]\cdots\right]J_{h_1}(u). \tag{A.18}$$

That is, for each $i = 1, \ldots, m$, $\nabla f_i(u)$ can be obtained by recursively computing the *vector Jacobian product* given by

$$v_\ell^\top := v_{\ell-1}^\top J_{h_{L-\ell+1}} \left(g_{L-\ell}(u)\right),$$

for $\ell = 1, \ldots, L$, starting with $v_0 = e_i$ and $g_0(u) = u$.

A.4 Taylor's Theorem

Once again consider a multivariate real-valued function $f : \mathbb{R}^n \to \mathbb{R}$. If all the k-order derivatives of f are continuous at a point $u \in \mathbb{R}^n$, then *Taylor's theorem* offers an approximation for f within a neighborhood of u in terms of these derivatives. We are particularly interested in cases where $k = 1$ and $k = 2$ as they are crucial in the implementation of, respectively, the first-order and the second-order optimization methods. It is easy to understand the theorem when the function f is univariate. Hence we start with the univariate case and then move to the general multivariate case. We omit the proof of Taylor's theorem as it is a well-known result that can be found in any standard multivariate calculus textbook.

Univariate Case

Suppose that $n = 1$, that is, f is a univariate real-valued function. We say that f is *k-times continuously differentiable* on an open interval $\mathcal{U} \subseteq \mathbb{R}$ if f is k times differentiable at every point on $\mathcal{U}$ (i.e., the k-th order derivative $\frac{d^k f(u)}{du^k}$ exists for all $u \in \mathcal{U}$) and $\frac{d^k f(u)}{du^k}$ is continuous on $\mathcal{U}$. If $k = 0$, we interpret $\frac{d^k f(u)}{du^k}$ simply as $f(u)$.

Theorem A.1 (Taylor's Theorem in $\mathbb{R}$). *Let $f : \mathbb{R} \to \mathbb{R}$ be k-times continuously differentiable on an open interval $\mathcal{U} \subseteq \mathbb{R}$. Then, for any $u, v \in \mathcal{U}$,*

$$f(u) = \sum_{i=0}^{k} \frac{(u-v)^i}{i!} \frac{d^i f(v)}{du^i} + O\left(|u-v|^{k+1}\right). \tag{A.19}$$

The polynomial,

$$P_k(u) = \sum_{i=0}^{k} \frac{(u-v)^i}{i!} \frac{d^i f(v)}{du^i},$$

appeared in (A.19) is called k-th order *Taylor polynomial*. Since the remainder

$$R_k(u) = f(u) - P_k(u) \longrightarrow 0, \quad \text{as } x \to a,$$

$f(u)$ is approximately equal to $P_k(u)$ for u within a small neighborhood of a. Particularly, for a point u near v, $P_1(u)$ is *linear approximation* of $f(u)$ and $P_2(u)$ is *quadratic approximation* of $f(u)$.

Multivariate Case

Now consider the multivariate case, that is, f is a multivariate real-valued function. In order to state Taylor's theorem for this case, we need some new notion relevant only here.

An n-tuple $\alpha = (\alpha_1, \ldots, \alpha_n)$ is called *multi-index* if each α_i is a non-negative integer. For a multi-index α, let

$$|\alpha| = \sum_{i=1}^{n} \alpha_i, \quad \alpha! = \alpha_1! \cdots \alpha_n!, \quad \text{and} \quad u^\alpha = u_1^{\alpha_1} \cdots u_n^{\alpha_n},$$

for any $u \in \mathbb{R}^n$. Then, the higher-order partial derivatives are expressed as

$$D^\alpha f(u) = \frac{\partial^{|\alpha|} f(u)}{\partial u_1^{\alpha_1} \cdots \partial u_n^{\alpha_n}}.$$

We say that f is *k-times continuously differentiable* on an open set $\mathcal{U} \subseteq \mathbb{R}^n$ if all the higher order partial derivatives $D^\alpha f(u)$ exists and are continuous on $\mathcal{U}$ for all multi-index α such that $|\alpha| \leq k$.

Theorem A.2 (Taylor's Theorem in $\mathbb{R}^n$). *Let $f : \mathbb{R}^n \to \mathbb{R}$ be a k-times continuously differentiable on an open set $\mathcal{U} \subseteq \mathbb{R}^n$. Then, for any $u, v \in \mathcal{U}$,*

$$f(u) = \sum_{\alpha:|\alpha|\leq k} D^\alpha f(v) \frac{(u-v)^\alpha}{\alpha!} + O\left(\|u-v\|^{k+1}\right). \tag{A.20}$$

The polynomial

$$P_k(u) = \sum_{\alpha:|\alpha|\leq k} D^\alpha f(v) \frac{(u-v)^\alpha}{\alpha!}$$

is called k-th order Taylor's polynomial. In particular,

$$P_1(u) = \sum_{\alpha:|\alpha|\leq 1} D^\alpha f(v) \frac{(u-v)^\alpha}{\alpha!} = f(v) + (u-v)^\top \nabla f(a), \tag{A.21}$$

for u near v, provides *linear approximation*, also called *first-order Taylor's approximation*, to $f(u)$, while

$$P_2(u) = \sum_{\alpha:|\alpha|\leq 2} D^\alpha f(v) \frac{(u-v)^\alpha}{\alpha!}$$

$$= f(v) + (u-v)^\top \nabla f(v) + \frac{1}{2}(u-v)^\top \nabla^2 f(v)(u-v) \tag{A.22}$$

provides *quadratic approximation*, also called *second-order Taylor's approximation*, to $f(u)$.

Linear Approximation with Jacobians and Hessians

Consider a differentiable function $f : \mathbb{R}^n \to \mathbb{R}^m$ with the $m \times n$ Jacobian $J_f(\cdot)$. Then with Theorem (A.2) we may construct a first-order linear approximation to $f(\cdot)$ around any $u_0 \in \mathbb{R}^n$,

$$\tilde{f}(u) = f(u_0) + J_f(u_0)(u - u_0), \tag{A.23}$$

where $\tilde{f}(u) \approx f(u)$.

Now consider a twice differentiable $g : \mathbb{R}^n \to \mathbb{R}$ with gradient $\nabla g(\cdot)$ and Hessian matrix $\nabla^2 g(\cdot)$. We can set $f(u) = \nabla g(u)$ with $f : \mathbb{R}^n \to \mathbb{R}^n$. Since the Hessian of $g(\cdot)$ is the Jacobian of $f(\cdot)$, from (A.23) we obtain a first-order linear approximation for the gradient around $u_0 \in \mathbb{R}^n$,

$$\widetilde{\nabla} g(u) = \nabla g(u_0) + \nabla^2 g(u_0)(u - u_0), \tag{A.24}$$

where $\widetilde{\nabla} g(u) \approx \nabla g(u)$.

B Cross Entropy and Other Expectations with Logarithms

This appendix expands on basic properties of cross entropy, the KL-divergence, and related concepts, also in the context of the multivariate normal distribution. It is not meant to be an extensive review of these concepts but rather provides key definitions, properties, and results needed for the content of the book.

B.1 Divergences and Entropies

We first define the relative entropy (KL-divergence), cross entropy, and entropy in the context of discrete probability distributions. We then provide a definition of the KL-divergence for continuous random variables. Finally we define the Jensen–Shannon divergence.

The KL-Divergence for Discrete Distributions

Assume two probability distributions $p(\cdot)$ and $q(\cdot)$ over elements in some discrete sets $\mathcal{X}_p$ and $\mathcal{X}_q$, respectively. That is, $p(x)$ or $q(x)$ denote the respective probabilities, which are strictly positive unless $x \notin \mathcal{X}_p$ for which $p(x) = 0$ (or similarly $x \notin \mathcal{X}_q$ for which $q(x) = 0$).

A key measure for the proximity between the distributions $p(\cdot)$ and $q(\cdot)$ is the *Kullback–Leibler divergence*, also shortened as *KL-divergence*, and also known as the *relative entropy*. It is denoted $D_{\mathrm{KL}}(p \parallel q)$ and as long as $\mathcal{X}_p \subseteq \mathcal{X}_q$ it is the expected value of $\log p(X)/q(X)$ where X is a random variable following the probability law $p(\cdot)$. Namely,

$$D_{\mathrm{KL}}(p \parallel q) = \sum_{x \in \mathcal{X}_p} p(x) \log \frac{p(x)}{q(x)}. \tag{B.1}$$

Further if $\mathcal{X}_p \not\subseteq \mathcal{X}_q$, that is if there is some element in $\mathcal{X}_p$ that is not in $\mathcal{X}_q$, then by definition $D_{\mathrm{KL}}(p \parallel q) = +\infty$. This definition as infinity is natural since we would otherwise divide by 0 for some $q(x)$.

Observe that the expression for $D_{\mathrm{KL}}(p \parallel q)$ from (B.1) can be decomposed into the difference of $H(p)$ from $H(p, q)$ via

$$D_{\mathrm{KL}}(p \parallel q) = \underbrace{\sum_{x \in \mathcal{X}} p(x) \log \frac{1}{q(x)}}_{H(p,q)} - \underbrace{\sum_{x \in \mathcal{X}} p(x) \log \frac{1}{p(x)}}_{H(p)}.$$

DOI: 10.1201/9781003298687-B

Here,

$$H(p, q) = -\sum_{x \in \mathcal{X}} p(x) \log q(x) \tag{B.2}$$

is called the *cross entropy* of p and q and

$$H(p) = -\sum_{x \in \mathcal{X}} p(x) \log p(x) \tag{B.3}$$

is called the entropy of p. Hence in words, the KL-divergence or relative entropy of p and q is the cross entropy of p and q with the entropy of p subtracted. Note that in case where there are only two values in $\mathcal{X}$, say 0 and 1, where we denote $p(1) = p_1$ and $q(1) = q_1$, we have

$$H(p) = -\big(p_1 \log p_1 + (1 - p_1) \log(1 - p_1)\big), \tag{B.4}$$
$$H(p, q) = -\big(p_1 \log q_1 + (1 - p_1) \log(1 - q_1)\big). \tag{B.5}$$

Some observations are in order. First observe that $D_{\mathrm{KL}}(p \parallel q) \geq 0$. Further note that in general $D_{\mathrm{KL}}(p \parallel q) \neq D_{\mathrm{KL}}(q \parallel p)$ and similarly $H(p, q) \neq H(q, p)$. Hence as a "distance measure" the KL-divergence is not a true metric since it is not symmetric over its arguments. Nevertheless, when $p = q$ the KL-divergence is 0 and similarly the cross entropy equals the entropy. In addition, it can be shown that $D_{\mathrm{KL}}(p \parallel q) = 0$ only when $p = q$. Hence the KL-divergence may play a role similar to a distance metric in certain applications. In fact, one may consider a sequence $q^{(1)}, q^{(2)}, \ldots$ which has decreasing $D_{\mathrm{KL}}(p \parallel q^{(t)})$ approaching 0 as $t \to \infty$. For such a sequence, the probability distributions $q^{(t)}$ approach[1] the target distribution p since the KL-divergence convergences to 0.

The KL-divergence for Continuous Distributions

The KL-divergence in (B.1) naturally extends to arbitrary probability distributions that are not necessarily discrete. In our case let us consider continuous multi-dimensional distributions. In this case $p(\cdot)$ and $q(\cdot)$ are probability densities, and the sets $\mathcal{X}_p$, and $\mathcal{X}_q$ are their respective supports. Now very similarly to (B.1), as long as $\mathcal{X}_p \subseteq \mathcal{X}_q$ we define

$$D_{\mathrm{KL}}(p \parallel q) = \int_{x \in \mathcal{X}_p} p(x) \log \frac{p(x)}{q(x)} \, dx. \tag{B.6}$$

The Jensen-Shannon Divergence

A related measure to the KL-divergence which is symmetric in arguments is the *Jensen-Shannon divergence* denoted $\mathrm{JSD}(p \parallel q)$. Either for the discrete or continuous case, it is defined by considering a mixture distribution with support $\mathcal{X}_p \cup \mathcal{X}_q$,

$$m(x) = \frac{1}{2}(p + q),$$

and then averaging the KL-divergence between each of the distributions and $m(\cdot)$, namely,

$$\mathrm{JSD}(p \parallel q) = \frac{D_{\mathrm{KL}}(p \parallel m) + D_{\mathrm{KL}}(q \parallel m)}{2}. \tag{B.7}$$

---

[1] There are multiple ways to define convergence of such a sequence of probability distributions. The exact form is out of our scope.

The square root of $\text{JSD}(p \parallel q)$, sometimes called the *Jensen-Shannon distance* is a metric in the mathematical sense.

B.2 Computations for Multivariate Normal Distributions

A univariate (single variable) *normal*, or *Gaussian*, distribution has a probability density function,

$$\mathcal{N}(x\,;\,\mu,\sigma^2) = \frac{1}{\sigma\sqrt{2\pi}}e^{-\frac{(x-\mu)^2}{2\sigma^2}}, \qquad \text{for} \qquad x \in \mathbb{R},$$

and is parameterized by $\mu \in \mathbb{R}$ and $\sigma^2 > 0$ which are the mean and variance of the distribution, respectively. The *standard normal* case has $\mu = 0$ and $\sigma^2 = 1$.

An m-dimensional multivariate normal distribution is characterized by a mean vector $\mu \in \mathbb{R}^m$ and a covariance matrix $\Sigma \in \mathbb{R}^{m \times m}$ which is assumed to be symmetric and positive definite. The probability density function (pdf) of a multivariate normal distribution is

$$\mathcal{N}(x\,;\,\mu,\Sigma) = \frac{1}{(\det \Sigma)^{1/2}(2\pi)^{m/2}}e^{-\frac{1}{2}(x-\mu)^\top \Sigma^{-1}(x-\mu)}, \qquad \text{for} \qquad x \in \mathbb{R}^m,$$

where $\det \Sigma$ stands for the determinant of a matrix Σ. There are many useful formulas associated with this distribution with one particular case being the *log-density*,

$$\log \mathcal{N}(x\,;\,\mu,\Sigma) = -\frac{1}{2}(x-\mu)^\top \Sigma^{-1}(x-\mu) - \frac{m}{2}\log(2\pi) - \frac{1}{2}\log(\det \Sigma). \qquad \text{(B.8)}$$

It is also useful to consider the KL-divergence between two multivariate normal distributions. For short, denote such a distribution as $\mathcal{N}_{\mu,\Sigma}$ when the mean vector is μ and the covariance matrix is Σ. Then if we consider two such distributions on $\mathbb{R}^m$ with corresponding mean vectors μ_1 and μ_2, and corresponding covariance matrices Σ_1 and Σ_2, then it is possible to show that

$$D_{\text{KL}}(\mathcal{N}_{\mu_1,\Sigma_1} \parallel \mathcal{N}_{\mu_2,\Sigma_2}) = \frac{1}{2}\Big((\mu_1-\mu_2)^\top \Sigma_2^{-1}(\mu_1-\mu_2) - m + \text{tr}(\Sigma_2^{-1}\Sigma_1) + \log\frac{\det(\Sigma_2)}{\det(\Sigma_1)}\Big). \text{ (B.9)}$$

A particularly useful case is one where $\Sigma_2 - \sigma_2^2 I$ for some constant $\sigma_2^2 > 0$. In this case,

$$D_{\text{KL}}(\mathcal{N}_{\mu_1,\Sigma_1} \parallel \mathcal{N}_{\mu_2,\sigma_2^2 I}) = \frac{1}{2\sigma_2^2}\|\mu_1-\mu_2\|^2 - \frac{m}{2} + \frac{\text{tr}(\Sigma_1)}{2\sigma_2^2} + \frac{m\log\sigma_2^2}{2} - \frac{\log\det(\Sigma_1)}{2}. \text{ (B.10)}$$

Furthermore, if the second distribution is standard, i.e., $\mu_2 = 0$ and $\sigma_2^2 = 1$, then

$$D_{\text{KL}}(\mathcal{N}_{\mu_1,\Sigma_1} \parallel \mathcal{N}_{0,I}) = \frac{1}{2}\|\mu_1\|^2 - \frac{m}{2} + \frac{\text{tr}(\Sigma_1)}{2} - \frac{\log\det(\Sigma_1)}{2}. \qquad \text{(B.11)}$$

Bibliography

[1] M. Abadi, A. Agarwal, P. Barham, E. Brevdo, Z. Chen, C. Citro, et al. TensorFlow: Large-scale machine learning on heterogeneous distributed systems. *arXiv:1603.04467*, 2016.

[2] D. H. Ackley, G. E. Hinton, and T. J. Sejnowski. A learning algorithm for Boltzmann machines. *Cognitive Science*, 1985.

[3] J. Adler and S. Lunz. Banach Wasserstein GAN. *Advances in Neural Information Processing Systems*, 2018.

[4] C. C. Aggarwal. Neural networks and deep learning. *Springer*, 2018.

[5] A. Agresti. *Categorical data analysis*. John Wiley & Sons, 2003.

[6] A. Agresti. *Analysis of ordinal categorical data*. John Wiley & Sons, 2010.

[7] H. Akaike. A new look at the statistical model identification. *IEEE Transactions on Automatic Control*, 1974.

[8] K. Akuzawa, Y. Iwasawa, and Y. Matsuo. Expressive speech synthesis via modeling expressions with variational autoencoder. *arXiv:1804.02135*, 2018.

[9] P. Albertos and I. Mareels. *Feedback and control for everyone*. Springer, 2010.

[10] J. J. Allaire. *Deep Learning with R*. Simon and Schuster, 2018.

[11] A. Alotaibi. Deep generative adversarial networks for image-to-image translation: A review. *Symmetry*, 2020.

[12] S. Amari. A theory of adaptive pattern classifiers. *IEEE Transactions on Electronic Computers*, 1967.

[13] S. Amari. Learning patterns and pattern sequences by self-organizing nets of threshold elements. *IEEE Transactions on Computers*, 1972.

[14] P. J. Antsaklis and A. N. Michel. *Linear systems*. Springer, 1997.

[15] M. Arjovsky, S. Chintala, and L. Bottou. Wasserstein generative adversarial networks. In *International Conference on Machine Learning*, 2017.

[16] L. Armijo. Minimization of functions having Lipschitz continuous first partial derivatives. *Pacific Journal of mathematics*, 1966.

[17] S. Arora, Z. Li, and K. Lyu. Theoretical analysis of auto rate-tuning by batch normalization. *arXiv:1812.03981*, 2018.

[18] J. Atwood and D. Towsley. Diffusion-convolutional neural networks. *Advances in neural information processing systems*, 2016.

[19] J. L. Ba, J. R. Kiros, and G. E. Hinton. Layer normalization. *arXiv:1607.06450*, 2016.

[20] D. Bahdanau, K. Cho, and Y. Bengio. Neural machine translation by jointly learning to align and translate. *arXiv:1409.0473*, 2014.

[21] P. Baldi and K. Hornik. Neural networks and principal component analysis: Learning from examples without local minima. *Neural networks*, 1989.

[22] W. Bao, J. Yue, and Y. Rao. A deep learning framework for financial time series using stacked autoencoders and long-short term memory. *PLoS ONE*, 2017.

[23] D. Bau, B. Zhou, A. Khosla, A. Oliva, and A. Torralba. Network dissection: Quantifying interpretability of deep visual representations. In *Proceedings of the IEEE conference on computer vision and pattern recognition*, 2017.

[24] A. G. Baydin, B. A. Pearlmutter, A. A. Radul, and J. M. Siskind. Automatic differentiation in machine learning: A survey. *Journal of Machine Learning Research*, 2018.

[25] L. M. Beda, L. N. Korolev, N. V. Sukkikh, and T. S. Frolova. Programs for automatic differentiation for the machine BESM (in Russian). *Technical report, Institute for Precise Mechanics and Computation Techniques, Academy of Science, Moscow, USSR*, 1959.

[26] M. Belkin, D. Hsu, S. Ma, and S. Mandal. Reconciling modern machine-learning practice and the classical bias–variance trade-off. *Proceedings of the National Academy of Sciences*, 2019.

[27] A. Ben-Tal and A. Nemirovski. *Lectures on modern convex optimization.* SIAM, Philadelphia, PA; MPS, Philadelphia, PA, 2001.

[28] Y. Bengio. Learning deep architectures for AI. *Foundations and Trends® in Machine Learning*, 2009.

[29] Y. Bengio. Practical recommendations for gradient-based training of deep architectures. In *Neural Networks: Tricks of the Trade.* 2012.

[30] Y. Bengio, P. Lamblin, D. Popovici, and H. Larochelle. Greedy layer-wise training of deep networks. *Advances in Neural Information Processing Systems*, 2006.

[31] J. O. Berger and R. L. Wolpert. The Likelihood Principle: A Review, Generalizations, and Statistical Implications. *Lecture Notes—Monograph Series*, 1988.

[32] D. Bertsekas, A. Nedić, and A. E. Ozdaglar. *Convex analysis and optimization.* Athena Scientific, 2003.

[33] D. P. Bertsekas. *Dynamic Programming and Optimal Control, Volume. II.* Athena Scientific, 3rd edition, 2007.

[34] D. P. Bertsekas. *Dynamic programming and optimal control: Volume I.* Athena Scientific, 2012.

[35] D. P. Bertsekas. *Nonlinear programming.* Athena Scientific, Third edition, 2016.

[36] D. P. Bertsekas and J. N. Tsitsiklis. *Neuro-dynamic programming.* Athena Scientific, 1996.

[37] D. Bertsimas and J. N. Tsitsiklis. *Introduction to linear optimization.* Athena Scientific, 1997.

[38] C. Biever. ChatGPT broke the Turing test-the race is on for new ways to assess AI. *Nature*, 2023.

[39] C. M. Bishop. *Pattern Recognition and Machine learning.* Springer, 2006.

[40] A. Bjerhammar. *Application of calculus of matrices to method of least squares: with special reference to geodetic calculations.* Elander, 1951.

[41] D. M. Blei, A. Kucukelbir, and J. D. McAuliffe. Variational inference: A review for statisticians. *J. Amer. Statist. Assoc.*, 2017.

[42] C. I. Bliss. The method of probits. *Science*, 1934.

[43] S. Bock and Weiß. A Proof of Local Convergence for the Adam Optimizer. In *2019 International Joint Conference on Neural Networks (IJCNN)*, 2019.

[44] D. Böhning. Multinomial logistic regression algorithm. *Annals of the institute of Statistical Mathematics*, 1992.

[45] D. Bolya, C. Zhou, F. Xiao, and Y. J. Lee. YOLACT: Real-Time Instance Segmentation. In *Proceedings of the IEEE/CVF International Conference on Computer Vision*, 2019.

[46] S. Bond-Taylor, A. Leach, Y. Long, and C. G. Willcocks. Deep generative modelling: A comparative review of vaes, gans, normalizing flows, energy-based and autoregressive models. *IEEE Transactions on Pattern Analysis and Machine Intelligence*, 2021.

[47] J. F. Bonnans, J. C. Gilbert, C. Lemaréchal, and C. A. Sagastizábal. *Numerical optimization: Theoretical and practical aspects.* Springer Science & Business Media, 2006.

[48] L. Bottou. Online algorithms and stochastic approximations. *Online learning in neural networks*, 1998.

[49] L. Bottou. Large-scale machine learning with stochastic gradient descent. In *Proceedings of COMPSTAT'2010: 19th International Conference on Computational Statistics*, 2010.

[50] L. Bottou. Stochastic gradient descent tricks. In *Neural Networks: Tricks of the Trade*. 2012.

[51] L. Bottou, F. E. Curtis, and J. Nocedal. Optimization methods for large-scale machine learning. *SIAM review*, 2018.

[52] L. Bottou and Y. LeCun. Large scale online learning. In *Advances in Neural Information Processing Systems*, 2003.

[53] H. Bourlard and Y. Kamp. Auto-association by multilayer perceptrons and singular value decomposition. *Biological cybernetics*, 1988.

[54] S. R. Bowman, L. Vilnis, O. Vinyals, A. M. Dai, R. Jozefowicz, and S. Bengio. Generating sentences from a continuous space. In *20th SIGNLL Conference on Computational Natural Language Learning, CoNLL 2016*, 2016.

[55] S. Boyd and L. Vandenberghe. *Convex optimization*. Cambridge University Press, 2004.

[56] S. Boyd and L. Vandenberghe. *Introduction to applied linear algebra: Vectors, matrices, and least squares*. Cambridge university press, 2018.

[57] J. Bradbury, R. Frostig, P. Hawkins, M. J. Johnson, C. Leary, D. Maclaurin, G. Necula, A. Paszke, J. VanderPlas, S. Wanderman-Milne, and Q. Zhang. JAX: Composable Transformations of Python+NumPy Programs. *http://github.com/google/jax*, 2018.

[58] A. P. Bradley. The use of the area under the ROC curve in the evaluation of machine learning algorithms. *Pattern Recognition*, 1997.

[59] L. Breiman. Bagging predictors. *Machine Learning*, 1996.

[60] L. Breiman. Random forests. *Machine Learning*, 2001.

[61] L. Breiman. Statistical Modeling: The Two Cultures (with Comments and a Rejoinder by the Author). *Statistical Science*, 2001.

[62] L. Breiman, J. H. Friedman, R. A. Olshen, and C. J. Stone. Classification and regression trees. Wadsworth & Brooks. *Cole Statistics/Probability Series*, 1984.

[63] J. S. Bridle. Probabilistic interpretation of feedforward classification network outputs, with relationships to statistical pattern recognition. In *Neurocomputing*. 1990.

[64] P. J. Brockwell and R. A. Davis. *Time Series: Theory and Methods*. Springer Science & Business Media, 1991.

[65] J. Bromley, I. Guyon, Y. LeCun, E. Säckinger, and R. Shah. Signature verification using a "Siamese" time delay neural network. *Advances in neural information processing systems*, 1993.

[66] T. Brown, B. Mann, N. Ryder, M. Subbiah, J. D. Kaplan, P. Dhariwal, A. Neelakantan, P. Shyam, G. Sastry, A. Askell, et al. Language models are few-shot learners. *Advances in Neural Information Processing Systems*, 2020.

[67] J. Bruna, W. Zaremba, A. Szlam, and Y. LeCun. Spectral networks and locally connected networks on graphs. *arXiv:1312.6203*, 2013.

[68] A. Burkov. *The Hundred-Page Machine Learning Book*. Andriy Burkov Quebec City, QC, Canada, 2019.

[69] O. Calin. *Deep Learning Architectures*. Springer, 2020.

[70] A. Canziani, A. Paszke, and E. Culurciello. An Analysis of Deep Neural Network Models for Practical Applications. *arXiv:1605.07678*, 2016.

[71] H. Cao, C. Tan, Z. Gao, Y. Xu, G. Chen, P. A. Heng, and S. Z. Li. A survey on generative diffusion model. *arXiv:2209.02646*, 2022.

[72] A. Cauchy. Méthode générale pour la résolution des systèmes d'équations simultanées. *Comp. Rend. Sci. Paris*, 1847.

[73] Y. Chang, X. Wang, J. Wang, Y. Wu, K. Zhu, H. Chen, L. Yang, X. Yi, C. Wang, Y. Wang, et al. A survey on evaluation of large language models. *arXiv:2307.03109*, 2023.

[74] D. Charte, F. Charte, S. García, M. J. del Jesus, and F. Herrera. A practical tutorial on autoencoders for nonlinear feature fusion: Taxonomy, models, software and guidelines. *Information Fusion*, 2018.

[75] N. V. Chawla, K. W. Bowyer, L. O. Hall, and W. P. Kegelmeyer. Smote: synthetic minority over-sampling technique. *Journal of Artificial Intelligence Research*, 2002.

[76] L. Chen, S. Li, Q. Bai, J. Yang, S. Jiang, and Y. Miao. Review of image classification algorithms based on convolutional neural networks. *Remote Sensing*, 2021.

[77] T. Chen and C. Guestrin. Xgboost: A scalable tree boosting system. In *Proceedings of the 22nd ACM SIGKDD International Conference on Knowledge Discovery and Data Mining*, 2016.

[78] X. Chen, Y. Duan, R. Houthooft, J. Schulman, I. Sutskever, and P. Abbeel. Infogan: Interpretable representation learning by information maximizing generative adversarial nets. *Advances in Neural Information Processing Systems*, 2016.

[79] P. S. Chib and P. Singh. Recent advancements in end-to-end autonomous driving using deep learning: A survey. *IEEE Transactions on Intelligent Vehicles*, 2023.

[80] K. Cho, B. Van Merriënboer, D. Bahdanau, and Y. Bengio. On the properties of neural machine translation: Encoder-decoder approaches. *arXiv:1409.1259*, 2014.

[81] K. Cho, B. Van Merriënboer, C. Gulcehre, D. Bahdanau, F. Bougares, H. Schwenk, and Y. Bengio. Learning phrase representations using RNN encoder-decoder for statistical machine translation. *arXiv:1406.1078*, 2014.

[82] D. Choi, C. J. Shallue, Z. Nado, J. Lee, C. J. Maddison, and G. E. Dahl. On empirical comparisons of optimizers for deep learning. *arXiv:1910.05446*, 2019.

[83] Y. Choi, M. Choi, M. Kim, J. Ha, S. Kim, and J. Choo. Stargan: Unified generative adversarial networks for multi-domain image-to-image translation. In *Proceedings of the IEEE conference on computer vision and pattern recognition*, 2018.

[84] S. Chopra, R. Hadsell, and Y. LeCun. Learning a similarity metric discriminatively, with application to face verification. In *2005 IEEE computer society conference on computer vision and pattern recognition (CVPR'05)*, 2005.

[85] J. Chung, C. Gulcehre, K. Cho, and Y. Bengio. Empirical evaluation of gated recurrent neural networks on sequence modeling. *arXiv:1412.3555*, 2014.

[86] D. C. Cireşan, U. Meier, L. M. Gambardella, and J. Schmidhuber. Deep, big, simple neural nets for handwritten digit recognition. *Neural Computation*, 2010.

[87] G. Claeskens and N. L. Hjort. Model selection and model averaging. *Cambridge Books*, 2008.

[88] W. S. Cleveland. Robust locally weighted regression and smoothing scatterplots. *Journal of the American Statistical Association*, 1979.

[89] A. K. Cline and I. S. Dhillon. Computation of the singular value decomposition. In *Handbook of linear algebra*. 2006.

[90] N. Cohen, O. Sharir, and A. Shashua. On the Expressive Power of Deep Learning: A Tensor Analysis, 2016.

[91] D. Commenges and H. Jacqmin-Gadda. *Dynamical biostatistical models*. CRC Press, 2015.

[92] P. Congdon. *Bayesian Statistical Modelling*. John Wiley & Sons, 2007.

[93] T. H. Cormen, C. E. Leiserson, R. L. Rivest, and C. Stein. *Introduction to algorithms*. MIT press, 2022.

[94] D. R. Cox and D. V. Hinkley. *Theoretical statistics*. CRC Press, 1979.

[95] J. S. Cramer. The origins of logistic regression. *Tinbergen Institute Working Paper No. 2002-119/4*, 2002.

[96] F. A. Croitoru, V. Hondru, R. T. Ionescu, and M. Shah. Diffusion models in vision: A survey. *IEEE Transactions on Pattern Analysis and Machine Intelligence*, 2023.

[97] G. Cybenko. Approximation by superpositions of a sigmoidal function. *Mathematics of Control, Signals, and Systems*, 1989.

[98] Y. H. Dai. Convergence properties of the BFGS algorithm. *SIAM Journal on Optimization*, 2002.

[99] W. C. Davidon. Variable metric method for minimization. Technical report, Argonne National Lab., Lemont, Ill., 1959.

[100] W. C. Davidon. Variable metric method for minimization. *SIAM Journal on Optimization*, 1991.

[101] M. P. Deisenroth, A. A. Faisal, and C. S. Ong. *Mathematics for Machine Learning*. Cambridge University Press, 2020.

[102] J. Deng, W. Dong, R. Socher, L. Li, K. Li, and L. Fei-Fei. ImageNet: A Large-Scale Hierarchical Image Database. In *2009 IEEE Conference on Computer Vision and Pattern Recognition*, 2009.

[103] J. Devlin, M. W. Chang, K. Lee, and K. Toutanova. BERT: Pre-training of deep bidirectional transformers for language understanding. *arXiv:1810.04805*, 2018.

[104] P. Dhariwal and A. Nichol. Diffusion models beat GANs on image synthesis. *Advances in Neural Information Processing Systems*, 2021.

[105] L. DO Q. Numerically efficient methods for solving least squares problems. *Pennsylvania State University*, 2012.

[106] A. J. Dobson and A. G. Barnett. *An introduction to generalized linear models*. Chapman and Hall/CRC, 2018.

[107] A. Dosovitskiy, L. Beyer, A. Kolesnikov, D. Weissenborn, X. Zhai, T. Unterthiner, M. Dehghani, M. Minderer, G. Heigold, S. Gelly, et al. An Image Is Worth 16x16 Words: Transformers for Image Recognition at Scale. *arXiv:2010.11929*, 2020.

[108] J. C. Douma and J. T. Weedon. Analysing continuous proportions in ecology and evolution: A practical introduction to beta and Dirichlet regression. *Methods in Ecology and Evolution*, 2019.

[109] T. Dozat. Incorporating Nesterov momentum into Adam. *International Conference on Learning Representations (ICLR) Workshop*, 2016.

[110] N. Du, Y. Huang, A. M. Dai, S. Tong, D. Lepikhin, Y. Xu, M. Krikun, Y. Zhou, A. W. Yu, O. Firat, et al. Glam: Efficient scaling of language models with mixture-of-experts. In *International Conference on Machine Learning*, 2022.

[111] S. R. Dubey, S. K. Singh, and B. B. Chaudhuri. Activation functions in deep learning: A comprehensive survey and benchmark. *Neurocomputing*, 2022.

[112] J. Duchi, E. Hazan, and Y. Singer. Adaptive subgradient methods for online learning and stochastic optimization. *Journal of Machine Learning Research*, 2011.

[113] V. Dumoulin, J. Shlens, and M. Kudlur. A learned representation for artistic style. *arXiv:1610.07629*, 2016.

[114] C. Eckart and G. Young. The approximation of one matrix by another of lower rank. *Psychometrika*, 1936.

[115] B. Efron and T. Hastie. *Computer age statistical inference*. Cambridge University Press, 2016.

[116] R. Eldan and O. Shamir. The power of depth for feedforward neural networks. In *Conference on learning theory*, 2016.

[117] J. L. Elman. Finding structure in time. *Cognitive science*, 1990.

[118] F. Emmert-Streib, S. Moutari, and M. Dehmer. A comprehensive survey of error measures for evaluating binary decision making in data science. *Wiley Interdisciplinary Reviews: Data Mining and Knowledge Discovery*, 2019.

[119] D. Erhan, Y. Bengio, A. Courville, and P. Vincent. Visualizing higher-layer features of a deep network. *University of Montreal*, 2009.

[120] A. E. Ezugwu, A. M. Ikotun, O. O. Oyelade, L. Abualigah, J. O. Agushaka, C. I. Eke, and A. A. Akinyelu. A comprehensive survey of clustering algorithms: State-of-the-art machine learning applications, taxonomy, challenges, and future research prospects. *Engineering Applications of Artificial Intelligence*, 2022.

[121] J. J. Faraway. *Extending the Linear Model with R: Generalized Linear, Mixed Effects and Nonparametric Regression Models*. Chapman and Hall/CRC, 2016.

[122] C. Feichtenhofer, A. Pinz, and A. Zisserman. Convolutional Two-Stream Network Fusion for Video Action Recognition. In *Proceedings of the IEEE Conference on Computer Vision and Pattern Recognition*, 2016.

[123] R. Fletcher. *Practical methods of optimization*. John Wiley & Sons, 2013.

[124] R. Fletcher and M. J. D. Powell. A rapidly convergent descent method for minimization. *The Computer Journal*, 1963.

[125] V. François-Lavet, P. Henderson, R. Islam, M. G. Bellemare, and J. Pineau. An introduction to deep reinforcement learning. *Foundations and Trends in Machine Learning*, 2018.

[126] K. Fukushima. Visual feature extraction by a multilayered network of analog threshold elements. *IEEE Transactions on Systems Science and Cybernetics*, 1969.

[127] K. Fukushima. c: A self-organizing neural network model for a mechanism of pattern recognition unaffected by shift in position. *Biological cybernetics*, 1980.

[128] Y. Gal and Z. Ghahramani. Dropout as a Bayesian approximation: Representing model uncertainty in deep learning. In *International conference on machine learning*, 2016.

[129] A. Garcia-Garcia, S. Orts-Escolano, S. Oprea, V. Villena-Martinez, P. Martinez-Gonzalez, and J. Garcia-Rodriguez. A Survey on Deep Learning Techniques for Image and Video Semantic Segmentation. *Applied Soft Computing*, 2018.

[130] S. Garg and G. Ramakrishnan. Advances in Quantum Deep Learning: An Overview. *arXiv:2005.04316*, 2020.

[131] L. A. Gatys, A. S. Ecker, and M. Bethge. A neural algorithm of artistic style. *arXiv:1508.06576*, 2015.

[132] Gauss, C. F. Theoria Motus Corporum Coelestium. Perthes, Hamburg. *Translation reprinted as Theory of the Motions of the Heavenly Bodies Moving about the Sun in Conic Sections. Dover, New York, 1963*, 1809.

[133] A. Géron. *Hands-On Machine Learning with Scikit-Learn, Keras, and TensorFlow: Concepts, Tools, and Techniques to Build Intelligent Systems*. O'Reilly Media, 2019.

[134] J. Gilmer, S. S. Schoenholz, P. F. Riley, O. Vinyals, and G. E. Dahl. Neural message passing for quantum chemistry. In *International conference on machine learning*, 2017.

[135] R. Girshick. Fast R-CNN. In *Proceedings of the IEEE International Conference on Computer Vision*, 2015.

[136] R. Girshick, J. Donahue, T. Darrell, and J. Malik. Rich feature hierarchies for accurate object detection and semantic segmentation. In *Proceedings of the IEEE conference on computer vision and pattern recognition*, 2014.

[137] X. Glorot and Y. Bengio. Understanding the difficulty of training deep feedforward neural networks. In *Proceedings of the thirteenth international conference on artificial intelligence and statistics*, 2010.

[138] Y. Goldberg. *Neural Network Methods for Natural Language Processing*. Springer Nature, 2022.

[139] G. H. Golub. Least squares, singular values and matrix approximations. *Aplikace matematiky*, 1968.

[140] G. H. Golub and C. Reinsch. Singular value decomposition and least squares solutions. In *Linear algebra*. 1971.

[141] G. H. Golub and C. F. Van Loan. *Matrix computations*. JHU Press, 2013.

[142] I. Goodfellow, Y. Bengio, and A. Courville. *Deep Learning*. MIT Press, 2016.

[143] I. Goodfellow, J. Pouget-Abadie, M. Mirza, B. Xu, D. Warde-Farley, S. Ozair, A. Courville, and Y. Bengio. Generative adversarial nets. *Advances in Neural Information Processing Systems*, 2014.

[144] M. Gori, G. Monfardini, and F. Scarselli. A new model for learning in graph domains. In *Proceedings. 2005 IEEE International Joint Conference on Neural Networks, 2005.*, 2005.

[145] A. Graves. Generating sequences with recurrent neural networks. *arXiv:1308.0850*, 2013.

[146] A. Graves, A. R. Mohamed, and G. Hinton. Speech recognition with deep recurrent neural networks. In *2013 IEEE international conference on acoustics, speech and signal processing*, 2013.

[147] A. Griewank and A. Walther. *Evaluating derivatives: Principles and techniques of algorithmic differentiation*. SIAM, 2008.

[148] A. Grover and J. Leskovec. node2vec: Scalable feature learning for networks. In *Proceedings of the 22nd ACM SIGKDD international conference on Knowledge discovery and data mining*, 2016.

[149] R. Gueorguieva, R. Rosenheck, and D. Zelterman. Dirichlet component regression and its applications to psychiatric data. *Computational Statistics & Data Analysis*, 2008.

[150] J. Gui, Z. Sun, Y. Wen, D. Tao, and J. Ye. A review on generative adversarial networks: Algorithms, theory, and applications. *IEEE Transactions on Knowledge and Data Engineering*, 2021.

[151] I. Gulrajani, F. Ahmed, M. Arjovsky, V. Dumoulin, and A. C. Courville. Improved training of Wasserstein GANs. *Advances in Neural Information Processing Systems*, 2017.

[152] I. Gulrajani, K. Kumar, F. Ahmed, A. A. Taiga, F. Visin, D. Vazquez, and A. Courville. PixelVAE: A Latent Variable Model for Natural Images. In *International Conference on Learning Representations*, 2016.

[153] H. Guo, Y. Li, J. Shang, M. Gu, Y. Huang, and B. Gong. Learning from class-imbalanced data: Review of methods and applications. *Expert systems with applications*, 2017.

[154] I. Guyon, P. Albrecht, Y. LeCun, J. Denker, and W. Hubbard. Design of a neural network character recognizer for a touch terminal. *Pattern Recognition*, 1991.

[155] M. U. Hadi, R. Qureshi, A. Shah, M. Irfan, A. Zafar, M. B. Shaikh, N. Akhtar, J. Wu, S. Mirjalili, et al. Large language models: A comprehensive survey of its applications, challenges, limitations, and future prospects. *Authorea Preprints*, 2023.

[156] W. Hamilton, Z. Ying, and J. Leskovec. Inductive representation learning on large graphs. *Advances in neural information processing systems*, 2017.

[157] W. L. Hamilton. *Graph Representation Learning*. Morgan & Claypool Publishers, 2020.

[158] D. J. Hand. Assessing the performance of classification methods. *International Statistical Review*, 2012.

[159] K. Hara, D. Saitoh, and H. Shouno. Analysis of dropout learning regarded as ensemble learning. In *Artificial Neural Networks and Machine Learning–ICANN 2016: 25th*

International Conference on Artificial Neural Networks, Barcelona, Spain, September 6-9, 2016, Proceedings, Part II 25, 2016.

[160] M. A. Hardy. *Regression with dummy variables*. Sage, 1993.

[161] D. Harrison Jr and D. L. Rubinfeld. Hedonic Housing Prices and the Demand for Clean Air. *Journal of Environmental Economics and Management*, 1978.

[162] G. M. Harshvardhan, M. K. Gourisaria, M. Pandey, and S. S. Rautaray. A comprehensive survey and analysis of generative models in machine learning. *Computer Science Review*, 2020.

[163] W. Harvey, S. Naderiparizi, V. Masrani, C. Weilbach, and F. Wood. Flexible diffusion modeling of long videos. *Advances in Neural Information Processing Systems*, 2022.

[164] D. Hassabis. Artificial Intelligence: Chess Match of the Century. *Nature*, 2017.

[165] D. Hassabis, D. Kumaran, C. Summerfield, and M. Botvinick. Neuroscience-Inspired Artificial Intelligence. *Neuron*, 2017.

[166] T. Hastie, R. Tibshirani, J. H. Friedman, and J. H. Friedman. *The elements of statistical learning: Data mining, inference, and prediction*. Springer, 2009.

[167] T. Hastie, R. Tibshirani, and M. Wainwright. Statistical learning with sparsity. *Monographs on statistics and applied probability*, 2015.

[168] T. J. Hastie and R. J. Tibshirani. *Generalized Additive Models*. Routledge, 2017.

[169] S. Haykin. *Neural networks: a comprehensive foundation*. Prentice Hall, 1998.

[170] K. He, G. Gkioxari, P. Dollár, and R. Girshick. Mask R-CNN. In *Proceedings of the IEEE International Conference on Computer Vision*, 2017.

[171] K. He, X. Zhang, S. Ren, and J. Sun. Delving deep into rectifiers: Surpassing human-level performance on imagenet classification. In *Proceedings of the IEEE international conference on computer vision*, 2015.

[172] K. He, X. Zhang, S. Ren, and J. Sun. Deep residual learning for image recognition. In *Proceedings of the IEEE conference on computer vision and pattern recognition*, 2016.

[173] R. Hecht-Nielsen. Theory of the backpropagation neural network. In *Neural Networks for Perception*. 1992.

[174] M. Henaff, J. Bruna, and Y. LeCun. Deep convolutional networks on graph-structured data. *arXiv:1506.05163*, 2015.

[175] S. Herculano-Houzel. The human brain in numbers: a linearly scaled-up primate brain. *Frontiers in human neuroscience*, 2009.

[176] A. Hertz, R. Mokady, J. Tenenbaum, K. Aberman, Y. Pritch, and D. Cohen-Or. Prompt-to-prompt image editing with cross attention control. *arXiv:2208.01626*, 2022.

[177] M. R. Hestenes and E. Stiefel. Methods of conjugate gradients for solving Linear Systems. *Journal of research of the National Bureau of Standards*, 1952.

[178] R. H. Hijazi and R. W. Jernigan. Modelling compositional data using Dirichlet regression models. *Journal of Applied Probability & Statistics*, 2009.

[179] J. M. Hilbe. *Logistic regression models*. Chapman and Hall/CRC, 2009.

[180] G. E. Hinton, S. Osindero, and Y. W. Teh. A fast learning algorithm for deep belief nets. *Neural Computation*, 2006.

[181] G. E. Hinton, N. Srivastava, A. Krizhevsky, I. Sutskever, and R. R. Salakhutdinov. Improving neural networks by preventing co-adaptation of feature detectors. *arXiv:1207.0580*, 2012.

[182] J. Ho, W. Chan, C. Saharia, J. Whang, R. Gao, A. Gritsenko, D. P. Kingma, B. Poole, M. Norouzi, D. J. Fleet, et al. Imagen video: High definition video generation with diffusion models. *arXiv:2210.02303*, 2022.

[183] J. Ho, A. Jain, and P. Abbeel. Denoising diffusion probabilistic models. *Advances in Neural Information Processing Systems*, 2020.

[184] S. Hochreiter and J. Schmidhuber. Long short-term memory. *Neural computation*, 1997.

[185] J. J. Hopfield. Neural networks and physical systems with emergent collective computational abilities. *Proceedings of the national academy of sciences*, 1982.

[186] K. Hornik. Approximation capabilities of multilayer feedforward networks. *Neural Networks*, 1991.

[187] D. W. Hosmer Jr, S. Lemeshow, and R. X. Sturdivant. *Applied logistic regression*. John Wiley & Sons, 2013.

[188] M. Z. Hossain, F. Sohel, M. Shiratuddin, and H. Laga. A comprehensive survey of deep learning for image captioning. *ACM Computing Surveys (CSUR)*, 2019.

[189] H. Hotelling. Analysis of a complex of statistical variables into principal components. *Journal of educational psychology*, 1933.

[190] J. Howard and S. Gugger. *Deep Learning for Coders with fastai and PyTorch*. O'Reilly Media, 2020.

[191] C. C. Hsu, H. T. Hwang, Y. C. Wu, Y. Tsao, and H. M. Wang. Voice conversion from unaligned corpora using variational autoencoding wasserstein generative adversarial networks. *Interspeech 2017*, 2017.

[192] G. Hu, Y. Yang, D. Yi, J. Kittler, W. Christmas, S. Z. Li, and T. Hospedales. When Face Recognition Meets with Deep Learning: An Evaluation of Convolutional Neural Networks for Face Recognition. In *Proceedings of the IEEE International Conference on Computer Vision Workshops*, 2015.

[193] Z. Hu, Z. Yang, X. Liang, R. Salakhutdinov, and E. P. Xing. Toward controlled generation of text. In *International Conference on Machine Learning*, 2017.

[194] L. Huang, J. Qin, Y. Zhou, F. Zhu, L. Liu, and L. Shao. Normalization techniques in training dnns: Methodology, analysis and application. *IEEE Transactions on Pattern Analysis and Machine Intelligence*, 2023.

[195] D. H. Hubel and T. N. Wiesel. Receptive fields of single neurons in the cat's striate cortex. *The Journal of physiology*, 1959.

[196] D. H. Hubel and T. N. Wiesel. Receptive fields, binocular interaction and functional architecture in the cat's visual cortex. *The Journal of Physiology*, 1962.

[197] P. J. Huber. Robust regression: asymptotics, conjectures and Monte Carlo. *The Annals of Statistics*, 1973.

[198] R. J. Hyndman and G. Athanasopoulos. *Forecasting: Principles and Practice*. OTexts, 3rd edition, 2021.

[199] F. N. Iandola, S. Han, M. W. Moskewicz, K. Ashraf, W. J. Dally, and K. Keutzer. Squeezenet: Alexnet-level accuracy with 50x fewer parameters and< 0.5 mb model size. *arXiv:1602.07360*, 2016.

[200] G. Iglesias, E. Talavera, and A. Díaz-Álvarez. A survey on GANs for computer vision: Recent research, analysis and taxonomy. *Computer Science Review*, 2023.

[201] M. Innes. Flux: Elegant Machine Learning with Julia. *Journal of Open Source Software*, 2018.

[202] S. Ioffe. Batch renormalization: Towards reducing minibatch dependence in batch-normalized models. *Advances in Neural Information Processing Systems*, 2017.

[203] S. Ioffe and C. Szegedy. Batch normalization: Accelerating deep network training by reducing internal covariate shift. In *International conference on machine learning*, 2015.

[204] P. Isola, J. Zhu, T. Zhou, and A. A. Efros. Image-to-image translation with conditional adversarial networks. In *Proceedings of the IEEE conference on computer vision and pattern recognition*, 2017.

[205] A. G. Ivakhnenko. Polynomial theory of complex systems. *IEEE Transactions on Systems, Man, and Cybernetics*, 1971.

[206] A. G. Ivakhnenko and V. G. Lapa. Cybernetic predicting devices. *Purdue Univ Lafayette Ind School of Electrical Engineering, appearing in The Defense Technical Information Center*, 1966.

[207] B. Jähne. *Digital image processing*. Springer Science & Business Media, 2005.

[208] A. K. Jain, M. N. Murty, and P. J. Flynn. Data clustering: a review. *ACM computing surveys (CSUR)*, 1999.

[209] G. Jin, Y. Liang, Y. Fang, Z. Shao, J. Huang, J. Zhang, and Y. Zheng. Spatio-temporal graph neural networks for predictive learning in urban computing: A survey. *IEEE Transactions on Knowledge and Data Engineering*, 2023.

[210] W. Jin, R. Barzilay, and T. Jaakkola. Junction tree variational autoencoder for molecular graph generation. In *International Conference on Machine Learning*, 2018.

[211] J. M. Johnson and T. M. Khoshgoftaar. Survey on deep learning with class imbalance. *Journal of Big Data*, 2019.

[212] I. T. Jolliffe. *Principal component analysis for special types of data*. Springer, 2002.

[213] L. V. Jospin, W. Buntine, F. Boussaid, H. Laga, and M. Bennamoun. Hands-on Bayesian Neural Networks–A Tutorial for Deep Learning Users. *arXiv:2007.06823*, 2020.

[214] R. Jozefowicz, W. Zaremba, and I. Sutskever. An empirical exploration of recurrent network architectures. In *International conference on machine learning*, 2015.

[215] Y. Jung. Multiple predicting K-fold cross-validation for model selection. *Journal of Nonparametric Statistics*, 2018.

[216] D. Jurafsky and J. H. Martin. *Speech and Language Processing*. Pearson, 2000.

[217] L. P. Kaelbling, M. L. Littman, and A. W. Moore. Reinforcement learning: A survey. *Journal of Artificial Intelligence Research*, 4, 1996.

[218] N. Kalchbrenner and P. Blunsom. Recurrent continuous translation models. In *Proceedings of the 2013 conference on empirical methods in natural language processing*, 2013.

[219] M. Kang, J. Y. Zhu, R. Zhang, J. Park, E. Shechtman, S. Paris, and T. Park. Scaling up GANs for text-to-image synthesis. In *Proceedings of the IEEE/CVF Conference on Computer Vision and Pattern Recognition*, 2023.

[220] Z. Karevan and J. A. K. Suykens. Transductive LSTM for time-series prediction: An application to weather forecasting. *Neural Networks*, 2020.

[221] A. Karpathy, G. Toderici, S. Shetty, T. Leung, R. Sukthankar, and L. Fei-Fei. Large-Scale Video Classification with Convolutional Neural Networks. In *Proceedings of the IEEE Conference on Computer Vision and Pattern Recognition*, 2014.

[222] T. Karras, T. Aila, S. Laine, and J. Lehtinen. Progressive growing of GANs for improved quality, stability, and variation. *arXiv:1710.10196*, 2017.

[223] T. Karras, M. Aittala, J. Hellsten, S. Laine, J. Lehtinen, and T. Aila. Training generative adversarial networks with limited data. *Advances in Neural Information Processing Systems*, 2020.

[224] T. Karras, S. Laine, and T. Aila. A style-based generator architecture for generative adversarial networks. In *Proceedings of the IEEE/CVF conference on computer vision and pattern recognition*, 2019.

[225] T. Karras, S. Laine, M. Aittala, J. Hellsten, J. Lehtinen, and T. Aila. Analyzing and improving the image quality of stylegan. In *Proceedings of the IEEE/CVF conference on computer vision and pattern recognition*, 2020.

[226] L. Kaufman and P. J. Rousseeuw. *Finding groups in data: an introduction to cluster analysis*. John Wiley & Sons, 2009.

[227] S. Khan, M. Naseer, M. Hayat, S. W. Zamir, F. S. Khan, and M. Shah. Transformers in vision: A survey. *ACM Computing Surveys (CSUR)*, 2022.

[228] D. Khurana, A. Koli, K. Khatter, and S. Singh. Natural language processing: State of the art, current trends and challenges. *Multimedia Tools and Applications*, 2023.

[229] J. Kiefer. Sequential minimax search for a maximum. *Proceedings of the American mathematical society*, 1953.

[230] J. Kiefer. Optimum experimental designs. *Journal of the Royal Statistical Society: Series B (Methodological)*, 1959.

[231] J. Kiefer and J. Wolfowitz. Stochastic estimation of the maximum of a regression function. *The Annals of Mathematical Statistics*, 1952.

[232] J. H. Kim. Estimating classification error rate: Repeated cross-validation, repeated hold-out and bootstrap. *Computational statistics & data analysis*, 2009.

[233] D. P. Kingma and J. Ba. Adam: A method for stochastic optimization. *arXiv:1412.6980*, 2014.

[234] D. P. Kingma and M. Welling. Auto-encoding variational Bayes. *arXiv:1312.6114*, 2013.

[235] D. P. Kingma and M. Welling. An introduction to variational autoencoders. *Foundations and Trends® in Machine Learning*, 2019.

[236] T. N. Kipf and M. Welling. Variational graph auto-encoders. *arXiv:1611.07308*, 2016.

[237] S. Kiranyaz, T. Ince, and M. Gabbouj. Optimization techniques: An overview. *Multidimensional Particle Swarm Optimization for Machine Learning and Pattern Recognition*, 2014.

[238] M. J. Kochenderfer and T. A. Wheeler. *Algorithms for optimization*. MIT Press, 2019.

[239] A. Krizhevsky, I. Sutskever, and G. E. Hinton. Imagenet classification with deep convolutional neural networks. *Advances in neural information processing systems*, 2012.

[240] D. P. Kroese, Z. Botev, T. Taimre, and R. Vaisman. *Data science and machine learning: Mathematical and statistical methods.* CRC Press, 2019.

[241] A. Krogh and J. Hertz. A Simple Weight Decay Can Improve Generalization. In *Advances in Neural Information Processing Systems*, 1991.

[242] J. Kuehn, S. Abadie, B. Liquet, and V. Roeber. A deep learning super-resolution model to speed up computations of coastal sea states. *Applied Ocean Research*, 2023.

[243] J. Kukačka, V. Golkov, and D. Cremers. Regularization for deep learning: A taxonomy. *arXiv:1710.10686*, 2017.

[244] H. Kwakernaak and R. Sivan. *Modern signal and systems.* Prentice Hall, 1991.

[245] P. Lafaye de Micheaux, R. Drouilhet, and B. Liquet. *The R software: Fundamentals of programming and statistical analysis.* Springer, 2013.

[246] H. Lai, S. Xiao, Y. Pan, Z. Cui, J. Feng, C. Xu, J. Yin, and S. Yan. Deep recurrent regression for facial landmark detection. *IEEE Transactions on Circuits and Systems for Video Technology*, 2016.

[247] K. J. Lang. A time-delay neural network architecture for speech recognition. *Technical Report, Carnegie-Mellon University*, 1988.

[248] H. Larochelle, Y. Bengio, J. Louradour, and P. Lamblin. Exploring strategies for training deep neural networks. *Journal of Machine Learning Research*, 2009.

[249] Y. LeCun, Y. Bengio, and G. Hinton. Deep learning. *Nature*, 2015.

[250] Y. LeCun, B. Boser, J. Denker, D. Henderson, W. Hubbard, and L. Jackel. Handwritten digit recognition with a back-propagation network. *Advances in neural information processing systems*, 1989.

[251] Y. LeCun, B. Boser, J. S. Denker, D. Henderson, R. E. Howard, W. Hubbard, and L. D. Jackel. Backpropagation applied to handwritten zip code recognition. *Neural Computation*, 1989.

[252] Y. LeCun, L. Bottou, Y. Bengio, and P. Haffner. Gradient-based learning applied to document recognition. *Proceedings of the IEEE*, 1998.

[253] C. Lemaréchal. Cauchy and the gradient method. *Doc Math Extra*, 2012.

[254] K. Levenberg. A method for the solution of certain non-linear problems in least squares. *Quarterly of applied mathematics*, 1944.

[255] H. Levitt. Transformed Up-Down Methods in Psychoacoustics. *The Journal of the Acoustical Society of America*, 1971.

[256] Q. Li, W. Cai, X. Wang, Y. Zhou, D. D. Feng, and M. Chen. Medical Image Classification with Convolutional Neural Network. In *2014 13th International Conference on Control Automation Robotics & Vision (ICARCV)*, 2014.

[257] Y. Li, R. Yu, C. Shahabi, and Y. Liu. Diffusion convolutional recurrent neural network: Data-driven traffic forecasting. *arXiv:1707.01926*, 2017.

[258] Z. Li, F. Liu, W. Yang, S. Peng, and J. Zhou. A survey of convolutional neural networks: Analysis, applications, and prospects. *IEEE Transactions on Neural Networks and Learning Systems*, 2021.

[259] H. W. Lin, M. Tegmark, and D. Rolnick. Why does deep and cheap learning work so well? *Journal of Statistical Physics*, 2017.

[260] M. Lin, Q. Chen, and S. Yan. Network in network. *arXiv:1312.4400*, 2013.

[261] T. Lin, S. U. Stich, K. K. Patel, and M. Jaggi. Don't Use Large Mini-batches, Use Local SGD. In *International Conference on Learning Representations*, 2020.

[262] T. Lin, Y. Wang, X. Liu, and X. Qiu. A survey of transformers. *AI Open*, 2022.

[263] A. Lindholm, N. Wahlström, F. Lindsten, and T. B. Schön. *Machine Learning - A First Course for Engineers and Scientists*. Cambridge University Press, 2022.

[264] H. Ling, K. Kreis, D. Li, S. W. Kim, A. Torralba, and S. Fidler. Editgan: High-precision semantic image editing. *Advances in Neural Information Processing Systems*, 2021.

[265] S. Linnainmaa. The representation of the cumulative rounding error of an algorithm as a taylor expansion of the local rounding errors. *Master's Thesis (in Finnish), Univ. Helsinki*, 1970.

[266] D. C. Liu and J. Nocedal. On the limited memory BFGS method for large-scale optimization. *Mathematical Programming*, 1989.

[267] S. Liu, L. Qi, H. Qin, J. Shi, and J. Jia. Path Aggregation Network for Instance Segmentation. In *Proceedings of the IEEE Conference on Computer Vision and Pattern Recognition*, 2018.

[268] Y. Liu, M. Ott, N. Goyal, J. Du, M. Joshi, D. Chen, O. Levy, M. Lewis, L. Zettlemoyer, and V. Stoyanov. RoBERTa: A robustly optimized BERT pretraining approach. *arXiv:1907.11692*, 2019.

[269] Y. Liu, Y. Zhang, Y. Wang, F. Hou, J. Yuan, J. Tian, Y. Zhang, Z. Shi, J. Fan, and Z. He. A survey of visual transformers. *IEEE Transactions on Neural Networks and Learning Systems*, 2023.

[270] I. Loshchilov and F. Hutter. Decoupled weight decay regularization. *arXiv:1711.05101*, 2017.

[271] M. Lucic, K. Kurach, M. Michalski, S. Gelly, and O. Bousquet. Are GANs created equal? A large-scale study. *Advances in Neural Information Processing Systems*, 2018.

[272] D. G. Luenberger and Y. Ye. *Linear and nonlinear programming*. Springer, Cham, Fifth edition, 2021.

[273] Y. Ma and J. Tang. *Deep Learning on Graphs*. Cambridge University Press, 2021.

[274] A. Maas, R. E. Daly, P. T. Pham, D. Huang, A. Y. Ng, and C. Potts. Learning word vectors for sentiment analysis. In *Proceedings of the 49th Annual Meeting of the Association for Computational Linguistics: Human Language Technologies*, 2011.

[275] A. L. Maas, R. E. Daly, P. T. Pham, D. Huang, A. Y. Ng, and C. Potts. Learning Word Vectors for Sentiment Analysis. In *Proceedings of the 49th Annual Meeting of the Association for Computational Linguistics: Human Language Technologies*, 2011.

[276] A. L. Maas, A. Y. Hannun, and A. Y. Ng. Rectifier nonlinearities improve neural network acoustic models. In *International Conference on Machine Learning*, 2013.

[277] J. MacQueen. Some methods for classification and analysis of multivariate observations. In *Proceedings of the 5th Berkeley symposium on mathematical statistics and probability*, 1967.

[278] A. Mahendran and A. Vedaldi. Understanding deep image representations by inverting them. In *Proceedings of the IEEE conference on computer vision and pattern recognition*, 2015.

[279] K. M. Malan. A survey of advances in landscape analysis for optimization. *Algorithms*, 2021.

[280] C. Manning and H. Schutze. *Foundations of Statistical Natural Language Processing*. MIT press, 1999.

[281] E. R. Mansfield and B. P. Helms. Detecting multicollinearity. *The American Statistician*, 1982.

[282] N. Mazyavkina, S. Sviridov, S. Ivanov, and E. Burnaev. Reinforcement learning for combinatorial optimization: A survey. *Computers & Operations Research*, 2021.

[283] W. S. McCulloch and W. Pitts. A logical calculus of the ideas immanent in nervous activity. *The Bulletin of Mathematical Biophysics*, 1943.

[284] S. Mei, Y. Bai, and A. Montanari. The landscape of empirical risk for nonconvex losses. *The Annals of Statistics*, 2018.

[285] G. Menghani. Efficient deep learning: A survey on making deep learning models smaller, faster, and better. *ACM Computing Surveys*, 2023.

[286] L. Mero, D. Yi, M. Dianati, and A. Mouzakitis. A survey on imitation learning techniques for end-to-end autonomous vehicles. *IEEE Transactions on Intelligent Transportation Systems*, 2022.

[287] A. Micheli. Neural network for graphs: A contextual constructive approach. *IEEE Transactions on Neural Networks*, 2009.

[288] T. Mikolov, K. Chen, G. Corrado, and J. Dean. Efficient estimation of word representations in vector space. *arXiv:1301.3781*, 2013.

[289] G. A. Miller. *WordNet: An Electronic Lexical Database.* MIT Press, 1998.

[290] S. Minaee, Y. Y. Boykov, F. Porikli, A. J. Plaza, N. Kehtarnavaz, and D. Terzopoulos. Image Segmentation Using Deep Learning: A Survey. *IEEE Transactions on Pattern Analysis and Machine Intelligence*, 2021.

[291] M. Minsky and S. A. Papert. *Perceptrons: An Introduction to Computational Geometry.* MIT Press, 1969.

[292] L. Mirsky. Symmetric gauge functions and unitarily invariant norms. *The quarterly journal of mathematics*, 1960.

[293] M. Mirza and S. Osindero. Conditional generative adversarial nets. *arXiv:1411.1784*, 2014.

[294] V. Mnih, K. Kavukcuoglu, D. Silver, A. Graves, I. Antonoglou, D. Wierstra, and M. Riedmiller. Playing atari with deep reinforcement learning. *arXiv:1312.5602*, 2013.

[295] A. Moghar and M. Hamiche. Stock market prediction using LSTM recurrent neural network. *Procedia Computer Science*, 2020.

[296] D. C. Montgomery. *Design and Analysis of Experiments.* John Wiley & Sons, 2017.

[297] E. H. Moore. On the reciprocal of the general algebraic matrix. *Bull. Am. Math. Soc.*, 1920.

[298] K. P. Murphy. *Machine learning: a probabilistic perspective.* MIT Press, 2012.

[299] K. P. Murphy. *Probabilistic Machine Learning: An Introduction.* MIT Press, 2022.

[300] F. Murtagh and P. Contreras. Algorithms for hierarchical clustering: an overview. *Wiley Interdisciplinary Reviews: Data Mining and Knowledge Discovery*, 2012.

[301] V. Nair and G. E. Hinton. Rectified linear units improve restricted Boltzmann machines. In *International conference on machine learning*, 2010.

[302] S. C. Narula and J. F. Wellington. The minimum sum of absolute errors regression: A state of the art survey. *International Statistical Review/Revue Internationale de Statistique*, 1982.

[303] Y. Nazarathy and H. Klok. *Statistics with Julia.* Springer, 2021.

[304] J. A. Nelder and R. W. M. Wedderburn. Generalized Linear Models. *Journal of the Royal Statistical Society: Series A*, 1972.

[305] Y. E. Nesterov. A method for solving the convex programming problem with convergence rate $O(1/k^2)$. *Dokl. Akad. Nauk SSSR*, 1983.

[306] A. Ng. Machine Learning Yearning. *https://info.deeplearning.ai/machine-learning-yearning-book*, 2017.

[307] A. Nichol, P. Dhariwal, A. Ramesh, P. Shyam, P. Mishkin, B. McGrew, I. Sutskever, and M. Chen. Glide: Towards photorealistic image generation and editing with text-guided diffusion models. *arXiv:2112.10741*, 2021.

[308] M. A. Nielsen. *Neural networks and deep learning*. Determination press San Francisco, CA, 2015.

[309] M. Niepert, M. Ahmed, and K. Kutzkov. Learning convolutional neural networks for graphs. In *International conference on machine learning*, 2016.

[310] J. Nocedal. Updating quasi-Newton matrices with limited storage. *Mathematics of Computation*, 1980.

[311] J. Nocedal and S. J. Wright. *Numerical optimization*. Springer, 1999.

[312] H. Noh, S. Hong, and B. Han. Learning Deconvolution Network for Semantic Segmentation. In *Proceedings of the IEEE International Conference on Computer Vision*, 2015.

[313] J. F. Nolan. *Analytical differentiation on a digital computer*. PhD thesis, Massachusetts Institute of Technology, 1953.

[314] A. B. Novikoff. On convergence proofs for perceptrons. Technical report, Stanford Research Inst. Menlo Park CA, 1963.

[315] N. A. Obuchowski and J. A. Bullen. Receiver Operating Characteristic (ROC) Curves: Review of Methods with Applications in Diagnostic Medicine. *Physics in Medicine & Biology*, 2018.

[316] A. Odena, C. Olah, and J. Shlens. Conditional image synthesis with auxiliary classifier GANs. In *International Conference on Machine Learning*, 2017.

[317] M. Ou, P. Cui, J. Pei, Z. Zhang, and W. Zhu. Asymmetric transitivity preserving graph embedding. In *Proceedings of the 22nd ACM SICKDD international conference on Knowledge discovery and data mining*, 2016.

[318] L. Ouyang, J. Wu, X. Jiang, D. Almeida, C. Wainwright, P. Mishkin, C. Zhang, S. Agarwal, K. Slama, A. Ray, et al. Training language models to follow instructions with human feedback. *Advances in Neural Information Processing Systems*, 2022.

[319] A. S. Pandya and R. B. Macy. *Pattern Recognition with Neural Networks in C++*. CRC Press, 1995.

[320] J. Papa. *PyTorch Pocket Reference: Building and Deploying Deep Learning Model*. O'Reilly Media, 2021.

[321] J. M. Papakonstantinou and R. A. Tapia. Origin and evolution of the secant method in one dimension. *The American Mathematical Monthly*, 2013.

[322] G. Papamakarios, E. Nalisnick, D. J. Rezende, S. Mohamed, and B. Lakshminarayanan. Normalizing flows for probabilistic modeling and inference. *The Journal of Machine Learning Research*, 2021.

[323] R. Pascanu, T. Mikolov, and Y. Bengio. On the difficulty of training recurrent neural networks. In *International conference on machine learning*, 2013.

[324] A. Paszke, S. Gross, S. Chintala, G. Chanan, E. Yang, Z. DeVito, Z. Lin, A. Desmaison, L. Antiga, and A. Lerer. Automatic differentiation in PyTorch. *31st Conference on Neural Information Processing Systems (NIPS2017)*, 2017.

[325] J. Pearl and D. Mackenzie. *The Book of Why: The New Science of Cause and Effect.* Basic Books, 2018.

[326] K. Pearson. LIII. On lines and planes of closest fit to systems of points in space. *The London, Edinburgh, and Dublin philosophical magazine and journal of science*, 1901.

[327] J. Pennington, R. Socher, and C. D. Manning. Glove: Global vectors for word representation. In *Proceedings of the 2014 conference on empirical methods in natural language processing (EMNLP)*, 2014.

[328] R. Penrose. A generalized inverse for matrices. In *Mathematical proceedings of the Cambridge Philosophical Society*, 1955.

[329] B. Perozzi, R. Al-Rfou, and S. Skiena. Deepwalk: Online learning of social representations. In *Proceedings of the 20th ACM SIGKDD international conference on Knowledge discovery and data mining*, 2014.

[330] K. B. Petersen and M. S. Pedersen. The matrix cookbook. *Technical University of Denmark*, 2012.

[331] M. Phuong and M. Hutter. Formal algorithms for transformers. *arXiv:2207.09238*, 2022.

[332] E. Plaut. From principal subspaces to principal components with linear autoencoders. *arXiv:1804.10253*, 2018.

[333] T. Poggio, A. Banburski, and Q. Liao. Theoretical issues in deep networks. *Proceedings of the National Academy of Sciences*, 2020.

[334] T. Poggio, H. Mhaskar, L. Rosasco, B. Miranda, and Q. Liao. Why and When Can Deep – but Not Shallow – Networks Avoid the Curse of Dimensionality: a Review, 2017.

[335] B. T. Polyak. Some methods of speeding up the convergence of iteration methods. *USSR Computational Mathematics and Mathematical Physics*, 1964.

[336] S. J. D. Prince. *Understanding Deep Learning.* MIT Press, 2023.

[337] M. L. Puterman. *Markov decision processes: discrete stochastic dynamic programming.* John Wiley & Sons, 2014.

[338] C. Rackauckas, Y. Ma, J. Martensen, C. Warner, K. Zubov, R. Supekar, D. Skinner, A. Ramadhan, and A. Edelman. Universal differential equations for scientific machine learning. *arXiv:2001.04385*, 2020.

[339] A. Radford, L. Metz, and S. Chintala. Unsupervised representation learning with deep convolutional generative adversarial networks. *arXiv:1511.06434*, 2015.

[340] A. Radford, J. Wu, R. Child, D. Luan, D. Amodei, I. Sutskever, et al. Language models are unsupervised multitask learners. *OpenAI blog*, 2019.

[341] J. W. Rae, S. Borgeaud, T. Cai, K. Millican, J. Hoffmann, F. Song, J. Aslanides, S. Henderson, R. Ring, S. Young, et al. Scaling language models: Methods, analysis & insights from training gopher. *arXiv:2112.11446*, 2021.

[342] L. Ramalho. *Fluent Python*. O'Reilly Media, Incorporated, 2021.

[343] R. Ranganath, D. Tran, and D. Blei. Hierarchical variational models. In *International conference on machine learning*. PMLR, 2016.

[344] W. Rawat and Z. Wang. Deep convolutional neural networks for image classification: A comprehensive review. *Neural Computation*, 2017.

[345] S. J. Reddi, S. Kale, and S. Kumar. On the Convergence of Adam and Beyond. *arXiv:1904.09237*, 2019.

[346] J. Redmon, S. Divvala, R. Girshick, and A. Farhadi. You only look once: Unified, real-time object detection. In *Proceedings of the IEEE conference on computer vision and pattern recognition*, 2016.

[347] D. Rezende and S. Mohamed. Variational inference with normalizing flows. In *International Conference on Machine Learning*. PMLR, 2015.

[348] S. Rezvani and X. Wang. A broad review on class imbalance learning techniques. *Applied Soft Computing*, 2023.

[349] H. Robbins. Some aspects of the sequential design of experiments. *Bulletin of the American Mathematical Society*, 1952.

[350] H. Robbins and S. Monro. A stochastic approximation method. *The annals of mathematical statistics*, 1951.

[351] J. Rodriguez-Perez, C. Leigh, B. Liquet, C. Kermorvant, E. Peterson, D. Sous, and K. Mengersen. Detecting technical anomalies in high-frequency water-quality data using artificial neural networks. *Environmental Science & Technology*, 2020.

[352] O. Ronneberger, P. Fischer, and T. Brox. U-net: Convolutional networks for biomedical image segmentation. In *Medical Image Computing and Computer-Assisted Intervention–MICCAI 2015: 18th International Conference, Munich, Germany, October 5-9, 2015, Proceedings, Part III 18*, 2015.

[353] F. Rosenblatt. The perceptron: a probabilistic model for information storage and organization in the brain. *Psychological review*, 1958.

[354] H. H. Rosenbrock. An automatic method for finding the greatest or least value of a function. *The computer journal*, 1960.

[355] S. M. Ross. *A first course in probability*. Pearson, 2014.

[356] S. Ruder. An overview of gradient descent optimization algorithms. *arXiv:1609.04747*, 2016.

[357] D. E. Rumelhart, G. E. Hinton, and R. J. Williams. Learning representations by back-propagating errors. *Nature*, 1986.

[358] C. Saharia, W. Chan, H. Chang, C. Lee, J. Ho, T. Salimans, D. Fleet, and M. Norouzi. Palette: Image-to-image diffusion models. In *ACM SIGGRAPH 2022 Conference Proceedings*, 2022.

[359] C. Saharia, W. Chan, S. Saxena, L. Li, J. Whang, E. L. Denton, K. Ghasemipour, R. G. Lopes, B. K. Ayan, T. Salimans, et al. Photorealistic text-to-image diffusion models with deep language understanding. *Advances in Neural Information Processing Systems*, 2022.

[360] C. Saharia, J. Ho, W. Chan, T. Salimans, D. J. Fleet, and M. Norouzi. Image super-resolution via iterative refinement. *IEEE Transactions on Pattern Analysis and Machine Intelligence*, 2022.

[361] R. Salakhutdinov and G. Hinton. Deep Boltzmann machines. In *Artificial Intelligence and Statistics*, 2009.

[362] S. Sarao Mannelli and P. Urbani. Analytical study of momentum-based acceleration methods in paradigmatic high-dimensional non-convex problems. *Advances in Neural Information Processing Systems*, 2021.

[363] M. Sarigül and M. Avci. Performance comparison of different momentum techniques on deep reinforcement learning. *Journal of Information and Telecommunication*, 2018.

[364] N. Savage. How AI and Neuroscience Drive Each Other Forwards. *Nature*, 2019.

[365] F. Scarselli, M. Gori, A. C. Tsoi, M. Hagenbuchner, and G. Monfardini. The graph neural network model. *IEEE transactions on neural networks*, 2008.

[366] R. E. Schapire and Y. Freund. *Boosting: Foundations and Algorithms*. The MIT Press, 2012.

[367] J. Schmidhuber. Annotated history of modern AI and Deep learning. *arXiv:2212.11279*, 2022.

[368] F. Schroff, D. Kalenichenko, and J. Philbin. Facenet: A unified embedding for face recognition and clustering. In *Proceedings of the IEEE conference on computer vision and pattern recognition*, 2015.

[369] M. Schuster and K. K. Paliwal. Bidirectional recurrent neural networks. *IEEE transactions on Signal Processing*, 1997.

[370] T. J. Sejnowski. *The Deep Learning Revolution*. MIT Press, 2018.

[371] Y. Seo, M. Defferrard, P. Vandergheynst, and X. Bresson. Structured sequence modeling with graph convolutional recurrent networks. In *Neural Information Processing: 25th International Conference, ICONIP, 2018, Proceedings, Part I 25*, 2018.

[372] P. Sermanet and Y. LeCun. Traffic Sign Recognition with Multi-Scale Convolutional Networks. In *The 2011 International Joint Conference on Neural Networks*, 2011.

[373] V. Sharma, M. Gupta, A. Kumar, and D. Mishra. Video Processing Using Deep Learning Techniques: A Systematic Literature Review. *IEEE Access*, 2021.

[374] A. Shashua and A. Levin. Ranking with large margin principle: Two approaches. *Advances in neural information processing systems*, 2002.

[375] Z. Shen, W. Bao, and D. S. Huang. Recurrent neural network for predicting transcription factor binding sites. *Scientific reports*, 2018.

[376] R. Shwartz-Ziv and N. Tishby. Opening the black box of deep neural networks via information. *arXiv:1703.00810*, 2017.

[377] D. Silver, A. Huang, C. J. Maddison, A. Guez, L. Sifre, G. Van Den Driessche, J. Schrittwieser, I. Antonoglou, V. Panneershelvam, M. Lanctot, et al. Mastering the game of Go with deep neural networks and tree search. *Nature*, 2016.

[378] J. S. Simonoff. *Smoothing Methods in Statistics*. Springer Science & Business Media, 2012.

[379] K. Simonyan and A. Zisserman. Two-Stream Convolutional Networks for Action Recognition in Videos. *Advances in Neural Information Processing Systems*, 2014.

[380] K. Simonyan and A. Zisserman. Very deep convolutional networks for large-scale image recognition. *arXiv:1409.1556*, 2014.

[381] S. A. Sisson, Y. Fan, and M. A. Beaumont, editors. *Handbook of Approximate Bayesian Computation*. CRC Press, Boca Raton, FL, 2019.

[382] S. Smith, M. Patwary, B. Norick, P. LeGresley, S. Rajbhandari, J. Casper, Z. Liu, S. Prabhumoye, G. Zerveas, V. Korthikanti, et al. Using deepspeed and megatron to train megatron-turing nlg 530b, a large-scale generative language model. *arXiv:2201.11990*, 2022.

[383] I. Sobel. History and definition of the sobel operator. *Retrieved from the World Wide Web*, 2014.

[384] J. Sohl-Dickstein, E. Weiss, N. Maheswaranathan, and S. Ganguli. Deep unsupervised learning using nonequilibrium thermodynamics. In *International Conference on Machine Learning*, 2015.

[385] C. K. Sønderby, T. Raiko, L. Maaløe, S. K. Sønderby, and O. Winther. Ladder variational autoencoders. *Advances in neural information processing systems*, 2016.

[386] B. Speelpenning. *Compiling fast partial derivatives of functions given by algorithms*. University of Illinois at Urbana-Champaign, 1980.

[387] A. Sperduti and A. Starita. Supervised neural networks for the classification of structures. *IEEE Transactions on Neural Networks*, 1997.

[388] N. Srivastava, G. Hinton, A. Krizhevsky, I. Sutskever, and R. Salakhutdinov. Dropout: a simple way to prevent neural networks from overfitting. *The journal of machine learning research*, 2014.

[389] S. M. Stigler. Gauss and the invention of least squares. *The Annals of Statistics*, 1981.

[390] P. Stoica and Y. Selen. Model-order selection: a review of information criterion rules. *IEEE Signal Processing Magazine*, 2004.

[391] G. Strang. *Linear algebra and learning from data*. Wellesley-Cambridge Press Cambridge, 2019.

[392] I. Sutskever, O. Vinyals, and Q. V. Le. Sequence to sequence learning with neural networks. In *Advances in neural information processing systems*, 2014.

[393] R. S. Sutton and A. G. Barto. *Reinforcement learning: An introduction*. MIT press, 2018.

[394] C. Szegedy, W. Liu, Y. Jia, P. Sermanet, S. Reed, D. Anguelov, D. Erhan, V. Vanhoucke, and A. Rabinovich. Going deeper with convolutions. In *Proceedings of the IEEE conference on computer vision and pattern recognition*, 2015.

[395] M. Tan and Q. Le. EfficientNet: Rethinking Model Scaling for Convolutional Neural Networks. In *International Conference on Machine Learning*, 2019.

[396] M. Tan and Q. V. Le. EfficientNetV2: Smaller Models and Faster Training. *International Conference on Machine Learning, PMLR*, 2021.

[397] J. Tang, M. Qu, M. Wang, M. Zhang, J. Yan, and Q. Mei. Line: Large-scale information network embedding. In *Proceedings of the 24th international conference on world wide web*, 2015.

[398] A. Tavanaei, M. Ghodrati, S. R. Kheradpisheh, T. Masquelier, and A. Maida. Deep Learning in Spiking Neural Networks. *Neural Networks*, 2019.

[399] M. Telgarsky. Benefits of depth in neural networks. In *Conference on learning theory*, 2016.

[400] R. Thoppilan, D. D. Freitas, J. Hall, N. Shazeer, A. Kulshreshtha, H. Cheng, A. Jin, T. Bos, L. Baker, Y. Du, et al. Lamda: Language models for dialog applications. *arXiv:2201.08239*, 2022.

[401] A. N. Tikhonov. On the stability of inverse problems. In *Comptes Rendus de l'Académie des Sciences de l'URSS*, 1943.

[402] A. C. Tsoi. Face recognition: A convolutional neural-network approach. *IEEE Transactions on Neural Networks*, 1997.

[403] L. Tunstall, L. V. Werra, and T. Wolf. *Natural Language Processing with Transformers*. O'Reilly Media, Inc., 2022.

[404] A. M. Turing and J. Haugeland. *Computing Machinery and Intelligence*. MIT Press Cambridge, MA, 1950.

[405] I. Ulku and E. Akagündüz. A Survey on Deep Learning-Based Architectures for Semantic Segmentation on 2D Images. *Applied Artificial Intelligence*, 2022.

[406] D. Ulyanov, A. Vedaldi, and V. Lempitsky. Instance normalization: The missing ingredient for fast stylization. *arXiv:1607.08022*, 2016.

[407] R. van de Schoot, S. Depaoli, R. King, B. Kramer, K. Märtens, M. G. Tadesse, M. Vannucci, A. Gelman, D. Veen, J. Willemsen, and C. Yau. Bayesian Statistics and Modelling. *Nature Reviews Methods Primers*, 2021.

[408] C. Van Rijsbergen. Information Retrieval (Book 2nd ed), 1979.

[409] V. N. Vapnick. *Statistical Learning Theory*. Wiley, New York, 1998.

[410] A. Vaswani, N. Shazeer, N. Parmar, J. Uszkoreit, L. Jones, A. N. Gomez, L. Kaiser, and I. Polosukhin. Attention is all you need. In *Advances in neural information processing systems*, 2017.

[411] P. Velickovic, G. Cucurull, A. Casanova, A. Romero, P. Lio, and Y. Bengio. Graph Attention Networks. *International Conference on Learning Representations (ICLR)*, 2018.

[412] A. Waibel, T. Hanazawa, G. Hinton, K. Shikano, and K. J. Lang. Phoneme recognition using time-delay neural networks. In *Backpropagation*. 2013.

[413] L. Waikhom and R. Patgiri. A survey of graph neural networks in various learning paradigms: methods, applications, and challenges. *Artificial Intelligence Review*, 2023.

[414] J. Walker, C. Doersch, A. Gupta, and M. Hebert. An uncertain future: Forecasting from static images using variational autoencoders. In *Computer Vision–ECCV 2016: 14th European Conference, 2016, Proceedings, Part VII 14*, 2016.

[415] C. Y. Wang, A. Bochkovskiy, and H. Y. M. Liao. YOLOv7: Trainable Bag-of-Freebies Sets New State-of-the-Art for Real-Time Object Detectors. In *Proceedings of the IEEE/CVF Conference on Computer Vision and Pattern Recognition*, 2023.

[416] L. Wang, Y. Xiong, Z. Wang, Y. Qiao, D. Lin, X. Tang, and L. Van Gool. Temporal Segment Networks: Towards Good Practices for Deep Action Recognition. In *European Conference on Computer Vision*, 2016.

[417] T. Wang, M. Liu, J. Zhu, A. Tao, J. Kautz, and B. Catanzaro. High-resolution image synthesis and semantic manipulation with conditional GANs. In *Proceedings of the IEEE conference on computer vision and pattern recognition*, 2018.

[418] Z. Wang, L. Zhao, and W. Xing. Stylediffusion: Controllable disentangled style transfer via diffusion models. In *Proceedings of the IEEE/CVF International Conference on Computer Vision*, 2023.

[419] C. J. Watkins and P. Dayan. Q-learning. *Machine learning*, 1992.

[420] R. E. Wengert. A simple automatic derivative evaluation program. *Communications of the ACM*, 1964.

[421] P. J. Werbos. Applications of advances in nonlinear sensitivity analysis. In *System Modeling and Optimization: Proceedings of the 10th IFIP Conference, 1981*, 2005.

[422] P. Wolfe. Convergence conditions for ascent methods. *SIAM Review*, 1969.

[423] P. Wolfe. Convergence conditions for ascent methods. II: Some corrections. *SIAM Review*, 1971.

[424] T. Wong and P. Yeh. Reliable accuracy estimates from k-fold cross-validation. *IEEE Transactions on Knowledge and Data Engineering*, 2019.

[425] H. Wu, Z. Xu, J. Zhang, W. Yan, and X. Ma. Face Recognition Based on Convolution Siamese Networks. In *2017 10th International Congress on Image and Signal Processing, BioMedical Engineering and Informatics (CISP-BMEI)*, 2017.

[426] Y. Wu and K. He. Group normalization. In *Proceedings of the European conference on computer vision (ECCV)*, 2018.

[427] Z. Wu, S. Pan, F. Chen, G. Long, C. Zhang, and P. S. Yu. A comprehensive survey on graph neural networks. *IEEE transactions on neural networks and learning systems*, 2020.

[428] P. Xu, X. Zhu, and D. A. Clifton. Multimodal learning with transformers: A survey. *IEEE Transactions on Pattern Analysis and Machine Intelligence*, 2023.

[429] A. Yang, B. Xiao, B. Wang, B. Zhang, C. Bian, C. Yin, C. Lv, D. Pan, D. Wang, D. Yan, et al. Baichuan 2: Open large-scale language models. *arXiv:2309.10305*, 2023.

[430] L. Yang, Z. Zhang, Y. Song, S. Hong, R. Xu, Y. Zhao, W. Zhang, B. Cui, and M. Yang. Diffusion models: A comprehensive survey of methods and applications. *ACM Computing Surveys*, 2023.

[431] T. Yang, Q. Lin, and Z. Li. Unified convergence analysis of stochastic momentum methods for convex and non-convex optimization. *arXiv:1604.03257*, 2016.

[432] Y. Yang. Can the strengths of AIC and BIC be shared? A conflict between model identification and regression estimation. *Biometrika*, 2005.

[433] Z. Yang, Z. Dai, Y. Yang, J. Carbonell, R. R. Salakhutdinov, and Q. V. Le. XLNet: Generalized autoregressive pretraining for language understanding. *Advances in neural information processing systems*, 2019.

[434] G. Yao, T. Lei, and J. Zhong. A Review of Convolutional-Neural-Network-Based Action Recognition. *Pattern Recognition Letters*, 2019.

[435] F. Yu and V. Koltun. Multi-scale context aggregation by dilated convolutions. *arXiv:1511.07122*, 2015.

[436] F. Yu, V. Koltun, and T. Funkhouser. Dilated residual networks. In *Proceedings of the IEEE conference on computer vision and pattern recognition*, pages 472–480, 2017.

[437] L. Yu, W. Zhang, J. Wang, and Y. Yu. Seqgan: Sequence generative adversarial nets with policy gradient. In *Proceedings of the AAAI conference on artificial intelligence*, 2017.

[438] Y. Yu, X. Si, C. Hu, and J. Zhang. A review of recurrent neural networks: LSTM cells and network architectures. *Neural Computation*, 2019.

[439] Y. Yuan. Recent advances in trust region algorithms. *Mathematical Programming*, 2015.

[440] M. D. Zeiler. Adadelta: An adaptive learning rate method. *arXiv:1212.5701*, 2012.

[441] M. D. Zeiler and R. Fergus. Visualizing and understanding convolutional networks. In *European conference on computer vision*, 2014.

[442] M. D. Zeiler, G. W. Taylor, and R. Fergus. Adaptive deconvolutional networks for mid and high level feature learning. In *2011 International Conference on Computer Vision*, 2011.

[443] X. Zeng and T. R. Martinez. Distribution-balanced stratified cross-validation for accuracy estimation. *Journal of Experimental & Theoretical Artificial Intelligence*, 2000.

[444] C. Zhang, J. Bütepage, H. Kjellström, and S. Mandt. Advances in variational inference. *IEEE Transactions on Pattern Analysis and Machine Intelligence*, 2019.

[445] M. Zhang and Y. Chen. Link prediction based on graph neural networks. *Advances in neural information processing systems*, 2018.

[446] N. Zhang, S. Shen, A. Zhou, and Y. Jin. Application of LSTM approach for modelling stress–strain behaviour of soil. *Applied Soft Computing*, 2021.

[447] Q. Zhang and S. Zhu. Visual interpretability for deep learning: a survey. *Frontiers of Information Technology & Electronic Engineering*, 2018.

[448] W. Zhao, J. P. Queralta, and T. Westerlund. Sim-to-real transfer in deep reinforcement learning for robotics: a survey. In *2020 IEEE symposium series on computational intelligence (SSCI)*, 2020.

[449] W. X. Zhao, K. Zhou, J. Li, T. Tang, X. Wang, Y. Hou, et al. A survey of large language models. *arXiv:2303.18223*, 2023.

[450] Y. Zhao, X. Li, W. Zhang, S. Zhao, M. Makkie, M. Zhang, Q. Li, and T. Liu. Modeling 4D fMRI Data via Spatio-Temporal Convolutional Neural Networks (ST-CNN). In *Medical Image Computing and Computer Assisted Intervention–MICCAI 2018: 21st International Conference, 2018, Proceedings, Part III 11*, 2018.

[451] J. Zhou, G. Cui, S. Hu, Z. Zhang, C. Yang, Z. Liu, L. Wang, C. Li, and M. Sun. Graph neural networks: A review of methods and applications. *AI open*, 2020.

[452] Q. Zhou, W. Chen, S. Song, J. Gardner, K. Weinberger, and Y. Chen. A reduction of the elastic net to support vector machines with an application to GPU computing. In *Proceedings of the AAAI Conference on Artificial Intelligence*, 2015.

[453] H. Zou and T. Hastie. Regularization and variable selection via the elastic net. *Journal of the Royal Statistical Society: Series B (Statistical Methodology)*, 2005.

[454] Z. Zou, K. Chen, Z. Shi, Y. Guo, and J. Ye. Object Detection in 20 Years: A Survey. *Proceedings of the IEEE*, 2023.

Index

For Product Safety Concerns and Information please contact our
EU representative GPSR@taylorandfrancis.com Taylor & Francis
Verlag GmbH, Kaufingerstraße 24, 80331 München, Germany